Classics in Mathematics

Lars Hörmander

The Analysis of Linear Partial Differential Operators II

Lars Hörmander

The Analysis of Linear Partial Differential Operators II

Differential Operators with Constant Coefficients

Reprint of the 1983 Edition

Springer

Lars Hörmander
University of Lund
Department of Mathematics
Box 118
SE-22100 Lund
Sweden
email: lvh@maths.lth.se

Originally published as Vol. 257 in the series:
Grundlehren der mathematischen Wissenschaften

Library of Congress Control Number: 2004097173

Mathematics Subject Classification (2000):
35B, 35C, 35E, 35G, 35L, 35H10, 35P25, 44A35

ISSN 1431-0821
ISBN 3-540-22516-1 Springer Berlin Heidelberg New York

Springer is a part of Springer Science+Business Media
springeronline.com

Printed in Germany

Printed on acid-free paper 41/3142YL-5 4 3 2 1 0

Lars Hörmander

The Analysis of Linear Partial Differential Operators II

Differential Operators with Constant Coefficients

Springer-Verlag
Berlin Heidelberg New York
London Paris Tokyo Hong Kong

Lars Hörmander

Lunds Universitet, Matematiska Institutionen
Box 118
S-22100 Lund
Sweden

With 7 Figures

Second revised printing 1990

AMS Subject Classification: 35 E; 35 G, 35 H, 35 L, 35 P, 44 A 35

ISBN 3-540-12139-0 Springer-Verlag Berlin Heidelberg New York Tokyo
ISBN 0-387-12139-0 Springer-Verlag New York Heidelberg Berlin Tokyo

Library of Congress Cataloging-in-Publication Data
Hörmander, Lars. The analysis of linear partial differential operators / Lars Hörmander. – Rev. ed. p. cm. – (Grundlehren der mathematischen Wissenschaften ; 257) Includes bibliographical references. Contents: 2. Differential operators with constant coefficients.
ISBN (invalid) 0-387-12139-0 (U.S. : v. 2)
1. Differential equations, Partial. 2. Partial differential operators. I. Title. II. Series.
QA377.H578 1990 515'.353–dc20 89-26134 CIP

Typesetting: Universitätsdruckerei H. Stürtz AG, 8700 Würzburg
Printing: Reprotechnik Deutschland GmbH, Heidelberg
Bookbinding: Schäffer, Grünstadt
2141/3111 - 543 Printed on acid-free paper

Preface

This volume is an expanded version of Chapters III, IV, V and VII of my 1963 book "Linear partial differential operators". In addition there is an entirely new chapter on convolution equations, one on scattering theory, and one on methods from the theory of analytic functions of several complex variables. The latter is somewhat limited in scope though since it seems superfluous to duplicate the monographs by Ehrenpreis and by Palamodov on this subject.

The reader is assumed to be familiar with distribution theory as presented in Volume I. Most topics discussed here have in fact been encountered in Volume I in special cases, which should provide the necessary motivation and background for a more systematic and precise exposition. The main technical tool in this volume is the Fourier-Laplace transformation. More powerful methods for the study of operators with variable coefficients will be developed in Volume III. However, the constant coefficient theory has given the guidelines for all that work. Although the field is no longer very active - perhaps because of its advanced state of development - and although it is possible to pass directly from Volume I to Volume III, the material presented here should not be neglected by the serious student who wants to gain a balanced perspective of the theory of linear partial differential equations.

I would like to thank all who have helped me in various ways during the preparation of this volume. As in the case of the first Volume I am particularly indebted to Niels Jørgen Kokholm of the University of Copenhagen who has read all the proofs and in doing so suggested many improvements of the text.

Lund in February 1983

Lars Hörmander

Contents

Introduction

Differential operators with constant coefficients acting on distributions of compact support are diagonalized by the Fourier-Laplace transformation. Already in Chapter VII we showed how this observation leads to general results on existence of fundamental solutions and approximation of solutions. In Chapter X we first examine the regularity properties of the fundamental solutions more closely, including the location of the wave front set. Existence and approximation of solutions is then studied systematically. Hypoelliptic operators are characterized in Chapter XI; the typical examples of elliptic operators or the heat operator were discussed in Volume I. We also prove a general result on propagation of regularity of solutions which is similar to Holmgren's uniqueness theorem. In Volume I we gave explicit formulas for the solution of the Cauchy problem for the wave equation. The hyperbolic operators which have similar properties are investigated in Chapter XII which also includes a study of the characteristic Cauchy problem modelled on say the heat or Schrödinger equation. The precision of the results in these chapters pays off in Chapter XIII where it allows us to treat a fairly large class of operators with variable coefficients locally as perturbations of constant coefficient operators. The new features which appear even for such operators are emphasized by a discussion of non-uniqueness for the Cauchy problem. Chapter XIV is devoted to perturbation theory in $\mathbb{R}^n$ (short range scattering theory).

The study of general overdetermined systems of differential operators with constant coefficients requires more prerequisites from the theory of analytic functions of several complex variables than we wish to assume here. As mentioned in the preface several monographs have already been devoted to this topic. In Chapter XV we shall therefore only develop some of the basic analytic techniques in a way which simplifies the existing treatments. Their use is illustrated in the case of a single differential equation. As in the preceding chapters it would cause no additional difficulties to consider determined systems.

Chapter XVI is devoted to convolution equations. The translation invariance which they share with differential operators with constant coefficients makes their study so closely related that this enlargement of our main theme seems natural if not unavoidable.

Chapter X. Existence and Approximation of Solutions of Differential Equations

Summary

In the preceding chapters we have constructed a number of explicit fundamental solutions. In Chapter VII we also gave a construction which is applicable to any differential operator with constant coefficients. We return to it in Section 10.2 to discuss the regularity properties of the fundamental solutions obtained in greater detail. First we determine when a fundamental solution is in one of the spaces introduced in Section 10.1. These generalize the $H_{(s)}$ spaces of Section 7.9 and are defined essentially as inverse Fourier transforms of L^p spaces with respect to appropriate densities. In Section 10.2 we also examine how the fundamental solution $E(P)$ of $P(D)$ given by the construction depends on P, and we estimate $WF(E(P))$ for a fixed P.

A differential equation $P(D)u=f$ with $f\in\mathscr{E}'$ can immediately be solved if one has a fundamental solution of $P(D)$ available. In Section 10.3 we determine the properties of the solution rather precisely by means of the results on fundamental solutions proved in Section 10.2. This leads to a comparison of the relative strengths of differential operators (polynomials); P is said to be stronger than Q if $Q(D)u$ is at least as regular as $P(D)u$ when $u\in\mathscr{E}'$. This notion, in a more precise form, is studied at some length in Section 10.4. In Section 10.5 we give a brief discussion of approximation theorems of Runge's type for solutions of the homogeneous differential equation $P(D)u=0$. This prepares for the study in Section 10.6 of the differential equation $P(D)u=f$ when f is a distribution of finite order. The same problem is discussed in Section 10.7 when f is an arbitrary distribution. Finally, Section 10.8 is devoted to the search for a geometrical form of the conditions for solvability encountered in Sections 10.6 and 10.7.

10.1. The Spaces $B_{p,k}$

In an existence theory for partial differential equations it is important to give precise statements concerning the regularity of the solutions obtained. Now a condition on the regularity of a distribution or function u (with compact support) can also be regarded as a condition on the behavior of the Fourier transform $\hat{u}$ at infinity. To classify this behavior one may for example ask for which weight functions k it is true that $k\hat{u}\in L^p$. The set of all such temperate distributions u is denoted by $B_{p,k}$ here. Only the cases $p=2$, $p=\infty$ and $p=1$ are really interesting. Concerning k we shall make a hypothesis which ensures that $B_{p,k}$ is a module over C_0^∞:

Definition 10.1.1. A positive function k defined in $\mathbb{R}^n$ will be called a temperate weight function if there exist positive constants C and N such that

(10.1.1) $$k(\xi+\eta)\leqq(1+C|\xi|)^N k(\eta); \quad \xi,\eta\in\mathbb{R}^n.$$

The set of all such functions k will be denoted by $\mathscr{K}$.

Remark. In harmonic analysis a weight function usually means a function K such that

$$K(\xi+\eta)\leqq K(\xi)\,K(\eta).$$

To avoid confusion we shall call such functions submultiplicative here.

From (10.1.1) it follows that

(10.1.1)′ $$(1+C|\xi|)^{-N}\leqq k(\xi+\eta)/k(\eta)\leqq(1+C|\xi|)^N; \quad \xi,\eta\in\mathbb{R}^n.$$

In fact, the left-hand inequality is obtained if η is replaced by $\xi+\eta$ and ξ is replaced by $-\xi$ in (10.1.1). If we let $\xi\to 0$ in (10.1.1)′ it follows that k is continuous, and when $\eta=0$ we obtain the useful estimates

(10.1.2) $$k(0)(1+C|\xi|)^{-N}\leqq k(\xi)\leqq k(0)(1+C|\xi|)^N.$$

If $k\in\mathscr{K}$ we shall write

(10.1.3) $$M_k(\xi)=\sup_\eta k(\xi+\eta)/k(\eta).$$

This means that M_k is the smallest function such that

(10.1.4) $$k(\xi+\eta)\leqq M_k(\xi)\,k(\eta).$$

It is clear that M_k is submultiplicative,

(10.1.5) $$M_k(\xi+\eta)\leqq M_k(\xi)\,M_k(\eta),$$

and since $M_k(\xi)\leqq(1+C|\xi|)^N$ this implies that $M_k\in\mathscr{K}$. From (10.1.5) we can deduce that

(10.1.6) $$1=M_k(0)\leqq M_k(\xi), \quad \xi\in\mathbb{R}^n.$$

In fact, for every positive integer ν we have

$$1=M_k(0)\leqq M_k(\xi)^\nu M_k(-\nu\xi)\leqq M_k(\xi)^\nu(1+C\nu|\xi|)^N,$$

and if we take ν^{th} roots in this inequality and let $\nu\to\infty$, the estimate (10.1.6) follows.

Example 10.1.2. The spaces $H_{(s)}$ in Section 7.9 correspond to

$$k_s(\xi)=(1+|\xi|^2)^{s/2}.$$

That $k_s\in\mathscr{K}$ and that $M_{k_s}(\eta)\leqq(1+|\eta|)^{|s|}$ follows from the estimates

$$1+|\xi+\eta|^2\leqq 1+|\xi|^2+2|\xi||\eta|+|\eta|^2\leqq(1+|\xi|^2)(1+|\eta|)^2.$$

Example 10.1.3. The basic example of a function in $\mathscr{K}$, which is the reason why we introduce it here, is the function $\tilde{P}$ defined by

$$\tilde{P}(\xi)^2=\sum_{|\alpha|\geqq 0}|P^{(\alpha)}(\xi)|^2 \tag{10.1.7}$$

where P is a polynomial so that the sum is finite. Here $P^{(\alpha)}=\partial^\alpha P$. It follows immediately from Taylor's formula that

$$\tilde{P}(\xi+\eta)\leqq(1+C|\xi|)^m\tilde{P}(\eta), \tag{10.1.8}$$

where m is the degree of P and C a constant depending only on m and the dimension n. From this example other functions in $\mathscr{K}$ are obtained in the theory as a result of the operations described in the next theorem.

Theorem 10.1.4. *If k_1 and k_2 belong to $\mathscr{K}$, it follows that k_1+k_2, k_1k_2, $\sup(k_1,k_2)$ and $\inf(k_1,k_2)$ are also in $\mathscr{K}$. If $k\in\mathscr{K}$ we have $k^s\in\mathscr{K}$ for every real s, and if μ is a positive measure we have either $\mu*k\equiv\infty$ or else $\mu*k\in\mathscr{K}$.*

Proof. It follows from (10.1.1)′ that $1/k\in\mathscr{K}$ if $k\in\mathscr{K}$. The statements are thus all trivial except perhaps the last. To prove that one, we note that (10.1.1) gives

$$(\mu*k)(\xi+\eta)\leqq(1+C|\xi|)^N(\mu*k)(\eta).$$

If $\mu*k$ is finite for some η it follows that $\mu*k$ is finite everywhere and belongs to $\mathscr{K}$.

Occasionally it is useful to know that functions in $\mathscr{K}$ can be replaced by equivalent functions which vary very slowly indeed:

Theorem 10.1.5. *If $k\in\mathscr{K}$ we can for every $\delta>0$ find a function $k_\delta\in\mathscr{K}$ and a constant C_δ such that*

(10.1.9) $$1 \leqq k_\delta(\xi)/k(\xi) \leqq C_\delta, \qquad \xi \in \mathbb{R}^n,$$

(10.1.10) $$M_{k_\delta}(\xi) \leqq (1+C|\xi|)^N, \qquad \xi \in \mathbb{R}^n,$$

where C and N are independent of δ, and

(10.1.11) $M_{k_\delta} \to 1$ *uniformly on compact subsets of* $\mathbb{R}^n$ *when* $\delta \to 0$.

Proof. We shall set

$$k_\delta(\xi) = \sup_\eta e^{-\delta|\eta|} k(\xi - \eta).$$

(Note the analogy with the definition of a convolution.) Then we have in view of (10.1.1)

$$k(\xi) \leqq k_\delta(\xi) \leqq \sup_\eta e^{-\delta|\eta|}(1+C|\eta|)^N k(\xi) = C_\delta k(\xi)$$

where the last equality is a definition of C_δ. This proves (10.1.9). To prove (10.1.10) we note that

$$\begin{aligned} k_\delta(\xi+\xi') &= \sup_\eta e^{-\delta|\eta|} k(\xi+\xi'-\eta) \\ &\leqq \sup_\eta e^{-\delta|\eta|}(1+C|\xi'|)^N k(\xi-\eta) \\ &= (1+C|\xi'|)^N k_\delta(\xi); \qquad \xi, \xi' \in \mathbb{R}^n. \end{aligned}$$

To prove (10.1.11) we first rewrite the definition of k_δ by introducing $\xi - \eta$ as a variable instead of η. This gives

$$k_\delta(\xi) = \sup_\eta e^{-\delta|\xi-\eta|} k(\eta).$$

Hence

$$\begin{aligned} k_\delta(\xi+\xi') &= \sup_\eta e^{-\delta|\xi+\xi'-\eta|} k(\eta) \\ &\leqq e^{\delta|\xi'|} \sup_\eta e^{-\delta|\xi-\eta|} k(\eta) = e^{\delta|\xi'|} k_\delta(\xi), \end{aligned}$$

which proves that

$$1 \leqq M_{k_\delta}(\xi) \leqq e^{\delta|\xi|}$$

The proof is complete.

Remark. It might have been more natural to require in Definition 10.1.1 only that k is continuous and that

$$k(\xi+\eta) \leqq C(1+|\xi|)^N k(\eta).$$

We have not done so since this would not guarantee the continuity of M_k. Now the proof of Theorem 10.1.5 shows that for such functions we still obtain (10.1.9) and $k_\delta \in \mathscr{K}$. Our choice of definition has therefore not led to any significant loss of generality.

We can now give the formal definition of the spaces we need:

Definition 10.1.6. If $k\in\mathscr{K}$ and $1\leqq p\leqq\infty$, we denote by $B_{p,k}$ the set of all distributions $u\in\mathscr{S}'$ such that $\hat{u}$ is a function and

$$\|u\|_{p,k}=((2\pi)^{-n}\int|k(\xi)\,\hat{u}(\xi)|^p d\xi)^{1/p}<\infty. \tag{10.1.12}$$

When $p=\infty$ we shall interpret $\|u\|_{p,k}$ as ess. $\sup|k(\xi)\,\hat{u}(\xi)|$.

The factor $(2\pi)^{-n}$ is included for convenience in the results and is motivated by the fact that the measure $(2\pi)^{-n}d\xi$ occurs in Parseval's formula. For example, we have thanks to this normalization

$$\|u\|_{2,\tilde{P}}=(\sum_\alpha\|P^{(\alpha)}(D)\,u\|_{L^2}^2)^{\frac{1}{2}}.$$

Theorem 10.1.7. $B_{p,k}$ *is a Banach space with the norm* (10.1.12). *We have*

$$\mathscr{S}\subset B_{p,k}\subset\mathscr{S}',$$

also in the topological sense.[1] C_0^∞ *is dense in* $B_{p,k}$ *if* $p<\infty$.

Proof. Let $L_{p,k}$ be the Banach space of all measurable functions v such that the norm $(2\pi)^{-n/p}\|kv\|_p<\infty$. Then we have

$$\mathscr{S}\subset L_{p,k}\subset\mathscr{S}',$$

also in the topological sense, and $\mathscr{S}$ is dense in $L_{p,k}$ if $p<\infty$. In fact, that $\mathscr{S}\subset L_{p,k}$ follows from the second inequality in (10.1.2), and if $p<\infty$ it follows from Theorem 1.3.2 that even C_0^∞ is dense in $L_{p,k}$, for C_0^0 is dense there. To prove that $L_{p,k}\subset\mathscr{S}'$ we note that Hölder's inequality gives

$$\int|\phi v|\,d\xi\leqq\|kv\|_p\,\|\phi/k\|_{p'}$$

where $1/p+1/p'=1$. This proves our assertion since $\|\phi/k\|_{p'}$ is a continuous semi-norm in $\mathscr{S}$ in view of the first inequality in (10.1.2). If we now use the fact that the Fourier transformation is an automorphism of $\mathscr{S}$ and of $\mathscr{S}'$, it follows that $B_{p,k}$ is complete, that $\mathscr{S}\subset B_{p,k}\subset\mathscr{S}'$ (topologically) and that $\mathscr{S}$ is dense in $B_{p,k}$ if $p<\infty$. Since C_0^∞ is dense in $\mathscr{S}$ by Lemma 7.1.8, this completes the proof.

Theorem 10.1.8. *If* $k_1, k_2\in\mathscr{K}$ *and*

$$k_2(\xi)\leqq C k_1(\xi),\qquad \xi\in\mathbb{R}^n, \tag{10.1.13}$$

it follows that $B_{p,k_1}\subset B_{p,k_2}$. *Conversely, if there exists an open set* $X\neq\emptyset$ *such that* $B_{p,k_1}\cap\mathscr{E}'(X)\subset B_{p,k_2}$, *then* (10.1.13) *is valid.*

Proof. The first part of the theorem is trivial. To prove the second we let F be a compact subset of X with non-empty interior, and set

[1] This means that the topology in $\mathscr{S}$ is stronger than that induced there by $B_{p,k}$ and that the topology in $B_{p,k}$ is stronger than the one induced by $\mathscr{S}'$.

$B=B_{p,k_1}\cap\mathscr{E}'(F)$. Since the topology in each of the spaces B_{p,k_j} is stronger than that in $\mathscr{S}'$ (Theorem 10.1.7), it is clear that B is a closed subspace of B_{p,k_1} and that the inclusion mapping of B into B_{p,k_2} is closed. Hence it follows from the closed graph theorem that

$$\|u\|_{p,k_2}\leqq C_1\|u\|_{p,k_1},\quad u\in B, \tag{10.1.14}$$

where C_1 is a constant. With a fixed function $u\in C_0^\infty(F)$ such that $u\not\equiv 0$, we shall apply (10.1.14) to $u_\eta(x)=u(x)e^{i\langle x,\eta\rangle}$ where $\eta\in\mathbb{R}^n$. Since $\hat{u}_\eta(\xi)=\hat{u}(\xi-\eta)$, the estimates

$$|k_1(\xi)\hat{u}(\xi-\eta)|\leqq k_1(\eta)|M_{k_1}(\xi-\eta)\hat{u}(\xi-\eta)|,$$
$$|k_2(\xi)\hat{u}(\xi-\eta)|\geqq k_2(\eta)|\hat{u}(\xi-\eta)/M_{k_2}(\eta-\xi)|$$

give

$$\|u_\eta\|_{p,k_1}\leqq k_1(\eta)\|u\|_{p,M_{k_1}};\qquad \|u_\eta\|_{p,k_2}\geqq k_2(\eta)\|u\|_{p,1/\check{M}_{k_2}}. \tag{10.1.15}$$

If we combine these inequalities with (10.1.14) we obtain (10.1.13) with the constant $C=C_1\|u\|_{p,M_{k_1}}/\|u\|_{p,1/\check{M}_{k_2}}$. The proof is complete.

Corollary 10.1.9. *If $k_1,k_2\in\mathscr{K}$, it follows that*

$$B_{p,k_1}\cap B_{p,k_2}=B_{p,k_1+k_2}$$

and that

$$\max_{j=1,2}\|u\|_{p,k_j}\leqq\|u\|_{p,k_1+k_2}\leqq\|u\|_{p,k_1}+\|u\|_{p,k_2},\qquad u\in B_{p,k_1}\cap B_{p,k_2}.$$

Proof. Since $k_j\leqq k_1+k_2$ we have

$$B_{p,k_1+k_2}\subset B_{p,k_j}\quad\text{and}\quad\|u\|_{p,k_j}\leqq\|u\|_{p,k_1+k_2}$$

for $j=1,2$. On the other hand, if $u\in B_{p,k_1}\cap B_{p,k_2}$, it follows from Minkowski's inequality that $u\in B_{p,k_1+k_2}$ and that the second part of the inequality is valid. This proves the corollary.

We next examine when the inclusion mapping in Theorem 10.1.8 is compact.

Theorem 10.1.10. *If K is a compact set in $\mathbb{R}^n$, the inclusion mapping of $B_{p,k_1}\cap\mathscr{E}'(K)$ into B_{p,k_2} is compact if*

$$k_2(\xi)/k_1(\xi)\to 0,\quad \xi\to\infty. \tag{10.1.16}$$

Conversely, if the mapping is compact for one set K with interior points, it follows that (10.1.16) *is valid.*

Proof. a) Assuming that (10.1.16) is valid, we take a sequence $u_\nu \in B_{p,k_1} \cap \mathscr{E}'(K)$ such that $\|u_\nu\|_{p,k_1} \leqq 1$. We have to prove that there exists a subsequence converging in B_{p,k_2}. Let ϕ be a function in $C_0^\infty(\mathbb{R}^n)$ which equals 1 in a neighborhood of K. Since $u_\nu = \phi u_\nu$, we have

$$\hat{u}_\nu(\xi) = (2\pi)^{-n} \int \hat{\phi}(\xi - \eta)\, \hat{u}_\nu(\eta)\, d\eta. \tag{10.1.17}$$

Multiplying (10.1.17) by $k_1(\xi)$ and using the inequality

$$k_1(\xi) \leqq M_{k_1}(\xi - \eta)\, k_1(\eta)$$

and Hölder's inequality, we obtain

$$\begin{aligned} |k_1(\xi)\, \hat{u}_\nu(\xi)| &\leqq (2\pi)^{-n} \|M_{k_1} \hat{\phi}\|_{p'} \|k_1 \hat{u}_\nu\|_p \\ &\leqq (2\pi)^{-n/p'} \|M_{k_1} \hat{\phi}\|_{p'}. \end{aligned}$$

Similarly, application of the same argument after differentiation of (10.1.17) gives

$$|k_1(\xi)\, D^\alpha \hat{u}_\nu(\xi)| \leqq (2\pi)^{-n/p'} \|M_{k_1} D^\alpha \hat{\phi}\|_{p'}.$$

This proves that the sequence $\hat{u}_\nu$ is uniformly bounded and equicontinuous on every compact set. Thus we can find a subsequence converging uniformly on all compact sets; changing notations, if necessary, we may assume that the sequence $\hat{u}_\nu$ itself is uniformly convergent on compact sets, Given any $\varepsilon > 0$ we now choose a ball S so large that $k_2(\xi)/k_1(\xi) < \varepsilon$ when $\xi \notin S$. Using Minkowski's inequality we obtain

$$\|u_\mu - u_\nu\|_{p,k_2} \leqq \varepsilon \|u_\mu - u_\nu\|_{p,k_1} + ((2\pi)^{-n} \int_S |k_2(\hat{u}_\mu - \hat{u}_\nu)|^p\, d\xi)^{1/p},$$

with the usual interpretation when $p = \infty$. The second term on the right-hand side tends to 0 when μ and $\nu \to \infty$, and the first is always $\leqq 2\varepsilon$. Hence the sequence u_ν is a Cauchy sequence in B_{p,k_2}, which proves the compactness.

b) Assuming the compactness of the inclusion mapping for some compact set K with interior points, we shall prove (10.1.16). To do so it is sufficient to prove that if a sequence $\xi_\nu \to \infty$ then $k_2(\xi_\nu)/k_1(\xi_\nu) \to 0$. Let $C_0^\infty(K) \ni u \not\equiv 0$, and set

$$u_\nu(x) = u(x)\, e^{i\langle x, \xi_\nu \rangle}/k_1(\xi_\nu).$$

From (10.1.15) we then obtain

$$\begin{aligned} &\|u_\nu\|_{p,k_1} \leqq \|u\|_{p,M_{k_1}}, \\ &\|u_\nu\|_{p,k_2} \geqq \|u\|_{p,1/\check{M}_{k_2}}\, k_2(\xi_\nu)/k_1(\xi_\nu). \end{aligned} \tag{10.1.18}$$

From the first of these inequalities it follows that the sequence u_ν is bounded in B_{p,k_1}, hence precompact in B_{p,k_2}. Now $u_\nu \to 0$ in $\mathscr{S}'$ for if

$\phi \in \mathscr{S}$ we have

$$\int \phi u_\nu \, dx = \widehat{(\phi u)}(-\xi_\nu)/k_1(\xi_\nu) \to 0 \qquad \text{as } \nu \to \infty$$

because $\widehat{(\phi u)} \in \mathscr{S}$. Since the topology in $\mathscr{S}'$ is weaker than that in B_{p,k_2}, it follows that the only limit point of the sequence u_ν in B_{p,k_2} is 0, hence $\|u_\nu\|_{p,k_2} \to 0$ because the sequence is precompact. But then the statement follows from the second inequality in (10.1.18). The proof is complete.

We shall now study how differential operators with constant coefficients act in the spaces $B_{p,k}$. Recall that if $P(\xi)$ is a polynomial in n variables $\xi_1, \dots, \xi_n$, with complex coefficients, then a differential operator $P(D)$ is defined by replacing ξ_j by $D_j = -i\partial/\partial x_j$, and a function $\tilde{P}$ is defined by (10.1.7).

Theorem 10.1.11. *If $u \in B_{p,k}$ it follows that $P(D)u \in B_{p,k/\tilde{P}}$.*

Proof. The statement is obvious since the Fourier transform of $P(D)u$ is $P(\xi)\hat{u}(\xi)$, with absolute value $\leqq \tilde{P}(\xi)|\hat{u}(\xi)|$.

The next theorem contains Theorem 10.1.11 and is equally simple.

Theorem 10.1.12. *If $u_1 \in B_{p,k_1} \cap \mathscr{E}'$ and $u_2 \in B_{\infty,k_2}$, it follows that $u_1 * u_2 \in B_{p,k_1 k_2}$, and we have the estimate*

$$\|u_1 * u_2\|_{p,k_1 k_2} \leqq \|u_1\|_{p,k_1} \|u_2\|_{\infty,k_2}. \tag{10.1.19}$$

Proof. The Fourier transform of $u_1 * u_2$ is $\hat{u}_1 \hat{u}_2$ by Theorem 7.1.15, and $k_2 |\hat{u}_2| \leqq \|u_2\|_{\infty,k_2}$.

To pass from results involving the spaces $B_{p,k}$ to statements of a classical form, the following result is needed.

Theorem 10.1.13. *If $k \in \mathscr{K}$ and j is a non-negative integer such that*

$$(1+|\xi|)^j/k(\xi) \in L^{p'}, \qquad 1/p + 1/p' = 1, \tag{10.1.20}$$

we have $B_{p,k} \subset C^j$. Conversely, if $B_{p,k} \cap \mathscr{E}'(X) \subset C^j$ for some open set $X \neq \emptyset$, it follows that (10.1.20) *is valid.*

Proof of the sufficiency of (10.1.20). If $u \in B_{p,k}$ it follows from (10.1.20) and Hölder's inequality that $\xi^\alpha \hat{u}(\xi) = (\xi^\alpha/k(\xi))(k(\xi)\hat{u}(\xi))$ is integrable if $|\alpha| \leqq j$. Hence the inverse Fourier transform

$$u(x) = (2\pi)^{-n} \int e^{i\langle x,\xi\rangle} \hat{u}(\xi) \, d\xi$$

and the integrals obtained by at most j differentiations with respect to x are absolutely convergent, which proves that $u \in C^j$. The proof of the necessity of (10.1.20) will be given after Theorem 10.1.15.

We shall now determine the dual space of $B_{p,k}$ when $p<\infty$. Since $\mathscr{S}$ is dense in $B_{p,k}$ if $p<\infty$ (Theorem 10.1.7), a continuous linear form on $B_{p,k}$ is uniquely determined in that case by its restriction to $\mathscr{S}$.

Theorem 10.1.14. *If L is a continuous linear form on $B_{p,k}$, $p<\infty$, we have for some $v \in B_{p',1/k}$, $1/p'+1/p=1$,*

$$L(u)=\check{v}(u), \qquad u \in \mathscr{S}.$$

The norm of this linear form is $\|v\|_{p',1/k}$. Hence $B_{p',1/k}$ is the dual space of $B_{p,k}$ and the canonical bilinear form in $B_{p,k} \times B_{p',1/k}$ is the continuous extension of the bilinear form $\check{v}(u)$; $v \in B_{p',1/k}$, $u \in \mathscr{S}$.

Proof. The Fourier transformation reduces the theorem to the fact that a continuous linear form on $\mathscr{S}$ with respect to the norm $\|kU\|_p$, $p<\infty$, is a scalar product with a function V such that $V/k \in L^{p'}$, and that the norm of the linear form is $\|V/k\|_{p'}$.

Remark. Similarly, it is clear that every anti-linear form on $B_{p,k}$ is an extension of a form $u \to v(\bar{u})$, $u \in \mathscr{S}$, and that the norm is $\|v\|_{p',1/k}$.

We now come to a most important property of $B_{p,k}$:

Theorem 10.1.15. *If $u \in B_{p,k}$ and $\phi \in \mathscr{S}$, it follows that $\phi u \in B_{p,k}$ and that*

$$\|\phi u\|_{p,k} \leqq \|\phi\|_{1,M_k} \|u\|_{p,k}. \tag{10.1.21}$$

Proof. We know from Theorems 7.1.10 and 7.1.15 that the Fourier transform of $v=\phi u$ is the convolution

$$\hat{v}(\xi)=(2\pi)^{-n} \int \hat{\phi}(\xi-\eta)\hat{u}(\eta)\,d\eta,$$

provided that $\hat{\phi} \in C_0^\infty$. Multiplying by $k(\xi)$ and noting the inequality $k(\xi) \leqq M_k(\xi-\eta)\,k(\eta)$, we obtain

$$|k\hat{v}| \leqq (2\pi)^{-n} |M_k \hat{\phi}| * |k\hat{u}|.$$

Hence Minkowski's inequality in integral form gives

$$\|k\hat{v}\|_p \leqq (2\pi)^{-n} \|M_k \hat{\phi}\|_1 \|k\hat{u}\|_p,$$

which is equivalent to (10.1.21). Since C_0^∞ is dense in $\mathscr{S}$, the result immediately extends to an arbitrary $\phi \in \mathscr{S}$.

End of proof of Theorem 10.1.13. We must prove that (10.1.20) is valid if $B_{p,k}\cap\mathscr{E}'(X)\subset C^j$. Assuming as we may that $0\in X$, we choose a function $\phi\in C_0^\infty(X)$ such that $\phi=1$ in a neighborhood of 0. If $v\in B_{p,k}$, it follows from Theorem 10.1.15 that $\phi v\in B_{p,k}\cap\mathscr{E}'(X)$, hence $\phi v\in C^j$ in virtue of the hypothesis. From the closed graph theorem it now follows as in the proof of Theorem 10.1.8 that

$$\sum_{|\alpha|\leqq j}\sup|D^\alpha(\phi v)|\leqq C\|v\|_{p,k},\qquad v\in B_{p,k}.$$

In particular, we have if $|\alpha|\leqq j$

$$|(2\pi)^{-n}\int\xi^\alpha\hat{v}(\xi)d\xi|=|D^\alpha v(0)|\leqq C(2\pi)^{-n/p}\|k\hat{v}\|_p,\qquad \hat{v}\in\mathscr{S},$$

so the converse of Hölder's inequality gives

$$\|\xi^\alpha/k(\xi)\|_{p'}\leqq C(2\pi)^{n/p'},\qquad |\alpha|\leqq j,$$

and this implies (10.1.20).

From Theorem 10.1.15 we also obtain a method for approximating by elements with compact support:

Theorem 10.1.16. *Let $\psi\in C_0^\infty$ and assume that $\psi(0)=1$. Set $\psi^\varepsilon(x)=\psi(\varepsilon x)$. If $u\in B_{p,k}$ and $p<\infty$, it then follows that $\psi^\varepsilon u\to u$ in $B_{p,k}$ when $\varepsilon\to 0$.*

Proof. From Theorem 10.1.15 we have

$$\|\psi^\varepsilon u\|_{p,k}\leqq C\|u\|_{p,k}\quad\text{if } u\in B_{p,k},\ 0<\varepsilon<1,$$

where

$$\begin{aligned}C&=\sup_{0<\varepsilon<1}(2\pi)^{-n}\int\varepsilon^{-n}|\hat{\psi}(\xi/\varepsilon)|M_k(\xi)d\xi\\&=\sup_{0<\varepsilon<1}(2\pi)^{-n}\int|\hat{\psi}(\xi)|M_k(\varepsilon\xi)d\xi\end{aligned}$$

is finite in virtue of (10.1.1). This proves the statement since it is true if $u\in C_0^\infty$, which is a dense subset of $B_{p,k}$.

Similarly we can approximate by the usual regularization:

Theorem 10.1.17. *Let $\phi\in C_0^\infty$ and assume that $\int\phi dx=1$. Set $\phi_\varepsilon(x)=\varepsilon^{-n}\phi(x/\varepsilon)$. If $u\in B_{p,k}$ and $p<\infty$, the regularizations $u*\phi_\varepsilon$ then converge to u in $B_{p,k}$ when $\varepsilon\to 0$.*

Proof. The Fourier transform of $u*\phi_\varepsilon$ is $\hat{u}(\xi)\hat{\phi}(\varepsilon\xi)$. Since $\hat{\phi}(\varepsilon\xi)\to\hat{\phi}(0)=1$ boundedly and uniformly on every compact set when $\varepsilon\to 0$, this proves the statement.

Theorem 10.1.15 makes it possible to define for every open set $X \subset \mathbb{R}^n$ a subspace of $\mathscr{D}'(X)$ whose elements behave locally just like the elements of $B_{p,k}$ but have unrestricted growth at the boundary and at infinity. In doing so, we use the following terminology.

Definition 10.1.18. A linear subspace $\mathscr{F}$ of $\mathscr{D}'(X)$ is called semi-local if $\phi u \in \mathscr{F}$ when $u \in \mathscr{F}$ and $\phi \in C_0^\infty(X)$. It is called local if, in addition, $\mathscr{F}$ contains every distribution u such that $\phi u \in \mathscr{F}$ for every $\phi \in C_0^\infty(X)$.

Examples of local spaces are $\mathscr{D}'(X)$, $C^k(X)$, $L^p_{\text{loc}}(X)$ whereas $\mathscr{D}'_F(X)$, $\mathscr{E}'(X)$, $L^p(X)$ are semi-local but not local. It follows from Theorem 10.1.15 that the set of restrictions to X of distributions in $B_{p,k}$ is semi-local.

It is obvious that $L^p_{\text{loc}}(X)$ is the smallest local space containing $L^p(X)$. More generally, we have

Theorem 10.1.19. *If $\mathscr{F}$ is semi-local, the smallest local space containing $\mathscr{F}$ is the space*

$$\mathscr{F}^{\text{loc}} = \{u;\, u \in \mathscr{D}'(X),\ \phi u \in \mathscr{F} \text{ for every } \phi \in C_0^\infty(X)\}.$$

Proof. Since $\mathscr{F}$ is semi-local, we have $\mathscr{F} \subset \mathscr{F}^{\text{loc}}$. It is also clear that $\mathscr{F}^{\text{loc}}$ is semi-local. To prove that $\mathscr{F}^{\text{loc}}$ is local, we take a distribution u such that $\phi u \in \mathscr{F}^{\text{loc}}$ for every $\phi \in C_0^\infty(X)$. Choose $\psi \in C_0^\infty(X)$ so that $\psi = 1$ in the support of ϕ. Then it follows that $\phi u = \psi(\phi u) \in \mathscr{F}$ in view of the definition of $\mathscr{F}^{\text{loc}}$. Hence $u \in \mathscr{F}^{\text{loc}}$, so that $\mathscr{F}^{\text{loc}}$ is a local space. It is obvious that it is the smallest local space containing $\mathscr{F}$.

For future reference we note the following useful property of local spaces:

Theorem 10.1.20. *Let $\mathscr{F}$ be a local subspace of $\mathscr{D}'(X)$. If $u \in \mathscr{D}'(X)$ and to every point $x_0 \in X$ there exists a function $\phi \in C_0^\infty(X)$ such that $\phi u \in \mathscr{F}$ and $\phi(x_0) \neq 0$, it follows that $u \in \mathscr{F}$.*

Proof. Let $\phi \in C_0^\infty(X)$. In view of the Borel-Lebesgue lemma we can then find a finite number of functions $\phi_1, \dots, \phi_k \in C_0^\infty$ such that $\phi_j u \in \mathscr{F}$ when $j = 1, \dots, k$ and $\Phi = \sum_1^k |\phi_j|^2 > 0$ in the support of ϕ. Then it follows that $\psi = \phi/\Phi$, defined as 0 outside $\operatorname{supp} \phi$, is in $C_0^\infty(X)$, and since $\mathscr{F}$ is semi-local we obtain $\phi u = \sum_1^k \psi \bar{\phi}_j \phi_j u \in \mathscr{F}$. Since $\mathscr{F}$ is local this implies that $u \in \mathscr{F}$.

We shall also use the notation $\mathscr{F}^c$ for the set of distributions in $\mathscr{F}$ with compact support in X. If $\mathscr{F}$ is semi-local, it is obvious that

$$\mathscr{F}^c = \mathscr{F}^{\text{loc}} \cap \mathscr{E}'(X), \qquad \mathscr{F}^{\text{loc}} = (\mathscr{F}^c)^{\text{loc}}. \tag{10.1.22}$$

Most of the results proved for the spaces $B_{p,k}$ above carry over immediately to the local spaces $B_{p,k}^{\text{loc}}(X)$ corresponding to the set of restrictions to X of distributions in $B_{p,k}$ (or, equivalently, corresponding to $B_{p,k} \cap \mathscr{E}'(X)$).

Theorem 10.1.21. *In order that* $B_{p,k_1}^{\text{loc}}(X) \subset B_{p,k_2}^{\text{loc}}(X)$ *it is necessary and sufficient that* (10.1.13) *be valid.*

Proof. The sufficiency is obvious in view of Theorem 10.1.8. To prove the necessity we note that (10.1.22) and the assumption imply that

$$B_{p,k_1} \cap \mathscr{E}'(X) = B_{p,k_1}^{\text{loc}}(X) \cap \mathscr{E}'(X) \subset B_{p,k_2}^{\text{loc}}(X) \cap \mathscr{E}'(X) \subset B_{p,k_2}.$$

Hence the necessity also follows from Theorem 10.1.8.

Theorem 10.1.22. *If* $u \in B_{p,k}^{\text{loc}}(X)$ *we have* $P(D)u \in B_{p,k/\tilde{P}}^{\text{loc}}(X)$.

Proof. To any $\phi \in C_0^\infty(X)$ we can choose $\psi \in C_0^\infty(X)$ so that $\psi = 1$ in a neighborhood of the support of ϕ. Since $\psi u \in B_{p,k}$, it then follows from Theorems 10.1.11 and 10.1.15 that

$$\phi P(D)u = \phi P(D)(\psi u) \in B_{p,k/\tilde{P}},$$

which proves the theorem.

Theorem 10.1.23. *If* $u \in B_{p,k}^{\text{loc}}(X)$ *and* $\phi \in C^\infty(X)$ *then* $\phi u \in B_{p,k}^{\text{loc}}(X)$.

Proof. This is trivially true for every local space.

Theorem 10.1.24. *Let* $u_1 \in B_{p,k_1}(\mathbb{R}^n) \cap \mathscr{E}'(\mathbb{R}^n)$ *and* $u_2 \in B_{\infty,k_2}^{\text{loc}}(\mathbb{R}^n)$. *Then it follows that* $u_1 * u_2 \in B_{p,k_1 k_2}^{\text{loc}}(\mathbb{R}^n)$.

Proof. If $\phi \in C_0^\infty(\mathbb{R}^n)$ we have to prove that $\phi(u_1 * u_2) \in B_{p,k_1 k_2}$. Let K be the set of all x such that $\{x\} + \operatorname{supp} u_1$ intersects $\operatorname{supp} \phi$. Since K is a compact set we can choose $\psi \in C_0^\infty(\mathbb{R}^n)$ such that $\psi = 1$ in a neighborhood of K. Then the support of

$$u_1 * u_2 - u_1 * (\psi u_2) = u_1 * ((1-\psi) u_2)$$

does not meet $\operatorname{supp} \phi$ in view of (4.2.2). Hence $\phi(u_1 * u_2) = \phi(u_1 * (\psi u_2))$ and the statement follows immediately from Theorems 10.1.15 and 10.1.12.

From Theorem 10.1.13 we also obtain immediately

Theorem 10.1.25. *In order that $B_{p,k}^{\text{loc}}(X) \subset C^j(X)$ where j is a non-negative integer, it is necessary and sufficient that* (10.1.20) *be valid.*

Theorem 10.1.7 leads to the following

Theorem 10.1.26. *$B_{p,k}^{\text{loc}}(X)$ is a Fréchet space with the topology defined by the semi-norms $u \to \|\phi u\|_{p,k}$, $\phi \in C_0^\infty(X)$, and we have*

$$C^\infty(X) \subset B_{p,k}^{\text{loc}}(X) \subset \mathscr{D}'^j(X)$$

for some j also in the topological sense.

Proof. To prove the metrizability we take an increasing sequence of compact sets $K_\nu \subset X$ such that every compact subset of X is contained in some K_ν. If we choose $\phi_\nu \in C_0^\infty(X)$ such that $\phi_\nu = 1$ in K_ν, the topology in $B_{p,k}^{\text{loc}}(X)$ is then defined by the countably many semi-norms $u \to \|\phi_\nu u\|_{p,k}$ alone. In fact, for an arbitrary $\phi \in C_0^\infty(X)$ we can take ν so large that $\operatorname{supp}\phi \subset K_\nu$. Then we have $\phi u = \phi\phi_\nu u$ so it follows from Theorem 10.1.15 that

$$\|\phi u\|_{p,k} \leqq \|\phi\|_{1,M_k} \|\phi_\nu u\|_{p,k}.$$

Hence the topology is metrizable. To prove the completeness we only have to prove the convergence of every sequence u_j such that ϕu_j is a Cauchy sequence in $B_{p,k}$ for every $\phi \in C_0^\infty(X)$. By Theorem 10.1.7 ϕu_j has a limit in $B_{p,k}$, hence in $\mathscr{D}'$. This proves that $u = \lim u_j$ exists in $\mathscr{D}'(X)$ and that $\|\phi u_j - \phi u\|_{p,k} \to 0$ for every $\phi \in C_0^\infty(X)$ when $j \to \infty$. The proof is complete, for the last statement follows immediately from the proof of Theorem 10.1.7.

If $k_\mu \in \mathscr{K}$ and $1 \leqq p_\mu \leqq \infty$, $\mu = 1, 2, \ldots$ we shall also study the space $\bigcap_1^\infty B_{p_\mu,k_\mu}^{\text{loc}}(X)$ with the topology which is the least upper bound of the topologies in the spaces $B_{p_\mu,k_\mu}^{\text{loc}}(X)$, that is, defined by the semi-norms

$$u \to \|\phi u\|_{p_\mu,k_\mu}, \qquad \phi \in C_0^\infty(X), \qquad \mu = 1, 2, \ldots.$$

It is clear that this is also a Fréchet space; it is metrizable because the topology is defined by the semi-norms $\|\phi_\nu u\|_{p_\mu,k_\mu}$ with the same ϕ_ν as in the proof of Theorem 10.1.26. Note that it follows from Theorem 10.1.25 that

$$C^\infty(X) = \bigcap_1^\infty B_{p_\mu,k_\mu}^{\text{loc}}(X) \quad \text{if } k_\mu(\xi) = (1+|\xi|)^\mu. \tag{10.1.23}$$

It is obvious that the intersection of more than countably many spaces $B_{p,k}^{\mathrm{loc}}(X)$ is also a locally convex space although not necessarily a Fréchet space.

Having introduced a topology in $B_{p,k}^{\mathrm{loc}}(X)$ we can now also extend Theorem 10.1.10.

Theorem 10.1.27. *In order that every bounded set in* $B_{p,k_1}^{\mathrm{loc}}(X)$ *shall be precompact in* $B_{p,k_2}^{\mathrm{loc}}(X)$ *it is necessary and sufficient that* (10.1.16) *hold.*

Proof. The necessity of (10.1.16) follows immediately from Theorem 10.1.10 since

$$\{u;\, u\in B_{p,k_1}\cap\mathscr{E}'(K),\ \|u\|_{p,k_1}\leqq 1\}$$

is a bounded set in $B_{p,k_1}^{\mathrm{loc}}(X)$ if K is any compact subset of X. (Note that the topologies of B_{p,k_2} and $B_{p,k_2}^{\mathrm{loc}}(X)$ coincide on this set.) To prove the sufficiency we take a bounded sequence u_j in $B_{p,k_1}^{\mathrm{loc}}(X)$. This means that the sequence $\|\phi_\nu u_j\|_{p,k_1}$ is bounded when $j\to\infty$ for every fixed ν, if ϕ_ν are the functions in the proof of Theorem 10.1.26. Hence it follows from Theorem 10.1.10 by application of Cantor's diagonal process that there is a subsequence u_{j_μ} such that $\phi_\nu u_{j_\mu}$ is convergent in B_{p,k_2} for every ν. But this means that u_{j_μ} converges in $B_{p,k_2}^{\mathrm{loc}}(X)$, which proves the theorem.

10.2. Fundamental Solutions

If P is a differential operator with constant coefficients then $E\in\mathscr{D}'(\mathbb{R}^n)$ is called a fundamental solution of P if

$$PE=\delta, \tag{10.2.1}$$

where δ is the Dirac measure at 0. (See Section 3.3 and also Section 4.4 for the importance of fundamental solutions.) In a number of cases we have given explicit fundamental solutions. In Example 3.1.2 we saw that the Heaviside function $H(x)$ is a fundamental solution of d/dx on $\mathbb{R}$. More generally (3.2.17)′ and (3.2.2)″ show that χ_+^{k-1} is a fundamental solution of $(d/dx)^k$ on $\mathbb{R}$. Thus $\chi_+^{\alpha_1-1}(x_1)\dots\chi_+^{\alpha_n-1}(x_n)$ is a fundamental solution of ∂^α (see the proof of Theorem 4.4.7). The fundamental solution $1/\pi z$ of the Cauchy-Riemann operator $\partial/\partial\bar z$ was obtained in (3.1.12), and in Section 3.3 we constructed fundamental solutions for the Laplacean and the heat operator as well as some related operators such as the Schrödinger operator. Fundamental solutions of real homogeneous second order operators were given in

Section 6.2. The general form of fundamental solutions of homogeneous elliptic operators was determined in Theorem 7.1.20, and in Theorem 7.3.10 we gave a general construction of fundamental solutions. Finally we gave in Theorems 7.1.22 and 8.3.7 a parametrix for elliptic operators and operators of real principal type (see Definition 7.1.21). We shall now examine the construction in Theorem 7.3.10 more carefully.

Theorem 10.2.1. *If $P(D)$ is a partial differential operator with constant coefficients which is not equal to 0, then (7.3.22) defines a fundamental solution $E \in B^{\text{loc}}_{\infty, \tilde{P}}$ of $P(D)$.*

Proof. We can write (7.3.22) in the form

$$E(\phi) = \int \langle g_\zeta, e^{i\langle \cdot, \zeta \rangle} \phi \rangle \, d\lambda(\zeta), \qquad \phi \in C_0^\infty, \tag{10.2.2}$$

where g_ζ has the Fourier transform

$$\hat{g}_\zeta(\xi) = \Phi(P_\xi, \zeta)/P(\xi + \zeta).$$

By (7.3.21) with $Q = P_\xi = P(\xi + .)$ we have

$$\tilde{P}(\xi)\, |\hat{g}_\zeta(\xi)| \leqq C \sup |\Phi| = C_1$$

that is, $\|g_\zeta\|_{\infty, \tilde{P}} \leqq C_1$. For given $\varepsilon > 0$ we can choose the set Z in Lemma 7.3.12 so that $|\zeta| < \varepsilon/2$ when $\zeta \in Z$, and then we have $g_\zeta = 0$ when $|\zeta| > \varepsilon/2$. The dual of $B_{1,k}$ when $\check{k} = 1/\tilde{P}$ is $B_{\infty, \tilde{P}}$ (Theorem 10.1.14). When $|\zeta| < \varepsilon/2$ it is clear that $f_{\varepsilon,\zeta}(x) = e^{i\langle x, \zeta \rangle}/\cosh |\varepsilon x|$ is in a bounded subset of $\mathscr{S}$. Since $M_k(\xi) \leqq (1 + C|\xi|)^m$ by (10.1.8) it follows that

$$\|f_{\varepsilon,\zeta}\|_{1, M_k} \leqq C_2, \qquad |\zeta| < \varepsilon/2.$$

From (10.2.2) we now obtain in view of Theorem 10.1.15

$$\begin{aligned} |\langle E/\cosh|\varepsilon .|, \phi \rangle| &\leqq \left| \int \langle g_\zeta, \phi f_{\varepsilon,\zeta} \rangle \, d\lambda(\zeta) \right| \\ &\leqq \int \|g_\zeta\|_{\infty, \tilde{P}} \|f_{\varepsilon,\zeta}\|_{1, M_k} \|\phi\|_{1,k} \, d\lambda(\zeta) \\ &\leqq C_3 \|\phi\|_{1,k}. \end{aligned}$$

Hence Theorem 10.1.14 gives

$$\|E/\cosh|\varepsilon .|\|_{\infty, \tilde{P}} \leqq C_3 \tag{10.2.3}$$

where C_3 only depends on n, ε and the degree m of P. The theorem is proved.

In terms of the spaces $B^{\text{loc}}_{p,k}$ of Section 10.1 it is not possible to make a better statement than Theorem 10.2.1. Indeed, suppose that there is a fundamental solution E of $P(D)$ with $E \in B^{\text{loc}}_{p,k}(\mathbb{R}^n)$. In view of

Theorem 10.1.22 we have then

$$\delta = P(D)E \in B^{loc}_{p,k/\tilde{P}}.$$

But this means that $k/\tilde{P} \in L^p$, so $k\hat{u} \in L^p$ whenever $\tilde{P}\hat{u} \in L^\infty$, that is,

$$B^{loc}_{\infty,\tilde{P}} \subset B^{loc}_{p,k}.$$

Definition 10.2.2. A fundamental solution E of $P(D)$ is called regular if $E \in B^{loc}_{\infty,\tilde{P}}$.

With this terminology Theorem 10.2.1 means that there always exists a regular fundamental solution. By (10.2.3) it can be chosen with small exponential growth and a bound which is uniform when $P \in \mathrm{Pol}^\circ(m,n)$, the space of polynomials of degree $\leqq m$ in n variables with 0 removed. We shall now denote by $E(P)$ the fundamental solution defined by (7.3.22) and examine its dependence on P. Let $Q \in \mathrm{Pol}(m,n)$ and replace P by $P+tQ$ in (7.3.22), where t is a small real number. Then

$$\frac{d}{dt}(\Phi(P_\xi + tQ_\xi, \zeta)/(P(\xi+\zeta)+tQ(\xi+\zeta)))|_{t=0}$$
$$= \Phi'(P_\xi, \zeta; Q_\xi)/P(\xi+\zeta) - \Phi(P_\xi,\zeta)Q(\xi+\zeta)/P(\xi+\zeta)^2.$$

Here $\Phi'(P,\zeta;Q)$ is the differential of Φ with respect to P in the direction Q. Since the derivatives of Φ with respect to P are homogeneous in P of degree -1, we have

$$|\Phi'(P,\zeta;Q)| \leqq C\tilde{Q}(0)/\tilde{P}(0).$$

Since $\tilde{Q}(\xi)/\tilde{P}(\xi)$ ist at most of polynomial growth, it follows when $\phi \in C_0^\infty$ that $\langle E(P),\phi\rangle$ is an infinitely differentiable function of $P \in \mathrm{Pol}^\circ(m,n)$, that is, $\mathrm{Pol}^\circ(m,n) \ni P \to E(P) \in \mathscr{D}'(\mathbb{R}^n)$ is a C^∞ map. The differential is obtained by formal differentiation of (7.3.22), thus

$$\text{(10.2.4)} \quad \langle E'(P;Q),\phi\rangle = (2\pi)^{-n} \int d\xi \int \hat{\phi}(-\xi-\zeta)(\Phi'(P_\xi,\zeta;Q_\xi)/P(\xi+\zeta)$$
$$-\Phi(P_\xi,\zeta)\,Q(\xi+\zeta)/P(\xi+\zeta)^2)\,d\lambda(\zeta).$$

The parenthesis here is bounded by a constant times $\tilde{Q}(\xi)/\tilde{P}(\xi)^2$. For the differential $E^{(j)}(P;Q_1,\dots,Q_j)$ we get a similar expression with the parenthesis replaced by a sum starting with

$$\Phi^{(j)}(P_\xi,\zeta;Q_{1\xi},\dots,Q_{j\xi})/P(\xi+\zeta)$$

and ending with

$$(-1)^j j!\,\Phi(P_\xi,\zeta)Q_1(\xi+\zeta)\dots Q_j(\xi+\zeta)/P(\xi+\zeta)^{j+1}.$$

It has a bound of the form $C_j\tilde{Q}_1(\xi)\dots\tilde{Q}_j(\xi)/\tilde{P}(\xi)^{j+1}$, so the proof of Theorem 10.2.1 gives immediately

Theorem 10.2.3. *The fundamental solution $E(P)$ defined by* (7.3.22) *is a C^∞ function from* $\mathrm{Pol}^\circ(m,n)$ *to* $\mathscr{D}'(\mathbb{R}^n)$. *If* $\sup_Z|\zeta|<\varepsilon$, *then one can for every j find a constant C_j such that the differential $E^{(j)}(P;Q_1,\dots,Q_j)$ of order j in the directions $Q_1,\dots,Q_j$ satisfies*

$$\|E^{(j)}(P;Q_1,\dots,Q_j)/\cosh|\varepsilon.|\,\|_{\infty,F}\leqq C_j \quad \text{if } F=\tilde{P}^{j+1}/\tilde{Q}_1\dots\tilde{Q}_j. \tag{10.2.5}$$

The space $B_{\infty,F}$ here is again optimal in the sense that it is contained in every space $B_{p,k}$ which could be used in (10.2.5). In fact, assume that $E(P)$ is any fundamental solution of P which is (locally) a C^j function of P. Then we have

$$P(D)^{j+1}E^{(j)}(P;Q_1,\dots,Q_j)=(-1)^j j!\,Q_1(D)\dots Q_j(D)\delta. \tag{10.2.6}$$

The proof is by induction starting for $j=0$ where (10.2.6) is the definition of a fundamental solution. Assume $j\geqq 1$ and that (10.2.6) has been proved for smaller values of j. Then

$$\begin{aligned} P(D)^j E^{(j-1)}(P;Q_1,\dots,Q_{j-1}) \\ =(-1)^{j-1}(j-1)!\,Q_1(D)\dots Q_{j-1}(D)\delta. \end{aligned} \tag{10.2.7}$$

Differentiation in the direction Q_j gives

$$jP(D)^{j-1}Q_j(D)E^{(j-1)}(P;Q_1,\dots,Q_{j-1})+P(D)^jE^{(j)}(P;Q_1,\dots,Q_j)=0.$$

If we multiply by $P(D)$ and use (10.2.7), then (10.2.6) follows.

If $E^{(j)}(P;Q_1,\dots,Q_j)\in B_{p,k}^{\mathrm{loc}}$ for some $k\in\mathscr{K}$ and $p\in[1,\infty]$ then (10.2.6) and Theorem 10.1.22 give that

$$Q_1(D)\dots Q_j(D)\delta\in B^{\mathrm{loc}}_{p,k/\tilde{P}^{j+1}}.$$

We may assume that all Q_j are different from 0 and choose a regular fundamental solution F_j for each of them. Then

$$Q_{i+1}(D)\dots Q_j(D)\delta=F_i*Q_i(D)\dots Q_j(D)\delta,\qquad i=1,\dots,j,$$

so by repeated use of Theorem 10.1.24 we obtain

$$\delta\in B^{\mathrm{loc}}_{p,k/F}$$

with F as in Theorem 10.2.3. Hence $k/F\in L^p$ which as above shows that

$$B^{\mathrm{loc}}_{p,k}\supset B^{\mathrm{loc}}_{\infty,F}.$$

We shall now study the location of the singularities of $E(P)$. As a preparation we begin with a simple result on the support. Set

$$\Lambda(P)=\{\eta\in\mathbb{R}^n;\ P(\xi+t\eta)\equiv P(\xi)\}, \tag{10.2.8}$$

which is obviously a linear space, and set

$$\Lambda'(P)=\{x\in\mathbb{R}^n;\ \langle x,\eta\rangle=0 \text{ if } \eta\in\Lambda(P)\}. \tag{10.2.9}$$

Note that if $\Lambda'(P)=\{x\in\mathbb{R}^n; x_{k+1}=\ldots=x_n=0\}$, then $P(\xi)$ is independent of $\xi_{k+1},\ldots,\xi_n$ so we have a differential operators in the variables $x_1,\ldots,x_k$ only.

Theorem 10.2.4. *For the fundamental solution $E(P)$ defined by* (7.3.22) *we have* $\operatorname{supp} E(P)\subset\Lambda'(P)$.

Proof. By a linear change of variables (and the corresponding change of Φ) we can reduce the proof to the situation where $\Lambda'(P)$ is defined by $x_{k+1}=\ldots=x_n=0$. Then $\Phi(P_\xi,\zeta)/P(\xi+\zeta)$ is independent of $\xi''=(\xi_{k+1},\ldots,\xi_n)$. By Fourier's inversion formula we can then write (7.3.22) in the form

$$\langle E(P),\phi\rangle=(2\pi)^{-k}\int d\xi'\int\hat\phi_0(-\xi'-\zeta')/P(\xi'+\zeta')\,\Phi(P_{\xi'},\zeta)\,d\lambda(\zeta)$$

where $\phi_0(x')=\phi(x',0)$. This proves the theorem and moreover that $E(P)$ is the tensor product of a distribution in the plane $x''=0$ with $\delta(x'')$.

There is no fundamental solution of P with support in a smaller *linear* space:

Theorem 10.2.5. *If $P(D)$ has a fundamental solution E with* $\operatorname{supp} E\subset V$ *then $V\supset\Lambda'(P)$ if V is a linear space.*

Proof. It suffices to prove that if $x_n=0$ in V then $P(D)$ is a polynomial in $D'=(D_1,\ldots,D_{n-1})$. We can write

$$P(D)=\sum_0^m a_j(D')\,D_n^j$$

and by Theorem 2.3.5 there is a similar decomposition

$$E=\sum_0^r E_j\otimes D_n^j\delta$$

in a neighborhood of 0. Here δ denotes the δ function in x_n and E_j is a distribution in x'. The equation $P(D)E=\delta$ gives a system of equations

$$\sum_{j+k=i} a_j(D')\,E_k=0 \quad\text{if } i=1,\ldots,m+r;\qquad a_0(D')\,E_0=\delta(x').$$

If $m>0$ we obtain $a_m(D')\,E_r=0$ by taking $i=m+r$. Multiplication of the equation with $i=m+r-1$ by $a_m(D')$ then eliminates E_r and gives $a_m(D')^2E_{r-1}=0$. Continuing in this way we obtain finally

$$a_m(D')^{r+1}E_0=0,\qquad\text{thus } a_m(D')^{r+1}\delta(x')=0.$$

This implies $a_m=0$ which proves the theorem.

Let us write (7.3.22) formally as

$$E(P)(x)=(2\pi)^{-n}\int d\xi \int e^{i\langle x,\xi+\zeta\rangle}\Phi(P_\xi,\zeta)/P(\xi+\zeta)\,d\lambda(\zeta).$$

The integral with respect to ξ over a compact set gives a C^∞ function so it is clear that only the behavior of P at ∞ influences the singularities of $E(P)$. Moreover, the integral over a large neighborhood of a point far away should be rather closely related to the fundamental solution of a limit of P_η, or rather of the normalized polynomial $P_\eta/\tilde{P}(\eta)$, as $\eta\to\infty$. We shall therefore discuss such limits now.

Definition 10.2.6. The set of limits of the normalized polynomial

$$\xi\to P(\xi+\eta)/\tilde{P}(\eta)$$

in $\mathrm{Pol}^\circ(m,n)$ as $\eta\to\infty$ in $\mathbb{R}^n$ is denoted by $L(P)$. If $\theta\in\mathbb{R}^n\setminus 0$ then the set of limits with $\eta/|\eta|\to\theta/|\theta|$ is denoted by $L_\theta(P)$.

It is clear that $L(P)$ and $L_\theta(P)$ are closed subsets of the unit sphere in $\mathrm{Pol}^\circ(m,n)$. The elements in $L(P)$ and $L_\theta(P)$ as well as their non-zero multiples will be called *localizations* at ∞ (in the direction θ). They can be reached along polynomial curves:

Proposition 10.2.7. *If $Q\in L_\theta(P)$ then one can find a polynomial*

$$\eta(t)=\sum_0^k \theta_j t^j,$$

with $\theta_j\in\mathbb{R}^n$, $\theta_k=\theta$ and $k>0$, such that

$$P(\xi+\eta(t))/\tilde{P}(\eta(t))\to Q(\xi)\quad \text{as } t\to\infty. \tag{10.2.10}$$

Proof. The Tarski-Seidenberg theorem or rather Corollary A.2.6 shows that

$$\begin{aligned}c(t)=\inf\{&(|\theta-b\eta|^2+\sum|Q^{(\alpha)}(0)-aP^{(\alpha)}(\eta)|^2);\\ &a>0,\ a^2\tilde{P}(\eta)^2=1,\ |\eta|^2=|\theta|^2t^2,\ bt=1\}\end{aligned}$$

is an algebraic function of t for large t, and $\liminf_{t\to\infty} c(t)=0$ since $Q\in L_\theta(P)$. Hence $c(t)\to 0$ as $t\to\infty$. The infimum is clearly attained, and by Theorem A.2.8 it is for large t attained with η equal to an algebraic function of t,

$$\eta(t)=\sum_{-\infty}^k \theta_j t^{j/k}$$

where $\theta_k=\theta$ since $\theta-\eta(t)/t\to 0$ as $t\to\infty$. The sum from $-\infty$ to -1 tends to 0 as $t\to+\infty$. If we set

$$\eta_0(t)=\sum_0^k \theta_j t^{j/k}$$

it is therefore clear that $P(\xi+\eta_0(t))/\tilde{P}(\eta_0(t)) \to Q(\xi)$ as $t\to\infty$. Replacing t by t^k we have proved the proposition.

Theorem 10.2.8. *If* $Q\in L_\theta(P)$ *then* $\theta\in\Lambda(Q)$ *and* $\deg Q<\deg P$ *if* $\theta\notin\Lambda(P)$.

Proof. By (10.2.10) we have

$$P(\xi+\eta(t))/at^\sigma \to Q(\xi)$$

for some $a>0$ and $\sigma\geqq 0$. If s is real then

$$\eta(t+st^{1-k}/k)=\eta(t)+s\theta+O(1/t)$$

so replacing t by $t+st^{1-k}/k$ gives $Q(\xi+s\theta)=Q(\xi)$ and proves the first statement. It is clear that $\deg Q<\deg P$ if $\tilde{P}(\eta(t))\to\infty$, so the last statement follows from

Proposition 10.2.9. $\tilde{P}(\xi)\to\infty$ *if* $\xi\in\mathbb{R}^n$ *and the distance from* ξ *to* $\Lambda(P)$ *tends to* ∞.

Proof. Let M be a set in $\mathbb{R}^n$ where $\tilde{P}$ is bounded. Write $P=p+r$ where p is homogeneous of order m and not identically 0 while r is of order $<m$. Then $p^{(\alpha)}-P^{(\alpha)}$ is a constant if $|\alpha|=m-1$, so $p^{(\alpha)}$ is then bounded in M. The linear space

$$N=\{\eta\in\mathbb{R}^n;\ p^{(\alpha)}(\eta)=0 \text{ when } |\alpha|=m-1\}$$

is contained in $\Lambda(p)$. In fact, since p is homogeneous we have

$$(m-|\alpha|)\, p^{(\alpha)}(\eta)=\sum \eta_j \partial p^{(\alpha)}(\eta)/\partial\eta_j$$

so induction for decreasing $|\alpha|$ shows that $p^{(\alpha)}(\eta)=0$ if $\eta\in N$ and $|\alpha|<m$. Hence Taylor expansion at η gives $p(\xi+\eta)=p(\xi)$ if $\eta\in N$, so $N\subset\Lambda(p)$. Now M is bounded modulo N so $\tilde{p}$ is bounded in M, hence $\tilde{r}$ is bounded in M. If the proposition is already proved for polynomials of order $<m$, as we may assume, it follows that M is bounded modulo $\Lambda(r)$. Since $\Lambda(P)\supset\Lambda(r)\cap\Lambda(p)$ it follows that M is bounded modulo $\Lambda(P)$, which proves the proposition.

Proposition 10.2.7 shows how well the elements of $L(P)$ can be approximated by polynomials of the form $P(.+\eta)/\tilde{P}(\eta)$. We shall now examine how well $P(.+\eta)/\tilde{P}(\eta)$ can be approximated by elements of $L(P)$ for large $|\eta|$.

Proposition 10.2.10. *There exist positive constants* C *and* b *such that for sufficiently large* $|\eta|$

$$\inf_{Q,\theta}\{\sum_\alpha |P^{(\alpha)}(\eta)/\tilde{P}(\eta)-Q^{(\alpha)}(0)|^2+|\theta-\eta/|\eta||^2;\ Q\in L_\theta(P)\} \leqq C^2|\eta|^{-2b}. \tag{10.2.11}$$

Proof. It follows from Theorem A.2.2 that

$$\bigcup_\theta \{\theta\} \times L_\theta(P) \subset (\mathbb{R}^n \setminus 0) \times \mathrm{Pol}^\circ(m, n)$$

is a semi-algebraic set. In fact, this set is defined as

$$\{(\theta, Q); \forall \varepsilon > 0 \; \exists \eta, \varepsilon|\eta| > 1, |\theta - \eta/|\eta|| < \varepsilon,$$
$$\Sigma |P^{(\alpha)}(\eta)/\tilde{P}(\eta) - Q^{(\alpha)}(0)|^2 < \varepsilon\}$$

(It is easy to eliminate denominators and square roots by just introducing new variable names for them.) By Corollary A.2.6 the supremum when $|\eta| = t$ of the infimum in (10.2.11) is therefore an algebraic function of t for large t. It tends to 0 as $t \to \infty$, for otherwise we could find a sequence $\eta_j \to \infty$ for which (10.2.11) does not converge to 0. This would give a contradiction if we take for Q a limit of $P(.+\eta_j)/\tilde{P}(\eta_j)$ which belongs to $L_\theta(P)$ when θ is a corresponding limit of $\eta_j/|\eta_j|$. Thus we have proved Proposition 10.2.10, for an algebraic function $f(t)$ with $f(t) \to 0$ as $t \to \infty$ is always $O(t^{-2b})$ for some $b > 0$.

We are now prepared for the proof of bounds for the wave front set and singular support of the fundamental solution $E(P)$. Let F be the closure of

(10.2.12) $$\{(x, \theta) \in \mathbb{R}^n \times (\mathbb{R}^n \setminus 0); \; x \in \Lambda'(Q) \text{ for some } Q \in L_\theta(P)\},$$

and let F_0 be the projection of F in $\mathbb{R}^n$, which is closed since we can always take θ to be a unit vector. Note that Theorem 10.2.8 gives

(10.2.13) $$\langle x, \theta \rangle = 0 \quad \text{if } (x, \theta) \in F.$$

Theorem 10.2.11. *If $E(P)$ is the fundamental solution defined by* (7.3.22) *then*

(10.2.14) $$WF(E(P)) \subset F, \qquad \operatorname{sing\,supp} E(P) \subset F_0.$$

Before the proof we state and prove a weakened converse result:

Theorem 10.2.12. *Let E be any regular fundamental solution of P. If $Q \in L_\theta(P)$ it follows that Q has a regular fundamental solution e_Q such that*

(10.2.15) $$\operatorname{supp} e_Q \times \{\theta\} \subset WF(E), \qquad \operatorname{supp} e_Q \subset \operatorname{sing\,supp} E.$$

Proof of Theorem 10.2.12. The second part of (10.2.15) follows from the first by projection in $\mathbb{R}^n$. From the equation $P(D)E = \delta$ it follows that

$$P(D+\eta)(E e^{-i\langle .,\eta \rangle}) = e^{-i\langle .,\eta \rangle} P(D) E = \delta.$$

Hence

(10.2.16) $$P(D+\eta)/\tilde{P}(\eta)(E\tilde{P}(\eta)\,e^{-i\langle .,\eta\rangle})=\delta.$$

Choose a sequence $\eta_j\to\infty$ such that

(10.2.17) $$P(D+\eta_j)/\tilde{P}(\eta_j)\to Q(D), \qquad \eta_j/|\eta_j|\to\theta/|\theta|.$$

We shall show that $E\tilde{P}(\eta_j)\,e^{-i\langle .,\eta_j\rangle}$ has a convergent subsequence.

If $\phi\in C_0^\infty$ we have by hypothesis $E_\phi=\phi E\in B_{\infty,\tilde{P}}$, that is,

$$|\hat{E}_\phi(\xi)|\,\tilde{P}(\xi)\leqq C_\phi.$$

This means that for the Fourier transform $\tilde{P}(\eta_j)\,\hat{E}_\phi(\xi+\eta_j)$ of $E_\phi\tilde{P}(\eta_j)e^{-i\langle .,\eta_j\rangle}$ we have the estimate

$$\tilde{P}(\eta_j)|\hat{E}_\phi(\xi+\eta_j)|\leqq C_\phi\tilde{P}(\eta_j)/\tilde{P}(\xi+\eta_j)\leqq C_\phi(1+C|\xi|)^m.$$

(See (10.1.8).) By Theorem 10.1.10 a subsequence converges in $B_{\infty,k}$ if $k(\xi)=(1+|\xi|)^{-m-1}$. Passing to a subsequence we may therefore assume that $E_\phi\tilde{P}(\eta_j)\,e^{-i\langle .,\eta_j\rangle}$ converges in $B_{\infty,k}$ for every ϕ in a partition of unity in $\mathbb{R}^n$. Then the limit

$$G=\lim E\tilde{P}(\eta_j)\,e^{-i\langle .,\eta_j\rangle}$$

exists in $\mathscr{D}'(\mathbb{R}^n)$, and (10.2.16) implies that $Q(D)\,G=\delta$. If $G_\phi=\phi G$ then

$$|\hat{G}_\phi(\xi)|=\lim|\hat{E}_\phi(\xi+\eta_j)\,\tilde{P}(\eta_j)|\leqq C_\phi\lim\tilde{P}(\eta_j)/\tilde{P}(\xi+\eta_j)=C_\phi/\tilde{Q}(\xi)$$

so $G\in B^{\text{loc}}_{\infty,\tilde{Q}}$. To prove the theorem it remains to show that if (x,θ) is not in $WF(E)$ then $G=0$ in a neighborhood of x. To do so we choose ϕ equal to 1 in a neighborhood of x so that $\hat{E}_\phi$ is rapidly decreasing in a conic neighborhood of θ. Then it follows that

$$\hat{E}_\phi(\xi+\eta_j)\,\tilde{P}(\eta_j)\to 0$$

uniformly on every compact set as $j\to\infty$. Hence $\phi G=0$ so $G=0$ in a neighborhood of x and (10.2.15) is valid for $e_Q=G$.

Proof of Theorem 10.2.11. We shall use a partition of unity to split up the integral in (7.3.22). It is convenient to let it depend on a continuous parameter. Thus we choose a function $\chi\in C_0^\infty(\mathbb{R}^n)$ with support in the unit ball so that $\int\chi(\xi)\,d\xi=1$ and set

$$\chi_\varepsilon(\xi,\eta)=(1+|\xi|^2)^{-\varepsilon n/2}\chi((\eta-\xi)(1+|\xi|^2)^{-\varepsilon/2})$$

where $\varepsilon\in(0,1)$ will be chosen quite small later on. Then

(10.2.18) $$\int\chi_\varepsilon(\xi,\eta)\,d\eta=1$$

and $|\xi-\eta|<(1+|\xi|^2)^{\varepsilon/2}$ in $\operatorname{supp}\chi_\varepsilon$. For large η it follows that ξ is large and that $|\xi-\eta|<|\xi|/2$, hence $3|\xi|/2>|\eta|>|\xi|/2$. This gives

(10.2.19) $$|\xi-\eta|<C_\varepsilon(1+|\eta|)^\varepsilon \quad \text{if } (\xi,\eta)\in\operatorname{supp}\chi_\varepsilon.$$

Finally it is clear that for an arbitrary multi-index α

(10.2.20) $$|D_\xi^\alpha \chi_\varepsilon(\xi,\eta)|\leqq C_{\varepsilon,\alpha}(1+|\eta|)^{-\varepsilon(n+|\alpha|)}.$$

With ε to be chosen later we set

(10.2.21) $$E(x,\eta)=(2\pi)^{-n}\int \chi_\varepsilon(\xi,\eta)\,d\xi \int e^{i\langle x,\xi+\zeta\rangle}\Phi(P_\xi,\zeta)/P(\xi+\zeta)\,d\lambda(\zeta)$$

and note that (10.2.18) gives

$$\langle E,\phi\rangle=\int\langle E(.,\eta),\phi\rangle\,d\eta.$$

We shall prove Theorem 10.2.11 by making appropriate estimates of $E(x,\eta)$.

The definition of F as the closure of the set (10.2.12) means that if $(x_0,\theta_0)\notin F$ then there is an open neighborhood ω_0 of x_0 and an open conic neighborhood Γ_0 of θ_0 such that

$$\omega_0\cap\Lambda'(Q)=\emptyset \quad \text{if } Q\in L_\theta(P) \quad \text{for some } \theta\in\Gamma_0.$$

We must show that if $\phi\in C_0^\infty(\omega_0)$ it follows that $(\widehat{\phi E})(\tau)\to 0$ rapidly when $\tau\to\infty$ in a closed cone $\Gamma\subset\Gamma_0$. Let Γ_1 and Γ_2 be open cones with

$$\Gamma\subset\Gamma_2\Subset\Gamma_1\Subset\Gamma_0$$

where $\Subset$ means inclusion for the closure (in $\mathbb{R}^n\setminus 0$). For every N we have

(10.2.22) $$|\widehat{\phi E}(\tau,\eta)|\leqq C_N(1+|\tau|)^{-N}(1+|\eta|)^{-N}$$

if $\tau\in\Gamma$ and $\eta\notin\Gamma_1$. In fact,

$$\widehat{\phi E}(\tau,\eta)=(2\pi)^{-n}\int\chi_\varepsilon(\xi,\eta)\,d\xi\int\hat\phi(\tau-\xi-\zeta)\,\Phi(P_\xi,\zeta)/P(\xi+\zeta)\,d\lambda(\zeta).$$

When $\Phi(P_\xi,\zeta)\neq 0$ we can estimate $|\hat\phi(\tau-\xi-\zeta)|$ by $C_N(1+|\tau-\xi|)^{-N}$ for any N. If $(\xi,\eta)\in\operatorname{supp}\chi_\varepsilon$ then (10.2.19) implies that $\xi\notin\Gamma_2$ when $\eta\notin\Gamma_1$ and $|\eta|>C_0$. For $\tau\in\Gamma$ and $\xi\notin\Gamma_2$ we have $|\tau|+|\xi|\leqq C_1|\tau-\xi|$ since this is true when $|\tau|+|\xi|=1$. Hence $|\tau|+|\eta|\leqq C_2|\tau-\xi|$ when $\tau\in\Gamma$, $\eta\notin\Gamma_1$, $|\eta|>C_0$ and $(\xi,\eta)\in\operatorname{supp}\chi_\varepsilon$. This proves (10.2.22). If we show that

$$\int_{\Gamma_1}\phi(x)\,E(x,\eta)\,d\eta$$

is a C^∞ function it will follow that $(\widehat{\phi E})(\tau)\to 0$ rapidly when $\tau\to\infty$ in Γ.

By Proposition 10.2.10 we can for large $\eta\in\Gamma_1$ find $Q^\eta\in L_\theta(P)$ for some $\theta\in\Gamma_0$ so that

(10.2.23) $$P(\xi+\eta)/\tilde P(\eta)=Q^\eta(\xi)+R^\eta(\xi), \qquad \tilde R^\eta(0)\leqq C|\eta|^{-b}.$$

For any ξ with $|\xi|<K|\eta|^{\varepsilon}$ and $t\in\Lambda(Q^{\eta})$ we have in view of (10.1.8)

$$\begin{aligned}|D_{\xi}^{\alpha}\langle t, D_{\xi}\rangle^{j}P(\xi+\eta)| &= \tilde{P}(\eta)|D^{\alpha}\langle t, D\rangle^{j}R^{\eta}(\xi)| \\ &\leqq C\tilde{P}(\xi+\eta)|\eta|^{m\varepsilon-b+(m-j)\varepsilon}|t|^{j},\end{aligned}$$

if $j>0$ so that $\langle t, D_{\xi}\rangle$ annihilates Q^{η}. If ε is so small that $2m\varepsilon\leqq b$ we obtain for large $\eta\in\Gamma_1$ and $t\in\Lambda(Q^{\eta})$, now for $j\geqq 0$,

(10.2.24) $$|D_{\xi}^{\alpha}\langle t, D_{\xi}\rangle^{j}P(\xi+\eta)|\leqq C\tilde{P}(\xi+\eta)|\eta|^{-\varepsilon j}|t|^{j}, \qquad |\xi|\leqq K|\eta|^{\varepsilon}.$$

This will give bounds for the derivatives of $\Phi(P_{\xi+\eta}, \zeta)/P(\xi+\eta+\zeta)$ if we prove the following

Lemma 10.2.13. *Let f be a C^{∞} function which is homogeneous of degree μ in $Y\setminus\{0\}$ where Y is a finite dimensional vector space over $\mathbb{R}$. If $s\to y(s)$ is a C^{k} function from $\mathbb{R}$ to Y then*

(10.2.25) $$|d^{k}f(y(s))/ds^{k}|\leqq C_{k}\|y(s)\|^{\mu}\sup_{0<j\leqq k}(\|y^{(j)}(s)\|/\|y(s)\|)^{k/j}.$$

Proof. We may assume that $s=0$, and multiplication of y and s by constant factors reduces the proof to the case where $\|y(0)\|=1$ and $\sup_{0<j\leqq k}\|y^{(j)}(0)\|=1$. But then the inequality (10.2.25) is an immediate consequence of the rules for differentiation of composite functions. Note that C_{k} only depends on k and the derivatives of f of order $\leqq k$ on the unit sphere.

End of proof of Theorem 10.2.11. We shall apply Lemma 10.2.13 to $Y=\operatorname{Pol}(m, n)$ with the norm $\|Q\|=\tilde{Q}(0)$ and $f(P)=\Phi(P, \zeta)/P(\zeta)$, $\mu=-1$, $y(s)=P_{st+\xi+\eta}$. The estimate (10.2.24) means that

$$\|y^{(j)}(0)\|=(\sum|D_{\xi}^{\alpha}\langle t, D_{\xi}\rangle^{j}P(\xi+\eta)|^{2})^{\frac{1}{2}}\leqq C\|y(0)\|(|t|/|\eta|^{\varepsilon})^{j}.$$

Hence

$$\begin{aligned}|\langle t, D_{\xi}\rangle^{j}\Phi(P_{\xi+\eta}, \zeta)/P(\xi+\eta+\zeta)| &\leqq C_{j}\tilde{P}(\xi+\eta)^{-1}|\eta|^{-\varepsilon j}|t|^{j}; \\ |\xi|\leqq K|\eta|^{\varepsilon}, &\quad \eta\in\Gamma_1,\ t\in\Lambda(Q^{\eta}).\end{aligned}$$

Replacing ξ by $\xi-\eta$ and taking (10.2.20) into account gives

$$|\langle t, D_{\xi}\rangle^{j}\chi_{\varepsilon}(\xi, \eta)\,\Phi(P_{\xi}, \zeta)/P(\xi+\zeta)|\leqq C_{j}|\eta|^{-\varepsilon(n+j)}|t|^{j}.$$

If we multiply (10.2.21) by $\langle t, x\rangle^{j}$, write

$$\langle t, x\rangle^{j}e^{i\langle x, \xi+\zeta\rangle}=\langle t, D_{\xi}\rangle^{j}e^{i\langle x, \xi+\zeta\rangle}$$

and integrate by parts, it follows now that for large $\eta\in\Gamma_1$ and $t\in\Lambda(Q^{\eta})$

(10.2.26) $$|\langle t, x\rangle^{j}E(x, \eta)|\leqq C_{j}(|t|/|\eta|^{\varepsilon})^{j}.$$

(Note that the measure of the set $\{\xi;\ \chi_\varepsilon(\xi,\eta)\neq 0\}$ is $\leqq C|\eta|^{\varepsilon n}$.) The distance $d(x,\Lambda'(Q))$ from x to $\Lambda'(Q)$ is

$$\sup_{t\in\Lambda(Q)} |\langle t,x\rangle|/|t|$$

so the previous estimate means that for large $\eta\in\Gamma_1$

$$|E(x,\eta)|\leqq C_j(d(x,\Lambda'(Q^\eta))|\eta|^\varepsilon)^{-j},\quad j=0,1,\ldots. \tag{10.2.27}$$

When $x\in\operatorname{supp}\phi$ there is a positive lower bound for $d(x,\Lambda'(Q^\eta))$. Since differentiation of $E(x,\eta)$ with respect to x_j only introduces a factor $\xi_j+\zeta_j$ in (10.2.21) it is clear that the same proof gives for all α

$$|D_x^\alpha E(x,\eta)|\leqq C_{j,\alpha}|\eta|^{|\alpha|-\varepsilon j},\quad j=0,1,\ldots,x\in\operatorname{supp}\phi,\quad \eta\in\Gamma_1. \tag{10.2.28}$$

If we choose j so large that $j\varepsilon-|\alpha|>n$, then the right-hand side is integrable so it follows that

$$\int_{\Gamma_1}\phi(x)\,E(x,\eta)\,d\eta$$

is a C^∞ function. This completes the proof.

In Section 8.3 we have seen the importance of information on $WF(E)$ for a fundamental solution E. We shall return to this topic in Chapter XI. The main virtue of the construction of fundamental solutions studied in this section is its universal character. However, it does not always give a fundamental solution E with minimal $WF(E)$. For example, in Theorem 8.3.7 $WF(E_\pm)$ was roughly half of the set F in Theorem 10.2.11. This gap is related to the fact that the localized operators Q may have fundamental solutions with support much smaller than $\Lambda'(Q)$, for example a half line if $\Lambda'(Q)$ is a line.

Every differential operator with constant coefficients has fundamental solutions in $\mathscr{S}'$. Several proofs are known of this fact (see the notes at the end of the chapter). However, there does not always exist a fundamental solution which is both regular and temperate:

Theorem 10.2.14. *Assume that $P(D)$ has a fundamental solution $E\in\mathscr{S}'\cap B^{\mathrm{loc}}_{\infty,\tilde P}(\mathbb{R}^n)$. Then there is a constant C such that*

$$\int_{|\eta|<1} d\eta/Q(\eta)<C \tag{10.2.29}$$

for every $Q\in L(P)$ which is positive in $\mathbb{R}^n$.

Proof. By Proposition 10.2.7 we can choose a polynomial $t\to\eta(t)\in\mathbb{R}^n$ such that for large t

$$|Q(\xi)-P(\xi+\eta(t))/\tilde P(\eta(t))|\leqq Ct^{-1}(1+|\xi|)^m,\quad \xi\in\mathbb{R}^n. \tag{10.2.30}$$

Moreover, as shown in Example A.2.7 it follows from Corollary A.2.6 that

$$Q(\xi) > 2c(1+|\xi|)^{-\mu} \tag{10.2.31}$$

for some $c>0$ and μ. Hence

$$|P(\xi+\eta(t))/\tilde{P}(\eta(t))| > c(1+|\xi|)^{-\mu} \tag{10.2.32}$$

when

$$ct > C(1+|\xi|)^{m+\mu}.$$

Now consider the fundamental solution

$$E_t = \tilde{P}(\eta(t))(\exp - i\langle .,\eta(t)\rangle)\,E$$

of $P(D+\eta(t))/\tilde{P}(\eta(t))$. The Fourier transform $\hat{E}_t$ satisfies the equation $P(.+\eta(t))/\tilde{P}(\eta(t))\,\hat{E}_t = 1$, so

$$\hat{E}_t(\xi) = \tilde{P}(\eta(t))/P(\xi+\eta(t)) \tag{10.2.33}$$

when

$$(1+|\xi|)^{m+\mu} < ct/C.$$

It follows that $\hat{E}_t \to 1/Q$ in $\mathscr{D}'$ when $t\to\infty$; we shall prove that there is convergence in $\mathscr{S}'$. To do so we choose $\chi\in C_0^\infty$ with support in the unit ball and $\chi(\xi)=1$ when $|\xi|<1/2$. For $\phi\in\mathscr{S}$ we make the decomposition

$$\phi = \phi_1^T + \phi_2^T; \quad \phi_1^T(\xi) = \chi(\xi/T)\,\phi(\xi).$$

If $(1+T)^{m+\mu} < ct/C$ then (10.2.30)–(10.2.33) give

$$|\langle \hat{E}_t - Q^{-1}, \phi_1^T\rangle| \leqq Cc^{-2}t^{-1}\int (1+|\xi|)^{m+2\mu}|\phi_1^T(\xi)|\,d\xi \to 0, \qquad t\to\infty,$$
$$|\langle Q^{-1}, \phi_2^T\rangle| \leqq c^{-1}\int_{|2\xi|>T} (1+|\xi|)^{\mu}|\phi(\xi)|\,d\xi \to 0, \qquad T\to\infty.$$

We have

$$\langle \hat{E}_t, \phi_2^T\rangle = \tilde{P}(\eta(t))\langle \hat{E}, \phi_2^T(.-\eta(t))\rangle,$$

and

$$\sup|\xi^\beta D_\xi^\alpha \phi_2^T(\xi-\eta(t))| = \sup|(\xi+\eta(t))^\beta D^\alpha \phi_2^T(\xi)|$$
$$\leqq (1+|\eta(t)|)^{|\beta|}\sup(1+|\xi|)^{|\beta|}|D^\alpha\phi_2^T(\xi)|.$$

We can choose $T=at^b$ with $a,b>0$ so small that $(1+T)^{m+\mu}<ct/C$. Then it follows that $\tilde{P}(\eta(t))\,\phi_2^T(.-\eta(t))\to 0$ in $\mathscr{S}$ when $t\to\infty$, hence $\hat{E}_t\to 1/Q$ in $\mathscr{S}'$.

The hypothesis that $E\in B_{\infty,\tilde{P}}^{\mathrm{loc}}$ means that

$$\sup \tilde{P}(\xi)|\hat{E}*\hat{\phi}(\xi)| \leqq (2\pi)^n\|\phi E\|_{\infty,\tilde{P}} = C_\phi < \infty, \qquad \phi\in C_0^\infty(\mathbb{R}^n).$$

When $\zeta=\eta(t)$ we obtain

$$|\hat{E}_t*\hat{\phi}(0)|\leqq C_\phi,$$

and letting $t\to\infty$ it follows that

$$|\int\hat{\phi}(-\xi)/Q(\xi)\,d\xi|\leqq C_\phi. \tag{10.2.34}$$

We can choose ϕ so that $\hat{\phi}\geqq 0$ and $\hat{\phi}>0$ in the unit ball, for example by taking $\phi=\psi*\tilde{\psi}$ where ψ is close to the δ function and $\tilde{\psi}(x)=\psi(-x)$. (10.2.29) is then a consequence of (10.2.34). The proof is complete.

Example 10.2.15. For the polynomial $P(\xi)=\xi_1^2\xi_2^2+\xi_3^2+i\xi_4$ the localization in the direction $(0,1,\varepsilon,0)$ is $(\xi_1^2+\varepsilon^2)(4+\varepsilon^4)^{-\frac{1}{2}}$. Since $1/\xi_1^2$ is not locally integrable it is clear that there is no uniform estimate of the form (10.2.29). Hence the unique temperate fundamental solution

$$\begin{aligned}E(x)=(2\pi)^{-3}\int\exp(i(x_1\xi_1+x_2\xi_2+x_3\xi_3)\\ -x_4(\xi_1^2\xi_2^2+\xi_3^2))\,d\xi_1\,d\xi_2\,d\xi_3 \quad \text{if } x_4>0,\\ E(x)=0 \quad \text{if } x_4\leqq 0\end{aligned}$$

is not regular.

10.3. The Equation $P(D)u=f$ when $f\in\mathscr{E}'$

We shall now combine the general observations in Section 4.4 on the use of fundamental solutions with the construction of regular fundamental solutions given in Section 10.2. This leads to the following two theorems.

Theorem 10.3.1. *Let E be the regular fundamental solution given by (7.3.22). Then the solution of the equation*

$$P(D)u=f\in\mathscr{E}' \tag{10.3.1}$$

which in virtue of (4.4.3) *is given by*

$$u=E*f \tag{10.3.2}$$

belongs to $B^{\mathrm{loc}}_{p,k\tilde{P}}(\mathbb{R}^n)$ for every $k\in\mathscr{K}$ and $p\in[1,\infty]$ such that $f\in B_{p,k}(\mathbb{R}^n)$, and we have

$$WF(u)\subset\{(x+y,\xi);\,(x,\xi)\in WF(f)\text{ and }(y,\xi)\in F\}. \tag{10.3.3}$$

Here F is the closure of the set (10.2.12).

Proof. This is an immediate consequence of Theorem 10.1.24, Theorem 10.2.11 and (8.2.16).

Remark. Theorem 10.1.22 shows that if the equation $P(D)u=f$ has a solution $u\in B^{\mathrm{loc}}_{p,k\tilde{P}}$ it follows that $f\in B^{\mathrm{loc}}_{p,k}$. Hence the solution of (10.3.1) given by (10.3.2) has the best possible local properties as described by the spaces $B^{\mathrm{loc}}_{p,k}$. However, (10.3.3) is not always optimal.

The following theorem adds considerable precision to Theorem 7.3.2.

Theorem 10.3.2. *If* $u\in\mathscr{E}'$, $k\in\mathscr{K}$ *and* $1\leqq p\leqq\infty$, *we have* $u\in B_{p,k\tilde{P}}$ *if and only if* $P(D)u\in B_{p,k}$.

Proof. That $u\in B_{p,k\tilde{P}}$ implies $P(D)u\in B_{p,k}$ follows from Theorem 10.1.11, and the converse follows from (4.4.2) combined with Theorem 10.1.24 if we choose a regular fundamental solution E.

Corollary 10.3.3. *Let* $u\in\mathscr{E}'$, $k\in\mathscr{K}$ *and* $1\leqq p\leqq\infty$. *If* $P(D)u\in B_{p,k}$ *it then follows that* $P(D)(\phi u)\in B_{p,k}$ *for every* $\phi\in C_0^\infty$.

Proof. This follows immediately from Theorem 10.3.2 and Theorem 10.1.15.

It is often useful to rephrase the above results in terms of the concept introduced in the following definition.

Definition 10.3.4. If $P(D)$ and $Q(D)$ are differential operators such that

$$\tilde{Q}(\xi)/\tilde{P}(\xi)<C, \quad \xi\in\mathbb{R}^n, \tag{10.3.4}$$

we shall say that Q is weaker than P and write $Q\prec P$, or that P is stronger than Q and write $P\succ Q$. If $P\prec Q\prec P$, the operators are called equally strong.

Theorems 10.3.1 and 10.3.2 now assume the following form:

Theorem 10.3.5. *If* $f\in B^{c}_{p,k}$, *where* $k\in\mathscr{K}$ *and* $1\leqq p\leqq\infty$, *and if* $Q\prec P$, *we have* $Q(D)u\in B^{\mathrm{loc}}_{p,k}$ *for the solution of the equation* (10.3.1) *given in Theorem* 10.3.1. *Conversely, if for some* $k\in\mathscr{K}$ *and some* p *with* $1\leqq p\leqq\infty$ *the equation* (10.3.1) *has a solution* u *such that* $Q(D)u\in B^{\mathrm{loc}}_{p,k}$ *for every* $f\in B^{c}_{p,k}$, *it follows that* $Q\prec P$.

Proof. To prove the first part we only have to note that $u\in B^{\mathrm{loc}}_{p,k\tilde{P}}$ implies that $Q(D)u\in B^{\mathrm{loc}}_{p,k\tilde{P}/\tilde{Q}}$ (Theorem 10.1.22) and then apply Theorem 10.1.21. To prove the second part, we observe that $Q(D)u\in B_{p,k}^{\mathrm{loc}}$ implies that $Q(D)f=P(D)Q(D)u\in B_{p,k/\tilde{P}}^{\mathrm{loc}}$ (Theorem 10.1.22). Since $f\in\mathscr{E}'$ it follows from Theorem 10.3.2 that $f\in B_{p,k\tilde{Q}/\tilde{P}}$. Hence

$$B_{p,k}^c\subset B_{p,k\tilde{Q}/\tilde{P}}^c$$

which implies that $Q\prec P$ in view of Theorem 10.1.8.

Theorem 10.3.6. *Let $k\in\mathscr{K}$ and $1\leqq p\leqq\infty$. Then $Q\prec P$ if and only if $u\in\mathscr{E}'$ and $P(D)u\in B_{p,k}$ imply that $Q(D)u\in B_{p,k}$.*

Proof. Since Theorem 10.3.2 shows that $P(D)u\in B_{p,k}$ is equivalent to $u\in B_{p,k\tilde{P}}$ and that $Q(D)u\in B_{p,k}$ is equivalent to $u\in B_{p,k\tilde{Q}}$, when $u\in\mathscr{E}'$, this is an immediate consequence of Theorem 10.1.8.

We may summarize Theorems 10.3.5 and 10.3.6 roughly as follows: $Q(D)u$ is as regular as $P(D)u$ if and only if $Q\prec P$. Such conclusions are obviously important in the study of perturbations of differential operators with constant coefficients (see Chapter XIII). To illustrate the results we shall write out in detail the special case $p=2$, $k=1$, that is, $B_{p,k}=L^2$.

Theorem 10.3.7. *If X is a bounded open set in $\mathbb{R}^n$, there exists a bounded linear operator E in $L^2(X)$ such that*

(10.3.5) $\quad P(D)Ef=f, \quad f\in L^2(X),$

(10.3.6) $\quad EP(D)u=u \quad$ *if* $u\in\mathscr{E}'(X)$, $P(D)u\in L^2(X)$,

(10.3.7) $\quad Q(D)E$ *is a bounded operator in* L^2 *for every* $Q\prec P$.

Proof. If $f\in L^2(X)$ we denote by f_0 the function which equals f in X and vanishes in $\mathbb{R}^n\setminus X$. We then define Ef as the restriction of $u_0=E_0*f_0$ to X where E_0 is a regular fundamental solution. Then (10.3.5) and (10.3.6) follow immediately from (4.4.2) and (4.4.3). Further, it follows from Theorem 10.3.5 that $Q(D)E$ maps $L^2(X)$ into itself. To conclude that $Q(D)E$ is bounded we could therefore apply the closed graph theorem. However, we prefer to prove this fact by repeating the proof of Theorem 10.1.24. Thus let ψ be a function in $C_0^\infty(\mathbb{R}^n)$ which is equal to 1 in a neighborhood of the closure of $X-X=\{x-y;\ x,y\in X\}$, and set $F_0=\psi E_0$. Then we have $F_0\in B_{\infty,\tilde{P}}$ and $F_0*f_0=E_0*f_0=u_0$ in X in view of Theorem 4.2.4. Hence

$$\begin{aligned}\|Q(D)Ef\|_{L^2}&\leqq\|Q(D)(F_0*f_0)\|_{2,1}\leqq\|f_0\|_{2,1}\|Q(D)F_0\|_{\infty,1}\\&\leqq\|f\|_{L^2}\|F_0\|_{\infty,\tilde{P}}\sup|Q(\xi)|/\tilde{P}(\xi),\end{aligned}$$

which proves the theorem.

Another interesting application is the following:

Theorem 10.3.8. *If $Q \prec P$ and $P(\xi) \geqq 0$, $\xi \in \mathbb{R}^n$, we have for every bounded open set $X \subset \mathbb{R}^n$ with a constant C_X*

$$|\int (Q(D)u)\bar{u}\,dx| \leqq C_X \int (P(D)\,u)\,\bar{u}\,dx, \quad u \in C_0^\infty(X). \tag{10.3.8}$$

Proof. This follows from Theorem 10.3.6 with $p=1$, $k=1$, applied to the convolution $u * \tilde{u}$ where $\tilde{u}(x) = \overline{u(-x)}$ is the function with Fourier transform $|\hat{u}|^2$. The details of the proof are left for the reader.

10.4. Comparison of Differential Operators

We shall here study the formal rules for using the pre-order relation $Q \prec P$ introduced in Definition 10.3.4, and also discuss some examples. In the proofs we shall only use Definition 10.3.4 although some of our statements have simple proofs by means of Theorems 10.3.5 and 10.3.6.

Theorem 10.4.1. *If $Q_1 \prec P$ and $Q_2 \prec P$ it follows that $a_1 Q_1 + a_2 Q_2 \prec P$ for arbitrary complex constants a_1 and a_2. If $Q_1 \prec P_1$ and $Q_2 \prec P_2$ we have $Q_1 Q_2 \prec P_1 P_2$. On the other hand, if $Q_1 Q_2 \prec P_1 P_2$ and $Q_1 \succ P_1$, it follows that $Q_2 \prec P_2$.*

Proof. The first rule, for addition, is trivial. The others require a lemma.

Lemma 10.4.2. *There exists a constant C such that for every polynomial Q of degree $\leqq m$ in $\mathbb{R}^n$*

$$\tilde{Q}(\xi, t)/C \leqq \sup_{|\eta|<t} |Q(\xi+\eta)| \leqq C\tilde{Q}(\xi, t); \qquad \xi \in \mathbb{R}^n,\ t>0. \tag{10.4.1}$$

Here we have used the notation

$$\tilde{Q}(\xi, t) = (\sum |Q^{(\alpha)}(\xi)|^2 t^{2|\alpha|})^{\frac{1}{2}}. \tag{10.4.2}$$

Proof. If we replace Q by the polynomial

$$\eta \to Q(\xi + t\eta)$$

the proof is reduced to the case where $\xi = 0$ and $t = 1$. The statement is then obvious since $\sup_{|\eta|<1} |Q(\eta)|$ and $\tilde{Q}(0,1) = \tilde{Q}(0)$ are norms in the finite dimensional vector space $\mathrm{Pol}(n, m)$, hence equivalent.

End of proof of Theorem 10.4.1. It suffices to prove that

$$C'\tilde{P}(\xi)\tilde{Q}(\xi) \leqq (\widetilde{PQ})(\xi) \leqq C''\tilde{P}(\xi)\tilde{Q}(\xi) \tag{10.4.3}$$

for polynomials P and Q in $\mathbb{R}^n$ of degree $\leqq m$. The second inequality follows at once if the derivatives of PQ are computed by Leibniz' rule. To prove the first we use Lemma 10.4.2 to choose for given $\xi \in \mathbb{R}^n$ an element $\eta \in \mathbb{R}^n$ with $|\eta| \leqq 1$ and

$$|Q(\xi+\eta)| \geqq \tilde{Q}(\xi)/C.$$

By (10.1.8) we have $\tilde{Q}(\xi)/C \geqq \tilde{Q}(\xi+\eta)/C_1$, hence Taylor's formula gives

$$\begin{aligned}|Q(\xi+\eta+\theta)| &\geqq |Q(\xi+\eta)| - \sum_{\alpha \neq 0} |Q^{(\alpha)}(\xi+\eta)\theta^\alpha|/\alpha! \\ &\geqq \tilde{Q}(\xi+\eta)(1/C_1 - C_2|\theta|), \qquad |\theta|<1.\end{aligned}$$

When $|\theta| < 1/2C_1C_2$ we obtain

$$|Q(\xi+\eta+\theta)| \geqq \tilde{Q}(\xi+\eta)/2C_1$$

and since $P = PQ/Q$ we conclude that

$$|P(\xi+\eta+\theta)| \leqq 2C_1 |(PQ)(\xi+\eta+\theta)|/\tilde{Q}(\xi+\eta), \qquad |\theta| < 1/2C_1C_2.$$

Using (10.4.1) and (10.1.8) we now obtain

$$\tilde{P}(\xi) \leqq C_3 \tilde{P}(\xi+\eta) \leqq C_4 (\widetilde{PQ})(\xi+\eta)/\tilde{Q}(\xi+\eta) \leqq C_5 (\widetilde{PQ})(\xi)/\tilde{Q}(\xi)$$

which proves (10.4.3).

Theorem 10.4.1 makes it natural to extend the pre-order relation to rational functions although we shall not actually make use of this. In fact, it follows from Theorem 10.4.1 that if R_1 and R_2 are rational functions, we have

$$PR_1 \prec PR_2 \tag{10.4.4}$$

either for every polynomial $P \neq 0$ such that PR_1 and PR_2 are polynomials or else for no such P. If (10.4.4) is fulfilled when PR_1 and PR_2 are polynomials, we define $R_1 \prec R_2$ and the following rules follow at once:

$$\begin{aligned}&R_1 \prec R_2 \Leftrightarrow R_2^{-1} \prec R_1^{-1}; \\ &R_1 \prec S_1 \quad \text{and} \quad R_2 \prec S_2 \Rightarrow R_1 S_1 \prec R_2 S_2.\end{aligned} \tag{10.4.5}$$

We shall now give two alternative characterization of the relation $Q \prec P$.

Theorem 10.4.3. *Each of the following conditions is necessary and sufficient in order that $Q \prec P$:*

a) *There is a constant C' such that*

$$|Q(\xi)| \leqq C' \tilde{P}(\xi), \quad \xi \in \mathbb{R}^n. \tag{10.4.6}$$

b) *There is a constant C'' such that*

$$\tilde{Q}(\xi, t) \leqq C'' \tilde{P}(\xi, t); \quad \xi \in \mathbb{R}^n, \ t \geqq 1. \tag{10.4.7}$$

Proof. It is trivial that b) implies that $Q \prec P$, which implies a). Hence we only have to prove that a) implies b). To do so we use Lemma 10.4.2,

$$\begin{aligned}\tilde{Q}(\xi, t) &\leqq C \sup_{|\eta|<t} |Q(\xi+\eta)| \leqq C'C \sup_{|\eta|<t} \tilde{P}(\xi+\eta)\\ &\leqq C'C^2 \sup_{|\eta|<t+1} |P(\xi+\eta)| \leqq C'C^3 \tilde{P}(\xi, t+1)\\ &\leqq C'C^3(1+t^{-1})^m \tilde{P}(\xi, t).\end{aligned}$$

This proves the theorem.

We now introduce another order relation which is closely connected with the relation $Q \prec P$. The connection will be clarified in a moment.

Definition 10.4.4. We shall say that P dominates Q and write $P \gg Q$ or $Q \ll P$ if

$$\sup_{\xi} \tilde{Q}(\xi, t)/\tilde{P}(\xi, t) \to 0, \quad t \to \infty, \tag{10.4.8}$$

where ξ and t are real variables.

Note that $P^{(\alpha)} \ll P$ for every $\alpha \neq 0$. It is obvious that $Q \prec P$ if $Q \ll P$. The following are analogues of Theorems 10.4.1 and 10.4.3.

Theorem 10.4.5. *If $Q_1 \ll P$ and $Q_2 \ll P$ it follows that $a_1 Q_1 + a_2 Q_2 \ll P$ for arbitrary complex constants a_1 and a_2. If $Q_1 \ll P_1$ and $Q_2 \prec P_2$ we have $Q_1 Q_2 \ll P_1 P_2$. On the other hand, if $Q_1 Q_2 \ll P_1 P_2$ and $Q_1 \succ P_1$, it follows that $Q_2 \ll P_2$. Similarly, $Q_1 Q_2 \prec P_1 P_2$ and $Q_1 \gg P_1$ implies that $Q_2 \ll P_2$.*

Proof. If we apply (10.4.3) to $P(t\xi)$ and $Q(t\xi)$ we obtain

$$C' \tilde{P}(\xi, t) \tilde{Q}(\xi, t) \leqq (\widetilde{PQ})(\xi, t) \leqq C'' \tilde{P}(\xi, t) \tilde{Q}(\xi, t). \tag{10.4.3$'$}$$

No other change is required in the proof of Theorem 10.4.1.

Theorem 10.4.6. *In order that $Q \ll P$ it is sufficient that*

$$\inf_t (\sup_\xi |Q(\xi)|/\tilde{P}(\xi, t)) = 0. \tag{10.4.9}$$

Proof. With the notation $C(t) = \sup_\xi |Q(\xi)|/\tilde{P}(\xi, t)$ we obtain as in the proof of Theorem 10.4.3 if s and t are >0

$$\begin{aligned}\tilde{Q}(\xi, t) &\leqq C \sup_{|\eta|<t} |Q(\xi+\eta)| \leqq C\, C(s) \sup_{|\eta|<t} \tilde{P}(\xi+\eta, s)\\ &\leqq C^2\, C(s) \sup_{|\eta|<t+s} |P(\xi+\eta)| \leqq C^3\, C(s) \tilde{P}(\xi, t+s)\\ &\leqq C^3\, C(s)(1+s/t)^m \tilde{P}(\xi, t).\end{aligned}$$

Hence it follows that

$$\overline{\lim_{t\to\infty}} (\sup_\xi \tilde{Q}(\xi, t)/\tilde{P}(\xi, t)) \leqq C^3\, C(s)$$

and since $\inf C(s) = 0$ this proves (10.4.8). The proof is complete.

Let P_0 be a fixed polynomial in $\mathbb{R}^n$ of degree $m>0$ and let

$$W = \{P;\ P \prec P_0\}, \qquad W_0 = \{P;\ P \ll P_0\}. \tag{10.4.10}$$

The degrees of the polynomials in W and W_0 are $\leqq m$ and $\leqq m-1$ respectively, so these are finite dimensional complex vector spaces. If $P \in W$ then $P^{(\alpha)} \in W_0$ for every $\alpha \neq 0$, hence the polynomial $\xi \to P(\xi+\theta) - P(\xi)$ is in W_0.

Let E be the open subset of W which consists of all polynomials which are equally strong as P_0. If $P \in E$ then $R = P + Q \in E$ if $Q \in W_0$, for

$$\tilde{P}(\xi, t) \leqq \tilde{R}(\xi, t) + \tilde{Q}(\xi, t) \leqq \tilde{R}(\xi, t) + \tilde{P}(\xi, t)/2$$

if t is large enough, so $P_0 \prec P \prec R$. We shall prove that E is characterized by the non-vanishing of a family of linear forms on W; it is then clear that these must vanish on W_0.

If $\xi \in \mathbb{R}^n$ we define

$$L_\xi(Q) = Q(\xi)/\tilde{P}_0(\xi), \qquad Q \in W.$$

The set $\mathscr{L} = \{L_\xi;\ \xi \in \mathbb{R}^n\}$ is a bounded subset of the dual space W' of W. Since

$$\sum_\alpha |L_\xi(P_0^{(\alpha)})|^2 = 1$$

the closure $\bar{\mathscr{L}}$ is a compact subset of W' which does not contain 0. Let

$$\mathscr{L}_0 = \{L \in \bar{\mathscr{L}};\ L(W_0) = 0\}.$$

Note that if $L_{\xi_j} \to L$ and $\tilde{P}_0(\xi_j, t_j)/\tilde{P}_0(\xi_j)$ is bounded for some sequence $t_j \to \infty$, then $L \in \mathscr{L}_0$ because

$$|L(Q)| \leqq C \varlimsup |Q(\xi_j)|/\tilde{P}_0(\xi_j, t_j).$$

Theorem 10.4.7. *The set E of polynomials which are equally strong as P_0 is the subset of $W = \{Q;\, Q \prec P_0\}$ defined by*

$$E = \{Q \in W;\ L(Q) \neq 0 \text{ for every } L \in \mathscr{L}_0\}$$

where $\mathscr{L}_0$ is a compact subset of $W' \setminus \{0\}$ such that

$$W_0 = \{Q;\ Q \ll P_0\} = \{Q \in W;\ L(Q) = 0 \text{ for every } L \in \mathscr{L}_0\}.$$

Proof. a) Assume that $Q \in W$ is not equally strong as P_0. Then we can choose a sequence $\xi_j \to \infty$ such that

$$\tilde{Q}(\xi_j)/\tilde{P}_0(\xi_j) \to 0.$$

We can then choose $t_j \to \infty$ so that

$$\tilde{Q}(\xi_j, t_j)/\tilde{P}_0(\xi_j, t_j) \leqq (1+t_j)^m \tilde{Q}(\xi_j)/\tilde{P}_0(\xi_j) \to 0.$$

By Lemma 10.4.2 we can find η_j with $|\eta_j| \leqq t_j$ so that if $\theta_j = \xi_j + \eta_j$

$$\tilde{P}_0(\xi_j, t_j) \leqq C|P_0(\theta_j)|.$$

Hence

$$|Q(\theta_j)| \leqq C\tilde{Q}(\xi_j, t_j) \leqq C^2 |P_0(\theta_j)| \tilde{Q}(\xi_j, t_j)/\tilde{P}_0(\xi_j, t_j),$$

so $L_{\theta_j}(Q) \to 0$, $j \to \infty$. By (10.1.8) applied to $P_0(t_j \xi)$ we have

$$\tilde{P}_0(\theta_j, t_j) \leqq C' \tilde{P}_0(\xi_j, t_j) \leqq C' C |P_0(\theta_j)|.$$

This proves that every limit point L of L_{θ_j} is in $\mathscr{L}_0$,

$$W \setminus E \subset \{Q \in W;\ L(Q) = 0 \text{ for some } L \in \mathscr{L}_0\}.$$

b) Assume that $Q \in W$ and that $L(Q) = 0$ for some $L \in \mathscr{L}_0$. Choose $\xi_j \in \mathbb{R}^n$ so that $L_{\xi_j} \to L$. Then $Q(\xi_j)/\tilde{P}_0(\xi_j) \to 0$, and since $Q^{(\alpha)} \ll P_0$ for $\alpha \neq 0$ we have $Q^{(\alpha)}(\xi_j)/\tilde{P}_0(\xi_j) \to 0$ for every α. Hence

$$\tilde{Q}(\xi_j)/\tilde{P}_0(\xi_j) \to 0$$

so $Q \notin E$.

c) Assume that $Q \in W \setminus W_0$. We must then show that $L(Q) \neq 0$ for some $L \in \mathscr{L}_0$. By hypothesis we can choose $\xi_j \in \mathbb{R}^n$ and $t_j \to \infty$ so that for some $c > 0$

$$c \leqq |Q(\xi_j)|/\tilde{P}(\xi_j, t_j).$$

Since $|Q(\xi_j)| \leqq C\tilde{P}(\xi_j)$ it follows that

$$\tilde{P}(\xi_j, t_j)/\tilde{P}(\xi_j) \leqq C/c. \qquad (10.4.11)$$

Hence $|L_{\xi_j}(Q)| \geqq c$ and every limit L of L_{ξ_j} is in $\mathscr{L}_0$ by (10.4.11). Since $L(Q) \neq 0$ this proves the theorem.

Corollary 10.4.8. *In order that $P \prec P+aQ \prec P$ for every complex number a it is necessary and sufficient that $Q \prec\!\!\prec P$.*

Proof. This follows from Theorem 10.4.7 with $P_0=P$. In fact, $L(P)+aL(Q) \neq 0$ for every $a \in \mathbb{C}$ and $L \in \mathscr{L}_0$ if and only if $L(Q)=0$, $L \in \mathscr{L}_0$, which means that $Q \in W_0$.

We end this section with a discussion of two simple examples of the comparison relation. Recall the Definition 7.1.19 of an elliptic operator.

Theorem 10.4.9. *A differential operator $P(D)$ of order m is stronger than every operator of order $\leqq m$ if and only if it is elliptic.*

Proof. a) Ellipticity is necessary. In fact, if there is a real $\xi \neq 0$ such that $P_m(\xi)=0$, it follows that $\tilde{P}(t\xi)=O(t^{m-1})$ when $t \to \infty$. If we take Q homogeneous of degree m with $Q(\xi) \neq 0$, we have $Q(t\xi)=t^m Q(\xi)$ so that Q is not weaker than P.

b) Ellipticity is sufficient. For the infimum C of $|P_m|$ on the real unit sphere is positive if P is elliptic, and the homogeneity gives

$$C|\xi|^m \leqq |P_m(\xi)|.$$

Writing $P_m = P + (P_m - P)$ we obtain

$$C|\xi|^m \leqq |P(\xi)| + C_1(|\xi|^{m-1}+1)$$

where C_1 is another constant. When $|\xi| > 2C_1/C$ it follows that

$$C|\xi|^m \leqq 2|P(\xi)| + 2C_1,$$

which implies that with another constant C

$$1+|\xi|^m \leqq C\tilde{P}(\xi). \tag{10.4.12}$$

This proves the theorem.

If P is elliptic of order m then Q is equally strong as P if and only if Q is of order m and Q also is elliptic, that is,

$$Q_m(\xi) \neq 0 \quad \text{when } |\xi|=1,\ \xi \in \mathbb{R}^n.$$

Thus the forms $Q \to Q_m(\xi)$ are the linear forms in Theorem 10.4.7.

Theorem 10.4.10. *A differential operator $P(D)$ with principal part $P_m(D)$ dominates all differential operators of order $<m$ if and only if*

$$(10.4.13) \qquad \sum_1^n |\partial P_m(\xi)/\partial \xi_j|^2 \neq 0, \qquad 0 \neq \xi \in \mathbb{R}^n.$$

Proof. a) The necessity of (10.4.13). Assume that

$$\partial P_m(\xi)/\partial \xi_j = 0, \qquad j=1, \ldots, n,$$

for some real $\xi \neq 0$. In view of Euler's identity for homogeneous polynomials this implies that $P_m(\xi)=0$. If $P=P_m+P_{m-1}+\ldots$ it then follows that

$$\tilde{P}(s\xi, t)^2 = \sum_\alpha |P^{(\alpha)}(s\xi)|^2 t^{2|\alpha|} = s^{2(m-1)}|P_{m-1}(\xi)|^2 + O(s^{2m-3})$$

when $s \to \infty$ for fixed t. If Q is homogeneous of degree $m-1$ and $Q(\xi) = 1$, then

$$\tilde{P}(s\xi, t)^2/|Q(s\xi)|^2 \to |P_{m-1}(\xi)|^2, \qquad s \to \infty.$$

The right-hand side does not tend to ∞ when $t \to \infty$, which means that P does not dominate Q.

b) The sufficiency of (10.4.13). Arguing as in the proof of Theorem 10.4.9 we obtain from (10.4.13) that

$$1+|\xi|^{2(m-1)} \leqq C \sum_{\alpha \neq 0} |P^{(\alpha)}(\xi)|^2.$$

Hence

$$t^2(1+|\xi|^{2(m-1)}) \leqq C \sum_\alpha |P^{(\alpha)}(\xi)|^2 t^{2|\alpha|} = C\tilde{P}(\xi, t)^2$$

which proves that P dominates every Q of order $<m$. The proof is complete.

We now extend Definition 8.3.5 as follows:

Definition 10.4.11. P is said to be of principal type if the principal part P_m satisfies (10.4.13).

If P is of principal type then Q is weaker than P if and only if Q is of degree $\leqq m$ and $|Q_m(\xi)| \leqq C|P_m(\xi)|$ for some C. The set $\mathscr{L}_0$ in Theorem 10.4.7 is the closure of the set of forms defined by $Q \to Q_m(\xi)/P_m(\xi)$ for some $\xi \in \mathbb{R}^n$ with $P_m(\xi) \neq 0$.

10.5 Approximation of Solutions of Homogeneous Differential Equations

We shall begin by extending Theorem 7.3.6 to the spaces considered in this chapter. Let $k_\iota \in \mathscr{K}$ and $1 \leqq p_\iota < \infty$ where ι belongs to an index set I, countable or not, and let for an open set $X \subset \mathbb{R}^n$

$$\mathscr{F}(X) = \bigcap_{\iota \in I} B^{\mathrm{loc}}_{p_\iota, k_\iota}(X)$$

with the topology defined in Section 10.1. Note that we assume throughout this section that $p_\iota < \infty$.

Theorem 10.5.1. *Let X be convex. Then the closed linear hull in $\mathscr{F}(X)$ of the exponential solutions of the equation*

$$P(D)u = 0 \tag{10.5.1}$$

(see Definition 7.3.5) consists of all solutions of (10.5.1) *in $\mathscr{F}(X)$.*

Proof. Let L be a continuous linear form on $\mathscr{F}(X)$ which is orthogonal to all exponential solutions. Since the topology in $C^\infty(X)$ is stronger than that in $\mathscr{F}(X)$ (Theorem 10.1.26), the restriction of L to $C^\infty(\mathbb{R}^n)$ is a distribution $v \in \mathscr{E}'(X)$. The proof of Theorem 7.3.6 gives $L(u) = v(u) = 0$ if $u \in C_0^\infty(X)$ and $P(D)u = 0$ in a neighborhood of $K = ch \operatorname{supp} v$. More generally, $L(u) = 0$ for every $u \in \mathscr{F}(X) \cap \mathscr{E}'(X)$ such that $P(D)u = 0$ in a neighborhood of K. In fact the regularizations $u_\varepsilon = u * \phi_\varepsilon$ in Theorem 10.1.17 are then in $C_0^\infty(X)$ and satisfy the equation $P(D)u_\varepsilon = 0$ in a neighborhood of K when ε is small, so $L(u_\varepsilon) = 0$. Now Theorem 10.1.17 shows that $u_\varepsilon \to u$ in $\mathscr{F}(X)$ when $\varepsilon \to 0$, so $L(u) = \lim L(u_\varepsilon) = 0$ because L is continuous on $\mathscr{F}(X)$. By the definition of the topology in $\mathscr{F}(X)$ there exists a compact set $K' \subset X$ such that $L(v) = 0$ for every $v \in \mathscr{F}(X)$ with $\operatorname{supp} v \cap K' = \emptyset$. If we choose $\chi \in C_0^\infty(X)$ so that $\chi = 1$ in a neighborhood of $K \cup K'$, we obtain $L((1-\chi)u) = 0$ for every $u \in \mathscr{F}(X)$, and $L(\chi u) = 0$ if u also satisfies (10.5.1). Hence $L(u) = L((1-\chi)u) + L(\chi u) = 0$ if $u \in \mathscr{F}(X)$ satisfies (10.5.1). The proof is completed by the Hahn-Banach theorem.

Remark. Malgrange [1] has proved that in Theorem 7.3.6 and therefore also in Theorem 10.5.1 it is sufficient to use polynomial solutions if and only if every non-constant factor of $P(\zeta)$ vanishes at the origin. It also follows easily from the proof of Lemma 7.3.7 that it suffices to use solutions of the form $e^{i\langle x, \zeta \rangle}$ if and only if P has no multiple factors.

We shall now prove an analogue of Theorem 4.4.5 for general differential operators with constant coefficients. It will be important in Section 10.6. The pattern of the proof is unchanged but the absence of analytic continuation for the solutions of (10.5.1) forces one to make a more implicit statement.

Theorem 10.5.2. *Let X_1 and X_2 be open sets, $X_1 \subset X_2$, and assume that every $\mu \in \mathscr{E}'(\bar{X}_2)$ with* $\operatorname{supp} P(-D)\mu \subset X_1$ *is in $\mathscr{E}'(X_1)$. Let*

$$\mathscr{N}_j = \{u;\, u \in \mathscr{F}(X_j),\, P(D)u = 0\}, \qquad j = 1, 2,$$

and let $\mathscr{N}_2'$ be the set of all restrictions of elements in $\mathscr{N}_2$ to X_1. *Then it follows that $\mathscr{N}_2'$ is dense in $\mathscr{N}_1$ in the topology induced by that in $\mathscr{F}(X_1)$.*

Proof. Let L be a continuous linear form on $\mathscr{F}(X_1)$ which is orthogonal to $\mathscr{N}_2'$. It follows from Theorem 10.1.26 that the restriction of L to $C^\infty(X_1)$ is a distribution $v \in \mathscr{E}'(X_1)$. We shall prove that there exists a distribution μ such that

$$\mu \in \mathscr{E}'(X_1), \qquad P(-D)\mu = v. \tag{10.5.2}$$

To do so we first note that since v is orthogonal to all exponential solutions of (10.5.1) it follows from Lemma 7.3.7 and Theorem 7.3.2 that there exists a distribution $\mu \in \mathscr{E}'(\mathbb{R}^n)$ such that $P(-D)\mu = v$. (See the proof of Theorem 7.3.6.) If we prove that

$$\operatorname{supp} \mu \subset \bar{X}_2 \tag{10.5.3}$$

it will follow from the hypothesis in the theorem that (10.5.2) is valid. To prove (10.5.3) we take a fundamental solution E of $P(-D)$ and note that $\mu = E * v$. If $\psi \in C_0^\infty(\complement \bar{X}_2)$ we have

$$\mu(\psi) = \mu * \check{\psi}(0) = E * v * \check{\psi}(0) = v(\check{E} * \psi).$$

Now $P(D)(\check{E} * \psi) = \psi = 0$ in X_2 since $\check{E}$ is a fundamental solution of $P(D)$. The restriction of the C^∞ function $\check{E} * \psi$ to X_1 is thus in $\mathscr{N}_2'$ so that by assumption $v(\check{E} * \psi) = 0$, which proves (10.5.3).

It follows from (10.5.2) that

$$L(u) = v(u) = (P(-D)\mu)(u) = \mu(P(D)u) = 0$$

if $u \in C^\infty(X_1)$ and $P(D)u = 0$ in a neighborhood of $\operatorname{supp} \mu$. As in the proof of Theorem 10.5.1 it follows that $L(u) = 0$ for every $u \in \mathscr{F}(X_1)$ such that $P(D)u = 0$ in X_1. The proof is now complete in view of the Hahn-Banach theorem.

Corollary 10.5.3. *Let X be an open set in $\mathbb{R}^n$ such that every $\mu \in \mathscr{E}'(\mathbb{R}^n)$ with* $\operatorname{supp} P(-D)\mu \subset X$ *is in* $\mathscr{E}'(X)$. *Then the exponential solutions of* (10.5.1) *are dense in the set of all solutions of* (10.5.1) *in* $\mathscr{F}(X)$.

Proof. By Theorem 10.5.2 we can approximate by solutions in $\mathbb{R}^n$ and by Theorem 10.5.1 these can be approximated by exponential solutions.

10.6. The Equation $P(D)u=f$ When f is in a Local Space $\subset \mathscr{D}'_F$

In this section we shall study the equation $P(D)u=f$ when f belongs to some space $B^{\text{loc}}_{p,k}(X)$ but is not otherwise globally restricted. We shall prove that there exists a solution for an arbitrary f if and only if X is P-convex in the sense of the following definition.

Definition 10.6.1. An open set $X \subset \mathbb{R}^n$ is called P-convex for supports if to every compact set $K \subset X$ there exists another compact set $K' \subset X$ such that $\phi \in C_0^\infty(X)$ and $\operatorname{supp} P(-D)\phi \subset K$ implies $\operatorname{supp} \phi \subset K'$.

Before showing the importance of this condition we give some simple properties of P-convexity for supports.

Theorem 10.6.2. *Every open convex set is P-convex for supports, and if X is convex one can take K' equal to the convex hull of K in Definition 10.6.1.*

Proof. This is a consequence of Theorem 7.3.2.

Theorem 10.6.3. *X is P-convex for supports if and only if the distances from $\complement X$ to* $\operatorname{supp}\mu$ *and to* $\operatorname{supp} P(-D)\mu$ *are equal for every $\mu \in \mathscr{E}'(X)$.*

Proof. a) Sufficiency. If K is a compact subset of X, the distance δ from K to $\complement X$ is positive. The set F of points at distance $\geq \delta$ to $\complement X$ is closed. If $\mu \in \mathscr{E}'(X)$ and $\operatorname{supp} P(-D)\mu \subset K$ we have $\operatorname{supp}\mu \subset K' = F \cap ch(K)$ by hypothesis and Theorem 7.3.2. b) Necessity. Since $\operatorname{supp}\mu \supset \operatorname{supp} P(-D)\mu$ it is sufficient to prove that the distance from $\operatorname{supp} P(-D)\mu$ to $\complement X$ does not exceed the distance δ from $\operatorname{supp}\mu$ to $\complement X$. First let $\mu \in C_0^\infty(X)$. Then the function $\mu_a(x) = \mu(x-a)$ belongs to $C_0^\infty(X)$ if $|a| < \delta$ but there is no compact subset of X containing $\operatorname{supp}\mu_a$ for all such a. Hence there cannot exist any compact subset of

X containing the support of $P(-D)\mu_a$ for every a with $|a|<\delta$. Since $(P(-D)\mu_a)(x)=(P(-D)\mu)(x-a)$, we conclude that the distance from $\operatorname{supp} P(-D)\mu$ to $\complement X$ cannot exceed δ. This proves the theorem when $\mu\in C_0^\infty(X)$. To prove the theorem for an arbitrary $\mu\in\mathscr{E}'(X)$ we only need to apply what has already been proved to the regularized functions μ_ϕ in Theorem 4.1.4 and let $\operatorname{supp}\phi\to\{0\}$.

If X_ι, $\iota\in I$, is a family of open sets, we denote the set of interior points in the intersection by $\bigcap^\circ X_\iota$ and the set of points which have a neighborhood contained in all but a finite number of X_ι by $\underline{\lim}^\circ X_\iota$.

Theorem 10.6.4. *If X_ι is P-convex for supports for every $\iota\in I$, then $\bigcap^\circ X_\iota$ and $\underline{\lim}^\circ X_\iota$ are also P-convex for supports.*

Proof. Let $X=\bigcap^\circ X_\iota$ and let $\mu\in\mathscr{E}'(X)$. If δ is the distance from $\operatorname{supp}P(-D)\mu$ to $\complement X$ the distance from $\operatorname{supp}P(-D)\mu$ to $\complement X_\iota$ ist at least δ for every ι. Hence it follows from Theorem 10.6.3 that the distance from $\operatorname{supp}\mu$ to $\complement X_\iota$ ist at least δ for every ι. The distance from $\operatorname{supp}\mu$ to $\complement X$ is therefore at least δ which proves that X is P-convex for supports. A similar argument shows that $\lim^\circ X_\iota$ is P-convex for supports.

Corollary 10.6.5. *To every open set X there is a smallest open set containing X which is P-convex for supports.*

This result and Theorem 10.6.2 make it natural to use the term convexity. Further justification will be given in Section 10.8 where we discuss the geometric meaning of P-convexity for supports.

We next prove that it is necessary to require P-convexity for supports in order to obtain an existence theorem for the equation

$$P(D)u=f \tag{10.6.1}$$

when f has unrestricted growth at the boundary of X including infinity.

Theorem 10.6.6. *Suppose that the equation* (10.6.1) *has a solution $u\in\mathscr{D}'(X)$ for every $f\in C^\infty(X)$. Then it follows that X is P-convex for supports.*

Proof. We shall consider the bilinear form

$$(\phi, f)\to\int\phi f\,dx \tag{10.6.2}$$

when $f\in C^\infty(X)$, which is a Fréchet space with its usual topology (see Theorem 2.3.1), and ϕ is in a metrizable space Φ defined as follows. The elements of Φ are all functions $\phi\in C_0^\infty(X)$ such that $\operatorname{supp}P(-D)\phi$

$\subset K$, where K is a fixed compact subset of X. The topology in Φ is defined by the semi-norms $\sup|D^\alpha P(-D)\phi|$ for all multi-indices α. It is trivial that (10.6.2) is continuous in f for a fixed ϕ, since ϕ has compact support. On the other hand, for every $f\in C^\infty(X)$ we can by assumption find $u\in\mathscr{D}'(X)$ satisfying (10.6.1). Then we have

$$\int \phi f\,dx = u(P(-D)\phi), \qquad \phi\in\Phi,$$

which proves that (10.6.2) is also continuous with respect to ϕ when f is fixed. But a bilinear form in the product of a Fréchet space and a metrizable space is continuous if it is separately continuous. Hence there exists a compact set $K'\subset X$ and constants C, N_1, N_2 such that

$$(10.6.3)\qquad \Big|\int \phi f\,dx\Big| \leqq C \sum_{|\alpha|\leqq N_1} \sup|D^\alpha P(-D)\phi| \sum_{|\beta|\leqq N_2} \sup_{K'}|D^\beta f|;$$
$$\phi\in\Phi,\quad f\in C^\infty(X).$$

In particular (10.6.3) shows that $\operatorname{supp}\phi\subset K'$ if $\phi\in\Phi$, which proves the theorem.

Conversely, we shall prove

Theorem 10.6.7. *Let X be P-convex for supports and $f\in B^{\mathrm{loc}}_{p_j,k_j}(X)$, $j=1,2,\dots$ where $k_j\in\mathscr{K}$ and $1\leqq p_j<\infty$. Then the equation* (10.6.1) *has a solution $u\in B^{\mathrm{loc}}_{p_j,\tilde{P}k_j}(X)$, $j=1,2,\dots$.*

In view of (10.1.23) we have the following corollary.

Corollary 10.6.8. *If X is P-convex for supports, the equation* (10.6.1) *has a solution $u\in C^\infty(X)$ for every $f\in C^\infty(X)$.*

Proof of Theorem 10.6.7. Let X_ν, $\nu=1,2,\dots$ be the open set of all $x\in X$ such that $|x|<\nu$ and the distance from x to $\complement X$ is larger than $1/\nu$. It is clear that $X_{\nu-1}\Subset X_\nu\Subset X$. From Theorem 10.6.3 and Theorem 7.3.2 it follows that every $\mu\in\mathscr{E}'(X)$, such that $\operatorname{supp}P(-D)\mu\subset X_\nu$, is in fact in $\mathscr{E}'(\overline{X_\nu})$. Let $\phi_\nu\in C_0^\infty(X_\nu)$ be equal to 1 in a neighborhood of $\overline{X_{\nu-1}}$. The semi-norms $\|\phi_\nu u\|_{p,k}$, $\nu=1,2,\dots$ then suffice to define the topology in $B^{\mathrm{loc}}_{p,k}(X)$. (See the proof of Theorem 10.1.26.) The essential part of the proof is now the following lemma.

Lemma 10.6.9. *Assume in addition to the assumptions of Theorem 10.6.7 that $f=0$ in X_ν. For every $\varepsilon>0$ one can then find $u\in\bigcap_1^\infty B_{p_j,\tilde{P}k_j}$ such that $P(D)u=f$ in $X_{\nu+1}$ and*

$$(10.6.4)\qquad \|\phi_\mu u\|_{p_j,\tilde{P}k_j}<\varepsilon;\qquad \mu\leqq\nu,\quad j\leqq\nu.$$

Proof. Replacing f by $\phi_{\nu+2} f$ does not change either the hypothesis or the conclusion of the lemma, so we may assume that $f \in \bigcap_1^\infty B_{p_j,k_j} \cap \mathscr{E}'(X)$. Using Theorem 10.3.1 we can then find a solution $v \in \bigcap_1^\infty B^{\text{loc}}_{p_j,\tilde{p}k_j}(\mathbb{R}^n)$ of the equation $P(D)v = f$. In particular, this means that $P(D)v = 0$ in X_ν. Now the assumptions of Theorem 10.5.2 are fulfilled with X_1 and X_2 replaced by X_ν and $X_{\nu+2}$ respectively. Hence it is possible to choose $w \in \bigcap_1^\infty B^{\text{loc}}_{p_j,\tilde{p}k_j}(X_{\nu+2})$ such that $P(D)w = 0$ in $X_{\nu+2}$ and

(10.6.5) $$\|\phi_\mu(v-w)\|_{p_j,\tilde{p}k_j} < \varepsilon; \quad \mu \leqq \nu,\ j \leqq \nu.$$

It is then clear that $u = \phi_{\nu+2}(v-w)$ has all the required properties.

End of proof of Theorem 10.6.7. Using the lemma we can successively construct $u_\nu \in \bigcap_1^\infty B^{\text{loc}}_{p_j,\tilde{p}k_j}(X)$, $\nu = 1, 2, \dots$ such that $P(D)u_\nu = f$ in X_ν and

(10.6.6) $$\|\phi_\mu(u_{\nu+1} - u_\nu)\|_{p_j,\tilde{p}k_j} < 2^{-\nu}; \quad \mu \leqq \nu,\ j \leqq \nu.$$

In fact, u_1 can be chosen as in the beginning of the proof of Lemma 10.6.9. When u_ν is chosen we must find $u_{\nu+1} = u_\nu + u$ so that $P(D)u = f - P(D)u_\nu$ in $X_{\nu+1}$ and u satisfies (10.6.4) with $\varepsilon = 2^{-\nu}$. Now it follows from the preceding step in the construction that $f - P(D)u_\nu = 0$ in X_ν, and $f - P(D)u_\nu \in \bigcap_1^\infty B^{\text{loc}}_{p_j,k_j}(X)$ in virtue of Theorem 10.1.22. Hence Lemma 10.6.9 shows precisely that the conditions on u can be satisfied. But (10.6.6) implies that $\lim_{\nu\to\infty} u_\nu = u$ exists in $B^{\text{loc}}_{p_j,\tilde{p}k_j}(X)$ for every j, and u obviously satisfies the equation $P(D)u = f$ in the whole of X. This proves the theorem.

Remark. The proof of Theorem 10.6.7 is often called the Mittag-Leffler procedure because it follows the same patterns as the classical proof of the Mittag-Leffler theorem giving a meromorphic function with prescribed singularities. Note that in the inductive argument it was very important to have precise information on the regularity of the solution of (10.6.1) when $f \in \mathscr{E}'$. One might otherwise easily loose more and more derivatives.

It follows at once from (7.3.2) that $\mathscr{D}'^j(X) \subset B^{\text{loc}}_{p,k}(X)$ if we have $k(\xi)(1+|\xi|)^j \in L^p$. Hence the union of all spaces $B^{\text{loc}}_{p,k}(X)$ is equal to $\mathscr{D}'_F(X)$, which gives

Corollary 10.6.10. *If X is P-convex for supports, the equation* (10.6.1) *has a solution $u \in \mathscr{D}'_F(X)$ for every $f \in \mathscr{D}'_F(X)$.*

However, the methods used here are not applicable if $f\in\mathscr{D}'(X)$ unless P is hypoelliptic (see Section 11.1). In that case the proof of Theorem 10.6.7 shows that there is a solution $u\in\mathscr{D}'(X)$ of (10.6.1) for every $f\in\mathscr{D}'(X)$, for the solutions of the homogeneous equation $P(D)u=0$ are automatically in C^∞. In the next section we shall see that stronger conditions on X are in general required to guarantee the existence of a solution of (10.6.1) for every $f\in\mathscr{D}'(X)$.

10.7. The Equation $P(D)u=f$ When $f\in\mathscr{D}'(X)$

It follows from Theorem 10.6.6 and Corollary 10.6.8 that $P(D)\mathscr{D}'(X)\supset C^\infty(X)$ if and only if X is P-convex for supports. Hence $P(D)\mathscr{D}'(X)=\mathscr{D}'(X)$ if and only if X is P-convex for supports and in addition the map induced by $P(D)$ in $\mathscr{D}'(X)/C^\infty(X)$ is surjective. We shall prove that this is equivalent to a condition analogous to P-convexity for supports.

Definition 10.7.1. An open set $X\subset\mathbb{R}^n$ is called P-convex for singular supports if to every compact set $K\subset X$ there exists another compact set $K'\subset X$ such that $\mu\in\mathscr{E}'(X)$ and $\operatorname{sing\,supp} P(-D)\mu\subset K$ implies $\operatorname{sing\,supp}\mu\subset K'$.

The properties of P-convexity for supports proved in Section 10.6 carry over to P-convexity for singular supports:

Theorem 10.7.2. *Every open convex set X is P-convex for singular supports.*

Proof. This is a consequence of Theorem 7.3.9.

Theorem 10.7.3. *X is P-convex for singular supports if and only if for every $\mu\in\mathscr{E}'(X)$, the distances from $\complement X$ to $\operatorname{sing\,supp}\mu$ and to $\operatorname{sing\,supp} P(-D)\mu$ are equal.*

Proof. In Definition 10.7.1 the condition $\mu\in\mathscr{E}'(X)$ may be replaced by $\mu\in\mathscr{E}'(\mathbb{R}^n)$ and $\operatorname{sing\,supp}\mu\subset X$. In fact, given such a μ we can choose $\psi\in C_0^\infty(X)$ equal to 1 in a neighborhood of $\operatorname{sing\,supp}\mu$ and apply the condition in Definition 10.7.1 to $\psi\mu$ instead of μ. In this extended form the condition allows us to repeat the argument involving translations used in the proof of Theorem 10.6.3. The repetition of the details is left for the reader.

Theorem 10.7.4. *If X_ι is P-convex for singular supports, $\iota \in I$, then $\bigcap {}^\circ X_\iota$ and $\underline{\lim} {}^\circ X_\iota$ are also P-convex for singular supports.*

Proof. In the proof of Theorem 10.6.4 we just have to replace the reference to Theorem 10.6.3 by a reference to Theorem 10.7.3.

Corollary 10.7.5. *To every open set X there is a smallest open set containing X which is P-convex for singular supports.*

Theorem 10.7.6. *If $P(D)$ induces a surjective map in $\mathscr{D}'(X)/C^\infty(X)$ then X is P-convex for singular supports.*

In the proof we shall use the spaces $H_{(s)}^{\text{loc}} = B_{2,k_s}^{\text{loc}}$, $k_s = (1+|\xi|^2)^{s/2}$, to measure the regularity of distributions. (See also Section 7.9.) We need the following simple lemma.

Lemma 10.7.7. *Let $v \in \mathscr{D}'(Y)$ and assume that there exists a real number s such that $D^\alpha v \in H_{(s)}^{\text{loc}}(Y)$ for every α. Then it follows that $v \in C^\infty(Y)$.*

Proof. If $\phi \in C_0^\infty(Y)$ and $w = \phi v$, it follows from Leibniz' formula that $D^\alpha w \in H_{(s)}$ for every α. Hence $w \in H_{(s+k)}$ for every integer $k \geqq 0$ so $w \in C^\infty$ by Theorem 10.1.13. The lemma is proved.

Proof of Theorem 10.7.6. Let K be a compact subset of X, and assume that there is no compact subset K' of X with the property in Definition 10.7.1. Choose an increasing sequence of compact subsets K_j of X such that every compact subset of X is contained in some K_j. Then there exist sequences $\mu_j \in \mathscr{E}'(X)$ and $x_j \in X$, $j = 1, 2, \ldots$ such that

$$\operatorname{sing\,supp} P(-D)\mu_j \subset K; \quad x_j \in \operatorname{sing\,supp} \mu_j; \quad x_j \notin \operatorname{supp} \mu_k, \quad k<j; \quad x_j \notin K_j. \tag{10.7.1}$$

In fact, if $\mu_1, \ldots, \mu_{j-1}$ and $x_1, \ldots, x_{j-1}$ are already chosen we can by assumption find μ_j so that $\operatorname{sing\,supp} P(-D)\mu_j \subset K$ and $\operatorname{sing\,supp} \mu_j$ contains some point x_j outside the compact set $K_j \cup \operatorname{supp} \mu_1 \cup \ldots \cup \operatorname{supp} \mu_{j-1}$.

Since (10.7.1) shows that there are only a finite number of points x_j in any compact subset of X, we can find open symmetric neighborhoods Y_k of 0 such that Y_k decreases with k, $K + \bar{Y}_1$ is a compact subset of X and

$$x_j \notin \operatorname{supp} \mu_k + \bar{Y}_k, \quad j>k; \quad \operatorname{supp} \mu_k + \bar{Y}_k \subset X. \tag{10.7.2}$$

Let $\mu_k\in H_{(s_k)}$. Since $x_k\in\operatorname{sing\,supp}\mu_k$ it follows from Lemma 10.7.7 that we can choose a multi-index α_k so that

$$D^{\alpha_k}\mu_k\notin H^{\mathrm{loc}}_{(s_k-|\alpha_{k-1}|)}(Y_k+\{x_k\}). \tag{10.7.3}$$

This means in particular that

$$|\alpha_k|>|\alpha_{k-1}|. \tag{10.7.4}$$

Now set

$$f=\sum_1^\infty(-D)^{\alpha_j}\delta_{x_j} \tag{10.7.5}$$

where δ_{x_j} is the Dirac measure at x_j. The series converges in $\mathscr{D}'(X)$ since only a finite number of points x_j belong to any compact subset. We shall now prove that assuming the existence of $u\in\mathscr{D}'(X)$ and $g\in C^\infty(X)$ such that

$$P(D)u=f+g \tag{10.7.6}$$

leads to a contradiction. This will prove the theorem.

The equation (10.7.6) means that

$$u(P(-D)\psi)=f(\psi)+g(\psi),\qquad \psi\in C_0^\infty(X). \tag{10.7.7}$$

If $\phi\in C_0^\infty(Y_k)$ we have $\mu_k*\phi\in C_0^\infty(X)$ in view of (10.7.2). Hence we may apply (10.7.7) to $\psi=\mu_k*\phi$ which gives

$$u(P(-D)\,\mu_k*\phi)=\sum_{j=1}^\infty(D^{\alpha_j}\mu_k*\phi)(x_j)+g(\mu_k*\phi),\qquad \phi\in C_0^\infty(Y_k). \tag{10.7.8}$$

Since $\operatorname{supp}(\mu_k*\phi)\subset\operatorname{supp}\mu_k+\bar{Y}_k$, it follows from (10.7.2) that all terms in (10.7.8) with $j>k$ must vanish. Thus we can rewrite (10.7.8) in the form

$$\begin{aligned}(D^{\alpha_k}\mu_k*\phi)(x_k)&=u(P(-D)\,\mu_k*\phi)-g(\mu_k*\phi)\\&\quad-\sum_{j=1}^{k-1}(D^{\alpha_j}\mu_k*\phi)(x_j),\qquad \phi\in C_0^\infty(Y_k).\end{aligned} \tag{10.7.9}$$

To estimate the right-hand side we first note that

$$\begin{aligned}|D^\alpha\mu_k*\phi|&\leqq\|\mu_k\|_{(s_k)}\|D^\alpha\phi\|_{(-s_k)}\\&\leqq\|\mu_k\|_{(s_k)}\|\phi\|_{(|\alpha|-s_k)},\qquad \phi\in C_0^\infty(Y_k).\end{aligned} \tag{10.7.10}$$

Here α is an arbitrary multi-index. Replacing g by a C_0^∞ function which is equal to g in $\operatorname{supp}\mu_k+Y_k$ we obtain for any s

$$|g(\mu_k*\phi)|\leqq C_{k,s}\|\phi\|_{(s)}. \tag{10.7.11}$$

To estimate $u(P(-D)\mu_k*\phi)$ we first choose $\chi\in C_0^\infty(X)$ equal to 1 in a neighborhood of K. Since $K+\bar{Y}_1\subset X$ we may choose χ so that $\operatorname{supp}\chi+\bar{Y}_1=\tilde{K}$ is a compact subset of X. We have

$$v_k'=\chi P(-D)\mu_k\in H_{(s_k-m)}, \qquad v_k''=(1-\chi)P(-D)\mu_k\in C_0^\infty(X),$$

if m is the order of P. Every derivative of $v_k''*\phi$ can be estimated by $\|\phi\|_{(s)}$ for any s. Since $\operatorname{supp}(v_k''*\phi)$ is contained in the compact subset $\operatorname{supp}\mu_k+\bar{Y}_k$ of X, this proves that for suitable constants $C_{k,s}$

$$|u(v_k''*\phi)|\leqq C_{k,s}\|\phi\|_{(s)}, \qquad \phi\in C_0^\infty(Y_k). \tag{10.7.12}$$

To estimate $u(v_k'*\phi)$, finally, we note that $\operatorname{supp}(v_k'*\phi)\subset\tilde{K}$ when $\phi\in C_0^\infty(Y_k)$ and that in analogy with (10.7.10) we have

$$|D^\alpha v_k'*\phi|\leqq\|v_k'\|_{(s_k-m)}\|\phi\|_{(|\alpha|+m-s_k)}, \qquad \phi\in C_0^\infty(Y_k).$$

If σ is the order of u in $\tilde{K}$, we obtain therefore

$$|u(v_k'*\phi)|\leqq C\|\phi\|_{(\sigma+m-s_k)}, \qquad \phi\in C_0^\infty(Y_k). \tag{10.7.13}$$

Summing up (10.7.9)–(10.7.13) we have proved

$$|((D^{\alpha_k}\mu_k)*\phi)(x_k)|\leqq C\|\phi\|_{(|\alpha_{k-1}|-s_k)}, \qquad \phi\in C_0^\infty(Y_k), \tag{10.7.14}$$

if k is so large that $m+\sigma\leqq|\alpha_{k-1}|$. From (10.7.14) it follows that

$$D^{\alpha_k}\mu_k\in H^{\mathrm{loc}}_{(s_k-|\alpha_{k-1}|)}(Y_k+\{x_k\}). \tag{10.7.15}$$

When proving this we may simplify the notations by assuming that $x_k=0$, and then we obtain if $\psi\in C_0^\infty(Y_k)$

$$\begin{aligned}|(\check{\psi}D^{\alpha_k}\mu_k)(\check{\phi})|=|(D^{\alpha_k}\mu_k)(\check{\psi}\check{\phi})|&\leqq C\|\psi\phi\|_{(|\alpha_{k-1}|-s_k)}\\ &\leqq C'_\psi\|\phi\|_{(|\alpha_{k-1}|-s_k)}, \qquad \phi\in\mathscr{S}.\end{aligned}$$

Here we have used (10.7.14) and Theorem 10.1.15. From the estimate just proved and Theorem 10.1.14 it follows that $\check{\psi}D^{\alpha_k}\mu_k\in H_{(s_k-|\alpha_{k-1}|)}$, which implies (10.7.15). Since (10.7.15) and (10.7.3) contradict each other, the proof of the theorem is complete.

We shall now prove the converse of Theorem 10.7.6.

Theorem 10.7.8. *If X is an open set in $\mathbb{R}^n$ which is P-convex for singular supports, then $P(D)$ induces a surjective map in $\mathscr{D}'(X)/C^\infty(X)$.*

Proof. It is sufficient to show that for every $f\in\mathscr{D}'(X)$ there is a continuous semi-norm q in $C_0^\infty(X)$, of the form (2.1.3), and a sequence $\psi_r\in C_0^\infty(X)$ such that no compact set intersects $\operatorname{supp}\psi_r$ for infinitely many r and

$$|f(v)|\leqq C(q(P(-D)v)+\sum|\langle v,\psi_r\rangle|), \qquad v\in C_0^\infty(X). \tag{10.7.16}$$

In fact, we can then apply the Hahn-Banach theorem to extend the linear form

$$(P(-D)v, \langle v,\psi_1\rangle, \langle v,\psi_2\rangle, \ldots)\to f(v)$$

to a linear form on $C_0^\infty(X)\oplus l^1$. This gives a linear form u on C_0^∞ and complex numbers a_r such that

$$f(v)=u(P(-D)v)+\sum a_r\langle v,\psi_r\rangle, \qquad v\in C_0^\infty(X),$$
$$|u(w)+\sum a_r w_r|\leqq C(q(w)+\sum|w_r|); \quad w\in C_0^\infty(X), \quad \{w_r\}\in l^1.$$

Thus $u\in\mathscr{D}'(X)$, $|a_r|\leqq C$ and

$$f-P(D)u=\sum a_r\psi_r\in C^\infty(X).$$

For technical reasons we must prove a stronger estimate than (10.7.16),

$$(10.7.16)' \quad |f(v)|+\sup|v|\leqq C(q(P(-D)v)+\sum|\langle v,\psi_r\rangle|), \qquad v\in C_0^\infty(X).$$

Choose an increasing sequence K_j of compact sets in X with union X and $K_0=\emptyset$, and for every j a corresponding K_j' satisfying the condition in Definition 10.7.1 so that $K_0'=\emptyset$ and K_j' is in the interior of K_{j+1}'. Also define $K_{-1}=K_{-1}'=\emptyset$.

Lemma 10.7.9. *For $j\geqq 0$ let V_j be the set of all $v\in C_0(K_{j+1}')$ with $P(-D)v\in C^\infty(\complement K_{j-1})$, equipped with the topology defined by the semi-norms $\sup|v|$ and $\sup_K|D^\alpha P(-D)v|$ where K is compact and $K\cap K_{j-1}=\emptyset$. Then V_j is a Fréchet space and we have a continuous restriction map $V_j\to C^\infty(\complement K_{j-1}')$.*

Proof. If v_ν is a Cauchy sequence in V_j then v_ν has a uniform limit $v\in C_0(K_{j+1}')$ and $P(-D)v_\nu$ has a limit w in $C^\infty(\complement K_{j-1})$. Hence $P(-D)v=w$ in $\complement K_{j-1}$ so $v\in V_j$ and $v_\nu\to v$ in V_j. Thus V_j is complete. By hypothesis restriction to $\complement K_{j-1}'$ maps V_j into $C^\infty(\complement K_{j-1}')$. Since the restriction map is obviously closed it follows from the closed graph theorem that it is continuous.

End of proof of Theorem 10.7.8. Assume that (10.7.16)′ has already been proved when $v\in C_0^\infty(K_j')$, which is a vacuous assumption if $j=0$. Let $\varepsilon_j>0$. Then we claim that (10.7.16)′ remains valid for all $v\in C_0^\infty(K_{j+1}')$ if C is replaced by $C(1+\varepsilon_j)$, q is replaced by another semi-norm q' in $C_0^\infty(X)$ such that $q'(\psi)=q(\psi)$, $\psi\in C_0^\infty(K_{j-1})$, and the functions ψ_r are supplemented by a finite number of functions in $C_0^\infty(X\smallsetminus K_{j-1}')$. Iteration of this conclusion with a sequence ε_j such that $\prod(1+\varepsilon_j)<\infty$ will of course yield (10.7.16) for all $v\in C_0^\infty(X)$. To prove the claim we choose a dense sequence $\chi_1,\chi_2,\ldots$ in $C_0^\infty(X\smallsetminus K_{j-1}')$.

If the claim were false we could (as in the proof of Theorem 2.1.4 for example) find a sequence $v_N \in C_0^\infty(K'_{j+1})$ such that

$$|f(v_N)| + \sup|v_N| = C(1+\varepsilon_j), \tag{10.7.17}$$
$$q(P(-D)\,v_N) + \sum |\langle v_N, \psi_r\rangle| + N \sum_1^N |\langle v_N, \chi_s\rangle| < 1$$

and $P(-D)v_N \to 0$ in $C^\infty(\complement K_{j-1})$. Since v_N is bounded in V_j it follows from Lemma 10.7.9 that v_N is bounded, hence precompact, in $C^\infty(\complement K'_{j-1})$. Every limit must be 0 since it is orthogonal to all χ_s. Thus $v_N \to 0$ in $C^\infty(\complement K'_{j-1})$. Choose $\chi \in C_0^\infty(K'_j)$ equal to 1 in a neighborhood of K'_{j-1}. Then $(1-\chi)v_N \to 0$ in $C_0^\infty(X)$. Hence it follows from (10.7.17) that for large N

$$|f(\chi v_N)| + \sup|\chi v_N| > C(1+\varepsilon_j/2),$$
$$q(P(-D)(\chi v_N)) + \sum |\langle \chi v_N, \psi_r\rangle| < 1 + \varepsilon_j/2.$$

But $\chi v_N \in C_0^\infty(K'_j)$ so this contradicts the hypothesis that (10.7.16)′ is valid when $v \in C_0^\infty(K'_j)$. The contradiction completes the proof of the theorem.

Remark. The proof of Theorem 10.7.8 remains valid if P is a differential operator with C^∞ coefficients (on a manifold) and $P(-D)$ is replaced by the formal adjoint. However, this is not the case for Theorem 10.7.6 since the proof uses translation invariance in an essential way. A counter-example is given by the surjective ordinary differential operator $P = \sin \pi x\, d/dx$ on $\mathbb{R}$. In fact, ${}^tPv = 0$ for all measures v supported by the integers so the analogue of P-convexity for (singular) supports is not valid.

Finally we state a corollary of Theorems 10.6.6, 10.7.6, 10.7.8 and Corollary 10.6.8:

Corollary 10.7.10. *The equation $P(D)u = f$ has a solution $u \in \mathscr{D}'(X)$ for every $f \in \mathscr{D}'(X)$ if and only if X is P-convex for supports as well as for singular supports.*

10.8. The Geometrical Meaning of the Convexity Conditions

No complete geometric characterization of P-convexity for (singular) supports is known but we shall give some partial results. The first is a simple consequence of Theorem 8.6.7. In the statement we shall say that a function f in X satisfies the minimum principle in a closed set F if for every compact set $K \subset F \cap X$ we have

$$\min_{x \in K} f(x) = \min_{x \in \partial_F K} f(x)$$

where $\partial_F K$ is the boundary of K as a subset of F. We write $d_X(x)$ for the Euclidean distance from $x \in X$ to $\complement X$.

Theorem 10.8.1. *If X is an open set in $\mathbb{R}^n$ which is P-convex for supports, then $d_X(x)$ satisfies the minimum principle in any characteristic hyperplane.*

Proof. Assume that the minimum principle is not fulfilled in the characteristic hyperplane π. Then there is a compact set $K \subset \pi \cap X$ such that

$$2d = \min_K d_X(x) < \min_{\partial_\pi K} d_X(x). \tag{10.8.1}$$

Choose $y_0 \in K$ and $x_0 \in \partial X$ with $|x_0 - y_0| = 2d$, and set $t_0 = (y_0 - x_0)/2$. If $x \in K' = K - \{t_0\}$ then $d_X(x + s t_0) \geqq d(1+s)$ by the triangle inequality if $|s| < 1$. Let H be the half space with boundary $\pi - t_0 = \pi'$ which contains K, and choose using Theorem 8.6.7 $u \in C^\infty$ with $P(-D)u = 0$ and $\operatorname{supp} u = H$. Set $v = \chi u$ where $\chi \in C_0^\infty$ is 1 near K' and 0 outside another small neighborhood of K'. Then $\operatorname{supp} P(-D) v \subset H \cap \operatorname{supp} d\chi$. Since the distance from $\partial_{\pi'} K'$ to ∂X exceeds d by (10.8.1) there is a neighborhood V of $\partial_{\pi'} K'$ with distance $> d$ to ∂X. We can choose χ so that $H \cap \operatorname{supp} d\chi \subset V \cup (K' + (0,1) t_0)$. Then $\operatorname{supp} P(-D) v$ has distance $> d$ from $\complement X$ but $x_0 + t_0$ is in $\operatorname{supp} v$ and $d_X(x_0 + t_0) = d$. Hence Theorem 10.6.3 shows that X is not P-convex for supports.

Corollary 10.8.2. *Every open set $X \subset \mathbb{R}^n$ is P-convex for supports if and only if P is elliptic.*

Proof. If P is not elliptic (see Definition 7.1.19) there exist real characteristic planes and the condition in Theorem 10.8.1 is not vacuous. For example, if X is P-convex for supports and $\partial X \in C^2$ then ∂X cannot be strictly concave at any characteristic point. – On the other hand, if P is elliptic and $u \in \mathscr{E}'(X)$, $d(\operatorname{supp} P(-D) u, \complement X) \geqq d$, then $d(\operatorname{supp} u, \complement X) \geqq d$. In fact, if B is an open ball with radius d and center $x_0 \in \partial X$, then u is analytic in B by Theorem 8.6.1 and $u = 0$ near x_0 so $u = 0$ in B. In view of Theorem 10.6.3 it follows now that X is P-convex for supports.

The necessary condition obtained in Theorem 10.8.1 is also sufficient when $n = 2$. When proving this fact we may assume that X is connected, for it is obvious that an open set X is P-convex for supports if and only if every connected component of X has this property.

Theorem 10.8.3. *If P is non-elliptic then the following conditions on an open connected set $X \subset \mathbb{R}^2$ are equivalent:*

(i) *X is P-convex for supports.*

(ii) *Every characteristic line intersects X in an interval.*

(iii) *Every $x_0 \in \complement X$ is the vertex of a closed proper convex angle $A \subset \complement X$ such that no characteristic line intersects A only at x_0.*

Proof. (i) $\Rightarrow$ (ii) Assume that (i) is valid. It suffices to show that if $(\pm 1, 0) \in X$ and the x axis is characteristic, then $I = [-1, 1] \times \{0\} \subset X$. To do so we join $(-1, 0)$ and $(1, 0)$ by a polygon γ in X without self-intersections. We may assume that γ only intersects the x axis at $(\pm 1, 0)$ for otherwise we can decompose γ into several polygons meeting the x-axis only at the end points and discuss them separately. Then I and γ bound together a closed set F. Let

$$Y = \{y;\, (x, y) \in F \text{ for some } x\},$$
$$Y_0 = \{y \in Y;\, (x, y) \in F \Rightarrow (x, y) \in X\}.$$

Y is a closed interval and Y_0 is not empty since the end point of Y which is not 0 belongs to Y_0. If $y \in Y_0$ then $(x, y) \in F$ implies $d((x, y), \complement X) \geqq d(\gamma, \complement X)$ by Theorem 10.8.1 applied to $F \cap (\mathbb{R} \times \{y\})$. It follows that Y_0 is closed in Y, and Y_0 is obviously open in Y since X is open. Thus $Y_0 = Y$, which implies $0 \in Y_0$ and $I \subset X$.

(ii) $\Rightarrow$ (iii) If $x_0 \notin X$ and L is a characteristic line through x_0 then one half ray L_1 of L bounded by x_0 must be in $\complement X$ by (ii). If there is a characteristic line M through x_0 with $L_1 \cap M = \{x_0\}$ then one of its half rays M_1 is in $\complement X$. Since X is connected it can be chosen so that the convex hull A of L_1 and M_1 is contained in $\complement X$. Next we examine if there is a characteristic line meeting A only at x_0 and continue extending A until we have a set with the stated properties.

(iii) $\Rightarrow$ (i) Let K be a compact subset of X with distance $d > 0$ to $\complement X$. If $x_0 \in \complement X$ and A has the property in (iii) then $(A + \{y\}) \cap K = \emptyset$ if $|y| < d$. Let A' be an open proper conic neighborhood of $A \setminus \{0\}$ such that $(A' + \{y\}) \cap K = \emptyset$. If $u \in \mathscr{E}'(X)$ and $\operatorname{supp} P(-D)u \subset K$ it follows from Corollary 8.6.11 applied to $A' + \{y\}$ that $u = 0$ in $A' + \{y\}$. Hence $d(\operatorname{supp} u, \complement X) \geqq d$ which proves (i).

Example. If $P(D) = D_1 D_2$ then the union X of the half disc $\{x;\, x_2 > 0, x_1^2 + x_2^2 < 9\}$ and the discs $\{x;\, (x_1 \pm 2)^2 + x_2^2 < 1\}$ is not P-convex for

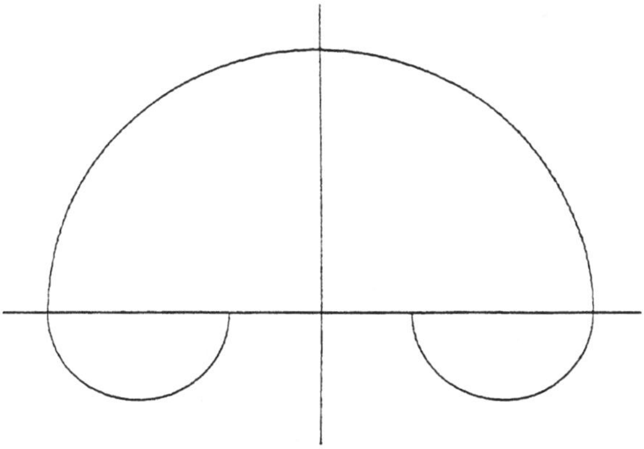

Fig. 1

supports but the intersection of X and any half plane $x_1 > a \geqq -1$ or $x_1 < a \leqq 1$ is P-convex for supports. This shows that P-convexity for supports is not a local property.

Corollary 10.8.4. *If an open connected set X in $\mathbb{R}^n$ is P-convex for supports when $P(\xi) = \langle t, \xi \rangle$, any $t \in \mathbb{R}^n \setminus 0$, then X is convex.*

Proof. The statement follows from Theorem 10.8.3 if $n=2$. If $n>2$ we can conclude from this two dimensional result that every component of the intersection of X and a two dimensional plane must be convex, for every C^∞ function there can be extended to a C^∞ function in X. Hence we obtain recursively that if a polygon with vertices x_0, $x_1, \dots, x_N$ lies in X, then the line segment between x_0 and x_j is contained in X for every $j \leqq N$. Since X is connected, this proves that X is convex.

Remark. An easy modification of the implications (ii) $\Rightarrow$ (iii) $\Rightarrow$ (i) in the proof of Theorem 10.8.3 shows that X is P-convex for supports if $X \cap \pi$ is convex for every characteristic plane π. The details are left for the reader. (A should be replaced by the product with a subspace of codimension 2.)

Theorem 10.8.5. *Let $P(D)$ be a differential operator in $\mathbb{R}^n$ which acts along a linear subspace V and is elliptic as an operator in V. Then an open set X in $\mathbb{R}^n$ is P-convex for supports if and only if $d_X(x)$ satisfies the minimum principle in every affine space parallel to V.*

Proof. The necessity is a simple modification of the proof of Theorem 10.8.1 since we can find u with $P(-D)u=0$ and $\operatorname{supp} u = V$. To prove the sufficiency, assume that $u \in C_0^\infty(X)$ and set $d = d(\operatorname{supp} P(-D)u, \complement X)$. Let $y_0 \in \operatorname{supp} u$ and set

$$K = (\{y_0\} + V) \cap \operatorname{supp} u.$$

If $y \in K$ and $d(y, \complement X) < d$ then $P(-D)u = 0$ in a neighborhood of y so $\operatorname{supp} u$ contains a neighborhood of y in $\{y_0\} + V$ since u is analytic in the V directions. Thus $d(y, \complement X) \geqq d$ if y is in the boundary of K in $\{y_0\} + V$, and by the minimum principle this implies that $d(y, \complement X) \geqq d$ in K. Hence $d(\operatorname{supp} u, \complement X) \geqq d$ which proves the theorem.

Assume now that P is of second order, with principal part

$$P_2(\xi) = -\xi_1^2 + \xi_2^2 + \ldots + \xi_n^2. \tag{10.8.2}$$

Write

$$A(x) = x_1^2 - x_2^2 - \ldots - x_n^2$$

for the Lorentz form and as usual $|x|^2$ for the Euclidean metric form. Sometimes we write $x' = (x_2, \ldots, x_n)$. Let X be an open set in $\mathbb{R}^n$.

Theorem 10.8.6. *X is not P-convex for supports if and only if it is possible to find points $x \neq y$ in X with $A(x-y) = 0$ and an open neighborhood W of the line segment $I = [x, y]$ so that*

$$W \cap \{z; A(z-x) \geqq 0\} \subset X \cup I, \quad I \cap \partial X \neq \emptyset. \tag{10.8.3}$$

Proof of Sufficiency. Assume that one can find x, y, W so that (10.8.3) is fulfilled. Denote by F the difference between the advanced and the retarded fundamental solutions of $P(-D)$ (Theorem 6.2.3). (First order terms in P can be eliminated and terms of order 0 can be handled by Theorem 12.5.3.) Then $P(-D)F = 0$, we have $A \geqq 0$ in $\operatorname{supp} F$ and $A(z) = 0$ implies $z \in \operatorname{supp} F$. Assume that $x = 0$ and choose $\chi \in C_0^\infty(W)$ equal to 1 near I. Then

$$v = \chi F$$

is in $\mathscr{E}'(\mathbb{R}^n)$ and

$$I \subset \operatorname{supp} v \subset I \cup X, \quad \operatorname{supp} P(-D)v \subset \operatorname{supp} d\chi \cap \operatorname{supp} F \subset X.$$

If $A(t) > 0$ and t is in the same Lorentz cone as y, then $\operatorname{supp} v(.-t\varepsilon) \subset X$ for small $\varepsilon > 0$ and $\operatorname{supp} P(-D)v(.-t\varepsilon)$ belongs to a fixed compact subset of X. Since $\operatorname{supp} v(.-t\varepsilon)$ can approach any point in $I \cap \partial X$ it follows that X is not P-convex for supports.

Before proving the necessity we note two consequences of Holmgren's uniqueness theorem.

Lemma 10.8.7. *If $P(-D)u = 0$ in $\Gamma_R = \{x; |x_1| + |x'| < R\sqrt{2}\}$ and $u = 0$ when $|x| < R$, then $u = 0$ in Γ_R.*

Proof. A characteristic plane which does not intersect the ball $|x|<R$ is of the form

$$ax_1+\langle a',x'\rangle=R, \quad |a|=|a'|, \quad |a|^2+|a'|^2\leqq 1.$$

Thus $|a|=|a'|\leqq 1/\sqrt{2}$ which proves that the plane contains no point with $|x_1|+|x'|<R\sqrt{2}$. The lemma follows now from Theorem 8.6.8.

Lemma 10.8.8. *If $P(-D)u=0$ in a convex set Y and we have two continuous curves $[0,1]\ni s\to x(s)$, $[0,1]\ni s\to y(s)$ in $Y\setminus \operatorname{supp} u$ such that $A(x(s)-y(s))>0$ for $0\leqq s\leqq 1$, then u vanishes near the line segment $[x(s),y(s)]$ for all $s\in[0,1]$ or for no such s.*

Proof. Let S be the set of all $s\in[0,1]$ such that u vanishes in a neighborhood of $[x(s),y(s)]$. From Theorem 8.6.8 it follows when $s\in S$ that u vanishes in the intersection of Y and the Lorentz half cones with vertices at $x(s)$ and $y(s)$ containing $y(s)$ and $x(s)$. This remains true if $s\in\bar{S}$. Since $u=0$ in a neighborhood of $x(s)$ and $y(s)$ it follows that S is open and closed, which proves the lemma.

Proof of the necessity in Theorem 10.8.6. If X is not P-convex for supports we can choose $u\in\mathscr{E}'(X)$ so that with distances in the Euclidean metric

$$d_1=d(\operatorname{supp} u, \complement X)<d(\operatorname{supp} P(-D)u, \complement X)=d_2.$$

Choose $x_0\in\partial X$ and $y_0\in\operatorname{supp} u$ so that $|t_0|=d_1$ if $t_0=y_0-x_0$. Then

$$v_\varepsilon=u(.+(1-\varepsilon)t_0)$$

belongs to $\mathscr{E}'(X)$ if $0<\varepsilon<1$, and

$$d(\operatorname{supp} P(-D)v_\varepsilon, \complement X)\geqq d_2-d_1 \quad \text{if } 0\leqq\varepsilon\leqq 1.$$

Choose a positive number R with

$$R<\min((d_2-d_1)/(1+\sqrt{2}), d_1).$$

Then

$$d(\operatorname{supp} P(-D)v_0, \complement X)>R+R\sqrt{2}, \tag{10.8.4}$$

$$x_0\in\operatorname{supp} v_0, \quad v_0=0 \text{ when } |x-x_0+t_0|<d_1. \tag{10.8.5}$$

Since $|t_0|=d_1>R$ we have $v_0=0$ when $|x-x_0+Rt_0/|t_0||<R$. Now $P(-D)v_0=0$ when $|x-x_0+Rt_0/|t_0||<R\sqrt{2}$, so it follows from Lemma 10.8.7 that $v_0=0$ in

$$\Gamma(x_0)=\Gamma_R+x_0-Rt_0/|t_0|. \tag{10.8.6}$$

Thus $x_0 \notin \Gamma(x_0)$ so $A(t_0)=0$. The generator

$$\{x_0\}+[-R,R]\,A'(t_0)/|A'(t_0)|$$

is in $\operatorname{supp} v_0$ by Corollary 8.6.14 or Theorems 8.5.9, 8.6.5. (On one side of x_0 this is also an easy consequence of Lemma 10.8.8.) The interval is therefore contained in $\bar{X}$.

Let J be the maximal compact interval on $\{x_0\}+\mathbb{R}A'(t_0)$ with $x_0 \in J \subset \partial X \cap \operatorname{supp} v_0$. Then $J+\{t_0\} \subset \operatorname{supp} u$ so

$$d(x, \operatorname{supp} u) \leqq d_1, \qquad x \in J,$$

and the opposite inequality follows from the definition of d_1 so equality must hold. All that has been said about x_0 is therefore equally true for every $x \in J$. Thus $v_0=0$ in the open set $\Gamma = \bigcup_{x \in J} \Gamma(x)$, and

$$\hat{J} = J+(-R,R)\,A'(t_0)/|A'(t_0)| \subset \operatorname{supp} v_0 \subset \bar{X}.$$

Thus there are points in $\hat{J} \cap X$ arbitrarily close to each end point of J.

Denote by Λ the open Lorentz half cone defined by $A(x)>0$ and with t_0 on its boundary. Set

$$W=\{x;\, d(x,J)<R\}.$$

(See Fig. 2; W is within dotted lines and $\Gamma(.)$ are the small double cones.) The essential step in the proof is now to show that

$$W \cap (\{x\}+\Lambda) \subset X \quad \text{if } x \in \hat{J}. \tag{10.8.7}$$

In the proof we assume that $x \in \hat{J}$, $y \in W$ and $y-x \in \Lambda$, and define $z=x-\varepsilon(y-x)$ for small $\varepsilon>0$. Then $z \in \Gamma$ and z is close to $\hat{J}$. Consider the

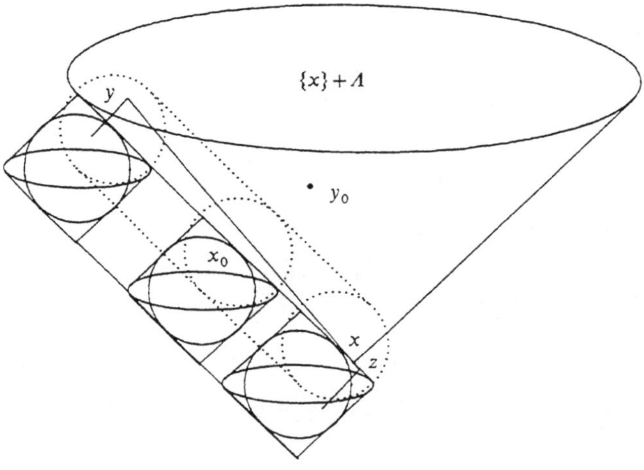

Fig. 2

maps

$$s \to y - st_0, \quad s \to z - st_0$$

with $0 < s < s_1$ and $s_1 < R/d_1$ chosen so that $y - s_1 t_0 \in \Gamma$. We have automatically $z - st_0 \in \Gamma$ for $0 \leqq s \leqq s_1$ if ε is small enough. Since $x \in \operatorname{supp} v_0$ it follows that $y - st_0 \in \operatorname{supp} v_0$ for some $s \in [0, s_1]$, for Lemma 10.8.8 would otherwise give $v_0 = 0$ near x. But then we have

$$y + \delta t_0 \in \operatorname{supp} v_{s+\delta} \subset X \quad \text{if } 0 < \delta < 1 - R/d_1,$$

which proves (10.8.7).

The cone $\{x\} + \Lambda$ in (10.8.7) increases when x approaches one end point J' of $\hat{J}$, so (10.8.7) can be strengthened to

$$W \cap (\{x\} + \bar{\Lambda}) \subset X \cup \hat{J} \quad \text{if } x \in \hat{J}. \tag{10.8.8}$$

If x is chosen in the component of $\hat{J} \smallsetminus J$ with J' on its boundary and y is in the other component of $\hat{J} \smallsetminus J$ then the condition (10.8.3) is fulfilled provided that x and y are taken in X and W is shrunk so that $(\{x\} - \bar{\Lambda}) \cap W \subset X$. The proof is complete.

Theorem 10.8.5 shows that a geometric description of P-convexity for supports may require a rather subtle examination of the curvature properties of ∂X. This explains why so few cases are completely understood. We leave this question now and pass to P-convexity for singular supports. Complete results are then known when P is of real principal type (Definition 8.3.5). If $P_m(\xi) = 0$ a line with direction $P'_m(\xi)$ is then called a bicharacteristic line.

Theorem 10.8.9. *If P is of real principal type then an open set $X \subset \mathbb{R}^n$ is P-convex for singular supports if and only if the boundary distance $d_X(x)$ satisfies the minimum principle in any bicharacteristic line.*

Proof. The necessity follows as in the proof of Theorem 10.8.1. Thus let $I \subset X$ be an interval on a bicharacteristic line L, and let $d(\partial_L I, \complement X) > d$. By Theorem 8.3.8 we can choose $u \in C^m(\mathbb{R}^n)$ so that $\operatorname{sing\,supp} u = L$ and $P(-D) u \in C^\infty(\mathbb{R}^n)$. Now we set $v = \chi u$ where $\chi \in C_0^\infty(X)$ is equal to 1 near I. Then $I \subset \operatorname{sing\,supp} v$ and $\operatorname{sing\,supp} P(-D) v \subset L \cap \operatorname{supp} d\chi$. If $\operatorname{supp} \chi$ is chosen close to I it follows that

$$d < d(\operatorname{sing\,supp} P(-D) v, \complement X).$$

By Theorem 10.7.3 we conclude if X is P-convex for singular supports that

$$d < d(\operatorname{sing\,supp} v, \complement X) \leqq d(I, \complement X)$$

which proves the minimum principle.

The proof of sufficiency is close to that of Theorem 10.8.5. Let $u \in \mathscr{E}'(X)$ and $y_0 \in \operatorname{sing\,supp} u \smallsetminus \operatorname{sing\,supp} P(-D) u$. Choose $\eta_0 \in \mathbb{R}^n \smallsetminus 0$ so

that
$$(y_0, \eta_0) \in WF(u).$$
Then $P_m(\eta_0)=0$ by Theorem 8.3.1. Let $L=y_0+\mathbb{R}P_m'(\eta_0)$ be the corresponding bicharacteristic line, and set
$$K=\{y\in L;\, (y,\eta_0)\in WF(u)\}.$$
This is a compact set, and $\partial_L K \subset \operatorname{sing\,supp} P(-D)u$ by Theorem 8.3.3′. Hence
$$d(K, \complement X)=d(\partial_L K, \complement X)\geqq d(\operatorname{sing\,supp} P(-D)u, \complement X)$$
by the minimum principle, and this completes the proof since $y_0\in K$.

Corollary 10.8.10. *If P is of real principal type and X an open set in $\mathbb{R}^n$, then $P(D)\mathscr{D}'(X)=\mathscr{D}'(X)$ if and only if the boundary distance $d_X(x)$ satisfies the minimum principle in any bicharacteristic line.*

Proof. The necessity follows from Theorem 10.7.6 and Theorem 10.8.9. The sufficiency will follow from Corollary 10.7.10 and Theorem 10.8.9 if we prove that X is P-convex for supports. To do so we take $u\in\mathscr{E}'(X)$ and assume that the minimum of $d_X(x)$ when $x\in\operatorname{supp} u$ is attained at y_0. If $x_0\in\partial X$ and $|y_0-x_0|=d_X(y_0)$ then $(y_0, \eta_0)\in WF_A(\check{u})$ if $\eta_0=y_0-x_0$ (Theorem 8.5.6). If $y_0\notin \operatorname{sing\,supp}_A P(-D)u$ then $P_m(\eta_0)=0$ (Theorem 8.6.1). The proof is now completed as that of Theorem 10.8.9 except that we consider analytic singularities now, noting that they are contained in the support.

Corollary 10.8.11. *If an open connected set X in $\mathbb{R}^n$ is P-convex for singular supports when $P(\xi)=\langle t,\xi\rangle$, any $t\in\mathbb{R}^n\setminus 0$, then X is convex.*

Proof. We can use the proof of Corollary 10.8.4 with the reference to Theorem 10.8.3 replaced by one to Theorem 10.8.9.

We shall return to P-convexity for singular supports in Chapter XI when we have additional results on singularities available. (See Corollaries 11.3.2 and 11.3.3.)

Notes

As mentioned in the notes for Chapter VII the existence of fundamental solutions was first proved in full generality by Ehrenpreis [1] and Malgrange [1] who used the Hahn-Banach theorem. The existence of regular fundamental solutions was proved similarly by Hörmander [2] who also proved that regular temperate fundamental solutions do not exist in general. (Example 10.2.15 is taken from Enqvist [1].) The constructive method used here was outlined in Hörmander [29]. It is very convenient for the study of fundamental solutions for operators depending on parameters and gives simple proofs of such results of Treves [6] based on a more primitive construction in the predecessor of this book. (See also Treves [1].) That fundamental solutions with the regularity properties in Theorem 10.2.11 exist was proved in Hörmander [24]; that the construction of Hörmander [29] yields these properties is proved here for the first time. Gabrielov [1] has shown that the set F_0 in Theorem 10.2.11 has positive codimension. It has not been possible to include his proof here nor the proof of the existence of temperate fundamental solutions first established by Łojasiewicz [1] and Hörmander [5]. A simplified proof based on Hironaka's resolution of singularities has been given by Atiyah [1] and by Bernstein-Gelfand [1]. A much more elementary construction is due to I.N. Bernstein [1] (see also Björk [1]).

The comparison of differential operators was introduced in Hörmander [1]; some results here are due to Fuglede [1]. Theorem 10.4.7 is probably new. It is still an open question if a fundamental solution $E(P)$ can be found which depends analytically on P when P is equally strong as P_0. Treves [7] has proved that constant strength is necessary. It is sufficient locally and the linear convexity of the set of such P proved in Theorem 10.4.7 should be helpful in proving a global result. (The term domination has been used differently by Treves [3].)

The proof of the approximation theorem in Section 10.5 follows Malgrange [1] and so does the existence theory in Section 10.6 apart from the simplifications resulting from the existence of regular fundamental solutions. The existence of solutions of $P(D)u=f$ for arbitrary $f\in\mathscr{D}'(X)$ was first proved by Ehrenpreis [2] when X is convex. His proof was simplified by Malgrange [3]. The condition in Section 10.7 is due to Hörmander [14]. The advantage of splitting existence theory in $\mathscr{D}'$ into existence theory in $\mathscr{D}'/C^\infty$ and in C^∞ was observed in Hörmander [29] which is followed closely here.

Malgrange [4] proved that if X is P-convex for supports and $\partial X\in C^2$, P of real principal type, then the normal curvature of ∂X is $\geqq 0$ at every characteristic boundary point in the corresponding bicharacteristic direction. The proof is based on existence of null solutions of the Cauchy problem constructed by means of Theorem 9.4.2. It was included in the predecessor of this book but has been replaced here by a complete characterisation of P-convexity for supports when P is the wave equation, essentially due to Persson [1]. General results are still missing on this problem. However, P-convexity for singular supports is well understood for large classes of operators. The results given in Section 10.8 are contained in those of Hörmander [29].

Chapter XI. Interior Regularity of Solutions of Differential Equations

Summary

Already in Section 4.4 we proved the smoothness of the distribution solutions of some differential equations for which we had constructed explicit fundamental solutions. In Section 11.1 we complete this discussion by characterizing the operators (with constant coefficients of course) such that all distribution solutions are in C^∞. They are called hypoelliptic. Their characterization is a simple corollary of the study in Section 10.2 of singularities of fundamental solutions. Another more elementary approach depending on the results in Section 10.3 is also given in Section 11.1, which concludes with some examples.

In Section 11.2 we study partially hypoelliptic equations, that is, equations for which the solutions are infinitely differentiable if infinite differentiability with respect to one group of variables is imposed. The discussion is quite parallel to that in Section 11.1 and could be bypassed by the reader without loss of the continuity.

The study of propagation of singularities initiated in Section 8.3 for operators of real principal type is resumed in Section 11.3. In particular we characterize the linear spaces which can carry the singularities of a solution of the differential equation $P(D)u=0$. The results are stated only in terms of $\operatorname{sing\,supp} u$. It is easy to modify them to results on $WF(u)$ instead. However, this has no particular advantage in general since P may have many localizations in the same direction at infinity. To obtain a perfect spectral resolution of the singularities one would have to introduce a notion of wave front set which is adapted to the operator at hand.

In Section 11.4 we estimate the growth of the derivatives of solutions of a hypoelliptic differential equation with the order of differentiation. Thus we obtain analogues of the fact that solutions of elliptic equations are real analytic as well as a converse of this and similar results.

11.1. Hypoelliptic Operators

In this section we shall study differential operators $P(D)$ such that the solutions u of the equation

$$P(D)u=f \tag{11.1.1}$$

are always smooth where f is smooth. They can easily be characterized by means of Theorem 10.2.11 and Theorem 10.2.12.

Theorem 11.1.1. *The following conditions on $P(D)$ are equivalent:*

(i) *For every open set $X\subset\mathbb{R}^n$ and $u\in\mathscr{D}'(X)$ we have*

$$WF(u)=WF(P(D)u).$$

(ii) *For every open set $X\subset\mathbb{R}^n$ and $u\in\mathscr{D}'(X)$ we have*

$$\operatorname{sing\,supp} u=\operatorname{sing\,supp} P(D)u.$$

(iii) *If X is an open set in $\mathbb{R}^n$ and $u\in\mathscr{D}'(X)$, $P(D)u=0$, then $u\in C^\infty(X)$.*

(iv) $P^{(\alpha)}(\xi)/P(\xi)\to 0$ as $\xi\to\infty$ in $\mathbb{R}^n$, if $\alpha\neq 0$.

(v) *$P(D)$ has a fundamental solution E with* $\operatorname{sing\,supp} E=\{0\}$.

Proof. (i) $\Rightarrow$ (ii) $\Rightarrow$ (iii) is trivial. Let E be a regular fundamental solution of $P(D)$ (Definition 10.2.2). If (iii) is fulfilled then $E\in C^\infty(\mathbb{R}^n\setminus 0)$ so it follows from Theorems 10.2.12 and 10.2.5 that the localizations of P at ∞ are constant. Since

$$P(\xi+\eta)/\tilde{P}(\eta)=\sum \xi^\alpha P^{(\alpha)}(\eta)/\tilde{P}(\eta)\alpha!$$

this shows that $P^{(\alpha)}(\eta)/\tilde{P}(\eta)\to 0$ when $\eta\to\infty$ if $\alpha\neq 0$, hence $|P(\eta)|/\tilde{P}(\eta)\to 1$ which proves (iv). Conversely, (iv) implies that the localizations at ∞ of P are constant so Theorem 10.2.11 gives a fundamental solution with $\operatorname{sing\,supp} E=\{0\}$, which proves (v). From (v) and (8.2.16) we obtain if $u\in\mathscr{E}'$

$$WF(u)=WF(E*(P(D)\,u))\subset WF(P(D)\,u)$$

and the opposite inclusion is valid by (8.1.11). If u is just in $\mathscr{D}'(X)$ we can choose $\phi\in C_0^\infty(X)$ equal to 1 in any given open subset $Y\Subset X$ and apply this result to ϕu. This shows that $WF(u)=WF(P(D)u)$ in Y, which completes the proof. (See also the proof of Theorem 4.4.1.)

Definition 11.1.2. The differential operator $P(D)$ (and the polynomial $P(\xi)$) are called hypoelliptic if the (equivalent) conditions in Theorem 11.1.1 are fulfilled.

The algebraic description (iv) of hypoelliptic polynomials can be given several equivalent forms which are sometimes useful.

Theorem 11.1.3. *Each of the following conditions is a necessary and sufficient condition for $P(\xi)$ to be hypoelliptic:*

Ia) *If $d(\xi)$ is the distance from $\xi \in \mathbb{R}^n$ to $\{\zeta; \zeta \in \mathbb{C}^n, P(\zeta)=0\}$ then $d(\xi) \to \infty$ when $\xi \to \infty$.*

Ib) *There exist positive constants c and C such that*

$$|\xi|^c \leqq C d(\xi)$$

if $\xi \in \mathbb{R}^n$ and $|\xi|$ is sufficiently large.

IIa) *$P^{(\alpha)}(\xi)/P(\xi) \to 0$ when $\xi \to \infty$ in $\mathbb{R}^n$ if $\alpha \neq 0$.*

IIb) *There exist positive constants c and C such that*

$$|P^{(\alpha)}(\xi)|/|P(\xi)| \leqq C|\xi|^{-|\alpha|c}$$

if $\xi \in \mathbb{R}^n$ and $|\xi|$ is sufficiently large.

Proof. IIa) is the condition (iv) in Theorem 11.1.1. It is obvious that IIb) implies IIa) and that Ib) implies Ia). That for example Ia) implies Ib) follows from the Tarski-Seidenberg theorem in the appendix. In fact, if

$$M(R) = \inf_{|\xi|=R} d(\xi)$$

we have

$$M(R)^2 = \inf |\xi - \zeta|^2$$

where the infimum is taken over all real ξ with $|\xi|^2 = R^2$ and all complex ζ with $P(\zeta)=0$. Thus it follows from Corollary A.2.6 that

$$M(R) = A R^a (1 + o(1)), \qquad R \to \infty.$$

Here $a>0$ and $A>0$ if Ia) is fulfilled, so Ib) follows then.

That Ia) and IIa), Ib) and IIb) are equivalent follows immediately from the next lemma, which will thus complete the proof of Theorem 11.1.3.

Lemma 11.1.4. *There exists a constant C such that for all polynomials P in $\mathbb{R}^n$ of degree $\leqq m$ we have*

$$C^{-1} \leqq d(\xi) \sum_{\alpha \neq 0} |P^{(\alpha)}(\xi)/P(\xi)|^{1/|\alpha|} \leqq C; \qquad \xi \in \mathbb{R}^n,\ P(\xi) \neq 0, \tag{11.1.2}$$

where $d(\xi)$ denotes the distance from ξ to the surface $\{\zeta \in \mathbb{C}^n; P(\zeta)=0\}$.

Proof. To prove the left-hand inequality we denote the sum by A. Then we have $|P^{(\alpha)}(\xi)| \leqq A^{|\alpha|}|P(\xi)|$, thus

$$|P(\xi+\zeta) - P(\xi)| \leqq |P(\xi)| \sum_{1 \leqq |\alpha| \leqq m} |A\zeta|^{|\alpha|}/\alpha!$$

by Taylor's formula. Choose a positive constant c such that

$$\sum_{0<|\alpha|\leqq m} c^{|\alpha|}/\alpha! \leqq 1.$$

When $A|\zeta|<c$ it then follows that $P(\xi+\zeta)\neq 0$. Hence $A\,d(\xi)\geqq c$ which proves that the left-hand inequality (11.1.2) is valid if $C\geqq 1/c$.

To prove the other part we shall first show that

$$|P(\xi+\zeta)|\leqq 2^m|P(\xi)| \quad \text{if } |\zeta|\leqq d(\xi). \tag{11.1.3}$$

For the proof we consider $g(t)=P(\xi+t\zeta)$ as a polynomial in t. The zeros t_i satisfy the inequality $|t_i|\,|\zeta|\geqq d(\xi)\geqq|\zeta|$, hence $|t_i|\geqq 1$. This gives

$$|P(\xi+\zeta)/P(\xi)|=|g(1)/g(0)|=|\prod(t_i-1)/t_i|\leqq 2^m$$

which proves (11.1.3). Application of Cauchy's inequality to the function $P(\xi+\zeta)$ which is analytic in the ball $|\zeta|\leqq d(\xi)$ now gives

$$|P^{(\alpha)}(\xi)|\leqq|\alpha|!\,2^m|P(\xi)|\,d(\xi)^{-|\alpha|}.$$

(We could also use (10.4.1) here.) This proves the lemma.

Using Theorem 11.1.3 one can eliminate the reference in the proof of Theorem 11.1.1 to the somewhat technical Theorems 10.2.11 and 10.2.12. Indeed, the proof of Theorem 7.1.22 shows that if IIb) in Theorem 11.1.3 is fulfilled then $P(D)$ has a parametrix E with $\operatorname{sing\,supp} E=\{0\}$ and this implies (i) in Theorem 11.1.1. On the other hand, Ia) in Theorem 11.1.3 follows very easily from a weaker form of (iii) in Theorem 11.1.1:

Theorem 11.1.5. *Assume that for some open set $X\neq\emptyset$ in $\mathbb{R}^n$ and some $k\in\mathscr{K}$, $p\in[1,\infty]$ the solutions $u\in B_{p,k}^{\mathrm{loc}}(X)$ of the equation $P(D)u=0$ are all in $C^\infty(X)$. Then it follows that*

$$\operatorname{Im}\zeta\to\infty \quad \textit{if} \quad \zeta\to\infty \quad \textit{on the surface } P(\zeta)=0, \tag{11.1.4}$$

so $P(D)$ is hypoelliptic.

Proof. The solution space in the theorem,

$$\mathscr{N}=\{u;\ u\in B_{p,k}^{\mathrm{loc}}(X),\ P(D)u=0\} \tag{11.1.5}$$

is a closed subspace of $B_{p,k}^{\mathrm{loc}}(X)$, hence a Fréchet space. By hypothesis we have an inclusion $\mathscr{N}\subset C^\infty(X)$. The inverse is continuous by Theorem 10.1.26 so the inclusion itself is continuous by Banach's theorem. A sequence which is bounded in $\mathscr{N}$ must therefore be bounded in $C^\infty(X)$.

Assume that (11.1.4) is not valid. We can then find a sequence $\zeta_\nu \in \mathbb{C}^n$ such that $P(\zeta_\nu)=0$ and $|\zeta_\nu| \to \infty$ while $|\operatorname{Im} \zeta_\nu|$ remains bounded. With real ξ_ν and η_ν we write $\zeta_\nu = \xi_\nu + i\eta_\nu$ and set

$$u_\nu(x) = e^{i\langle x, \zeta_\nu \rangle}/k(\xi_\nu).$$

It is clear that $u_\nu \in \mathcal{N}$, and if $\phi \in C_0^\infty(X)$ we have

$$\|\phi u_\nu\|_{p,k} = ((2\pi)^{-n} \int |\hat{\phi}(\xi - \zeta_\nu) k(\xi)/k(\xi_\nu)|^p d\xi)^{1/p} \tag{11.1.6}$$
$$\leqq ((2\pi)^{-n} \int |\hat{\phi}(\xi - i\eta_\nu) M_k(\xi)|^p d\xi)^{1/p}$$

(Cf. the first part of (10.1.18).) This is bounded in virtue of (7.3.3) since η_ν remains bounded when $\nu \to \infty$.

That u_ν is bounded in $C^\infty(X)$ means that for $K \Subset X$ and any α we have

$$\sup_K |D^\alpha u_\nu(x)| \leqq C_{K,\alpha}.$$

Since η_ν is bounded this means that for any integer $\mu > 0$

$$|\xi_\nu|^\mu / k(\xi_\nu) \leqq C_\mu$$

which contradicts (10.1.2) if $\mu > N$. The proof is complete, for (11.1.4) is just another way of phrasing I a) in Theorem 11.1.3.

When $P(D)$ is hypoelliptic then the set $\mathcal{N}$ defined by (11.1.5) is contained in $C^\infty(X)$ and consists of all solutions of $P(D)u=0$ in $\mathscr{D}'(X)$. By Theorem 4.4.2 the topologies of $\mathscr{D}'(X)$ and $C^\infty(X)$ coincide on $\mathcal{N}$. In particular, $\mathcal{N}$ is a Montel space, that is, one can extract a convergent subsequence from every bounded sequence. (This follows from Ascoli's theorem.) When $P(D)$ is the Cauchy-Riemann operator this is the classical Stieltjes-Vitali theorem. It is characteristic of hypoelliptic operators:

Theorem 11.1.6. *Assume that there is an open non-empty set $X \subset \mathbb{R}^n$ and some $k \in \mathscr{K}$, $p \in [1, \infty]$, such that the space* (11.1.5) *is a Montel space with the topology induced by $B_{p,k}^{\text{loc}}(X)$. Then* (11.1.4) *is valid so $P(D)$ is hypoelliptic.*

Proof. Assuming that (11.1.4) is false we define a bounded sequence $u_\nu \in \mathcal{N}$ just as in the proof of Theorem 11.1.5. It converges to 0 in $\mathscr{D}'(X)$, for if $\phi \in C_0^\infty(X)$ we have

$$\int \phi(x) u_\nu(x)\, dx = \hat{\phi}(-\zeta_\nu)/k(\xi_\nu) \to 0, \qquad \nu \to \infty,$$

in view of (7.3.3) and (10.1.2). Since $\mathcal{N}$ is by hypothesis a Montel space and 0 is the only limit point of the bounded sequence u_ν, it follows that $u_\nu \to 0$ in $B_{p,k}^{\text{loc}}(X)$ also. But using the inequality

$k(\xi_\nu)\leqq k(\xi)M_k(\xi_\nu-\xi)$ we obtain instead of (11.1.6) the estimate

$$\|\phi u_\nu\|_{p,k}\geqq((2\pi)^{-n}\int|\hat{\phi}(\xi-i\eta_\nu)/M_k(-\xi)|^p d\xi)^{1/p}.$$

Since $\|\phi u_\nu\|_{p,k}\to 0$ when $\nu\to\infty$ we obtain that $\hat{\phi}(\xi-i\eta_0)=0$ identically if η_0 is a limit point of the sequence η_ν. Hence $\phi=0$ which is a contradiction since ϕ was supposed to be an arbitrary function in $C_0^\infty(X)$. This proves the theorem.

Next we shall use the results of Section 10.3, which depend only on Theorem 10.2.1, to prove a general theorem on interior regularity. Afterwards we shall specialize it to the hypoelliptic case.

Theorem 11.1.7. *Let $u\in B_{p,k_1}^{\mathrm{loc}}(X)$ and $P(D)u=f\in B_{p,k_2}^{\mathrm{loc}}(X)$ where $k_j\in\mathscr{K}$. Then it follows that $u\in B_{p,k}^{\mathrm{loc}}(X)$ if $k\in\mathscr{K}$ and*

$$k(\xi)\leqq Ck_1(\xi)(\tilde{P}(\xi)/\tilde{P}'(\xi))^N,\qquad \xi\in\mathbb{R}^n, \tag{11.1.7}$$

$$k(\xi)\leqq C(k_1(\xi)+\tilde{P}(\xi)k_2(\xi)),\qquad \xi\in\mathbb{R}^n, \tag{11.1.8}$$

for some constants C and N. Here we have used the notation

$$\tilde{P}'(\xi)^2=\sum_{\alpha\neq 0}|P^{(\alpha)}(\xi)|^2. \tag{11.1.9}$$

Note that the function $\tilde{P}'$ defined by (11.1.9) is in $\mathscr{K}$.

Proof. We first prove the theorem when $N=1$. Assuming that k satisfies (11.1.7) and (11.1.8) with $N=1$ we have to show that $\phi u\in B_{p,k}$ if $\phi\in C_0^\infty(X)$. In virtue of Theorem 10.3.2 this will follow if we can prove that $P(D)(\phi u)\in B_{p,k/\tilde{P}}$. Now Leibniz' formula gives

$$P(D)(\phi u)=\phi P(D)u+\sum_{\alpha\neq 0}D^\alpha\phi P^{(\alpha)}(D)u/\alpha!.$$

Since $u\in B_{p,k_1}^{\mathrm{loc}}(X)$ it follows from Theorems 10.1.22 and 10.1.21 that the terms in the sum are all in $B_{p,k_1/\tilde{P}'}$, which is a subspace of $B_{p,k/\tilde{P}}$ if (11.1.7) holds with $N=1$. Furthermore, the same argument gives that $\phi P(D)u\in B_{p,k_1/\tilde{P}}$. Since $P(D)u\in B_{p,k_2}^{\mathrm{loc}}$ we have also $\phi P(D)u\in B_{p,k_2}$, hence $\phi P(D)u\in B_{p,(k_1/\tilde{P}+k_2)}$ (Corollary 10.1.9), so it follows from (11.1.8) that $\phi P(D)u\in B_{p,k/\tilde{P}}$. This proves the theorem when $N=1$. Iterating this result N times, where N is a positive integer, we obtain that $u\in B_{p,k_{(N)}}^{\mathrm{loc}}(X)$ where

$$k_{(N)}(\xi)=\inf(k_1(\xi)(\tilde{P}(\xi)/\tilde{P}'(\xi))^N,\ k_1(\xi)+\tilde{P}(\xi)k_2(\xi)),$$

This completes the proof.

Remark. It follows from Theorem 10.3.2 that (11.1.8) is necessary and sufficient for the conclusion of the theorem to be valid even when

$u \in \mathscr{E}'(X)$. Using exponential solutions of the equation $P(D)u=0$ as in the proof of Theorem 11.1.5 one can also show that if the conclusion is valid when $f=0$, the quotient $k(\xi)/k_1(\xi)$ must be bounded by a function of $\tilde{P}(\xi)/\tilde{P}'(\xi)$. When k and k_1 are both polynomials it follows by means of Corollary A.2.6 that (11.1.7) must be valid for some N. We shall not use this fact here and leave the proof to the reader.

The following theorem makes condition (ii) in Theorem 11.1.1 more precise.

Theorem 11.1.8. *Let $P(D)$ be hypoelliptic and $u \in \mathscr{D}'(X)$. If $P(D)u \in B^{\text{loc}}_{p,k}(X)$ it follows that $u \in B^{\text{loc}}_{p,\tilde{P}k}(X)$.*

Proof. If Y is an open set $\Subset X$, we have $u \in B^{\text{loc}}_{p,k_1}(Y)$ for some $k_1 \in \mathscr{K}$ because u is of finite order in Y. It follows from IIb) in Theorem 11.1.3 that

$$(1+|\xi|)^c \leqq C\tilde{P}(\xi)/\tilde{P}'(\xi)$$

for some positive constants c and C. Hence there are constants C and N such that

$$\tilde{P}(\xi)\,k(\xi) \leqq C k_1(\xi)(\tilde{P}(\xi)/\tilde{P}'(\xi))^N,$$

so it follows immediately from Theorem 11.1.7 that $u \in B^{\text{loc}}_{p,\tilde{P}k}(Y)$. Since Y is an arbitrary open set $\Subset X$, this proves the theorem.

We shall now study the algebraic conditions for hypoellipticity further and give some examples.

Theorem 11.1.9. *If $P_1(D)$ and $P_2(D)$ are equally strong and $P_1(D)$ is hypoelliptic, it follows that $P_2(D)$ is also hypoelliptic. If $d_j(\xi)$ is the distance from ξ to the zeros of P_j, we even have with a constant C*

$$C^{-1} \leqq (d_1(\xi)+1)/(d_2(\xi)+1) \leqq C, \quad \xi \in \mathbb{R}^n. \tag{11.1.10}$$

Proof. In view of (11.1.2) we have

$$(d_1(\xi))^{|\alpha|}\,|P_1^{(\alpha)}(\xi)| \leqq C^{|\alpha|}\,|P_1(\xi)|.$$

When $|\xi|$ is so large that $d_1(\xi) \geqq 1$, we may apply (10.4.7) to $Q=P_2$ and $P=P_1$ with $t=d_1(\xi)$. This gives with some constant C

$$\sum_\alpha |P_2^{(\alpha)}(\xi)|^2 (d_1(\xi))^{2|\alpha|} \leqq C\,|P_1(\xi)|^2.$$

Since P_1 is weaker than P_2 the estimate

$$\sum_\alpha |P_2^{(\alpha)}(\xi)|^2 (d_1(\xi))^{2|\alpha|} \leqq C \sum_\alpha |P_2^{(\alpha)}(\xi)|^2$$

follows with some other constant C. When $|\xi|$ is so large that $d_1(\xi)^2 > 2C$ also, we obtain

$$\sum_\alpha |P_2^{(\alpha)}(\xi)|^2 (d_1(\xi))^{2|\alpha|} \leqq 2C|P_2(\xi)|^2.$$

Application of (11.1.2) to P_2 now gives $d_2(\xi)/d_1(\xi) > c$ for large $|\xi|$ and some positive constant c. Hence P_2 satisfies conditions I) in Theorem 11.1.3 and is thus hypoelliptic. Since the roles of P_1 and P_2 in the above estimates may now be interchanged, we obtain the inequality (11.1.10).

Remark. If P is hypoelliptic, it is clear that the relation $Q \prec P$ is equivalent to boundedness of $Q(\xi)/P(\xi)$ at infinity in $\mathbb{R}^n$. The same reasoning as in the proof above also shows that the relation $Q \ll P$ is equivalent to $Q(\xi)/P(\xi) \to 0$ when $\xi \to \infty$ in $\mathbb{R}^n$, if and only if P is hypoelliptic. (The necessity follows from the fact that $P^{(\alpha)} \ll P$ for every P if $\alpha \neq 0$.)

Theorem 11.1.10. *Elliptic operators are hypoelliptic. The principal part of a hypoelliptic operator cannot have a simple real zero* $\neq 0$.

Proof. That elliptic operators are hypoelliptic follows immediately from (10.4.12). - If $P(D)$ is of order m and $\xi \in \mathbb{R}^n \setminus 0$ is a simple zero of the principal part P_m then $P_m(\xi) = 0$ but $\partial P_m(\xi)/\partial \xi_j \neq 0$ for some j. Since $P(t\xi)$ is then at most of order $m-1$ in t and the coefficient of t^{m-1} in $P^{(j)}(t\xi)$ is precisely $\partial P_m(\xi)/\partial \xi_j$, it follows that $P^{(j)}(t\xi)/P(t\xi)$ does not converge to 0 when $t \to \infty$ which proves that $P(D)$ is not hypoelliptic.

More general classes of hypoelliptic operators can be defined as follows. Let $m_1, \dots, m_n$ be positive integers and set

$$|\alpha : m| = \sum_1^n \alpha_k / m_k. \tag{11.1.11}$$

If $|\alpha : m| \leqq 1$ for all terms in $P(D)$,

$$P(D) = \sum_{|\alpha : m| \leqq 1} a_\alpha D^\alpha, \tag{11.1.12}$$

we set

$$P^0(D) = \sum_{|\alpha : m| = 1} a_\alpha D^\alpha. \tag{11.1.13}$$

Theorem 11.1.11. *If* $P^0(\xi) \neq 0$ *for* $0 \neq \xi \in \mathbb{R}^n$, *it follows that* $P(D)$ *is hypoelliptic. Such an operator is called semi-elliptic.*

Proof. We first note that for some constant C

$$\sum_1^n |\xi_j|^{m_j} \leqq C|P^0(\xi)|, \quad \xi \in \mathbb{R}^n. \tag{11.1.14}$$

In fact, the lower bound of P^0 when the left-hand side is equal to 1 must be positive. Denoting the lower bound by $1/C$, we have (11.1.14) when the left-hand side is equal to 1. But replacing ξ_j by $\xi_j t^{1/m_j}$, $j=1, \dots, n$, only means multiplying both sides by t, if $t>0$. Hence (11.1.14) is valid for every ξ. Now the trivial inequality $|\xi_k|^{m_k} \leqq \sum |\xi_j|^{m_j}$ implies that

$$|\xi^\alpha| \leqq \Big(\sum_1^n |\xi_j|^{m_j}\Big)^{|\alpha:m|}. \tag{11.1.15}$$

As in the proof of (10.4.12) we thus obtain the estimate

$$1+\sum_1^n |\xi_j|^{m_j} \leqq C(|P(\xi)|+1), \qquad \xi \in \mathbb{R}^n.$$

In view of (11.1.15) this implies that $|P^{(\alpha)}(\xi)|/|P(\xi)| \to 0$ when $\xi \to \infty$, which proves the theorem.

Note that when $m_j = m$ for every j, the semi-elliptic operators are just the elliptic operators of order m. If $m_1 = 1$ and $m_j = 2$ for $j > 1$, we find that the heat equation is hypoelliptic (see also Section 4.4). Also the p-parabolic operators of Petrowsky [2] are semi-elliptic.

We finally give a more complicated example of a hypoelliptic operator in order to show that the principal part may be chosen rather arbitrarily provided that the real characteristics have high multiplicities.

Theorem 11.1.12. *Let $Q(\xi)$ be a polynomial of degree m with real coefficients and k an integer $\geqq 2$. Let $R(\xi)$ be a homogeneous positive definite polynomial of degree $2km-2(k-1)$. Then the polynomial*

$$P(\xi) = Q(\xi)^{2k} + R(\xi) \tag{11.1.16}$$

is hypoelliptic.

Proof. We shall show that condition (iv) in Theorem 11.1.1 is satisfied. Writing $Q(\xi)^{2k} = S(\xi)$, we have $P^{(\alpha)}(\xi) = S^{(\alpha)}(\xi) + R^{(\alpha)}(\xi)$; and since

$$|R^{(\alpha)}(\xi)|/P(\xi) \leqq |R^{(\alpha)}(\xi)|/R(\xi) \to 0 \quad \text{when} \quad \xi \to \infty \qquad \text{if } \alpha \neq 0,$$

the only difficulty is to estimate $S^{(\alpha)}$. Now we can write

$$S^{(\alpha)}(\xi) = \sum_{j=1}^{\min(2k,|\alpha|)} Q(\xi)^{2k-j} F_j^\alpha(\xi) \tag{11.1.17}$$

where $F_j^\alpha(\xi)$ is a polynomial of degree $jm-|\alpha|$ at most. Estimating $Q(\xi)^{2k}$ and $R(\xi)$ by $P(\xi)$ gives

$$|Q(\xi)|^{2k-j} R(\xi)^{j/2k} \leqq P(\xi). \tag{11.1.18}$$

If μ is the degree of R we obtain for the terms in (11.1.17) the estimate

$$|Q(\xi)|^{2k-j}|F_j^\alpha(\xi)| \leqq CP(\xi)R(\xi)^{((jm-|\alpha|)/\mu)-j/2k}.$$

The exponent of R increases with j since $2km>\mu$. When $j\leqq|\alpha|$ it is therefore at most equal to $|\alpha|((m-1)/\mu-1/2k)=-|\alpha|/k\mu$. Hence we have proved that $S^{(\alpha)}(\xi)/P(\xi)\to 0$ when $\xi\to\infty$ if $\alpha\neq 0$, which completes the proof.

Note that in the operator (11.1.16) the principal part is the principal part of Q raised to the power $2k$.

11.2. Partially Hypoelliptic Operators

In this section we shall study differential operators $P(D)$ such that the solutions u of the equation $P(D)u=f$ have to be smooth if f is smooth and in addition we impose smoothness in some variables on u. The results are parallel to and contain those of Section 11.1, but we have preferred to study the more important question of hypoellipticity separately.

We suppose that the coordinates $x=(x_1,\dots,x_n)$ are split into two parts $x'=(x_1,\dots,x_j)$ and $x''=(x_{j+1},\dots,x_n)$. Multi-indices α will be split in the same way. If $\psi\in C_0^\infty(\mathbb{R}^j)$ then the tensor product $\psi(x')\otimes\delta(x'')$ is a measure in the plane $x''=0$. If X is an open set in $\mathbb{R}^n$ we set $X_\psi=\{x;\{x\}-\operatorname{supp}(\psi\otimes\delta)\subset X\}$. It is clear that the convolution $u*'\psi=u*(\psi\otimes\delta)$ is defined in X_ψ if $u\in\mathscr{D}'(X)$ (see Section 4.2). When $u\in B_{p,k_1}^{\text{loc}}(X)$ we have $u*'\psi\in B_{p,k_2}^{\text{loc}}(X_\psi)$ if

$$k_2(\xi)\leqq C(1+|\xi'|)^N k_1(\xi), \qquad \xi\in\mathbb{R}^n,$$

for some constants C and N. This follows from Theorem 10.1.24 if $X=\mathbb{R}^n$, and the simple modification which is otherwise required is left for the reader. The following is an analogue of Theorem 11.1.5.

Theorem 11.2.1. *Assume that there is an open non-void set $X\subset\mathbb{R}^n$ and some $k\in\mathscr{K}$, $p\in[1,\infty]$ such that for every $u\in B_{p,k}^{\text{loc}}(X)$ satisfying the equation $P(D)u=0$ and every $\psi\in C_0^\infty(\mathbb{R}^j)$ the convolution $u*'\psi\in C^\infty(X_\psi)$. Then it follows that*

(11.2.1) $$|\operatorname{Im}\zeta|+|\operatorname{Re}\zeta'|\to\infty$$

if $\zeta\to\infty$ on the surface $P(\zeta)=0$.

Proof. By the closed graph theorem the map

$$\mathcal{N}\ni u\to u*'\psi\in C^\infty(X_\psi)$$

must be continuous. (Here $\mathcal{N}$ is defined by (11.1.5).) Assuming that (11.2.1) is not valid, we choose a sequence of points $\zeta_\nu = \xi_\nu + i\eta_\nu \in \mathbb{C}^n$ such that $P(\zeta_\nu)=0$ and $|\zeta_\nu| \to \infty$ while $|\eta_\nu|$ and $|\xi'_\nu|$ remain bounded. Then the sequence

$$u_\nu(x) = e^{i\langle x, \zeta_\nu\rangle}/k(\xi_\nu)$$

is bounded in $\mathcal{N}$ (see the proof of Theorem 11.1.5), so $u_\nu *' \psi = \hat{\psi}(\zeta'_\nu)\, u_\nu$ must be bounded in $C^\infty(X_\psi)$. Choose ψ so close to the Dirac measure that X_ψ is not empty and $|\hat{\psi}(\zeta')| \geqq 1/2$ when $|\zeta'| \leqq \sup|\zeta'_\nu|$. Then it follows that u_ν must be bounded in $C^\infty(X_\psi)$ which gives a contradiction as at the end of the proof of Theorem 11.1.5.

There is a stronger version of Theorem 11.2.1 which is parallel to Theorem 11.1.6:

Theorem 11.2.2. *Assume that there is an open non-void set $X \subset \mathbb{R}^n$ and some $k \in \mathcal{K}$, $p \in [1, \infty]$ such that the map*

$$\mathcal{N} \ni u \to u *' \psi \in B^{\mathrm{loc}}_{p,k}(X_\psi)$$

is compact for every $\psi \in C_0^\infty(\mathbb{R}^j)$. (Here $\mathcal{N}$ is defined by (11.1.5).) Then it follows that (11.2.1) is valid.

Proof. If (11.2.1) is not valid we choose ζ_ν and u_ν as in the proof of Theorem 11.2.1. Then u_ν is bounded, and $u_\nu *' \psi$ converges to 0 in $\mathcal{D}'(X_\psi)$ (see the proof of Theorem 11.1.6) so $u_\nu *' \psi \to 0$ in $B^{\mathrm{loc}}_{p,k}(X_\psi)$ by the compactness. When ψ is close to the Dirac measure we conclude that $u_\nu \to 0$ in $B^{\mathrm{loc}}_{p,k}(X_\psi)$ which contradicts the end of the proof of Theorem 11.1.6.

Before proving the converse of Theorem 11.2.1 we give a number of equivalent forms of the condition (11.2.1).

Theorem 11.2.3. *The condition* (11.2.1) *is equivalent to each of the following six:*

I. *If $d(\xi)$ is the distance from $\xi \in \mathbb{R}^n$ to the surface $\{\zeta \in \mathbb{C}^n;\, P(\zeta)=0\}$, it follows that*

a) *$d(\xi) \to \infty$ if $\xi'' \to \infty$ while ξ' remains bounded.*

b) *There exist positive constants c and C such that*

$$(1+|\xi|^2)^c \leqq C(1+d(\xi)^2)(1+|\xi'|^2), \qquad \xi \in \mathbb{R}^n.$$

II. a) *$P^{(\alpha)}(\xi)/P(\xi) \to 0$ if $\alpha \neq 0$ and $\xi'' \to \infty$ while ξ' remains bounded.*

II. b) *There exist positive constants c and C such that*

$$\tilde{P}'(\xi)/\tilde{P}(\xi) \leqq C(1+|\xi'|)(1+|\xi|)^{-c}.$$

III. *P can be written as a finite sum*

$$P(\xi)=\sum_{\alpha''=0} P_\alpha(\xi'')\,\xi'^\alpha \tag{11.2.2}$$

where $P_0(\xi'')$ *is hypoelliptic (as a polynomial in* ξ''*) and*

a) $P_\alpha(\xi'')/P_0(\xi'')\to 0$ *when* $\xi''\to\infty$ *if* $\alpha\neq 0$.

b) *There exist positive constants c and C such that*

$$|P_\alpha(\xi'')|/(|P_0(\xi'')|+1)\leqq C(1+|\xi''|)^{-c}, \qquad \alpha\neq 0.$$

Proof. Condition I a) is obviously equivalent to (11.2.1), and it is clear that I b) implies I a). On the other hand, if I a) is fulfilled, it follows that $(1+d(\xi)^2)(1+|\xi'|^2)\to\infty$ when $\xi\to\infty$. Hence I b) follows if we use Corollary A.2.6 in the same way as in the proof of Theorem 11.1.3. From Lemma 11.1.4 it follows that I b) implies II b) and that II a) implies I a). (The constants C are not the same.) Since it is trivial that II b) implies II a), this proves the equivalence of the conditions I and II.

By Taylor's formula we have (11.2.2) with

$$P_\alpha(\xi'')=P^{(\alpha)}(0,\xi'')/\alpha!.$$

If P satisfies II, it follows that P_0 satisfies the conditions II in Theorem 11.1.3. Hence P_0 is hypoelliptic, and the remaining conditions in III follow immediately from II.

Since III b) implies III a) it only remains to prove that III a) implies II a). To do so we note that it follows from III a) that $P_\beta^{(\alpha)}\prec P_\beta \ll P_0$ if $\beta\neq 0$, hence (see the remark following Theorem 11.1.9)

$$P_\beta^{(\alpha)}(\xi'')/P_0(\xi'')\to 0, \qquad \xi''\to\infty \quad \text{if } \beta\neq 0 \text{ or } \alpha\neq 0. \tag{11.2.3}$$

Since II a) follows from (11.2.3), the proof is now complete.

Definition 11.2.4. The differential operator $P(D)$ is called partially hypoelliptic with respect to the plane $x''=0$ if the (equivalent) conditions in Theorem 11.2.3 are fulfilled. The polynomial $P(\xi)$ is then called partially hypoelliptic in ξ''.

Theorem 11.2.5. *Let* $P(D)$ *be partially hypoelliptic with respect to the plane* $x''=0$ *and let* $u\in\bigcap_{\nu=0}^{\infty} B_{p,k_\nu}^{\mathrm{loc}}(X)$ *where* $k_\nu(\xi)=k_0(\xi)(1+|\xi'|)^\nu$ *are functions in* $\mathscr{K}$. *If* $P(D)u=f\in C^\infty(X)$, *it follows that* $u\in C^\infty(X)$.

Proof. In virtue of Theorem 11.1.7 we have $u\in B_{p,k}^{\mathrm{loc}}(X)$ if for some C, N and ν

$$k(\xi)\leqq C k_\nu(\xi)(\tilde P(\xi)/\tilde P'(\xi))^N, \qquad \xi\in\mathbb{R}^n. \tag{11.2.4}$$

Now it follows from condition IIb) in Theorem 11.2.3 that (11.2.4) is valid if for some other constant C

$$k(\xi)/k_0(\xi) \leqq C(1+|\xi'|)^{\nu-N}(1+|\xi|)^{cN}.$$

Since this inequality is valid for an arbitrary $k \in \mathscr{K}$ if we choose $\nu = N$ sufficiently large, the theorem is proved.

Corollary 11.2.6. *Let $P(D)$ be partially hypoelliptic with respect to the plane $x''=0$ and let $\psi \in C_0^\infty(\mathbb{R}^j)$. If $P(D)u = f \in C^\infty(X)$, then $u *' \psi \in C^\infty(X_\psi)$.*

Proof. First assume that u is of finite order in X, hence that $u \in B_{p,k_0}^{\mathrm{loc}}(X)$ for some $k_0 \in \mathscr{K}$. By the remarks made before Theorem 11.2.1 the convolution $u *' \psi$ satisfies the hypothesis of Theorem 11.2.5 in X_ψ. Hence $u *' \psi \in C^\infty(X_\psi)$. Now if u is an arbitrary distribution it follows from what we have already proved that $u *' \psi \in C^\infty(Y_\psi)$ for every open set $Y \Subset X$. This proves the corollary.

Corollary 11.2.7. *Let $P(D)$ be partially hypoelliptic with respect to the plane $x''=0$. With the notations of Theorem 11.2.1 and* (11.1.5) *the mapping $\mathscr{N} \ni u \to u *' \psi$ is then continuous with values in $C^\infty(X_\psi)$, thus compact with values in $B_{p,k'}^{\mathrm{loc}}(X_\psi)$ for an arbitrary $k' \in \mathscr{K}$.*

Proof. The continuity follows from Corollary 11.2.6 and the closed graph theorem (or else by inspecting the proof of Corollary 11.2.6). Since bounded sets in $C^\infty(X_\psi)$ are precompact in the topology of $C^\infty(X_\psi)$ which is stronger than that in $B_{p,k'}^{\mathrm{loc}}(X_\psi)$, the proof is complete.

Example 11.2.8. The differential operator $P(D)$ is partially hypoelliptic with respect to the hyperplane $x_n = 0$ if and only if the part of $P(\xi)$ which is of highest order with respect to ξ_n is independent of the other variables. Theorem 4.4.8 implies Corollary 11.2.6 then. (In the non-characteristic case the same result follows from Theorem 8.3.1 and (8.2.16).)

Theorem 11.1.7 shows that if $u \in \mathscr{D}'(X)$ and $P(D)u \in C^\infty(X)$, then

$$|\widehat{\phi u}(\xi)| \leqq C_N(1+|\xi|)^{-N}, \qquad N = 1, 2, \ldots \tag{11.2.5}$$

if $\xi \in M$, $\phi \in C_0^\infty(X)$, and M is a subset of $\mathbb{R}^n$ such that for some positive constants C', c'

$$\tilde{P}'(\xi)/\tilde{P}(\xi) \leqq C'(1+|\xi|)^{-c'}, \qquad \xi \in M. \tag{11.2.6}$$

Condition IIb) in Theorem 11.2.3 means that we can take

$$M = \{\xi;\, |\xi'| < C|\xi''|^{c''}\}$$

for some $c''>0$. If $c''=1$ then (11.2.5) would mean that the wave front set of u does not meet the plane $\xi'=0$. When $c''<1$ the set M is much more narrow than a conic neighborhood of the plane $\xi'=0$. However, one can introduce modifications of $WF(u)$ which take the validity of (11.2.5) in such sets into account and rephrase the results of this section in terms of them. (See also Hörmander [29, Section 1.6].)

11.3. Continuation of Differentiability

In this section we shall prove analogues for singularities of the (non-) uniqueness theorems in Section 8.6. Those were essential when we studied P-convexity for supports in Section 10.8. The results in this section have similar applications to P-convexity for singular supports.

We shall first introduce an appropriate substitute for the characteristic hyperplanes in Theorem 8.6.8. In doing so we shall not only consider hyperplanes but linear spaces of arbitrary dimension. Let $P(D)$ be any differential operator in $\mathbb{R}^n$ with constant coefficients. If V is a linear subspace of $\mathbb{R}^n$ we set

$$\tilde{P}_V(\xi, t)=\sup\{|P(\xi+\theta)|;\ \theta\in V, |\theta|\leqq t\}. \tag{11.3.1}$$

Note that if $V=\mathbb{R}^n$ it follows from (10.4.1) that (11.3.1) is equivalent to our earlier definition of $\tilde{P}(\xi, t)$. We shall then drop the subscript V, and we shall switch back to (10.4.2) whenever this is more convenient. If V is the coordinate plane $\xi''=(\xi_{k+1}, \dots, \xi_n)=0$ it is clear in view of (10.4.1) that $\tilde{P}_V(\xi, t)$ is equivalent to

$$\Big(\sum_{\alpha''=0} |P^{(\alpha)}(\xi)|^2 t^{2|\alpha|}\Big)^{\frac{1}{2}}. \tag{11.3.1)'$$

Note that this shows that a change of norm in $\mathbb{R}^n$ does not affect $\tilde{P}_V(\xi, t)$ by more than a fixed factor.

Now set

$$\sigma_P(V)=\inf_{t>1}\ \varliminf_{\xi\to\infty} \tilde{P}_V(\xi, t)/\tilde{P}(\xi, t). \tag{11.3.2}$$

This is a continuous function of V in the sense that

$$\sigma_P(V)\leqq\sigma_P(W)+Cd(V, W), \tag{11.3.3}$$

where

$$d(V, W)=\sup_{x\in V, |x|=1}\Big(\inf_{y\in W, |y|=1} |x-y|\Big). \tag{11.3.4}$$

To prove (11.3.3) we first observe that by Taylor's formula

$$|P(\xi+\theta)-P(\xi+\eta)| \leqq C\tilde{P}(\xi,t)|\theta-\eta|/t \tag{11.3.5}$$

if $|\theta|\leqq t$, $|\eta|\leqq t$. If $\theta\in V$ and $|\theta|\leqq t$ we can choose $\eta\in W$ with $|\eta|=|\theta|$ and $|\theta-\eta|\leqq t d(V,W)$ by the definition (11.3.4). Hence

$$|P(\xi+\theta)|\leqq|P(\xi+\eta)|+Cd(V,W)\tilde{P}(\xi,t)\leqq\tilde{P}_W(\xi,t)+Cd(V,W)\tilde{P}(\xi,t)$$

which proves (11.3.3). From (11.3.3) it follows that

$$|\sigma_P(V)-\sigma_P(W)|\leqq C\max(d(V,W)+d(W,V)). \tag{11.3.3$'$}$$

Thus $\sigma_P(V)$ is a continuous function of V when V has fixed dimension.

Theorem 11.3.1. *Let V be a linear subspace of $\mathbb{R}^n$ such that $\sigma_P(V')=0$ if V' is the orthogonal space of V. For every non-negative integer μ one can then find $u\in C^\mu(\mathbb{R}^n)$ with $P(D)u=0$ and* $\operatorname{sing\,supp} u=V$. *More precisely, we can find u so that $u\notin C^{\mu+1}(N)$ for any open set N intersecting V.*

Later on we shall also prove a converse of Theorem 11.3.1. Before the proof we make an application.

Corollary 11.3.2. *If X is an open set in $\mathbb{R}^n$ which is P-convex for singular supports, then the minimum principle is valid for the boundary distance $d_X(x)$ on every affine subspace parallel to a linear subspace V with $\sigma_P(V')=0$.*

Proof. The statement follows if we use Theorem 11.3.1 instead of Theorem 8.3.8 in the first part of the proof of Theorem 10.8.9.

Corollary 11.3.3. *Every open set $X\subset\mathbb{R}^n$ is P-convex for singular supports if and only if $P(-D)$ is hypoelliptic.*

Proof. If $P(-D)$ is not hypoelliptic, then there is a non-constant localization Q at infinity (see the proof of Theorem 11.1.1). It is clear that

$$\tilde{Q}_V(0,t)/\tilde{Q}(0,t)\geqq\sigma_{\check{P}}(V), \qquad t>1.$$

If we take $V=\Lambda(Q)$ then the numerator is independent of t while the denominator tends to ∞ with t. Hence $\sigma_{\check{P}}(V)=0$, so Corollary 11.3.2 shows that $d_X(x)$ satisfies the minimum principle in every affine subspace parallel to $\Lambda'(Q)$. This condition is not fulfilled if ∂X is strictly concave at a point where $\Lambda'(Q)$ is a tangent of X. On the other hand, if $P(-D)$ is hypoelliptic, then $\operatorname{sing\,supp} u=\operatorname{sing\,supp} P(-D)u$ for every $u\in\mathscr{E}'(X)$ which implies that X is P-convex for singular supports.

To prove Theorem 11.3.1 we shall first give the hypothesis a seemingly stronger form.

Lemma 11.3.4. *If $\sigma_P(V')=0$, it follows that there are positive constants b, β, t_1, ρ such that for any $t>t_1$ and $r>t^\rho$ one can find $\xi\in\mathbb{R}^n$ with $|\xi|=r$ and*

(11.3.6) $$\tilde{P}_{V'}(\xi, t)<bt^{-\beta}\tilde{P}(\xi, t).$$

Proof. It follows from Theorem A.2.2 and Theorem A.2.5 that

$$a(t)=\varliminf_{\xi\to\infty} \tilde{P}_{V'}(\xi, t)/\tilde{P}(\xi, t)$$

is an algebraic function of t for large t. In fact, $y\geqq a(t)$ means that for every $\varepsilon>0$ there is some ξ with $\varepsilon^2|\xi|^2>1$ and

$$\tilde{P}_{V'}(\xi, t)^2<(y+\varepsilon)^2\tilde{P}(\xi, t)^2.$$

Here we use expressions of the form (11.3.1)′ for $\tilde{P}_{V'}$ and $\tilde{P}$. It is then also clear that

$$a(st)\leqq Cs^m a(t), \quad s\geqq 1.$$

By hypothesis $\inf_{t>1} a(t)=\sigma_P(V')=0$, so it follows that $\lim_{t\to\infty} a(t)=0$. Hence

$$a(t)t^\beta<b, \quad t>1,$$

for some rational number $\beta>0$ and constant $b>0$. The set

$$M=\{(r, t);\ t>1,\ \tilde{P}_{V'}(\xi, t)<bt^{-\beta}\tilde{P}(\xi, t) \text{ for some } \xi\in\mathbb{R}^n,\ |\xi|=r\}$$

is semi-algebraic by the Tarski-Seidenberg theorem, and if $t>1$, we have $(r, t)\in M$ for arbitrarily large values of r. Since the boundary of M is piecewise algebraic it follows that $(r, t)\in M$ if $r>t^\rho$, $t>t_1$, and t_1, ρ are large enough.

When we examine the consequences of (11.3.6) it is convenient to work with coordinates chosen so that V is defined by $x'=0$ if $x=(x', x'')$, $x'=(x_1, \dots, x_k)$, $x''=(x_{k+1}, \dots, x_n)$, is a splitting of the coordinates in two groups. The orthogonal space V' is then defined by $\xi''=0$. Consider now a polynomial Q with $\tilde{Q}(0, t)=1$ and $\tilde{Q}_{V'}(0, t)$ small. We can choose η with $|\eta|\leqq t$ so that $|Q(\eta)|=1$ and then find a whole circle where Q is bounded away from 0. In fact, Lemma 7.3.12 shows that if q is a polynomial in one variable of degree $\leqq m$, then

(11.3.7) $$\sup_{|z|\leqq 1}|q(z)|\leqq C_0 \sup_{0<r<1}\inf_{|z|=r}|q(z)|$$

where C_0 only depends on m. If we apply this to $q(z)=Q(\eta', z\eta'')$ we obtain

(11.3.8) $$|Q(\eta', z\eta'')|\geqq 1/C_0, \quad |z|=r,$$

for some $r \in [0,1]$. If $|Q(\eta',0)|<1/C_0$ it follows that the equation $Q(\eta', z\eta'')=0$ has a root with $|z|<r$. Let κ be the number of such roots. Since $\tilde{Q}(\zeta,t) \leqq C'$ when $|\zeta| \leqq t$, we obtain from (11.3.8) and Taylor's formula if γ is small enough

$$|Q(\eta'+\theta, z\eta'')| \geqq 1/2C_0 \quad \text{when } |z|=r,\ \theta \in \mathbb{C}^k,\ |\theta|<\gamma t. \tag{11.3.9}$$

Here γ depends only on n and m. The equation $Q(\eta'+\theta, z\eta'')=0$ must have κ roots with $|z|<r$ for every θ with $|\theta|<\gamma t$. If we set

$$U(\theta,x'')=(2\pi i\kappa)^{-1} \int_{|z|=r} e^{i\langle x'', z\eta''\rangle} \left(\frac{d}{dz} Q(\eta'+\theta, z\eta'')\right) / Q(\eta'+\theta, z\eta'')\, dz$$

it follows that when $|\theta|<\gamma t$

$$U(\theta,0)=1, \quad Q(\eta'+\theta, D'')U(\theta,x'')=0, \quad |U(\theta,x'')| \leqq Ce^{t|x''|} \tag{11.3.10}$$

where C is another constant depending only on n and m. It is clear that U is an analytic function of x'' and θ when $|\theta|<\gamma t$. The following lemma is an easy consequence:

Lemma 11.3.5. *There are constants $\varepsilon_0, c, C, \gamma$ (depending only on n and m) such that if P is of order $\leqq m$ and for some $\xi \in \mathbb{R}^n$ and $t>0$*

$$\tilde{P}_{V'}(\xi,t)/\tilde{P}(\xi,t) \leqq \varepsilon < \varepsilon_0$$

then one can find $\eta \in \mathbb{R}^n$ with $|\eta-\xi| \leqq t$, $\eta-\xi \in V'$, and an analytic function $U(\theta,x'')$, $\theta \in \mathbb{C}^k$, $|\theta|<\gamma t$, $x'' \in \mathbb{C}^{n-k}$, satisfying

$$U(\theta,0)=1, \quad P(\eta+(\theta,D''))\,U(\theta,x'')=0, \tag{11.3.11}$$

$$|U(\theta,x'')| \leqq C \exp(ct|x''|\varepsilon^{1/m}). \tag{11.3.12}$$

Proof. Without restriction we may assume that $\xi=0$. Set

$$Q(\zeta)=P(\zeta', \delta\zeta'')$$

where $0<\delta<1$. Then

$$\tilde{P}(0,t) \leqq C'\tilde{Q}(0,t)\,\delta^{-m}, \quad \tilde{P}_{V'}(0,t)=\tilde{Q}_{V'}(0,t),$$

hence

$$\tilde{Q}_{V'}(0,t)/\tilde{Q}(0,t) \leqq C'\varepsilon\delta^{-m}.$$

If $C'\varepsilon\delta^{-m}=1/2\,C_0$ where C_0 is the constant in (11.3.8) we obtain a function satisfying (11.3.10). Then $U(\theta,\delta x'')$ has the properties required in Lemma 11.3.5 if $(\eta',0)$ is now called η. The proof is complete.

Proof of Theorem 11.3.1. Banach's theorem and a condensation of singularities argument shows just as in the proof of Theorem 8.3.8 that if the statement is false then

(11.3.13) $$\sum_{|\alpha|=\mu+1} |D^\alpha u(0)| \leqq C\Big(\sum_{|\alpha|\leqq\mu} \sup_{K_1} |D^\alpha u| + \sum_{|\alpha|\leqq\nu} \sup_{K_2} |D^\alpha u|\Big)$$

for all $u\in C^\infty(\mathbb{R}^n)$ with $P(D)u=0$. Here K_1 and K_2 are compact sets, $K_2\subset\mathbb{R}^n\setminus V$. The theorem will be proved if we construct some u for which (11.3.13) is not true. To do so we start with two small positive numbers ε and δ to be chosen later and use Lemma 11.3.4 to find arbitrarily large t and $\xi\in\mathbb{R}^n$ with

(11.3.14) $$\log|\xi|=\delta t, \qquad \tilde{P}_{V'}(\xi,t)/\tilde{P}(\xi,t)<\varepsilon.$$

By Theorem 1.4.2 we can select a sequence $\chi_N\in C_0^\infty(\mathbb{R}^k)$ with support in $\{\theta;|\theta|<\gamma/2\}$ such that $\int\chi_N d\theta=1$ and

$$|D^\alpha\chi_N|\leqq C^{|\alpha|+1}N^{|\alpha|}, \quad |\alpha|\leqq N.$$

With N to be chosen later we set

(11.3.15) $$u(x)=e^{i\langle x,\eta\rangle}\int e^{i\langle x',\theta\rangle}U(\theta,x'')\chi_N(\theta/t)t^{-k}d\theta$$

where $U(\theta,x'')$ and η are given by Lemma 11.3.5. Then $P(D)u=0$, $u\in C^\infty$ and $u(0)=1$. Since

$$|D_{x''}^{\alpha''}U(\theta,x'')|\leqq Ce(ct\varepsilon^{1/m})^{|\alpha''|}\exp(ct|x''|\varepsilon^{1/m})|\alpha''|!$$

by Cauchy's inequalities, we have in K_1

(11.3.16) $$|D^\alpha u(x)|\leqq C_\alpha(1+|\eta|)^{|\alpha|}\exp(C_1t\varepsilon^{1/m})<C'_\alpha(1+|\xi|)^{|\alpha|+1/2}$$

if $C_1\varepsilon^{1/m}<\delta/2$, for $e^{\delta t}=|\xi|$. In the same way we obtain

(11.3.17) $$D^\alpha u(0)=\eta^\alpha+O(|\eta|^{|\alpha|-1/2}).$$

It remains to examine the last sum in (11.3.13). In K_2 we have $|x'|>c_0>0$ by hypothesis. We can write

$$u(x)=e^{i\langle x,\eta\rangle}\int e^{i\langle tx',\theta\rangle}U(t\theta,x'')\chi_N(\theta)d\theta,$$

and by Cauchy's inequalities

$$|D_\theta^\alpha U(t\theta,x'')|\leqq C^{|\alpha|+1}\exp(ct|x''|\varepsilon^{1/m})|\alpha|!, \qquad |\theta|<\gamma/2.$$

Hence the proof of Proposition 8.4.2, part a, gives

$$|u(x)|\leqq C_0(C'N/c_0t)^N\exp(C_2t\varepsilon^{1/m}), \qquad x\in K_2.$$

Differentiation of u can only give a factor $|\xi|$ and change C_0, hence

$$S=\sum_{|\alpha|\leqq\nu}\sup_{K_2}|D^\alpha u(x)|\leqq C_\nu(C'N/c_0t)^N|\xi|^\nu\exp(C_2t\delta/2C_1).$$

When $N=[c_0t/eC']$ this gives

(11.3.18) $$S\leqq eC_\nu\exp(-c_0t/eC'+\delta t(\nu+C_2/2C_1)).$$

Choose δ so that $\delta(v+C_2/2\,C_1)<c_0/e\,C'$ and then ε with $C_1\,\varepsilon^{1/m}<\delta/2$. When ξ, $t\to\infty$ it follows from (11.3.16), (11.3.17), (11.3.18) that (11.3.13) is not valid. The proof is complete.

Our next theorem is much more than a converse of Theorem 11.3.1.

Theorem 11.3.6. *Let X be an open set in $\mathbb{R}^n$, x^0 a point in X and $\phi_1,\dots,\phi_k\in C^1(X)$ real valued functions such that $d\phi_1(x^0),\dots,d\phi_k(x^0)$ are linearly independent. Assume that $\sigma_P(W)\neq 0$ for the space W spanned by $d\phi_1(x^0),\dots,d\phi_k(x^0)$ and set*

$$X_-=\{x\in X;\ \phi_j(x)<\phi_j(x^0)\ \text{for some}\ j=1,\dots,k\}.$$

If $u\in\mathscr{D}'(X)$, $P(D)u\in C^\infty(X)$ and $u\in C^\infty(X_-)$ then $u\in C^\infty$ in a neighborhood of x^0 which is independent of u.

The proof of the theorem is fairly long so we postpone it to list some consequences which may make the reader more willing to study it.

Corollary 11.3.7. *Let $X_1\subset X_2$ be open convex sets in $\mathbb{R}^n$ and $P(D)$ a differential operator with constant coefficients. Then the following conditions are equivalent:*

(i) *Every solution $u\in\mathscr{D}'(X_2)$ of the equation $P(D)u=0$ which has a C^∞ restriction to X_1 is in fact in $C^\infty(X_2)$.*

(ii) *Every hyperplane H with $\sigma_P(H')=0$ which intersects X_2 also intersects X_1.*

Proof. That (i) $\Rightarrow$ (ii) follows at once from Theorem 11.3.1. To prove that (ii) $\Rightarrow$ (i) we just have to copy the proof of Theorem 8.6.8 with the reference to Holmgren's uniqueness theorem (Theorem 8.6.5) replaced by a reference to Theorem 11.3.6 with $k=1$.

Minimal linear spaces V with $\sigma_P(V')=0$ are minimal carriers of singularities also among sets which are not linear:

Corollary 11.3.8. *Let V be a linear subspace of $\mathbb{R}^n$ such that $\sigma_P(V')=0$ but $\sigma_P(W')\neq 0$ for every linear subspace $W\subset V$, $W\neq V$. If $P(D)u\in C^\infty$ and* sing supp $u\subset V$, *it follows that either* sing supp $u=V$ *or that $u\in C^\infty$.*

Proof. Let V be defined by $x'=0$ and assume that for some $r>0$ we have $u\in C^\infty$ when $|x|<r$. Thus $u\in C^\infty$ in $\{x=(x',x'');\ |x''|<r$ or $x'\neq 0\}$.

If a is a linear function of x' it follows that $u \in C^\infty$ where

$$\phi(x) = |x''|^2 - r^2 - a(x') < 0,$$

for if $\phi(x) < 0$ then either $x' \neq 0$ or $|x''| < r$. We have $\operatorname{grad} \phi = (-a', 2x'')$ so if $x_0 = (0, x_0'')$, $|x_0''| = r$, the plane spanned by these gradients is $V' \oplus \mathbb{R} x_0$, which is strictly larger than V'. Hence $u \in C^\infty$ in a neighborhood of x_0. The set of all r such that $u \in C^\infty$ when $|x| < r$ is therefore both open and closed which proves the corollary.

Corollary 11.3.8 does *not* mean that for solutions of $P(D)u = 0$ the singularities must propagate along linear spaces in the sense that for every $x \in \operatorname{sing\,supp} u$ there is a linear space V with $\sigma_P(V') = 0$ such that $\{x\} + V \subset \operatorname{sing\,supp} u$. This is true for operators of real principal type by Theorem 8.3.3', but not in general. For consider the differential operator $P(D) = D_2 D_3$ in $\mathbb{R}^3$. Set $u = \delta(x_1)(f(x_2) - g(x_3))$ where f and g are the characteristic functions of $(-1, 1)$ and $(2, 3)$ respectively. Then we have $P(D)u = 0$ and $0 \in \operatorname{sing\,supp} u$ but no straight line through 0 is contained in $\operatorname{sing\,supp} u$.

The proof of Theorem 11.3.6 like that of Theorem 8.3.3' will depend on the construction of a fundamental solution for $P(D)$, or rather a family of parametrices. As a preparation we first give analogues of Lemmas 11.3.4 and 11.3.5 for the case where $\sigma_P(V') \neq 0$.

Lemma 11.3.9. *If $\sigma_P(V') \neq 0$ it follows that there are positive constants b, t_1, ρ such that for $t > t_1$ and $|\xi| > t^\rho$ we have*

$$\tilde{P}_{V'}(\xi, t) \geqq b \tilde{P}(\xi, t). \tag{11.3.19}$$

Proof. Let $0 < b < \sigma_P(V')$. Then the function

$$a(t) = \sup \{|\xi|;\ \tilde{P}_{V'}(\xi, t) < b \tilde{P}(\xi, t)\}$$

is finite for $t > 1$, and it is a piecewise algebraic function of t by Corollary A.2.4 and Theorem A.2.5. Hence there are positive constants ρ, t_1 such that $a(t) < t^\rho$ if $t > t_1$, which proves Lemma 11.3.9.

In particular (11.3.19) is applicable when $\log|\xi| = \delta t$ if $t > T_\delta$, and this is the form in which (11.3.19) will be used. (See also the proof of Theorem 11.3.1.) Here δ denotes a small positive number.

(11.3.19) means that one can find a large zero free region by moving in a direction belonging to V':

Lemma 11.3.10. *Let κ and b be fixed positive numbers, $\kappa < 1$. Then there exist positive constants b_1 and γ depending only on κ, b, n and m*

such that (11.3.19) *implies that if* $\eta^0 \in V'$, $|\eta^0|=1$, *we have for some* $\theta \in V'$ *with* $|\theta| \leqq \kappa t$

$$|P(\xi + it\eta^0 + z\theta + \zeta)| \geqq b_1 \tilde{P}(\xi, t) \tag{11.3.20}$$

if $z \in \mathbb{C}$, $|z|=1$; $\zeta \in \mathbb{C}^n$, $|\zeta| < 2\gamma t$.

Proof. By (11.3.19) and (10.1.8) we have

$$b\tilde{P}(\xi, t) \leqq \tilde{P}_{V'}(\xi, t) \leqq C\tilde{P}_{V'}(\xi + it\eta^0, t\kappa),$$

for $\eta^0 \in V'$. As in the proof of Lemma 11.3.5 we can choose $\theta \in V'$ with $0 \leqq |\theta| \leqq t\kappa$ so that

$$\tilde{P}_{V'}(\xi + it\eta^0, t\kappa) \leqq C_1 |P(\xi + it\eta^0 + z\theta)|, \qquad |z|=1.$$

(See (11.3.7).) If $|\zeta| \leqq 2\gamma t$, $\gamma < 1$, it follows that

$$\begin{aligned} |P(\xi + it\eta^0 + z\theta + \zeta)| &\geqq C_3 \tilde{P}(\xi, t) - C_2 \gamma \tilde{P}(\xi + it\eta^0 + z\theta, t) \\ &\geqq (C_3 - C_4\gamma)\tilde{P}(\xi, t). \end{aligned}$$

This proves the lemma.

Remark. Conversely the inequality (11.3.20) implies already for $\zeta = 0$ an estimate of the form (11.3.19) so (11.3.20) contains all information given by (11.3.19). Note that since $|\theta| \leqq \kappa t$ the scalar product of $t\eta^0 + \operatorname{Im} z\theta$ with η^0 is at least $t(1-\kappa)$ so the argument in (11.3.20) is moved effectively in the direction $i\eta^0$.

Before developing further technical details we shall indicate the main lines of the construction of a parametrix which we are aiming at. Our purpose is to interpret in the sense of (7.3.14) the integral

$$E = (2\pi)^{-n} \int e^{i\langle x, \zeta\rangle} P(\zeta)^{-1} d\zeta$$

taken over some cycle avoiding the zeros of P and which for large ξ is close to

$$\mathbb{R}^n \ni \xi \to \xi + i\delta^{-1}(\log|\xi|)\eta^0.$$

That this is useful for the study of singularities was already seen in the proof of Theorem 7.3.8. The reason is that the absolute value of the exponential is $|\xi|^{-\langle x, \eta^0\rangle/\delta}$ on the cycle so one can expect E to be roughly $\langle x, \eta^0\rangle/\delta$ times differentiable at x (thus a distribution of order $-\langle x, \eta^0\rangle/\delta$ when $\langle x, \eta^0\rangle < 0$). To avoid the zeros of P we shall use a device similar to the proof of Theorem 7.3.10 but based on (11.3.20). This forces us to work locally and decompose the integral by a partition of unity.

To make these ideas precise we set $t_\delta(\xi) = \delta^{-1}\log(\delta|\xi|+2)$ and choose a partition of unity with mesh width $\gamma t_\delta(\xi)$ at ξ where $0 < \gamma < 1$

is fixed so that (11.3.20) is valid. As in the proof of Theorem 11.3.1 we need good estimates for a number of derivatives proportional to $t_\delta(\xi)$. In the following lemma ε denotes a positive proportionality constant, not necessarily small, to be chosen later on.

Lemma 11.3.11. *There exists a partition of unity* $1=\sum \chi_j$ *in* $\mathbb{R}^n$ *and points* $\xi_j \in \operatorname{supp} \chi_j$ *such that for large* j

$$\operatorname{supp} \chi_j \subset B_j = \{\xi;\ |\xi - \xi_j| \leqq \gamma t_\delta(\xi_j)\}, \tag{11.3.21}$$

the intersection of more than C_0 *balls* B_j *is always empty and*

$$|D^\alpha \chi_j(\xi)| \leqq C_1^{|\alpha|+1} \varepsilon^{|\alpha|}, \quad |\alpha| \leqq \varepsilon t_\delta(\xi_j). \tag{11.3.22}$$

Here C_0 *and* C_1 *are independent of* ε *and* δ. *We have*

$$\sum t_\delta(\xi_j)^n |\xi_j|^{-n-1} < \infty. \tag{11.3.23}$$

Proof. The metric

$$\|\eta\|_\xi = |\eta| / (\gamma t_\delta(\xi))$$

is uniformly slowly varying for this is obvious when $\gamma = \delta = 1$, hence when $\gamma < 1$, and changing scales gives a general δ. The balls B_j can then be taken from Theorem 1.4.10, and the partition of unity is obtained from the remark following Theorem 1.4.10 if in B_j we take $d_i = 1/(\varepsilon t_j + 2)$ when $i/\varepsilon \leqq t_j = t_\delta(\xi_j)$. (11.3.23) follows since the sum is majorized by the convergent integral

$$\int_{|\xi|>1} C_0 |\xi|^{-n-1} d\xi.$$

For large j we have by Lemma 11.3.9

$$\tilde{P}_{V'}(\xi_j, t_j) \geqq b \tilde{P}(\xi_j, t_j),$$

and using Lemma 11.3.10 we choose $\theta_j \in V'$ so that $|\theta_j| \leqq \kappa t_j$ and

$$|P(\xi_j + i t_j \eta^0 + z\theta_j + \zeta)| \geqq b_1 \tilde{P}(\xi_j, t_j) \tag{11.3.24}$$

if $|z| = 1$, $|\zeta| < 2\gamma t_j$.

Here and in what follows z and ζ are complex variables. Set

$$E_j(x) = (2\pi)^{-n} \iint_{|z|=1} \chi_j(\xi) e^{i\langle x, \zeta + z\theta_j\rangle} / P(\zeta + z\theta_j) d\zeta\, dz / 2\pi i z, \tag{11.3.25}$$

where $\zeta(\xi) = \xi + i t_\delta(\xi) \eta^0$ and $d\zeta = J(\xi) d\xi$, $J(\xi) = D(\zeta(\xi))/D\xi$. Note that $J(\xi)$ is an analytic function of ξ in a conic neighborhood of $\mathbb{R}^n$ and tends to 1 at ∞. When $\xi \in \operatorname{supp} \chi_j$ we have $|\xi - \xi_j| \leqq \gamma t_j$, hence

$$|\zeta - \xi_j - i t_j \eta^0| \leqq \gamma t_j + |t_\delta(\xi) - t_\delta(\xi_j)| = \gamma t_j (1 + O(1/|\delta \xi_j|)) < 3\gamma t_j / 2$$

for large j. The denominator in (11.3.25) is therefore bounded from below in view of (11.3.24). Hence $E_j \in C^\infty$ is defined for $j > j_0$, and the discussion of (7.3.14) shows that the sum

$$E = \sum_{j > j_0} E_j$$

exists in $\mathscr{D}'(\mathbb{R}^n)$. Since

$$P(D) E_j(x) = (2\pi)^{-n} \int \chi_j(\xi) e^{i\langle x, \zeta\rangle} d\zeta$$

we have if $u \in C_0^\infty(\mathbb{R}^n)$

$$\langle P(D)E, u\rangle = \sum \langle P(D)E_j, u\rangle = (2\pi)^{-n} \int \sum_{j > j_0} \chi_j(\xi) \hat{u}(-\zeta) d\zeta.$$

Here the integral is equal to the integral of $\hat{u}(-\zeta)\, d\zeta_1 \wedge \ldots \wedge d\zeta_n$ over the chain $\xi \to \xi + i t_\delta(\xi)\, \eta^0$ outside $\bigcup_{j \leqq j_0} \operatorname{supp} \chi_j$. By Stokes' formula it can be replaced by integration over $\mathbb{R}^n$ and over a compact chain. Hence

$$\langle P(D)E, u\rangle = u(0) + \int \hat{u}(-\zeta)\, d\mu(\zeta)$$

where $d\mu$ is a measure of compact support. Equivalently

$$P(D)E - \delta_0 = \int e^{i\langle x, \zeta\rangle} d\mu(\zeta)$$

where the right-hand side is an entire function of x. Thus we have a parametrix.

It remains to estimate E_j and its derivatives in detail. We can write

$$E_j(x) = (2\pi)^{-n} e^{-t_j\langle x, \eta^0\rangle} \iint_{|z|=1} e^{i\langle x, \xi\rangle} \chi_j(\xi) F_j(\xi, z)\, d\xi\, dz / 2\pi i z$$

where

$$F_j(\xi, z) = (\exp i\langle x, i(t_\delta(\xi) - t_j)\eta^0 + z\theta_j\rangle) J(\xi) / P(\zeta(\xi) + z\theta).$$

F_j is an analytic function of ξ in the complex ball with center ξ_j and radius $3\gamma t_j/2$, and for $j > j_0$

(11.3.26) $$|F_j(\xi, z)| \leqq C \exp(|x'| + \kappa t_j |x'|)$$

if $\xi \in \mathbb{C}^n$, $|\xi - \xi_j| < 3\gamma t_j/2$, $|z| = 1$. In fact, for $j > j_0$

$$|t_\delta(\xi) - t_j| = |t_\delta(\xi) - t_\delta(\xi_j)| < C t_j / \delta |\xi_j| < 1, \quad |\xi - \xi_j| < 3\gamma t_j/2,$$

which gives a bound for the exponential while $P(\zeta(\xi) + z\theta)$ is bounded from below by (11.3.24) since

$$|\zeta(\xi) - \xi_j - i t_j \eta^0| < 3\gamma t_j/2 + 1 < 2\gamma t_j.$$

Now we can use part a) of the proof of Proposition 8.4.2 again to estimate

$$E_j(x)=(2\pi)^{-n}e^{i\langle x,\xi_j\rangle - t_j\langle x,\eta^0\rangle}\iint t_j^n\chi_j(\xi_j+t_j\xi)\cdot F_j(\xi_j+t_j\xi, z)e^{i\langle xt_j,\xi\rangle}\,d\xi\,dz/2\pi i z.$$

By (11.3.26) we have

$$|D_\xi^\alpha F_j(\xi_j+t_j\xi, z)|\leqq C_2^{|\alpha|+1}|\alpha|!\exp(|x'|+\kappa t_j|x'|), \quad |\xi|<\gamma, \tag{11.3.26'}$$

and (11.3.22) gives

$$|D^\alpha\chi_j(\xi_j+t_j\xi)|\leqq C_1^{|\alpha|+1}(\varepsilon t_j)^{|\alpha|}, \quad |\alpha|\leqq\varepsilon t_j. \tag{11.3.22'}$$

It follows that

$$|E_j(x)|\leqq C e^{|x'|}t_j^n(C_3\varepsilon t_j/|x t_j|)^{\varepsilon t_j-1}\exp(t_j(\kappa|x'|-\langle x,\eta^0\rangle)).$$

For derivatives of E_j we only have additional factors $(\zeta+z\theta_j)$ to take into account, so we have more generally

$$|D^\alpha E_j(x)|\leqq C_\alpha e^{|x'|}t_j^n(C_3\varepsilon t_j/|x t_j|)^{\varepsilon t_j-1}\exp(t_j(\kappa|x'|+\delta|\alpha|-\langle x,\eta^0\rangle)).$$

When $eC_3\varepsilon<|x|<2eC_3\varepsilon$ we can use the estimates

$$(C_3\varepsilon/|x|)^{\varepsilon t_j-1}\leqq 3e^{-\varepsilon t_j}\leqq 3e^{-\rho t_j|x|}$$

where we have written $\rho=1/2eC_3$. Hence

$$|D^\alpha E_j(x)|\leqq 3C_\alpha e^{|x'|}t_j^n\exp(t_j(\kappa|x'|+\delta|\alpha|-\langle x,\eta^0\rangle-\rho|x|))$$

so it follows from (11.3.23) that $\sum D^\alpha E_j(x)$ converges uniformly in the set where

$$\varepsilon/2<\rho|x|<\varepsilon, \quad \kappa|x'|+\delta|\alpha|-\langle x,\eta^0\rangle-\rho|x|<-\delta(n+1).$$

In the set defined by

$$\varepsilon/2<\rho|x|<\varepsilon, \quad \kappa|x'|-\langle x,\eta^0\rangle-\rho|x|/2<0 \tag{11.3.27}$$

we obtain $E\in C^\nu$ for any desired ν when $\delta<\delta_\nu$.

Now choose a cutoff function $\psi\in C_0^\infty(\mathbb{R}^n)$ which is 1 when $|x|<3\varepsilon/5\rho$ and 0 when $|x|>4\varepsilon/5\rho$, and set $F=\psi E$. Then we have $P(D)F=\delta_0+\omega$ where $\omega\in C^\infty$ for $|x|<3\varepsilon/5\rho$ and $\omega\in C^\nu$ for small δ in the set where $\kappa|x'|-\langle x,\eta^0\rangle-\rho|x|/2<0$. Replacing $2\varepsilon/5\rho$ by ε we have proved

Proposition 11.3.12. *Assume that V is a linear subspace of $\mathbb{R}^n$ with $\sigma_P(V')\neq 0$. Choose $\eta^0\in V'$ with $|\eta^0|=1$ and κ with $0<\kappa<1$. For $\varepsilon>0$ and any positive integer ν one can then find $F_{\varepsilon,\nu}\in\mathscr{E}'(\mathbb{R}^n)$ such that $|x|<2\varepsilon$ in $\operatorname{supp}F_{\varepsilon,\nu}$, the difference $P(D)F_{\varepsilon,\nu}-\delta_0\in C^\infty\{x;|x|<\varepsilon\}$ and*

$$P(D)F_{\varepsilon,\nu}\in C^\nu\{x;\kappa|x'|-\langle x,\eta^0\rangle-\rho|x|/2<0\}.$$

Here x' *is the residue class of* x mod V *and* ρ *is a positive constant independent of* ε *and* ν.

If u is a distribution with $P(D)u=f\in C^\infty$ when $|x|<3\varepsilon$, we have for $|x|<\varepsilon$

$$u=u*(\delta_0-P(D)F_{\varepsilon,\nu})+f*F_{\varepsilon,\nu}.$$

The last term is in $C^\infty(\{x;\,|x|<\varepsilon\})$. If $u\in C^\infty$ in a neighborhood of

$$\{x;\ \varepsilon\leqq|x|\leqq 2\varepsilon,\ \kappa|x'|+\langle x,\eta^0\rangle-\rho|x|/2\geqq 0\},\tag{11.3.28}$$

it follows by Theorem 4.2.6 when $\nu\to\infty$ that $u\in C^\infty$ in a neighborhood of 0.

End of proof of Theorem 11.3.6. We may assume that $x^0=0$ and that $\phi_j(x)=x_j+o(|x|)$ as $x\to 0$. Take $\eta^0=-(1,\dots,1,0,\dots,0)/\sqrt{k}$ with k coordinates equal to $-1/\sqrt{k}$, and choose $\kappa<1/\sqrt{k}$. When $x\in\operatorname{sing\,supp}u$ we have by hypothesis $x_j\geqq -o(|x|)$ for $j=1,\dots,k$. Hence $|x_j-|x_j||=o(|x|)$ then, which gives

$$\begin{aligned}\kappa|x'|+\langle x,\eta^0\rangle-\rho|x|/2&\leqq\kappa\sum_1^k|x_j|-\sum_1^k|x_j|/\sqrt{k}+o(|x|)-\rho|x|/2\\&\leqq o(|x|)-\rho|x|/2<0\end{aligned}$$

if $\varepsilon<|x|<2\varepsilon$ and ε is small. It follows that $u\in C^\infty$ in the set (11.3.28), hence in a neighborhood of 0, which completes the proof.

Proposition 11.3.12 also gives some useful global information:

Theorem 11.3.13. *Let* Γ *be a closed convex set in* $\mathbb{R}^n$ *and let* V *be the largest vector space with* $\Gamma+V=\Gamma$. *Then* $u\in\mathscr{D}'(\mathbb{R}^n)$, $P(D)u=0$ *and* $\operatorname{sing\,supp}u\subset\Gamma$ *implies* $u\in C^\infty(\mathbb{R}^n)$ *if and only if* $\sigma_P(V')\neq 0$.

Proof. If $\sigma_P(V')=0$ it follows from Theorem 11.3.1 that for any $y\in\Gamma$ we can find $u\in C^m(\mathbb{R}^n)$ with $P(D)u=0$ and $\operatorname{sing\,supp}u=\{y\}+V\subset\Gamma$. (It is easy to conclude that u can be chosen with $\operatorname{sing\,supp}u=F$ if F is any closed set with $F+V=F$.) On the other hand, assume that $\sigma_P(V')\neq 0$ and let $0\notin\Gamma$. Then there is a proper cone in $\mathbb{R}^n/V$ containing the image of Γ there. We may therefore assume that Γ is the inverse image of such a cone. The set of points where Proposition 11.3.12 can be used to show that $u\in C^\infty$ is obviously a cone since ε may be any positive number. In the proof of Theorem 11.3.6 we saw that it contains a neighborhood of 0 so it is all of $\mathbb{R}^n$.

The results of this section like those of Section 11.1 agree well with the classification of differential operators in Section 10.4:

Theorem 11.3.14. *If P' and P'' are equally strong then $\sigma_{P'}(V)=0$ if and only if $\sigma_{P''}(V)=0$.*

Proof. By Theorem 10.4.3 the hypothesis implies that

$$C_1 \leqq \tilde{P}''(\xi, t)/\tilde{P}'(\xi, t) \leqq C_2; \quad \xi \in \mathbb{R}^n, \ t>1. \tag{11.3.29}$$

From Lemmas 11.3.4 and 11.3.9 we know that $\sigma_{P'}(V)=0$ if and only if for some sequence $\xi_j \to \infty$ in $\mathbb{R}^n$

$$\tilde{P}'_V(\xi_j, t_j)/\tilde{P}'(\xi_j, t_j) \to 0,$$

where $t_j = \log|\xi_j|$. Passing to a subsequence we may assume that the limits

$$Q'(\xi) = \lim_{j\to\infty} P'(\xi_j + t_j \xi)/\tilde{P}'(\xi_j, t_j),$$
$$Q''(\xi) = \lim_{j\to\infty} P''(\xi_j + t_j \xi)/\tilde{P}'(\xi_j, t_j)$$

exist. In fact, the supremum of $|P'(\xi_j + t_j\xi)|/\tilde{P}'(\xi_j, t_j)$ when ξ varies over the unit ball is 1, and that of $|P''(\xi_j + t_j\xi)|/\tilde{P}'(\xi_j, t_j)$ lies between C_1 and C_2. The limits are therefore not identically 0, and (11.3.29) implies

$$C_1 \tilde{Q}'(\xi, t) \leqq \tilde{Q}''(\xi, t) \leqq C_2 \tilde{Q}'(\xi, t), \quad t>0.$$

Letting $t \to 0$ we conclude that $Q''=0$ when $Q'=0$. Since $Q'=0$ in V it follows that $Q''=0$ in V, so that $\tilde{P}''_V(\xi_j, t_j)/\tilde{P}''(\xi_j, t_j) \to 0$ as $j \to \infty$. Hence $\sigma_{P''}(V)=0$ and the theorem is proved.

11.4. Estimates for Derivatives of High Order

By refining the proof of Theorem 11.1.7 we shall now prove precise estimates for the growth of the derivatives of the solutions of the hypoelliptic differential equation

$$P(D)u = 0 \tag{11.4.1}$$

with the order of differentiation.

Theorem 11.4.1. *Let $y \in \mathbb{R}^n$ and let C, ρ be constants such that*

$$|\langle y, \xi\rangle| \leqq C(d(\xi)+1)^\rho, \quad \xi \in \mathbb{R}^n, \tag{11.4.2}$$

where $d(\xi)$ is the distance from ξ to the zeros of the hypoelliptic polynomial $P(\zeta)$. If u is a solution of (11.4.1) in an open neighborhood of a compact set K, then there exists another constant C such that

$$|\langle y, D\rangle^j u(x)| \leqq C^j j^{\rho j}, \quad j=1,2,\ldots; \ x \in K. \tag{11.4.3}$$

Remark. If $y \neq 0$ it follows from (11.4.2) that $\rho \geqq 1$, for if $P(\zeta_0)=0$ we have $d(\xi) \leqq |\xi - \zeta_0|$.

The proof of Theorem 11.4.1 follows readily from the following theorem where we use the notation

$$\|u, X\|_\varepsilon = \Big(\sum_{\alpha \neq 0} \varepsilon^{-2|\alpha|} \int_X |P^{(\alpha)}(D)u|^2 \, dx\Big)^{\frac{1}{2}}.$$

Theorem 11.4.2. *Let X be an open set in $\mathbb{R}^n$ and denote by X_δ the set of points in X with distance $>\delta$ to the boundary of X. If* (11.4.2) *holds, then there is a constant C, independent of u, δ and X such that for every solution of* (11.4.1) *in X and all δ with $0<\delta<1$ we have*

$$\|\langle y, D\rangle u, X_\delta\|_\delta \leqq C\delta^{-\rho} \|u, X\|_\delta. \tag{11.4.4}$$

Proof of Theorem 11.4.1. Using Theorem 11.4.2 it is easy to prove Theorem 11.4.1. In fact, let X be a neighborhood of K where $P(D)u=0$ and all derivatives of u are in L^2. Iteration of (11.4.4) gives

$$\|\langle y, D\rangle^j u, X_{j\delta}\|_\delta \leqq C^j \delta^{-\rho j} \|u, X\|_\delta, \tag{11.4.5}$$

for every positive integer j. Fix a constant c with $0<c<1$ such that $K \subset X_c$ and set $\delta = c/j$. Since $P^{(\alpha)}$ is a constant $\neq 0$ for some α with $|\alpha| = m$ and since $P^{(\alpha)}(D)u \in L^2(X)$ for every α, this gives with another C

$$\int_{X_c} |\langle y, D\rangle^j u|^2 \, dx \leqq C^j j^{2\rho j}, \quad j=1,2,\ldots. \tag{11.4.6}$$

To pass to the pointwise estimate (11.4.3) we now use the inequality

$$\sup_{x\in K} |u(x)|^2 \leqq C \sum_{|\beta| \leqq n} \int |D^\beta u|^2 \, dx, \quad u \in C^\infty(X_c), \tag{11.4.7}$$

which follows from Lemma 7.6.3 after a change of scales. If we apply (11.4.6) to $D^\beta u$ for all β with $|\beta| \leqq n$, add the estimates so obtained, and use (11.4.7) with u replaced by $\langle y, D\rangle^j u$, then (11.4.3) follows.

In the proof of Theorem 11.4.2 we shall use the norms

$$|||v|||_{s,\varepsilon} = \|u\|_{2, d_{s,\varepsilon}} \tag{11.4.8}$$

where $\varepsilon > 0$ and

$$d_{s,\varepsilon}(\xi) = (1+\varepsilon d(\xi))^s. \tag{11.4.9}$$

(Recall that $d(\xi)$ is the distance from ξ to the zeros of P.) Since by the triangle inequality $d(\xi+\eta) \leqq d(\xi) + |\eta|$, it follows that

$$d_{s,\varepsilon}(\xi+\eta) \leqq d_{s,\varepsilon}(\xi)(1+\varepsilon|\eta|)^{|s|} \tag{11.4.10}$$

which in particular proves that $d_{s,\varepsilon} \in \mathscr{K}$.

Lemma 11.4.3. *If $v \in C_0^\infty(\mathbb{R}^n)$ and $\varepsilon>0$, we have*

$$(11.4.11)\quad \sum_{\alpha\neq 0} \varepsilon^{-2|\alpha|} |||P^{(\alpha)}(D)v|||^2_{s+1,\varepsilon} \leqq C \sum_{\alpha} \varepsilon^{-2|\alpha|} |||P^{(\alpha)}(D)v|||^2_{s,\varepsilon}.$$

Proof. The estimate (11.4.11) follows from the inequality

$$(11.4.12)\quad (1+\varepsilon d(\xi))^2 \sum_{\alpha\neq 0} \varepsilon^{-2|\alpha|} |P^{(\alpha)}(\xi)|^2 \leqq C \sum_{\alpha} \varepsilon^{-2|\alpha|} |P^{(\alpha)}(\xi)|^2, \qquad \xi \in \mathbb{R}^n,$$

if we multiply by $(2\pi)^{-n}|d_{s,\varepsilon}(\xi)\hat{v}(\xi)|^2$ and integrate with respect to ξ. Now the inequality (11.4.12) is trivial when $\varepsilon d(\xi)<1$ if $C \geqq 4$. On the other hand, if $\varepsilon d(\xi) \geqq 1$ and $|\alpha| \geqq 1$, we have

$$(1+\varepsilon d(\xi))^2 \varepsilon^{-2|\alpha|} |P^{(\alpha)}(\xi)|^2 \leqq 4\varepsilon^2 d(\xi)^2 \varepsilon^{-2|\alpha|} |P^{(\alpha)}(\xi)|^2 \leqq 4 d(\xi)^{2|\alpha|} |P^{(\alpha)}(\xi)|^2.$$

Hence (11.4.12) follows from Lemma 11.1.4.

Estimates for solutions u of (11.4.1) will be obtained by applying Lemma 11.4.3 to the product of u with suitable functions in $C_0^\infty(X)$.

Lemma 11.4.4. *Let $\phi \in C_0^\infty(\mathbb{R}^n)$ and set $\phi^\varepsilon(x)=\phi(x/\varepsilon)$ when $\varepsilon>0$. Then we have for all multi-indices α and integers j*

$$(11.4.13)\qquad (2\pi)^{-n} \int |\xi^\alpha \hat{\phi}^\varepsilon(\xi)|(1+\varepsilon|\xi|)^j \, d\xi \leqq C_{\alpha,j}\varepsilon^{-|\alpha|}$$

where $C_{\alpha,j}$ is independent of ε.

Proof. Since $\hat{\phi}^\varepsilon(\xi)=\varepsilon^n \hat{\phi}(\varepsilon\xi)$, we obtain (11.4.13) by introducing $\varepsilon\xi$ as a new variable and noting that $\hat{\phi} \in \mathscr{S}$.

Lemma 11.4.5. *Let u be a solution of the equation* (11.4.1) *defined in the ball $B_\varepsilon=\{x;\, |x|<\varepsilon\}$, let $\phi \in C_0^\infty(B_1)$, and let s be an integer $\geqq 0$. Then we have*

$$(11.4.14)\quad \sum_{\alpha\neq 0} \varepsilon^{-2|\alpha|} |||P^{(\alpha)}(D)(\phi^\varepsilon u)|||^2_{s,\varepsilon} \leqq C \sum_{\alpha\neq 0} \varepsilon^{-2|\alpha|} \int_{B_\varepsilon} |P^{(\alpha)}(D)u|^2 \, dx,$$

where C is independent of ε and u.

Proof. To prove (11.4.14) for $s=0$ we note that $|||\cdot|||_{0,\varepsilon}$ is the L^2 norm. Since $\varepsilon^{|\alpha|} \sup|D^\alpha \phi^\varepsilon| \leqq \sup|D^\alpha \phi|$, the inequality (11.4.14) is then a consequence of the identity

$$\varepsilon^{-|\alpha|} P^{(\alpha)}(D)(\phi^\varepsilon u) = \sum \varepsilon^{|\beta|} D^\beta \phi^\varepsilon \varepsilon^{-|\alpha+\beta|} P^{(\alpha+\beta)}(D)u/\beta!.$$

In proving (11.4.14) when $s>0$ we may thus assume that it is already proved for an arbitrary $\phi \in C_0^\infty(B_1)$ when s is replaced by $s-1$. From (11.4.11) we obtain

$$(11.4.15)\quad \sum_{\alpha\neq 0} \varepsilon^{-2|\alpha|} |||P^{(\alpha)}(D)(\phi^\varepsilon u)|||^2_{s,\varepsilon} \leqq C \sum \varepsilon^{-2|\alpha|} |||P^{(\alpha)}(D)(\phi^\varepsilon u)|||^2_{s-1,\varepsilon}.$$

To study the term with $\alpha=0$ in the right-hand side, we choose a function $\psi\in C_0^\infty(B_1)$ such that $\psi=1$ in a neighborhood of $\operatorname{supp}\phi$. Since $\phi^\varepsilon u=\phi^\varepsilon(\psi^\varepsilon u)$ and $\phi^\varepsilon P(D)(\psi^\varepsilon u)=\phi^\varepsilon P(D)u=0$, we obtain by Leibniz' formula

$$P(D)(\phi^\varepsilon u)=\sum_{\alpha\neq 0} D^\alpha\phi^\varepsilon P^{(\alpha)}(D)(\psi^\varepsilon u)/\alpha!. \tag{11.4.16}$$

Now we can estimate the norms of the terms in the right-hand side of (11.4.16) by using Theorem 10.1.15, (11.4.10) and (11.4.13). This gives

$$|||P(D)(\phi^\varepsilon u)|||_{s-1,\varepsilon}\leqq\sum_{\alpha\neq 0} C_{\alpha,s-1}\varepsilon^{-|\alpha|}|||P^{(\alpha)}(D)(\psi^\varepsilon u)|||_{s-1,\varepsilon}. \tag{11.4.17}$$

(11.4.14) now follows from (11.4.15) and (11.4.17) since we have assumed that (11.4.14) is already proved when s is replaced by $s-1$.

Lemma 11.4.6. *Let* (11.4.2) *be valid. Then there is a constant C such that if $P(D)u=0$ in B_ε and $0<\varepsilon\leqq 1$ we have*

$$\varepsilon^{2\rho}\sum_{\alpha\neq 0}\varepsilon^{-2|\alpha|}\int_{B_{\varepsilon/2}}|P^{(\alpha)}(D)\langle y,D\rangle u|^2\,dx$$
$$\leqq C\sum_{\alpha\neq 0}\varepsilon^{-2|\alpha|}\int_{B_\varepsilon}|P^{(\alpha)}(D)u|^2\,dx.$$

Proof. Let s be the smallest integer such that $s\geqq\rho$, where ρ is the exponent in (11.4.2). It follows from (11.4.2) that

$$\varepsilon^\rho|\langle y,\xi\rangle|\leqq C(1+\varepsilon d(\xi))^\rho\leqq C d_{s,\varepsilon}(\xi)\quad \text{if } 0<\varepsilon\leqq 1 \text{ and } \xi\in\mathbb{R}^n.$$

Hence Parseval's formula gives

$$\varepsilon^{2\rho}\int|\langle y,D\rangle v|^2\,dx\leqq C^2|||v|||_{s,\varepsilon}^2;\qquad 0<\varepsilon\leqq 1,\ v\in C_0^\infty(\mathbb{R}^n).$$

Let $\phi\in C_0^\infty(B_1)$ be equal to 1 in $B_{\frac12}$. If we apply the estimate just proved to $v=P^{(\alpha)}(D)(\phi^\varepsilon u)$ and use (11.4.14), we obtain with another constant C

$$\varepsilon^{2\rho}\sum_{\alpha\neq 0}\varepsilon^{-2|\alpha|}\int|P^{(\alpha)}(D)\langle y,D\rangle(\phi^\varepsilon u)|^2\,dx\leqq C\sum_{\alpha\neq 0}\varepsilon^{-2|\alpha|}\int_{B_\varepsilon}|P^{(\alpha)}(D)u|^2\,dx.$$

Since $\phi^\varepsilon=1$ in $B_{\varepsilon/2}$, the lemma is proved.

Proof of Theorem 11.4.2. When the hypotheses of Theorem 11.4.2 are fulfilled and $z\in X_\varepsilon$, $0<\varepsilon\leqq 1$, we obtain by applying Lemma 11.4.6 to $u(.+z)$

$$\varepsilon^{2\rho}\sum_{\alpha\neq 0}\varepsilon^{-2|\alpha|}\int_{|x-z|<\varepsilon/2}|P^{(\alpha)}(D)\langle y,D\rangle u|^2\,dx$$
$$\leqq C\sum_{\alpha\neq 0}\varepsilon^{-2|\alpha|}\int_{|x-z|<\varepsilon}|P^{(\alpha)}(D)u|^2\,dx.$$

We now integrate the two sides with respect to z over X_ε and invert the order of integration. The integral with respect to z in the right-hand side is $\leqq m(B_\varepsilon)$ and vanishes if $x \notin X$. When $x \in X_{2\varepsilon}$ the integral with respect to z in the left-hand side is equal to $m(B_{\varepsilon/2}) = 2^{-n} m(B_\varepsilon)$. Hence we obtain

$$2^{-n}\varepsilon^{2\rho} \sum_{\alpha \neq 0} \varepsilon^{-2|\alpha|} \int_{X_{2\varepsilon}} |P^{(\alpha)}(D)\langle y, D\rangle u|^2\, dx$$
$$\leqq C \sum_{\alpha \neq 0} \varepsilon^{-2|\alpha|} \int_X |P^{(\alpha)}(D)u|^2 dx, \qquad 0<\varepsilon\leqq 1,$$

which implies (11.4.4) if we choose $\varepsilon=\delta/2$.

Remark. A simple modification of the proofs of Theorems 11.4.1 and 11.4.2 gives results for solutions of inhomogeneous equations $P(D)u = f$ also.

We shall now prove that Theorem 11.4.1 gives the best possible result.

Theorem 11.4.7. *Let $y \in \mathbb{R}^n$ and $x^0 \in X$, where X is an open set in $\mathbb{R}^n$. Assume that a sequence M_j has the property that for every solution of the hypoelliptic equation* (11.4.1) *in X there is a constant C such that*

$$|\langle y, D\rangle^j u(x^0)| \leqq C^j M_j, \quad j=1,2,\dots. \tag{11.4.18}$$

Then there is a number σ and positive constants c, C such that

$$M_j \geqq c^j j^{\sigma j}, \quad j=1,2,\dots \tag{11.4.19}$$

$$|\langle y, \zeta\rangle| \leqq C(1+|\operatorname{Im}\zeta|)^\sigma, \quad P(\zeta)=0. \tag{11.4.20}$$

Proof. Let $\mathcal{N}$ be the set of all solutions of (11.4.1) in X with the topology induced by $L^2_{\mathrm{loc}}(X) = B^{\mathrm{loc}}_{2,1}(X)$. According to Theorem 4.4.2 this topology is identical with that induced by $C^\infty(X)$, which proves that the set

$$F_r = \{u;\, u \in \mathcal{N}, |\langle y, D\rangle^j u(x^0)| \leqq r^j M_j, j=1,2,\dots\} \tag{11.4.21}$$

is closed for every integer r. By hypothesis we have $\bigcup_1^\infty F_r = \mathcal{N}$, and since $\mathcal{N}$ is a Fréchet space, it follows from Baire's theorem that F_r has an interior point for a suitable choice of r. Since F_r is convex and symmetric the origin must be an interior point. Hence there is a number $\delta>0$ and a compact set K such that F_r contains every $u \in \mathcal{N}$ with L^2 norm $\leqq \delta$ over K. But then we have for every u

$$|\langle y, D\rangle^j u(x^0)| \leqq r^j M_j \delta^{-1} \Big(\int_K |u|^2 dx\Big)^{\frac{1}{2}}, \quad j=1,2,\dots \tag{11.4.22}$$

for this inequality is homogeneous with respect to u and it is valid when the L^2 norm of u over K is equal to δ.

Now take any $\zeta\in\mathbb{C}^n$ with $P(\zeta)=0$ and set $u(x)=e^{i\langle x,\zeta\rangle}$ in (11.4.22). If A is an upper bound for $|x-x^0|$ when $x\in K$, this gives with a constant C

$$|\langle y,\zeta\rangle|^j \leqq C r^j M_j e^{A|\operatorname{Im}\zeta|}, \quad j=1,2,\ldots. \tag{11.4.23}$$

Let $\mu(\tau)$ be the supremum of $|\langle y,\zeta\rangle|$ when $P(\zeta)=0$ and $|\operatorname{Im}\zeta|\leqq\tau$. From Corollary A.2.6 it follows that there are constants a and b such that

$$\mu(\tau)=2b\tau^a(1+o(1)), \quad \tau\to\infty. \tag{11.4.24}$$

Thus the inequality (11.4.20) is valid with $\sigma=a$ if C is sufficiently large. On the other hand, since it follows from (11.4.24) that $\mu(\tau)\geqq b\tau^a$ for large τ, the inequality (11.4.23) implies

$$M_j \geqq C^{-1}(b/r)^j \tau^{aj} e^{-A\tau} \tag{11.4.25}$$

for large τ. If we choose $\tau=j$ the inequality (11.4.19) follows for large j with $\sigma=a$. The proof is complete.

We shall now compare the conditions (11.4.2) and (11.4.20) when $n>1$.

Theorem 11.4.8. *Let P be a hypoelliptic polynomial and denote the distance from ξ to the zeros of P by $d(\xi)$. Then the inequality*

$$|\langle y,\xi\rangle| \leqq C(d(\xi)+1)^\rho, \quad \xi\in\mathbb{R}^n, \tag{11.4.26}$$

is valid for some constant C if and only if for another constant C we have

$$|\langle y,\zeta\rangle| \leqq C(|\operatorname{Im}\zeta|+1)^\rho; \quad \zeta\in\mathbb{C}^n,\ P(\zeta)=0. \tag{11.4.27}$$

Proof. We may assume that $y\neq 0$. Each of the inequalities (11.4.26) and (11.4.27) then implies that $\rho\geqq 1$. If $\zeta=\xi+i\eta$ with real ξ, η, and if $P(\zeta)=0$, we have $d(\xi)\leqq|\eta|$, so it follows from (11.4.26) that

$$|\langle y,\zeta\rangle| \leqq |\langle y,\xi\rangle| + |\langle y,\eta\rangle| \leqq C(|\eta|+1)^\rho + |y|\,|\eta|,$$

which implies (11.4.27) since $\rho\geqq 1$. On the other hand, assume that (11.4.27) is valid. If $\xi\in\mathbb{R}^n$ we can choose ζ so that $P(\zeta)=0$ and $|\zeta-\xi| = d(\xi)$. Then we have $|\operatorname{Im}\zeta|\leqq d(\xi)$, hence

$$|\langle y,\xi\rangle| \leqq |\langle y,\xi-\zeta\rangle| + |\langle y,\zeta\rangle| \leqq |y| d(\xi) + C(d(\xi)+1)^\rho.$$

Since $\rho\geqq 1$, the inequality (11.4.26) follows. The proof is complete.

Definition 11.4.9. The smallest number ρ such that (11.4.26) or (11.4.27) is valid for some C will be denoted by $\rho(y)$, $y \in \mathbb{R}^n$.

Note that the existence of a smallest exponent ρ follows from the proof of Theorem 11.4.7. The function $\rho(y)$ is of a very special kind:

Theorem 11.4.10. *There exists a strictly increasing family of subspaces of* $\mathbb{R}^n$,

$$\{0\} = G_0 \subset G_1 \subset \ldots \subset G_k = \mathbb{R}^n$$

and a strictly increasing sequence of rational numbers $\rho_1, \ldots, \rho_k$ *all* $\geqq 1$, *so that* $\rho(y) = \rho_j$ *if* j *is the smallest integer such that* $y \in G_j$.

Proof. It is obvious that

$$\rho(t_1 y_1 + t_2 y_2) \leqq \max(\rho(y_1), \rho(y_2)). \tag{11.4.28}$$

But this means that for every real ρ the set

$$G_\rho = \{y;\ \rho(y) \leqq \rho\}$$

is a linear subspace of $\mathbb{R}^n$, increasing with ρ. The dimension of G_ρ is an increasing function with only a finite number of discontinuities $\rho_1, \ldots, \rho_k$. We have

$$\{0\} \subset G_{\rho_1} \subset G_{\rho_2} \subset \ldots \subset G_{\rho_k} = \mathbb{R}^n,$$

and $\rho(y) = \rho_j$ if G_{ρ_j} is the first of these spaces which contains y. Changing notations slightly, we have proved the theorem.

We now assume that the coordinate system is chosen so that the spaces G_j are defined by the equations

$$x_j = 0 \quad \text{if } j > \dim G_j. \tag{11.4.29}$$

If we write ρ_j for $\rho(e_j)$, where e_j is the unit vector along the jth coordinate axis, it is then clear that $\rho(y) = \rho_j$ where j is the largest index with $y_j \neq 0$.

Definition 11.4.11. By $\Gamma^{(\rho)}(X)$ we denote the set of functions $u \in C^\infty(X)$ such that for every compact set $K \subset X$ there is a constant C for which the inequality

$$|D^\alpha u(x)| \leqq C^{|\alpha|+1} \alpha_1^{\rho_1 \alpha_1} \ldots \alpha_n^{\rho_n \alpha_n}, \quad x \in K, \tag{11.4.30}$$

is valid for every multi-index α.

Theorem 11.4.12. *The solutions of* (11.4.1) *in* X *are of class* $\Gamma^{(\rho)}(X)$ *with* ρ *defined as above.*

The slight modifications of the proof of Theorem 11.4.1 which give this result are left for the reader. When $\rho_j=1$ for every j, the class $\Gamma^{(\rho)}(X)$ consists of all real analytic functions in X. This leads to the following corollary.

Corollary 11.4.13. *All solutions of the equation $P(D)u=0$ are (real) analytic if and only if $P(D)$ is elliptic.*

Proof. If P is elliptic, it follows from (10.4.12) that

$$P^{(\alpha)}(\xi)/P(\xi)=O(|\xi|^{-|\alpha|})$$

when $\xi\to\infty$ in $\mathbb{R}^n$. In view of Lemma 11.1.4 this implies that $|\xi|/d(\xi)$ is bounded when $\xi\to\infty$, hence $\rho(y)=1$ for every y. Thus the solutions of the equation $P(D)u=0$ are analytic according to Theorem 11.4.12. This is of course also a very special case of Theorem 8.6.1.

On the other hand, if all solutions are analytic, we have $\rho(y)=1$ for every y, hence $|\xi|/d(\xi)$ is bounded when $\xi\to\infty$. If m is the order of P, we can choose α so that $|\alpha|=m$ and $P^{(\alpha)}\neq 0$. Then it follows from Lemma 11.1.4 that $|\xi|^m/P(\xi)$ is bounded when $\xi\to\infty$. In particular, for every real $\xi\neq 0$ the quotient $|t\xi|^m/P(t\xi)$ is bounded when $t\to\infty$, which proves that $P_m(\xi)\neq 0$. Hence P is elliptic. (This is also a consequence of Theorem 8.6.7.)

Theorem 11.4.14. *Definition 11.4.9 assigns the same function ρ to any two equally strong hypoelliptic operators.*

Proof. This is an immediate consequence of (11.1.10).

Example 11.4.15. For the semi-elliptic operators in Theorem 11.1.11 we have $\rho_j=m/m_j$ where $m=\sup m_j$ is the order of the operator.

Notes

Petrowsky [3] proved that all classical solutions of a partial differential equation with constant coefficients are analytic if and only if the equation is elliptic (Corollary 11.4.13). He also gave a proof of the analyticity of the solutions of (non-linear) analytic elliptic differential equations, thus extending the results of S. Bernstein [1]. The differentiability of weak solutions is important in the applications of variational calculus (Dirichlet's principle) and was first proved for the Laplace equation by Weyl [1]. Such results have therefore often been called extensions of Weyl's lemma. In most early work the proofs depend on representation formulas involving fundamental solutions but in the 1950's alternative methods depending on precise estimates for solutions with compact supports of inhomogeneous equations were developed. (See e.g. Schwartz [1], John [2], Friedrichs [3], Lax [2], Nirenberg [1].)

Hypoelliptic equations with constant coefficients were characterized in Hörmander [1] by means of a construction of fundamental solutions. (The algebraic conditions are in fact suggested by the proofs of Petrowsky [3].) In Section 11.1 we first follow this approach, using the precise results on fundamental solutions from Section 10.2 though. Alternative more elementary proofs are then given which depend on the study of the operator $P(D)$ in $\mathscr{E}'$ given in Section 10.3 (cf. Malgrange [2], Hörmander [4], Peetre [1]). The results on partial hypoellipticity in Section 11.2 are due to Gårding and Malgrange [1] and to Ehrenpreis [3].

Theorem 7.3.9 implies that if $P(D)u \in C^\infty$ and $u \in C^\infty$ outside a strictly convex surface S then $u \in C^\infty$ in a neighborhood of S. This was first proved by Malgrange [3] and John [4]. The general results in Section 11.3 on propagation of differentiability across other surfaces are due to Hörmander [30].

As already mentioned the results proved in Section 11.4 go back to Petrowsky [3] for general elliptic equations. For the heat equation they are due to Holmgren [2]. The general results were obtained in Hörmander [1, 6]. The method of proof used here goes back to Gevrey [1].

Chapter XII. The Cauchy and Mixed Problems

Summary

To solve the Cauchy problem for a differential operator $P(D)$ with data on a hypersurface $S=\{x;\, \rho(x)=0\}$ means, roughly speaking, to find a solution u of the equation

$$P(D)u=f \tag{12.1}$$

with given f so that for another given function ϕ

$$u(x)-\phi(x)=O(\rho(x)^m) \quad \text{when } \rho(x)\to 0. \tag{12.2}$$

Here m is the order of P. If $u-\phi$ is sufficiently differentiable, the condition (12.2) is of course equivalent to the vanishing on S of $u-\phi$ and all its derivatives of order $<m$ in a direction transversal to S. This is the form in which the Cauchy problem is usually given.

The condition (12.2) means that we can write $u=\phi+\rho^m v$. When $\rho=0$ the equation (12.1) gives $m!P_m(\rho'/i)v=f-P(D)\phi$. If S is non-characteristic, that is, $P_m(\rho')\neq 0$ when $\rho=0$, we can then compute v when $\rho=0$. If all data are in C^∞ we can continue this process and determine the Taylor expansion of u on S in one and only one way so that (12.1) is fulfilled of infinite order on S. This is the formal reason for believing in the soundness of the Cauchy problem. However, we shall find that in contrast to the analytic case studied in Section 9.4 such faith is only justified for certain classes of operators $P(D)$ and surfaces S.

If we write $u=\phi+v$ then the Cauchy problem is reduced to one with homogeneous Cauchy data. Changing notation we may thus assume that the boundary condition has the homogeneous form

$$u(x)=O(\rho(x)^m). \tag{12.2)'$$

If we set $U_+=u$, $F_+=f$ when $\rho>0$, $U_+=F_+=0$ when $\rho<0$, it follows from (12.1) and (12.2)′ that $P(D)U_+=F_+$ in the distribution sense; conversely this implies (12.1) and (12.2)′ when $\rho(x)>0$ if S is non-characteristic. (The other side of S can be handled in the same way.) Thus the Cauchy problem with homogeneous Cauchy data is then equivalent to finding a solution of (12.1) vanishing on one side of S when f does. This is how the Cauchy problem should be stated when S is characteristic or the data are not very smooth.

To make the preceding vague outline precise we start in Section 12.1 with a detailed discussion of the wave equation. It culminates in the beautiful explicit solution formulas of M. Riesz in the four dimensional case. In Section 12.2 we discuss the Cauchy problem for the wave equation when the initial data are highly oscillatory. The asymptotic formulas of geometrical optics are given, including a discussion of the simplest kind of caustics. This will serve as a guide for a study of operators with variable coefficients later on, and is also of independent interest. However, the results of Section 12.2 and most of those in Section 12.1 are not required for the rest of this chapter or indeed this volume.

In Section 12.3 we begin the systematic study of the Cauchy problem by proving that the Cauchy problem for $P(D)$ with data on a non-characteristic hyperplane $\langle x, N\rangle=0$ cannot be solved in general unless P is hyperbolic with respect to N. This means that one can find τ_0 such that $P(\xi+i\tau N)\neq 0$ if $\xi\in\mathbb{R}^n$ and $\tau<\tau_0$. A detailed study of such polynomials is made in Section 12.4. We show that P is also hyperbolic with respect to $-N$ or any vector in a convex cone $\Gamma(P,N)$ containing N. Furthermore, P is hyperbolic if and only if the principal part P_m is hyperbolic, that is, the zeros τ of $P_m(\xi+\tau N)$ are real for real ξ, and in addition the lower order terms are weaker than P_m. With these results available it is easy to construct in Section 12.5 a fundamental solution of a hyperbolic operator with support in the dual cone $\Gamma^\circ(P,N)$. The Cauchy problem is then studied just as in Section 12.1. Section 12.6 is devoted to a study of the singularities of the fundamental solution and to the Herglotz-Petrowsky formulas for it outside the singular support. In Section 12.7 we prove a uniqueness theorem which roughly speaking shows that the Cauchy data for which a non-characteristic Cauchy problem can be solved are coherent if the principal part is not hyperbolic. This means that they are uniquely determined in an open set by the restriction to the complement of a compact subset. Section 12.8 is devoted to the Cauchy problem with data on a characteristic hyperplane. There is no uniqueness then, but we characterize the operators for which an existence theorem is valid. Simple examples are the heat equation and the Schrödinger equation. However, the general theory is quite technical, and the results will not be needed in the sequel. In Section 12.9, finally, we study mixed problems in a quarter space. Thus Cauchy boundary conditions are required on one boundary plane and some other differential boundary conditions with constant coefficients are imposed on the other boundary plane.

12.1. The Cauchy Problem for the Wave Equation

In Section 6.2 we constructed a fundamental solution E_+ of the wave operator in $\mathbb{R}^{n+1}$

$$\square = c^{-2}\partial^2/\partial t^2 - \Delta = c^{-2}\partial^2/\partial t^2 - \sum_1^n \partial^2/\partial x_j^2$$

which has support in the forward light cone

$$\{(t,x);\ ct \geqq |x|\}.$$

We shall now discuss how it can be used to solve the Cauchy problem in the half space

$$H = \{(t,x);\ t \geqq 0\}.$$

First we prove a uniqueness theorem; it is of course a special case of Corollary 8.6.9. The argument was already given in the proof of Theorem 6.2.4.

Theorem 12.1.1. *Suppose that $u \in C^2(H)$, $\square u = 0$ in H and $u = \partial u/\partial t = 0$ when $t=0$. Then $u=0$ in H.*

Proof. Set $U=u$ in H and $U=0$ in $\complement H$. Then $U \in C^1$ and $\square U = 0$ in $\mathbb{R}^{n+1}$ by Theorem 3.1.9. Hence the remarks at the end of Section 4.2 give

$$U = \delta * U = (\square E_+) * U = E_+ * \square U = 0$$

for the map

$$\operatorname{supp} E_+ \times H \ni ((t,x),(s,y)) \to (t+s, x+y)$$

is proper since $s+t \leqq C$ implies $s \leqq C$, $t \leqq C$, hence $|x| \leqq cC$, if $s \geqq 0$, $t \geqq 0$.

The proof of an existence theorem is similar:

Theorem 12.1.2. *Let ν be an integer $\geqq (n-3)/2$ and k an integer $\geqq 2$. Then the Cauchy problem*

$$\square u = f \text{ in } H, \quad u = u_0 \quad \text{and} \quad \partial u/\partial t = u_1 \qquad \text{when } t=0 \tag{12.1.1}$$

has a solution $u \in C^k(H)$ for every $f \in C^{k+\nu}(H)$ and $u_j \in C^{k+\nu+2-j}(\mathbb{R}^n)$. There is no other C^2 solution.

Proof. The last statement follows since the difference between two solutions must be 0 by Theorem 12.1.1. To prove the existence we assume first that $u_0 = u_1 = 0$ and that $\partial^\alpha f = 0$ when $t=0$ and $|\alpha| \leqq k+\nu$. If

$$F = f \text{ in } H, \qquad F = 0 \text{ in } \complement H,$$

it follows then that $F \in C^{k+\nu}(\mathbb{R}^{n+1})$. The convolution

$$U = E_+ * F$$

is well defined as we have seen in the proof of Theorem 12.1.1, we have $U=0$ in $\complement H$ and $\Box U = \delta * F = F$. Now $E_+ = 0$ when $ct < |x|$, and for $t>0$

$$E_+ = C\chi_+^{(1-n)/2}(R), \qquad R = c^2t^2 - |x|^2.$$

Since $(n-1)/2 \leqq \nu+1$, Theorem 6.1.2 shows that the order of E_+ in $\mathbb{R}^{n+1} \setminus 0$ is at most ν if $\nu \geqq 0$, and the proof of Theorem 3.2.3 extends this conclusion to $\mathbb{R}^{n+1}$. If $\nu = -1$ then $n=1$ (or 0) and the explicit formula for E_+ shows that E_+ and its first order derivatives are of order 0. Hence $U \in C^k(\mathbb{R}^{n+1})$ so the restriction of U to H has the required properties.

The proof of Theorem 12.1.2 in the general case is reduced to the special case already considered if we put $u = w + v$ where $v \in C^{k+\nu+2}$ and

$$\partial^\alpha(f - \Box v) = 0, \quad u_0 = v, \quad u_1 = \partial v/\partial t \quad \text{when } t=0, \ |\alpha| \leqq k+\nu.$$

The conditions on v mean that when $t=0$

(12.1.2) $$v = u_0, \quad \partial v/\partial t = u_1,$$
$$\partial^{j+2} v/\partial t^{j+2} = c^2 \partial^j (f + \Delta v)/\partial t^j, \quad 0 \leqq j \leqq k+\nu.$$

If we write

(12.1.3) $$\partial^j v/\partial t^j = u_j \quad \text{when } t=0, \ 0 \leqq j \leqq k+\nu+2$$

then (12.1.2) means that we know u_0 and u_1 and that

$$u_{j+2} = c^2(\partial^j f/\partial t^j + \Delta u_j) \quad \text{when } t=0, \ 0 \leqq j \leqq k+\nu.$$

This permits us to compute $u_2, u_3, \ldots$ successively. By induction we obtain $u_j \in C^{k+\nu+2-j}$, $0 \leqq j \leqq k+\nu+2$, and (12.1.2) is equivalent to (12.1.3) with these u_j. Now Corollary 1.3.4 guarantees the existence of a function $v \in C^{k+\nu+2}$ satisfying (12.1.3), which completes the proof.

Since we know E_+ explicitly we can use the preceding proof to derive explicit formulas for the solution of the Cauchy problem (12.1.1). For the sake of simplicity we restrict ourselves to the physically important case $n=3$. According to (6.2.7) the fundamental solution is then given by

$$\langle E_+, \phi \rangle = (4\pi)^{-1} \int_{\mathbb{R}^3} \phi(|x|/c, x)\, dx/|x|, \qquad \phi \in C_0(\mathbb{R}^4).$$

Let u be the solution of the Cauchy problem in Theorem 12.1.2, and set with $\chi \in C^\infty(\mathbb{R})$ equal to 0 on $(-\infty, 0)$ and 1 on $(1, \infty)$

$$u_\varepsilon(t, x) = \chi(t/\varepsilon)\, u(t, x).$$

To determine u in H we observe that

$$\Box u_\varepsilon(t,x)=f\chi(t/\varepsilon)+2c^{-2}\partial u/\partial t\,\chi'(t/\varepsilon)/\varepsilon+c^{-2}u\chi''(t/\varepsilon)/\varepsilon^2.$$

When $t>\varepsilon$ we have $u(t,x)=u_\varepsilon(t,x)=E_+*\Box u_\varepsilon(t,x)$ or explicitly in terms of polar coordinates in $\mathbb{R}^3$

$$\begin{aligned}u(t,x)=(4\pi)^{-1}\int(f(t-r/c,x-r\omega)\,\chi((t-r/c)/\varepsilon)\\+c^{-2}(2u_t'(t-r/c,x-r\omega)\,\chi'((t-r/c)/\varepsilon)\,\varepsilon^{-1}\\+u(t-r/c,x-r\omega)\,\chi''((t-r/c)/\varepsilon)\,\varepsilon^{-2}))r\,dr\,d\omega.\end{aligned}$$

In the last term we integrate by parts noting that

$$\chi''((t-r/c)/\varepsilon)(\varepsilon c)^{-1}=-(\partial/\partial r)\,\chi'((t-r/c)/\varepsilon),$$

and that $\chi'((t-r/c)/\varepsilon)/\varepsilon\to\delta(t-r/c)=c\,\delta(ct-r)$ as $\varepsilon\to 0$. When $r=ct$ we have

$$\begin{aligned}\frac{\partial}{\partial r}(ru(t-r/c,x-r\omega))=u_0(x-tc\omega)-tu_1(x-tc\omega)\\-ct\langle\omega,\partial/\partial x\rangle\,u_0(x-tc\omega).\end{aligned}$$

Letting $\varepsilon\to 0$ we obtain again Kirchoff's formula (cf. (7.3.8))

$$\begin{aligned}(12.1.4)\quad u(t,x)=(4\pi)^{-1}\int_{|y|<ct}f(t-|y|/c,x-y)\,dy/|y|\\+(4\pi)^{-1}\int_{|\omega|=1}(u_0(x+tc\omega)+tu_1(x+tc\omega)\\+ct\langle\omega,\partial/\partial x\rangle\,u_0(x+tc\omega))\,d\omega.\end{aligned}$$

Here $d\omega$ denotes the element of area on the unit sphere in $\mathbb{R}^3$.

The proof of Theorems 12.1.1 and 12.1.2 can easily be adapted to handle the Cauchy problem for any *spacelike* surface, that is, any smooth surface in $\mathbb{R}^{1+n}$ which has precisely one transversal intersection with every line with direction in the closed light cone. In particular, such a surface intersects the lines in the direction of the t axis, so the surface is defined by an equation

$$t=S(x).$$

Here $|S'|<1/c$ since there would otherwise be tangents belonging to the light cone. In addition

$$|x|-c|S(x)|\to\infty\quad\text{as } x\to\infty.$$

In fact, for any $T>0$ the intersection with the light cone placed at $(T,0)$ or $(-T,0)$ must be compact, since the intersection with a ray is a continuous function of the direction. Hence the sets

$$\{x;\,cT-cS(x)\geqq|x|\},\qquad\{x;\,cS(x)+cT\geqq|x|\}$$

are compact and therefore the union $\{x;\, |x|-c\,|S(x)|\leqq cT\}$ is compact.

Let H be the set defined by $t\geqq S(x)$. We shall study the Cauchy problem

(12.1.1)′ $\quad \Box u=f$ in H, $u=u_0$ and $\partial u/\partial t=u_1$ on ∂H.

Since the backward light cone with vertex at any point in H has a compact intersection with H, the proof of the uniqueness (Theorem 12.1.1) remains unchanged. We assume that the boundary ∂H is in $C^{k+\nu+2}$, that is, that $S\in C^{k+\nu+2}$. As in the proof of Theorem 12.1.2 we can then compute the derivatives of u on ∂H from the data in (12.1.1)′ and obtain on ∂H

$$\partial^j u/\partial t^j \in C^{k+\nu+2-j}, \qquad 0\leqq j\leqq k+\nu+2-j.$$

The existence proof then proceeds as before.

We shall now show that the discussion which led to Kirchoff's formula can also be modified and that it gives a formula of M. Riesz. Assuming that the origin is in the interior of H we shall compute u there. Let s be any smooth function such that H and ∂H are defined by $s\geqq 0$ and $s=0$ respectively and $ds\neq 0$ on ∂H. We could take $s=t-S(x)$ but it is better to leave the choice open for the time being. Choose χ as in the proof of (12.1.4) and set

$$\chi_\varepsilon(s)=\chi(s/\varepsilon), \qquad u_\varepsilon(x)=\chi_\varepsilon(s)\,u(t,x).$$

We shall use the notation ρ for the quadratic form $\tau^2/c^2-|\xi|^2$ dual to $R=c^2t^2-|x|^2$ and also for the corresponding symmetric bilinear form. Thus $\Box=\rho(\partial)$, and if $E=E_-$ we have for small ε

$$\begin{aligned} u(0)=u_\varepsilon(0)&=\langle E, \Box u_\varepsilon\rangle=-\langle \rho(E', u_\varepsilon'), 1\rangle\\ &=-\langle \rho(E', u'), \chi_\varepsilon\rangle-\langle \rho(E', s')u, \chi_\varepsilon'\rangle\\ &=\langle E, \chi_\varepsilon \Box u\rangle+\langle E\rho(s', u'), \chi_\varepsilon'\rangle-\langle u\rho(E', s'), \chi_\varepsilon'\rangle. \end{aligned}$$

Here $E=c\,\delta(R)/2\pi$ so

$$\rho(E', s')=c\,\delta'(R)\,\rho(R', s')/2\pi.$$

We fix the choice of s now so that

(12.1.5) $$-1=\rho(R', s')=2\left(\textstyle\sum x_j\,\partial s/\partial x_j+t\,\partial s/\partial t\right)$$

near the intersection Σ of ∂H and the support of E. This is consistent with s being positive above ∂H. Before letting $\varepsilon\to 0$ we must remove the term involving $\delta'(R)$. To do so we choose a vector field V near Σ so that

(12.1.6) $$\langle V, s'\rangle=0, \qquad \langle V, R'\rangle=1.$$

which is possible since s' and R' are linearly independent on Σ. Then we have

$$\langle\delta'(R)\chi_\varepsilon'(s), u\rangle = -\langle\delta(R)\chi_\varepsilon'(s), \operatorname{div}(Vu)\rangle,$$

for the right-hand side is

$$\langle(\delta(R)\chi_\varepsilon'(s))', Vu\rangle = \langle\delta'(R)\langle V, R'\rangle\chi_\varepsilon'(s), u\rangle + \langle\delta(R)\chi_\varepsilon''(s)\langle V, s'\rangle, u\rangle$$

and simplifies to the desired result by (12.1.6). Letting $\varepsilon\to 0$ we now obtain

(12.1.7)
$$u(0) = (4\pi)^{-1}\int_{(-|y|/c,\, y)\in H} \Box u(-|y|/c, y)\, dy/|y| + c(2\pi)^{-1}\langle\delta(R, s), \rho(s', u') - \operatorname{div}(Vu)\rangle.$$

Here $\delta(R, s)$ is the pullback of δ_0 in $\mathbb{R}^2$ so it is a positive density on Σ. Now Σ is a Riemannian manifold with the metric induced by $-R(dX)$, $X = (t, x_1, x_2, x_3)$ for this is positive even on the spacelike surface ∂H.

Lemma 12.1.3. *$c\delta(R, s)$ is the Riemannian surface measure on Σ.*

Proof. Near any point on Σ we can introduce local coordinates $y = (y_1, y_2, y_3, y_4)$ so that $y_1 = R$, $y_2 = s$. Then we have by (6.1.1)

$$c\langle\delta(R, s), \phi\rangle = \iint \phi(X(0, 0, y_3, y_4))c|\det \partial X/\partial y|\, dy_3\, dy_4$$

when ϕ has small support. Set $g^{jk} = -\rho(y_j', y_k')$. Then $g^{11} = 0$ on Σ, and $g^{12} = 1$ by (12.1.5). Thus $\det(g^{ij})_{i,j=1}^2 = -1 \neq 0$. We can therefore replace y_3, y_4 by new coordinates

$$\tilde{y}_k = y_k + \sum_1^2 a_{kj}(y)\, y_j, \qquad k = 3, 4,$$

equal to the old ones on Σ, such that on Σ

$$0 = \rho(\tilde{y}_k', y_i') = \rho(y_k', y_i') - \sum_1^2 a_{kj}(y)\, g^{ji}; \qquad i = 1, 2;\ k = 3, 4.$$

Hence we may assume that $g^{ij} = 0$ if $i \leqq 2$ and $j \geqq 3$. Then we have

$$\det(g^{jk})_{j,k=3}^4 = -\det(g^{jk})_{j,k=1}^4 = c^{-2}|\det \partial y/\partial X|^2.$$

The pseudo-Riemannian metric

$$-R(dX) = \sum g_{jk}\, dy_j\, dy_k$$

has the inverse matrix, so $\det(g_{jk})_{j,k=3}^4 = c^2|\det \partial X/\partial y|^2$. Since the Riemannian surface measure is $(\det(g_{jk})_{j,k=3}^4)^{1/2}\, dy_3\, dy_4$ the lemma is proved.

The vector field V is only determined by (12.1.6) up to a vector field which is tangential to Σ. To make a definite choice of V we require that

$$\langle V, \psi' \rangle = a\rho(s', \psi') + b\rho(R', \psi'), \qquad \psi \in C^1. \tag{12.1.8}$$

for some functions a and b to be determined. This is equivalent to

$$R(V, t) = a\langle t, s' \rangle + b\langle t, R' \rangle$$

so on Σ the meaning is that $R(V, t) = 0$ if t is a tangent. The conditions (12.1.6) now become

$$a\rho(s', s') = b, \qquad -a + 4bR = 1 \tag{12.1.9}$$

if we recall (12.1.5) and note that $\rho(R', R') = 4R$. Thus

$$\begin{aligned} a &= -1/(1 - 4R\rho(s', s')), \\ b &= -\rho(s', s')/(1 - 4R\rho(s', s')). \end{aligned} \tag{12.1.10}$$

On Σ we have $a = -1$ and $b = -\rho(s', s')$, so using (12.1.8) we can write the interesting quantity in the right hand side of (12.1.7) in the form

$$\begin{aligned} \rho(s', u') - \operatorname{div}(Vu) &= \rho(s', u') + \rho(s', u') + \rho(s', s')\rho(R', u') - u \operatorname{div} V \\ &= \rho(\theta, u') - u \operatorname{div} V, \qquad \theta = 2s' + \rho(s', s')R'. \end{aligned}$$

From (12.1.5) we obtain on Σ

$$\begin{aligned} &\rho(\theta, R') = 2\rho(s', R') = -2, \\ &\rho(\theta, \theta) = 4\rho(s', s') + 4\rho(s', s')\rho(s', R') = 0. \end{aligned}$$

If L is the vector corresponding to the covector θ by $\rho(\psi', \theta) = \langle L, \psi' \rangle$, then L is isotropic, that is, $R(L, L) = 0$; L is R-orthogonal to $T(\Sigma)$ since θ is a linear combination of s' and R', and $\langle L, R' \rangle = -2$. These conditions determine L on Σ; it is the reflection in Σ of the light ray coming from 0.

To compute $\operatorname{div} V$ we integrate (12.1.8) for a test function ψ and obtain

$$\operatorname{div} V = \rho(a', s') + a\Box s + \rho(b', R') + b\Box R.$$

When $R = s = 0$ we have by (12.1.10)

$$a' = -4R'\rho(s', s'), \qquad \text{thus } \rho(a', s') = 4\rho(s', s'), \ \rho(a', R') = 0.$$

Recalling that $b = a\rho(s', s')$ we obtain on Σ

$$\rho(b', R') = a\rho(\rho(s', s')', R') = -\rho(\rho(s', s')', R').$$

Now s' is homogeneous of degree -1 by (12.1.5) for $s + \log\sqrt{-t}$ is homogeneous of degree 0. Hence $\rho(s', s')$ is homogeneous of degree

-2 and (cf. (12.1.5))

$$-\rho(\rho(s', s')', R') = 4\rho(s', s').$$

Finally $b\Box R = -8\rho(s', s')$ so we obtain on Σ

$$-\operatorname{div} V = \Box s. \tag{12.1.11}$$

The restriction of $\Box s$ to Σ does not change if we replace s by another function $s+q$ which satisfies (12.1.5) and vanishes on Σ. For then we have $q=0$ when $R=0$ so we can write $q=fR$ where

$$\rho(f'R+R'f, R')=0, \qquad \text{that is,} \quad \rho(f', R')+4f=0.$$

When $R=0$ we obtain

$$\Box q = \Box(fR) = 2\rho(f', R') + f\Box R = -8f+8f=0.$$

This was of course to be expected because of the role played by $\Box s$ in (12.1.7).

With polar coordinates in $\mathbb{R}^3$ we can define Σ by $-tc=r=\exp\psi(\omega)$ and can then take

$$s = -(\log|ct| - \psi(x/|x|))/2$$

for this is a solution of (12.1.5). When $ct=-|x|$ we obtain

$$\Box s = (|x|^{-2} - \Delta\psi(x/|x|))/2.$$

With polar coordinates in $\mathbb{R}^3$ we have, if Δ_ω denotes the Laplacean in S^2,

$$\Delta = r^{-2}\Delta_\omega + 2r^{-1}\partial/\partial r + \partial^2/\partial r^2$$

where the last two terms annihilate every function which is homogeneous of degree 0. Hence

$$\Box s = (1-\Delta_\omega\psi)/2|x|^2. \tag{12.1.12}$$

The preceding description of Σ identifies Σ with S^2. The metric in Σ induced by $-R(dX)$ is then

$$(\omega\, dr + r\, d\omega, \omega\, dr + r\, d\omega) - dr^2 = r^2(d\omega, d\omega) = e^{2\psi}(d\omega, d\omega)$$

where $(d\omega, d\omega)$ is the standard line element in the sphere.

Lemma 12.1.4. *The Gaussian curvature of the line element*

$$e^{2\psi(\omega)}(d\omega, d\omega)$$

on S^2 is equal to $e^{-2\psi}(1-\Delta_\omega\psi)$.

Proof. We consider S^2 as embedded in $\mathbb{R}^3$ and prove the assertion at $(0,0,-1)$ where we take x_1 and x_2 as local coordinates. Since $\sum x_j dx_j$

$=0$ the standard line element is

$$dx_1^2+dx_2^2+(x_1\,dx_1+x_2\,dx_2)^2/(1-x_1^2-x_2^2).$$

Up to third order terms the metric tensor to consider is therefore

$$g_{11}=e^{2\psi}(1+x_1^2), \quad g_{12}=e^{2\psi}x_1x_2, \quad g_{22}=e^{2\psi}(1+x_2^2).$$

At 0 we have $g_{11}g_{22}-g_{12}^2=e^{4\psi}$, and the Christoffel symbols are

$$\begin{aligned}
&\Gamma_{111}=2^{-1}\partial g_{11}/\partial x_1=e^{2\psi}\,\partial\psi/\partial x_1,\\
&\Gamma_{222}=e^{2\psi}\,\partial\psi/\partial x_2,\\
&\Gamma_{112}=-2^{-1}\partial g_{11}/\partial x_2+\partial g_{12}/\partial x_1=-e^{2\psi}\,\partial\psi/\partial x_2,\\
&\Gamma_{221}=-e^{2\psi}\,\partial\psi/\partial x_1,\\
&\Gamma_{121}=2^{-1}\partial g_{11}/\partial x_2=e^{2\psi}\,\partial\psi/\partial x_2,\\
&\Gamma_{122}=e^{2\psi}\,\partial\psi/\partial x_1.
\end{aligned}$$

Hence we have at 0, if R is the Riemann curvature tensor,

$$\begin{aligned}
R_{1212}&=\partial^2 g_{12}/\partial x_1\,\partial x_2-(\partial^2 g_{11}/\partial x_2^2+\partial^2 g_{22}/\partial x_1^2)/2\\
&\quad+\sum(\Gamma_{12r}^2-\Gamma_{11r}\Gamma_{22r})e^{-2\psi}\\
&=e^{2\psi}-e^{2\psi}(2\,\Delta\psi+4|\psi'|^2)/2+e^{2\psi}(|\psi'|^2+|\psi'|^2)\\
&=e^{2\psi}(1-\Delta\psi).
\end{aligned}$$

At 0 the Laplacean in the unit sphere is the standard Laplacean in (x_1,x_2) which completes the proof of the lemma.

We have now computed all the quantities occurring in (12.1.7) and can sum up the result as the following theorem of M. Riesz (see Fig. 3):

Theorem 12.1.5. *Let H be the set in $\mathbb{R}^{3+1}$ above a smooth spacelike surface ∂H, let $(t,x)\in H\setminus\partial H$ and denote by Σ the intersection of ∂H and the light cone with vertex at (t,x). Define a vector field L on Σ so that on Σ*

$$R(L,L)=0, \quad \langle L, R'(.-(t,x))\rangle=-2, \quad R(L,T(\Sigma))=0.$$

Then the solution of the Cauchy problem (12.1.1′) is given by

$$u(t,x)=(4\pi)^{-1}\int_{(t-|y|/c,\,y)\in H} f(t-|y|/c,y)\,dy/|y| \tag{12.1.13}$$
$$+(4\pi)^{-1}\int_\Sigma u(\sigma)\,K(\sigma)\,d\sigma+(2\pi)^{-1}\int_\Sigma\langle L,u'\rangle\,d\sigma$$

where $d\sigma$ is the Riemannian surface element of the metric defined by $-R(dX)$ on Σ and K is the Gaussian curvature of Σ. Here $R=c^2t^2-|x|^2$ is the Lorentz form.

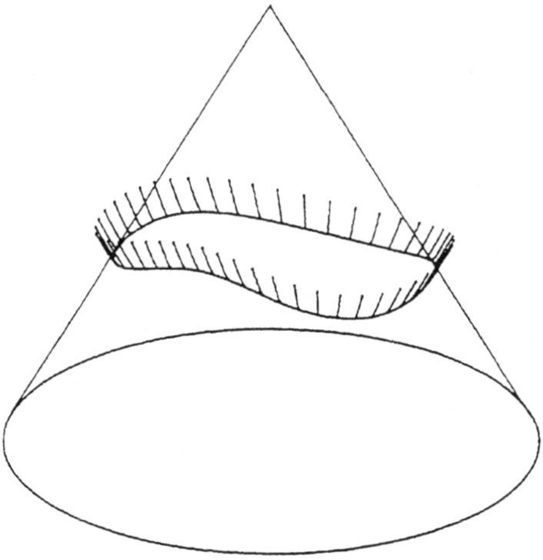

Fig. 3. The backward light cone, Σ and L

When $u=1$ identically we obtain

$$\int_\Sigma K(\sigma)\,d\sigma = 4\pi$$

which is of course a consequence of the Gauss-Bonnet formula.

12.2. The Oscillatory Cauchy Problem for the Wave Equation

Already the Riesz formula (Theorem 12.1.5) shows that the light rays of geometrical optics have a prominent role in the solution of the Cauchy problem for the wave equation. This becomes still more pronounced if one considers highly oscillatory initial data as we shall do in this section. (This is essentially the well known fact that radio waves propagate more and more along straight lines as the frequency increases.) We restrict ourselves to the homogeneous wave equation in $\mathbb{R}^{n+1}$ with unit speed of light, and pose the Cauchy data on the plane $t=0$,

$$\partial^2 u/\partial t^2 - \Delta u = 0, \tag{12.2.1}$$

$$u=0, \quad \partial u/\partial t = f \quad \text{when } t=0. \tag{12.2.2}$$

We assume that $f \in C_0^\infty(\mathbb{R}^n)$. The solution will then have compact support in any slab $|t| \leqq T$. Taking Fourier transforms with respect to

x in (12.2.1) and (12.2.2) we obtain an ordinary differential equation with initial conditions. After solving it we get from Fourier's inversion formula

$$u(t,x)=(2\pi)^{-n}\Big(\int e^{i(\langle x,\xi\rangle+t|\xi|)}\hat{f}(\xi)/2i|\xi|\,d\xi - \int e^{i(\langle x,\xi\rangle-t|\xi|)}\hat{f}(\xi)/2i|\xi|\,d\xi\Big). \tag{12.2.3}$$

We shall alway assume $n\geqq 2$ so the integrals converge, and it is of course elementary to verify that (12.2.3) satisfies (12.2.1) and (12.2.2).

Introducing the definition of $\hat{f}$ we can write (12.2.3) in the form

$$u(t,x)=(2\pi)^{-n}\Big(\iint e^{i(\langle x-y,\xi\rangle+t|\xi|)}f(y)/2i|\xi|\,dy\,d\xi - \iint e^{i(\langle x-y,\xi\rangle-t|\xi|)}f(y)/2i|\xi|\,dy\,d\xi\Big), \tag{12.2.4}$$

where the integrals exist as oscillatory integrals (see Section 7.8) since the differentials with respect to y of the phase functions $\langle x-y,\xi\rangle \pm t|\xi|$ do not vanish when $|\xi|\neq 0$. To identify (12.2.4) with (12.2.3) we just have to recall that the oscillatory integrals are limits as $\varepsilon\to 0$ of the absolutely convergent integrals obtained when a convergence factor $\chi(\varepsilon\xi)$ is introduced, with $\chi\in C_0^\infty$ and $\chi=1$ near 0. (See also Example 7.8.4.)

We shall now choose oscillatory initial data, that is,

$$f(x)=a(x)\,e^{i\omega\phi(x)}, \qquad \omega>0, \tag{12.2.5}$$

where $a\in C_0^\infty(\mathbb{R}^n)$, $\phi\in C^\infty(\mathbb{R}^n)$ is real and $\phi'_x\neq 0$ in $\mathbb{R}^n$. (Only minor modifications would be required if we just assume $\phi'_x\neq 0$ in $\operatorname{supp} a$.) The solution is then $u_\omega=u_\omega^+-u_\omega^-$ where

$$u_\omega^\pm(t,x)=(2\pi)^{-n}\iint e^{i(\langle x-y,\xi\rangle\pm t|\xi|+\omega\phi(y))}a(y)/2i|\xi|\,dy\,d\xi. \tag{12.2.6}$$

The method of stationary phase (Section 7.7) will now be used to examine the asymptotic behavior when $\omega\to\infty$.

Stationary points with respect to y and ξ occur for the function in the exponent when

$$\xi=\omega\phi'(y), \qquad y=x\pm t\xi/|\xi|. \tag{12.2.7}$$

Writing $\xi=\omega\phi'(y)+\eta$ we see that the stationary point is non-degenerate if and only if

$$0\neq D(\eta, y-x\mp t\xi/|\xi|)/D(\eta,y)=D(y\mp t\phi'(y)/|\phi'(y)|)/Dy.$$

Thus the stationary points are non-degenerate unless (t,x) belongs to

$$C_\pm=\{(t,x);\ x=y\mp t\phi'(y)/|\phi'(y)|;\ D(y\mp t\phi'(y)/|\phi'(y)|)/Dy=0 \text{ for some } y\in\mathbb{R}^n\}.$$

This is called the caustic set. $C_\pm$ is closed since $|x-y|\leqq t$ and the image of a compact set under a continuous map is compact. The measure is 0 for fixed t by an elementary case of the Morse-Sard theorem:

Lemma 12.2.1. *If K is a compact set in $\mathbb{R}^n$ and f a real valued C^1 map from a neighborhood of K to $\mathbb{R}^n$, then the set of critical values*

$$\{f(x);\ x\in K,\ \det f'(x)=0\}$$

is of measure 0.

Proof. For every $\varepsilon>0$ we can cover $\{x\in K;\ \det f'(x)=0\}$ by disjoint cubes I_j of diameter ε and $\sum m(I_j)<C$ where C is independent of ε. By Taylor's formula

$$f(x)=f(x_j)+f'(x_j)(x-x_j)+o(\varepsilon), \qquad x\in I_j,$$

where $x_j\in I_j$ and $\det f'(x_j)=0$. Thus $f(I_j)$ is at distance $o(\varepsilon)$ from a set of diameter $O(\varepsilon)$ in a hyperplane, so $m(f(I_j))=o(\varepsilon^n)=m(I_j)\,o(1)$. Hence

$$\sum m(f(I_j))\to 0 \quad \text{as } \varepsilon\to 0,$$

which proves the lemma.

When (t,x) varies in a component G of $\mathbb{R}^{n+1}\setminus C_\pm$ it follows from the implicit function theorem that the equation

$$y=x\pm t\,\phi'(y)/|\phi'(y)| \tag{12.2.8}$$

has a fixed number $v_\pm(G)$ of solutions y each of which is a C^∞ function of (t,x) (locally). The solution is unique when $t=0$ and therefore also in the component of $\mathbb{R}^{n+1}\setminus C_\pm$ containing $\{0\}\times\mathbb{R}^n$. In any component G we define $v_\pm(G)$ C^∞ functions $\phi^\pm(t,x)$ locally by setting

$$\phi^\pm(t,x)=\phi(y)$$

when (12.2.8) is valid. Then we have

$$\partial\phi^\pm/\partial t=\pm|\partial\phi^\pm/\partial x|; \qquad \partial\phi^\pm/\partial x=\partial\phi/\partial y \tag{12.2.9}$$

when (12.2.8) is valid. (The first equation is the eiconal equation of geometrical optics.) This is a consequence of the Hamilton-Jacobi integration theory (see Section 6.4) but a direct verification is elementary: When (12.2.8) is valid we have

$$d\phi(y)=\langle\phi'(y),dy\rangle=\langle\phi'(y),dx\rangle\pm|\phi'(y)|\,dt\pm t\langle\phi'(y),d(\phi'(y)/|\phi'(y)|)\rangle.$$

Here the last term vanishes since

$$d\langle\phi'(y)/|\phi'(y)|,\phi'(y)/|\phi'(y)|\rangle=d1=0.$$

The lines $t \to (t, y \mp t\,\phi'(y)/|\phi'(y)|)$ are the light rays corresponding to our initial value problem. What we have observed so far is that outside the caustics formed by them we obtain a finite number of smooth phase functions by keeping the phase constant along the light rays.

To study the asymptotic behavior of $u_\omega^\pm$ we shall first show that large values of ξ in (12.2.6) give small contributions.

Lemma 12.2.2. *If* $\chi \in C_0^\infty(\mathbb{R}^n \setminus 0)$ *is equal to* 1 *in a neighborhood of* $\{\phi'(y);\ y \in \operatorname{supp} a\}$, *then*

$$(2\pi)^{-n} \iint e^{i(\langle x-y,\xi\rangle \pm t|\xi| + \omega\phi(y))}\, a(y)\,(1-\chi(\xi/\omega))/2i|\xi|\, dy\, d\xi$$

is rapidly decreasing as $\omega \to \infty$.

Proof. If $y \in \operatorname{supp} a$ and $\xi \in \operatorname{supp}(1-\chi)$ we have $|\phi'(y) - \xi| > c(1+|\xi|)$ for some $c>0$ since the left hand side does not vanish and ϕ' is bounded. Hence $|\omega\phi'(y) - \xi| > c(|\omega| + |\xi|)$ in the support of the integrand. An application of Theorem 7.7.1 therefore shows that the integral with respect to y is bounded by any power of $(|\xi|+\omega)^{-1}$, which proves the lemma.

It remains to discuss

(12.2.10) $$v_\omega^\pm(t,x) = \omega^{n-1}(2\pi)^{-n} \iint e^{i\omega(\langle x-y,\xi\rangle \pm t|\xi| + \phi(y))}\, \chi(\xi)\, a(y)/2i|\xi|\, dy\, d\xi.$$

Here the critical points are non-degenerate when $(t,x) \notin C_\pm$. After a partition of unity to separate the finitely many critical points we can apply Theorem 7.7.6 to each term. Note that $\langle x-y,\xi\rangle \pm t|\xi| = 0$ at a critical point because this is a homogeneous function of ξ. Hence it follows that in each component G of $\mathbb{R}^{n+1} \setminus C_\pm$ we have locally

(12.2.11) $$u_\omega^\pm(t,x) = \omega^{-1} \sum e^{i\omega\phi^\pm(t,x)}\, a_\omega^\pm(t,x) + R_\omega^\pm(t,x)$$

where $a_\omega^\pm \in C^\infty$ has an asymptotic expansion in powers of $1/\omega$ and $R_\omega^\pm$ is rapidly decreasing. The summation is taken over the $\nu_\pm(G)$ solutions of the eiconal equation discussed above. We shall now examine the critical points more carefully in order to calculate the main terms in (12.2.11) and also determine the asymptotic behavior at the caustic set.

To simplify notation we assume that the phase function in (12.2.10)

(12.2.12) $$(\xi, y) \to \langle x-y,\xi\rangle \pm t|\xi| + \phi(y)$$

has a critical point $\xi = (\xi_1, 0, \dots, 0)$, $y=0$, thus

(12.2.13) $$\partial\phi(0)/\partial y = \xi, \qquad x = \mp t e_1$$

where $e_1=(1,0,\dots,0)$. We assume $\xi_1>0$ and note that for small θ

$$\begin{aligned}|\xi+\theta|&=(\xi_1+\theta_1)(1+|\theta'|^2/(\xi_1+\theta_1)^2)^{1/2}\\&=\xi_1+\theta_1+|\theta'|^2/2\xi_1-|\theta'|^2\theta_1/2\xi_1^2+\dots\end{aligned}$$

where $\theta'=(\theta_2,\dots,\theta_n)$ and terms of order four or higher are neglected. If ϕ_2 and ϕ_3 are the homogeneous parts of order 2 and 3 in the Taylor expansion of ϕ at 0, we obtain for (12.2.12) the Taylor expansion at ξ, 0

$$\begin{aligned}\langle x-y,\xi+\theta\rangle\pm t|\xi+\theta|+\phi(y)&\\=\phi(0)-\langle y,\theta\rangle\pm t|\theta'|^2/2|\xi|+\phi_2(y)&\\\mp t|\theta'|^2\theta_1/2|\xi|^2+\phi_3(y)+\dots.&\end{aligned}\tag{12.2.14}$$

By an orthogonal transformation in the variables $y_2,\dots,y_n$ we can always put $\phi_2(0,y')$ in diagonal form,

$$\phi_2(0,y')=\sum_2^n b_j y_j^2/2.$$

The second order part of (12.2.14) is then

$$\sum_2^n(b_j y_j^2\pm t\theta_j^2/|\xi|-2y_j\theta_j)/2+y_1(L(y)-\theta_1),$$
$$L(y)=(\phi_2(y_1,y')-\phi_2(0,y'))/y_1.$$

To study this quadratic form we take $\theta_1-L(y)$ as a new variable instead of θ_1. The form $\theta_1 y_1$ is non-degenerate in the variables θ_1, y_1 with signature 0. Thus the quadratic form is non-degenerate unless for some j we have

$$\pm t b_j/|\xi|=1.\tag{12.2.15}$$

The signature of a term in the sum is 0 when $t=0$ and changes to $2\operatorname{sgn}\pm t$ where (12.2.15) is valid. Now note that the level surface of ϕ through 0 has second order contact with the surface $|\xi|y_1+\phi_2(0,y')=0$, that is,

$$y_1=-\sum_2^n y_j^2 b_j/2|\xi|.$$

When the normal is oriented in the direction $\phi'(0)=\xi$, the radii of curvature are $-|\xi|/b_j$. Hence the quadratic form is non-singular unless $x=\mp t\xi/|\xi|$ is a center of curvature; for all other points the signature is $2N\operatorname{sgn}\pm t$ if N is the number of centers of curvature between $y=0$ and x. If we recall that the determinant of the quadratic form was actually calculated after (12.2.7) when we first examined the non-degeneracy, we have proved:

Theorem 12.2.3. *The solution of the Cauchy problem* (12.2.1), (12.2.2) *with data* (12.2.5) *is* $u_\omega = u_\omega^+ - u_\omega^-$ *where* $u_\omega^\pm$ *has the asymptotic expansion* (12.2.11) *outside the caustic set* $C_\pm$, *with the leading term*

$$(12.2.16)\qquad a_\infty^\pm(t,x) = \sum a(y)(2i|\phi'(y)|)^{-1} |D(y \mp t\phi'(y)/|\phi'(y)|)/Dy|^{-\frac{1}{2}} i^{N \operatorname{sgn} \pm t}.$$

Here y *is any solution of* (12.2.8) *and* N *is the number of caustic points (centers of curvature of the level surface of* ϕ *through* y) *between* $(0, y)$ *and* (t, x) (*between* y *and* x), *counted with multiplicity.*

Note that the square of the modulus of the amplitude times the cross section of an infinitesimal light beam is constant along the ray, and that the phase remains constant along the ray apart from a jump of $\pm\pi/2$ when a caustic is passed. This is a well known observable phenomenon in optics. However, (12.2.16) gives the incorrect impression that the amplitudes blow up at the caustics. What happens is of course that the asymptotic expansion (12.2.11) ceases to be valid when we approach the caustics. Thus we need to have a more refined expansion to describe what happens there, and this can be obtained from Theorem 7.7.19.

To be able to apply Theorem 7.7.19 at a caustic point (t, x) we need to know first of all that it is a simple caustic, that is, that (12.2.15) is fulfilled for just one j with our preceding choice of coordinates. This makes the rank of the Hessian of (12.2.12) equal to $2n-1$. The kernel is determined by $b_j y_j = \theta_j$ for this index, $y_i = \theta_i = 0$ for $2 \leqq i \neq j$, and $y_1 = 0$, $\theta_1 = L(y) = \partial\phi_2/\partial y_1$. In this direction the third order terms in the Taylor expansion of (12.2.12) are, if we use (12.2.15),

$$(12.2.17)\qquad \begin{aligned} &\mp t\theta_j^2\theta_1/2|\xi|^2 + \phi_3(y) \\ &= -(b_j/2|\xi|)\, y_j^2\, \partial\phi_2(0, y_j, 0)/\partial y_1 + \phi_3(0, y_j, 0). \end{aligned}$$

We need to assume that this quantity is not equal to 0. To interpret that condition we observe that the level surface of ϕ through $y=0$ is defined by

$$y_1\xi_1 + \phi_2(y) + \phi_3(y) + \ldots = 0$$

or if we use the approximation $y_1 = -\phi_2(0, y')/|\xi| + \ldots$ except in the first term

$$\begin{aligned} y_1 = -|\xi|^{-1}(&\phi_2(0, y') + \phi_3(0, y') \\ &- \phi_2(0, y')\,\partial\phi_2(0, y')/\partial y_1/|\xi|) + O(|y'|^4). \end{aligned}$$

Thus the non-vanishing of (12.2.17) means that the pertinent circle of curvature does not have higher order contact with the level surface of

ϕ. (When $n=2$ this means that the evolute is non-singular.) From Theorem 7.7.19 we now obtain:

Theorem 12.2.4. *Let $x_0=y_0\mp t_0\phi'(y_0)/|\phi'(y_0)|$ be a simple center of curvature for the level surface $\{y;\, \phi(y)=\phi(y_0)\}$, and assume that the corresponding osculating circle is not a third order tangent of the level surface. Then there exist C^∞ functions ψ and ρ near (t_0,x_0) with $\psi(t_0,x_0)=0$ such that the contribution to $u_\omega^\pm(t,x)$ from a neighborhood of y_0 can be written*

$$\omega^{-\frac{5}{6}}(w_\omega(t,x)\,Ai(\psi(t,x)\,\omega^{\frac{2}{3}}) + \tilde{w}_\omega(t,x)\,Ai'(\psi(t,x)\,\omega^{\frac{2}{3}})\,\omega^{-\frac{1}{3}})\,e^{i\omega\rho(t,x)}. \tag{12.2.18}$$

Here w_ω and $\tilde{w}_\omega$ have asymptotic expansions in powers of $1/\omega$.

Thus the blowup of the expansion (12.2.11) shown by (12.2.16) is caused by the fact that the leading term in (12.2.18) is of the order of magnitude $\omega^{-\frac{5}{6}}$ and not of the magnitude ω^{-1} as in (12.2.11). Note that (12.2.18) is exponentially decreasing where $\psi(t,x)>0$ but of the form (12.2.11) at a point with $\psi(t,x)<0$. Thus $\psi(t,x)=0$ defines the boundary of the set near (t_0,x_0) illuminated from a neighborhood of y_0. At a distance of the order of magnitude $\omega^{-\frac{2}{3}}$ from this surface we can practically neglect the term (12.2.18) in the shadow and replace it by (12.2.11) in the illuminated region.

12.3 Necessary Conditions for Existence and Uniqueness of Solutions to the Cauchy Problem

Let H be a half space defined by an inequality $\langle x,N\rangle\geqq 0$. As stated in the summary of this chapter, the Cauchy problem for $P(D)$ in H with homogeneous Cauchy data is to find a solution of the equation

$$P(D)u=f \quad \text{in } \mathbb{R}^n \tag{12.3.1}$$

with $\operatorname{supp} u\subset H$ when f is given with $\operatorname{supp} f\subset H$. By Theorem 8.6.7 and Corollary 8.6.9 we know that $f=0$ then implies $u=0$ if and only if ∂H is *non-characteristic*, that is, $P_m(N)\neq 0$ if P_m is the principal part of P. We shall now prove that P must satisfy a very restrictive algebraic condition in order that even a very weak existence theorem shall be valid.

Theorem 12.3.1. *Assume that the equation $P(D)u=f$ has a solution $u\in\mathscr{D}'(\mathbb{R}^n)$ with support in H for every $f\in C_0^\infty(H)$, and that the boundary* *of H is non-characteristic. Then there exists a number* τ_0 *such that*

$$P(\xi+i\tau N)\neq 0 \quad \text{if } \xi\in\mathbb{R}^n \text{ and } \tau<\tau_0. \tag{12.3.2}$$

In order to simplify the proof we shall first show that the hypotheses in the theorem are in fact very much stronger than they seem.

Lemma 12.3.2. *From the hypotheses in Theorem 12.3.1 it follows that the equation* $P(D)u=f$ *has one and only one solution* $u\in C^\infty(\mathbb{R}^n)$ *with* $\operatorname{supp} u\subset H$ *for every* $f\in C^\infty(\mathbb{R}^n)$ *with* $\operatorname{supp} f\subset H$.

Proof. Let $f\in C_0^\infty(H)$ and let $u\in\mathscr{D}'(\mathbb{R}^n)$ be the solution of the equation $P(D)u=f$ with $\operatorname{supp} u\subset H$ which exists by hypothesis. If V is a convex neighborhood of N such that $P_m(\theta)\neq 0$ when $\theta\in V$, and $\langle x,\theta\rangle\geqq -1$ when $x\in\operatorname{supp} f$ and $\theta\in V$, it then follows from Corollary 8.6.10 that

$$\operatorname{supp} u\subset \bigcap_{\theta\in V}\{x;\langle x,\theta\rangle\geqq -1\}.$$

In particular, this means that the intersection of $\operatorname{supp} u$ and each half space $\{x;\langle x,N\rangle\leqq a\}$ is compact. If $\chi\in C_0^\infty(\mathbb{R})$ is equal to 1 in $(-a,a)$, it follows that $v=\chi(\langle .,N\rangle)u\in\mathscr{E}'$ and that $P(D)v\in C^\infty$ when $\langle x,N\rangle<a$. Hence $v\in C^\infty$ when $\langle x,N\rangle<a$, by Theorem 7.3.9, so $u\in C^\infty$ when $\langle x,N\rangle<a$. Since a is arbitrary this shows that $u\in C^\infty(\mathbb{R}^n)$.

Now let f be any function in $C^\infty(\mathbb{R}^n)$ with $\operatorname{supp} f\subset H$. Choose a sequence of functions $\chi_\nu\in C_0^\infty(\mathbb{R}^n)$ such that $\chi_\nu(x)=1$ when $|x|<\nu$. Then the equation $P(D)u_\nu=\chi_\nu f$ has a solution $u_\nu\in C^\infty(\mathbb{R}^n)$ with support in H. Since

$$P(D)(u_{\nu+1}-u_\nu)=(\chi_{\nu+1}-\chi_\nu)f=0$$

when $|x|<\nu$, it follows from Corollary 8.6.11 that on any compact set we have $u_{\nu+1}-u_\nu=0$ for large ν. Hence $u=\lim_{\nu\to\infty} u_\nu$ exists, and it is obvious that $u\in C^\infty(\mathbb{R}^n)$, $\operatorname{supp} u\subset H$ and $P(D)u=f$. As already observed, the uniqueness follows from Corollary 8.6.9, so the proof of the lemma is complete.

Proof of Theorem 12.3.1. From Lemma 12.3.2 and Banach's theorem it follows that the differential operator $P(D)$ is an automorphism of the closed subspace

$$\{u;u\in C^\infty(\mathbb{R}^n),\operatorname{supp} u\subset H\}$$

of $C^\infty(\mathbb{R}^n)$. In particular, if y is a point with $\langle y,N\rangle>0$, this implies that for some compact set $K\subset\mathbb{R}^n$ and some constants C and k we have

(12.3.3)
$$|u(y)| \leqq C \sum_{|\alpha| \leqq k} \sup_K |D^\alpha P(D) u|;$$
$$u \in C^\infty(\mathbb{R}^n), \qquad \operatorname{supp} u \subset H.$$

Now choose a function $\chi \in C^\infty(\mathbb{R})$ with support on the positive half axis so that $\chi(t)=1$ when $t > \langle y, N\rangle/2 = c$. We can then apply (12.3.3) to $u(x) = e^{i\langle x-y, \zeta\rangle} \chi(\langle x, N\rangle)$. If $P(\zeta)=0$ we have $P(D)u=0$ when $\langle x, N\rangle > c$. Writing $K' = \{x; x \in K, \langle x, N\rangle \leqq c\}$, we thus obtain with another constant C

(12.3.4)
$$1 \leqq C(1+|\zeta|)^{k+m} \sup_{x \in K'} |e^{i\langle x-y, \zeta\rangle}|, \qquad P(\zeta)=0.$$

If $\zeta = \xi + i\tau N$ where ξ is real and $\tau < 0$, it follows from (12.3.4) that

(12.3.5)
$$1 \leqq C(1+|\zeta|)^{k+m} e^{c\tau}$$

if $P(\zeta)=0$ and $\operatorname{Im} \zeta = \tau N$, $\tau < 0$.

Now let
$$\mu(\sigma) = \sup(-\tau)$$
where the supremum is taken over all real ξ and τ such that
$$|\xi + i\tau N|^2 \leqq \sigma^2 \quad \text{and} \quad |P(\xi + i\tau N)|^2 = 0.$$
It follows from Corollary A.2.6 that $\mu(\sigma)=0$ or
$$\mu(\sigma) = A\sigma^a(1+o(1)), \qquad \sigma \to \infty,$$
for some constants $A \neq 0$ and a. Since (12.3.5) shows that
$$1 \leqq C(1+\sigma)^{k+m} e^{-c\mu(\sigma)} \qquad \text{if } \mu(\sigma) > 0,$$
it follows that $\mu(\sigma)$ must be bounded above when $\sigma \to \infty$. This completes the proof.

Definition 12.3.3. A polynomial P is called hyperbolic with respect to the real vector N if $P_m(N) \neq 0$ and (12.3.2) is valid for some τ_0.

We shall prove in Section 12.5 that the Cauchy problem can always be solved when P is hyperbolic. This could be done directly by means of (12.3.2) but it is convenient to study the properties of hyperbolic polynomials first as we shall do in Section 12.4.

12.4. Properties of Hyperbolic Polynomials

In this section we shall prove that seemingly much stronger conditions follow from the definition of a hyperbolic polynomial. Part of these could in fact have been obtained quite easily in the proof of Theorem 12.3.1 but we have preferred to give direct proofs.

Theorem 12.4.1. *If P is hyperbolic with respect to N, it follows that P is hyperbolic with respect to $-N$.*

Proof. The homogeneity of the principal part P_m shows that $P_m(-N) = (-1)^m P_m(N) \neq 0$. From (12.3.2) it follows that $P(\xi + i\tau N) \neq 0$ if $\xi \in \mathbb{R}^n$ and $\operatorname{Re}\tau < \tau_0$, for we can write $\xi + i\tau N = \xi - \operatorname{Im}\tau N + i \operatorname{Re}\tau N$. Now note that the coefficient of τ^{m-1} in the equation $P(\xi + i\tau N) = 0$ is a linear function of ξ, and that the coefficient of τ^m is equal to $P_m(iN) \neq 0$. If the zeros are denoted by τ_j, it thus follows that $\sum_1^m \tau_j$ is a linear function of ξ. Hence $\sum_1^m \operatorname{Re}\tau_j$ is a linear function of ξ when $\xi \in \mathbb{R}^n$, and it is bounded from below by $m\tau_0$ in view of (12.3.2). Consequently $\sum_1^m \operatorname{Re}\tau_j = c$ where c is a constant, and this implies that

$$\operatorname{Re}\tau_j = c - \sum_{k \neq j} \operatorname{Re}\tau_k \leqq c - (m-1)\tau_0.$$

Thus we have $P(\xi + i\tau N) \neq 0$ when $\tau > c - (m-1)\tau_0$, which proves the theorem.

Theorem 12.4.2. *If P is hyperbolic with respect to N, it follows that the principal part P_m is also hyperbolic with respect to N.*

Proof. By (12.3.2) the roots of the equation $P(\xi + i\tau N) = 0$ are located in the half plane $\operatorname{Re}\tau \geqq \tau_0$. Now we have

$$P_m(\xi + i\tau N) = \lim_{\sigma \to +\infty} \sigma^{-m} P(\sigma\xi + i\sigma\tau N).$$

Since the zeros of $P(\sigma\xi + i\sigma\tau N)$ are in the half plane $\sigma \operatorname{Re}\tau \geqq \tau_0$, we conclude that the zeros of $P_m(\xi + i\tau N)$ all lie in the half plane $\operatorname{Re}\tau \geqq 0$. The proof is complete.

For homogeneous polynomials the definition of hyperbolicity is particularly simple:

Theorem 12.4.3. *A homogeneous polynomial P is hyperbolic with respect to N if and only if $P(N) \neq 0$ and the equation*

$$P(\xi + \tau N) = 0 \tag{12.4.1}$$

has only real roots when ξ is real.

Proof. If $P(\xi+\tau N)=0$ it follows from the homogeneity that $P(\sigma\xi+\sigma\tau N)=0$ for every real σ. If (12.3.2) is fulfilled we must thus have $\sigma\,\mathrm{Im}\,\tau \geqq \tau_0$ for every real σ. This implies that $\mathrm{Im}\,\tau=0$. Conversely, if (12.4.1) has only real roots, we know that $P(\xi+i\tau N)\neq 0$ for real $\tau\neq 0$, so that P is hyperbolic.

By Lemma 8.7.3 a homogeneous hyperbolic polynomial is proportional to a polynomial with real coefficients. The component $\Gamma(P,N)$ of N in $\{\theta; P_m(\theta)\neq 0\}$ is a convex cone, and $\theta\in\Gamma(P,N)$ if and only if $P_m(\theta+tN)=0$ implies $t<0$. The following theorem is close to Lemma 8.7.4.

Theorem 12.4.4. *Let P be hyperbolic with respect to N and assume that (12.3.2) is valid. If $\theta\in\Gamma(P,N)$ then*

$$P(\xi+i\tau N+i\sigma\theta)\neq 0 \tag{12.4.2}$$

when $\mathrm{Re}\,\tau<\tau_0$ and $\mathrm{Re}\,\sigma\leqq 0$.

Proof. (12.4.2) follows from (12.3.2) when $\mathrm{Re}\,\sigma=0$. We shall study the zeros of $P(\xi+i\tau N+i\sigma\theta)$, considered as a polynomial in σ, when τ varies in the half plane $\mathrm{Re}\,\tau<\tau_0$. Note that the coefficient of σ^m is $i^m P_m(\theta)\neq 0$ which shows that the zeros vary continuously with τ. The number of zeros σ with negative real part must thus be a constant when $\mathrm{Re}\,\tau<\tau_0$ since we have seen that there is no zero with $\mathrm{Re}\,\sigma=0$. To prove the theorem it will thus suffice to prove that there are no zeros with $\mathrm{Re}\,\sigma<0$ when τ is a sufficiently large negative number. With the notation $\sigma=u\tau$, the equation $P(\xi+i\tau N+i\sigma\theta)=0$ can be written in the form

$$i^{-m}\tau^{-m}P(\xi+i\tau(N+u\theta))=0,$$

which converges to the equation $P_m(N+u\theta)=0$ when $\tau\to-\infty$. The roots of this equation are negative when $\theta\in\Gamma(P,N)$ as pointed out above. Hence all roots of the equation $P(\xi+i\tau N+i\sigma\theta)=0$ have a positive real part when $\mathrm{Re}\,\tau<\tau_0$. The proof is complete.

Corollary 12.4.5. *If P is hyperbolic with respect to N, it follows that P is hyperbolic with respect to every $\theta\in\Gamma(P,N)$.*

Proof. If we take σ and τ real in (12.4.2), $\tau=\varepsilon\sigma$, it follows that P is hyperbolic with respect to $\theta+\varepsilon N$ for every $\varepsilon>0$. Since $\Gamma(P,N)$ is open we have $\theta-\varepsilon N\in\Gamma(P,N)$ for small ε if $\theta\in\Gamma(P,N)$. Hence P is hyperbolic with respect to $\theta=(\theta-\varepsilon N)+\varepsilon N$.

We shall now discuss the converse of Theorem 12.4.2. Let

$$P = P_m + P_{m-1} + \dots + P_0$$

where P_j is homogeneous of degree j.

Theorem 12.4.6. *If the principal part P_m of P is hyperbolic with respect to N, then each of the following conditions is necessary and sufficient for P to be hyperbolic with respect to N:*

(i) $P(\xi + iN)/P_m(\xi + iN)$ *is bounded when* $\xi \in \mathbb{R}^n$.
(i)′ $P_j(\xi + iN)/P_m(\xi + iN)$ *is bounded when* $\xi \in \mathbb{R}^n$, $j = 0, \dots, m$.
(ii) P *is weaker than* P_m.
(ii)′ P_j *is weaker than* P_m *for* $j = 0, \dots, m$.
(iii) P_m *dominates the lower order part* $P - P_m$.
(iii)′ P_m *dominates* P_j *for* $j = 0, \dots, m-1$.
(iv) P *and* P_m *are equally strong.*
(v) $P(\sigma(\xi + iN)) \neq 0$ *if* $\xi \in \mathbb{R}^n$, $\sigma \in \mathbb{C}$ *and* $|\sigma|$ *is sufficiently large.*

Proof. (i) ⇒ (ii) for (i) implies $\tilde{P}(\xi + iN) \leqq C_1 \tilde{P}_m(\xi + iN)$ by Theorem 10.4.3, hence $\tilde{P}(\xi) \leqq C_2 \tilde{P}_m(\xi)$ by (10.1.8). (ii) ⇒ (ii)′. In fact, if P is weaker than P_m then $P(t\xi)$ is weaker than $P_m(t\xi) = t^m P_m(\xi)$ for every real $t \neq 0$. Thus

$$P(t\xi) = \sum t^j P_j(\xi) \prec P_m(\xi).$$

Choose $m+1$ different real t_j, $j = 0, \dots, m$, all different from 0. Since the matrix (t_j^k); $j, k = 0, \dots, m$; is non-singular, each polynomial P_k is a linear combination of the polynomials $P(t_j \xi)$, hence weaker than $P_m(\xi)$. (ii)′ ⇒ (i)′ Since $P_j(\xi + iN)$ is equally strong as $P_j(\xi)$ for every j, this follows if we show that there is a constant C such that

$$\tilde{P}_m(\xi + iN) \leqq C |P_m(\xi + iN)|, \quad \xi \in \mathbb{R}^n. \tag{12.4.3}$$

Now $P_m(\xi + iN + i\zeta) \neq 0$ if $N + \operatorname{Re}\zeta \in \Gamma(P_m, N)$, which is true when $|\zeta|$ is sufficiently small since $\Gamma(P_m, N)$ is open. Hence (12.4.3) is a consequence of Lemma 11.1.4. (i)′ ⇒ (i) is trivial, so the equivalence of the first four conditions is proved.

From (ii)′ it follows that

$$|\tau|^{m-j} \tilde{P}_j(\xi, \tau) \leqq C \tilde{P}_m(\xi, \tau), \tag{12.4.4}$$

$j = 0, \dots, m$, $(\xi, \tau) \in \mathbb{R}^{n+1}$, for this is precisely condition (ii)′ when $\tau = 1$ and both sides are homogeneous of degree m in (ξ, τ). Thus (ii)′ ⇒ (iii)′ ⇒ (iii), and (iii) ⇒ (iv) by Corollary 10.4.8 while (iv) ⇒ (ii) is trivial. The condition (v) means that the zeros of

$$\sigma^m P_m(\xi + iN) + \sigma^{m-1} P_{m-1}(\xi + iN) + \dots + P_0$$

have a bound independent of ξ. This is equivalent to the existence of a bound for $P_j(\xi+iN)/P_m(\xi+iN)$ which is independent of j and ξ. Thus (i)$'\Rightarrow$(v)$\Rightarrow$(i). Condition (v) with σ real implies that P is hyperbolic. What remains to prove is that hyperbolicity of P implies condition (i), which is the essential difficulty in the theorem.

Let $p(\xi,s)=s^m P(\xi/s)$ be the homogeneous polynomial corresponding to P. Thus $p(\xi,0)=P_m(\xi)$ and $p(\xi,1)=P(\xi)$. That P is hyperbolic with respect to N and $-N$ means that for some C

$$p(\xi+tN,s)=0, \quad (\xi,s)\in\mathbb{R}^{n+1} \Rightarrow |\operatorname{Im} t|\leqq C|s|, \tag{12.4.5}$$

for this is condition (12.3.2) when $s=1$. If we prove that

$$|p(\xi+itN,t)|\leqq C|p(\xi+itN,0)|, \quad (\xi,t)\in\mathbb{R}^{n+1}, \tag{12.4.6}$$

then condition (i) follows when $t=1$. By the homogeneity it suffices to prove (12.4.6) when $|\xi|=1$. Since $|t|^m\leqq C|p(\xi+itN,0)|$ the only question is if (12.4.6) is valid for small t. If not, then Theorem A.2.8 shows that one can find analytic functions $\xi(r)$ and $t(r)$ in a neighborhood of $0\in\mathbb{C}$ which are real for real r, such that $|\xi(r)|=1$ and

$$|p(\xi(r)+it(r)N,t(r))/p(\xi(r)+it(r)N,0)|\to\infty, \quad t(r)\to 0 \quad \text{as } r\to 0.$$

However, this contradicts the following lemma applied to

$$h(t,r,s)=p(\xi(r)+tN,s)$$

which completes the proof of Theorem 12.4.6.

Lemma 12.4.7. *Let h be an analytic function at 0 in $\mathbb{C}^3$ and assume that*

$$h(t,r,s)=0 \quad \textit{implies} \quad |\operatorname{Im} t|\leqq C|s| \quad \text{if } (r,s)\in\mathbb{R}^2, \tag{12.4.5$'$}$$

in a neighborhood of 0. Then there is a constant C such that

$$|h(it,r,t)|\leqq C|h(it,r,0)| \quad \textit{for small } (t,r)\in\mathbb{R}^2. \tag{12.4.6$'$}$$

Proof. By (12.4.5)$'$ we have $h(t,0,0)\neq 0$ if $\operatorname{Im} t\neq 0$ and t is small. Hence $h(t,0,0)$ has a zero of finite order m when $t=0$. We assume that the lemma is already proved for smaller values of m. (It is trivial when $m=0$.) Lemma 8.7.2 and the remark after its proof show that at 0

$$h(t,r,s)=h^0(t,r,s)+O(|t|+|r|+|s|)^{m+1} \tag{12.4.7}$$

where h^0 is a homogeneous polynomial of degree m. Hence

$$h_1(t,r,s)=h(rt,r,rs)/r^m$$

is analytic near 0 and

$$h_1(t,0,s)=h^0(t,1,s).$$

Thus $h_1(t,0,0)$ has a zero of order $\leqq m$ at 0. For zeros of h_1 near 0 with r and s real we have $|\mathrm{Im}\, rt| \leqq C|rs|$, hence $|\mathrm{Im}\, t| \leqq C|s|$, so h_1 also satisfies (12.4.5)'. Assume for a moment that (12.4.6)' is valid for h_1. Then we have for some $\delta > 0$ and C

$$|h(irt, r, rt)| \leqq C|h(irt, r, 0)| \quad \text{if } |t|+|r|<\delta.$$

Hence (12.4.6)' is valid if $|t| \leqq \delta|r|/2$ and $|r|<\delta/2$. On the other hand,

$$|h(it,r,t)| \leqq C_1|t|^m \quad \text{if } |t|>\delta|r|/2,$$

and if t, r are real then

$$|h(it,r,0)| \geqq C_2|t|^m \tag{12.4.8}$$

because the small zeros of $z \to h(z,r,0)$ are real. Hence (12.4.6)' is also valid when $|t|>|r|\delta/2$ and r is small enough.

If we introduce the power series expansion

$$h(t,r,s) = \sum a_{\alpha\beta\gamma} t^\alpha r^\beta s^\gamma$$

then $a_{\alpha\beta\gamma}=0$ when $\alpha+\beta+\gamma<m$, and

$$h_1(t,r,s) = \sum a_{\alpha\beta\gamma} t^\alpha r^{\beta+(\alpha+\gamma-m)} s^\gamma.$$

If $h_1(it,0,0)$ has a zero of order $<m$ at 0 then Lemma 12.4.7 follows from the inductive hypothesis and the arguments above. Otherwise we repeat the same procedure with h replaced by h_1. If the lemma is not yet proved after k such steps then

$$a_{\alpha\beta\gamma}=0 \quad \text{when } \beta+k(\alpha+\gamma-m)<0.$$

Hence the lemma follows for some k unless $a_{\alpha\beta\gamma}=0$ when $\alpha+\gamma<m$. This implies

$$|h(it,r,t)| \leqq C|t|^m$$

and (12.4.6)' is then an immediate consequence of (12.4.8). The proof is complete.

In (10.2.8) we defined the space $\Lambda(P)$ along which P is constant. Now condition (i) in Theorem 12.4.6 shows in particular that if $P_m(\xi+iN+t\eta)$ is independent of t then $P(\xi+iN+t\eta)$ is independent of t, when η is a real vector. Hence we obtain

Corollary 12.4.8. *If P is hyperbolic then $\Lambda(P)=\Lambda(P_m)$, if P_m denotes the principal part.*

Since the comparison of differential operators is not always so easy to make, we list some further corollaries and examples of Theorem 12.4.6.

Corollary 12.4.9. *Let $P=P_m+P_{m-1}+\dots$ be hyperbolic with respect to N. If $\xi\in\mathbb{R}^n$ is not proportional to N, and t_0 is a zero of $t\to P_m(\xi+tN)$ of order μ, then $\xi+t_0N$ is a zero of order $\mu-j$ of P_{m-j}, $j=0,\dots,m-1$.*

Proof. We may assume that $t_0=0$, $\xi\neq 0$, for ξ can be replaced by $\xi+t_0N$. By (i)′ in Theorem 12.4.6 and (10.1.8) we have for every α

$$|P_{m-j}^{(\alpha)}(\xi)|\leqq C|P_m(\xi+iN)|\leqq C\sum_0^m|\langle\partial,N\rangle^k P_m(\xi)|.$$

If we replace ξ by $s\xi$ and let $s\to\infty$, then the right-hand side is $O(s^{m-\mu})$ so $P_{m-j}^{(\alpha)}(\xi)=0$ when $m-j-|\alpha|>m-\mu$, that is, when $|\alpha|<\mu-j$.

Corollary 12.4.10. *Each of the following two conditions is necessary and sufficient in order that every polynomial P with principal part P_m shall be hyperbolic with respect to N:*

a) *P_m is hyperbolic with respect to N and of principal type (Definition 10.4.11).*

b) *$P_m(N)\neq 0$ and the polynomial $P_m(\xi+\tau N)$ has only simple real zeros for every real ξ which is not proportional to N.*

Proof. Corollary 12.4.9 shows that b) is necessary, and it is obvious that b) implies a). If P_m is of principal type then P_m dominates every Q of order $<m$ (Theorem 10.4.10) so P_m+Q is hyperbolic by Theorem 12.4.6, condition (iii).

Definition 12.4.11. P is called strictly hyperbolic (or hyperbolic in the sense of Petrowsky) with respect to N if the equivalent conditions a) and b) of Corollary 12.4.10 are satisfied.

An example is of course the wave operator corresponding to $P(\xi)=\xi_1^2-\xi_2^2-\dots-\xi_n^2$, which is strictly hyperbolic with respect to every N such that $N_1^2-N_2^2-\dots-N_n^2>0$. There is a very complete theory of strictly hyperbolic operators with variable coefficients also. We shall return to it in Chapter XXIII.

In general the condition in Corollary 12.4.9 is weaker than those in Theorem 12.4.6. We shall illustrate this by an example.

Example 12.4.12. a) Let $P_4(\xi)=(\xi_1^2-\xi_2^2-\xi_3^2)^2$ be the square of the Lorentz form. It is then clear that P_4 is hyperbolic. The condition in Corollary 12.4.9 is that $P_3(\xi)=0$ when $\xi_1^2-\xi_2^2-\xi_3^2=0$, that is,

$$P_3(\xi)=L(\xi)(\xi_1^2-\xi_2^2-\xi_3^2) \tag{12.4.9}$$

where L is a linear form. By Theorem 10.4.1 this implies $P_3 \prec P_4$. Since P_3 is stronger than every lower order operator if $L(\xi)=\xi_1$, say, (12.4.9) is the only restriction on the lower order terms of a hyperbolic operator with principal part P_4.

b) Let us now take the product of two Lorentz forms with tangential light cones,

$$P_4(\xi)=(\xi_1^2-\xi_2^2-\xi_3^2)(\xi_1^2-\xi_2^2-2\xi_3^2).$$

Then the condition in Corollary 12.4.9 is that $P_3(\xi)=0$ when $\xi_3=0$ and $\xi_1^2=\xi_2^2$, that is,

$$P_3(\xi)=(\xi_1^2-\xi_2^2)L(\xi)+\xi_3 q(\xi) \tag{12.4.10}$$

where L is linear and q is quadratic. By Theorem 10.4.1 P_4 is stronger than $(\xi_1^2-\xi_2^2-2\xi_3^2)L(\xi)$ and $(\xi_1^2-\xi_2^2-\xi_3^2)L(\xi)$ if L is a linear form. Hence P_4 is also stronger than $(\xi_1^2-\xi_2^2)L(\xi)$ and $\xi_3^2 L(\xi)$. It remains to examine if P_4 is stronger than $\xi_3 q(\xi_1,\xi_2,0)$, that is, if

$$|\xi_3 q(\xi_1,\xi_2,0)| \leqq C(|P_4(\xi)|+|\partial P_4(\xi)/\partial\xi_1|+|\xi|^2+1).$$

When $\xi_1=\pm\xi_2$ the right-hand side is $C(2\xi_3^4+6|\xi_1\xi_3^2|+2\xi_1^2+\xi_3^2+1)$. If we replace (ξ_1,ξ_2) by $s(\xi_1,\xi_2)$ it follows when we let $s\to\infty$ that $|\xi_3 q(\xi_1,\pm\xi_1,0)| \leqq 2C\xi_1^2$. Hence $q(\xi_1,\pm\xi_1,0)=0$. This proves that $P=P_4+P_3+\ldots$ is hyperbolic if and only if

$$P_3(\xi)=(\xi_1^2-\xi_2^2)L_1(\xi)+\xi_3^2 L_2(\xi) \tag{12.4.11}$$

for some linear L_1, L_2. Condition (12.4.10) does not imply (12.4.11).

c) Let us now take light cones intersecting transversally,

$$P_4(\xi)=(\xi_1^2-2\xi_2^2-\xi_3^2)(\xi_1^2-\xi_2^2-2\xi_3^2).$$

Then the condition in Corollary 12.4.9 is that $P_3(\xi)=0$ when $\xi_2^2=\xi_3^2$ and $\xi_1^2=3\xi_2^2$. If we write

$$P_3(\xi)=L_1(\xi)(\xi_1^2-2\xi_2^2-\xi_3^2)+\xi_1 a(\xi_2,\xi_3)+b(\xi_2,\xi_3)$$

where L is linear, a quadratic and b cubic, this means that $a(\xi_2,\xi_3)$ and $b(\xi_2,\xi_3)$ must vanish when $\xi_2^2=\xi_3^2$. Hence

$$P_3(\xi)=L_1(\xi)(\xi_1^2-2\xi_2^2-\xi_3^2)+(\xi_2^2-\xi_3^2)L_2(\xi) \tag{12.4.12}$$

where L_2 is also linear. Now Theorem 10.4.1 shows that $P_4(\xi)$ is stronger than $L(\xi)(\xi_1^2-2\xi_2^2-\xi_3^2)$ and $L(\xi)(\xi_1^2-\xi_2^2-2\xi_3^2)$, hence stronger than $L(\xi)(\xi_2^2-\xi_3^2)$, if L is a linear form, so (12.4.12) is the only condition that has to be fulfilled in order that P be hyperbolic.

12.5. The Cauchy Problem for a Hyperbolic Equation

In this section we shall prove that hyperbolicity implies existence of solutions of the Cauchy problem. The proof of the following theorem is motivated by (7.4.7).

Theorem 12.5.1. *Let $P(D)$ be hyperbolic with respect to N. Then there exists one and only one fundamental solution E of the operator $P(D)$ with support in the half space $H=\{x; \langle x, N\rangle \geqq 0\}$. This fundamental solution is regular (Definition 10.2.2), and* $\operatorname{supp} E$ *is contained in the dual cone*

$$\Gamma^\circ(P,N)=\{x; \langle x,\theta\rangle \geqq 0,\ \theta\in\Gamma(P,N)\} \tag{12.5.1}$$

but in no smaller closed convex cone with vertex at 0.

Proof. a) If E_1 and E_2 are two fundamental solutions with support in H, the difference $u=E_1-E_2$ is a solution of the equation $P(D)u=\delta-\delta=0$ with support in H. Thus it follows from Corollary 8.6.9 that $u=0$.

b) We shall now prove the existence of E. First note that (12.3.2) implies

$$|P(\xi+i\tau N)| \geqq |P_m(N)|\,|\tau-\tau_0|^m, \qquad \tau<\tau_0,\ \xi\in\mathbb{R}^n. \tag{12.5.2}$$

In fact, $P(\xi+i\tau N)=i^m P_m(N)\prod_1^m(\tau-\tau_j)$ where the zeros τ_j satisfy the inequality $\operatorname{Re}\tau_j\geqq\tau_0$. Hence $|\tau-\tau_j|\geqq \operatorname{Re}\tau_j-\tau\geqq\tau_0-\tau$ when τ is real, which gives (12.5.2).

With a fixed $\tau<\tau_0$ we now define a distribution by

$$\check{E}(\phi)=(2\pi)^{-n}\int \hat\phi(\xi+i\tau N)/P(\xi+i\tau N)\,d\xi, \qquad \phi\in C_0^\infty. \tag{12.5.3}$$

This definition is of the form (7.3.14) so it is legitimate by (12.5.2). Since $\hat\phi(\xi+i\tau N)$ is the Fourier transform of

$$\phi(x)e^{-i\langle x, i\tau N\rangle}=\phi(x)e^{\langle x,\tau N\rangle}=\psi(x),$$

the definition of E means that $F=e^{\langle x,\tau N\rangle}E$ is defined by

$$\check{F}(\psi)=(2\pi)^{-n}\int\hat\psi(\xi)/P(\xi+i\tau N)\,d\xi,$$

that is, $\hat F=1/P(.+i\tau N)$. Thus F is a distribution in $\mathscr{S}'$ in view of (12.5.2) (although E need not be in $\mathscr{S}'$). In virtue of Theorem 12.4.4 we have $P(\xi+i\tau N+i\zeta)\neq 0$ if ξ is real and $(\tau_0-\tau)N-\operatorname{Re}\zeta\in\Gamma(P,N)$, thus if $|\zeta|$ is sufficiently small. Hence the distance from $\xi+i\tau N$ to the zeros of P is bounded from below, so it follows from Lemma 11.1.4 that $\tilde P(\xi+i\tau N)\leqq C|P(\xi+i\tau N)|$, hence $\tilde P(\xi)\leqq C|P(\xi+i\tau N)|$

for some other constant C. By definition this means that $F \in B_{\infty,\tilde{P}}$ and that $E \in B^{\mathrm{loc}}_{\infty,\tilde{P}}$.

That E is a fundamental solution follows from the Fourier inversion formula

$$\check{E}(P(D)\phi) = (2\pi)^{-n} \int \hat{\phi}(\xi + i\tau N)\, d\xi$$
$$= (2\pi)^{-n} \int \hat{\psi}(\xi)\, d\xi = \psi(0) = \phi(0), \qquad \phi \in C_0^\infty.$$

We next show that E is in fact independent of the choice of τ as long as $\tau < \tau_0$. In doing so we may assume that N is directed along the ξ_1 axis. All we have to do then is to move a line of integration in the complex z_1 plane. That this is permissible follows from the analyticity of $\hat{\phi}(\zeta)/P(\zeta)$, the estimate (12.5.2) of P from below and the fact that $\hat{\phi}$ satisfies (7.3.3). Using the proof of Theorem 7.3.1 it is now easy to show that $\operatorname{supp} E \subset H$. First note that, if $\phi \in C_0^\infty(H)$ we have by (7.3.3)

$$|\xi + i\tau N|^{n+1} |\hat{\phi}(\xi + i\tau N)| \leqq C \quad \text{when } \tau \leqq 0,\ \xi \in \mathbb{R}^n. \tag{12.5.4}$$

Using (12.5.2) and (12.5.4) we now obtain for $\tau < \inf(\tau_0, 0)$

$$|\check{E}(\phi)| \leqq C(\tau_0 - \tau)^{-m} \int |\xi + i\tau N|^{-n-1}\, d\xi = C_1 (\tau_0 - \tau)^{-m} |\tau|^{-1}$$

where C and C_1 are constants. When $\tau \to -\infty$ it follows that $\check{E}(\phi) = 0$. Hence $\operatorname{supp} E \subset H$.

It remains to prove that $\operatorname{supp} E \subset \Gamma^\circ(P, N)$. This is in fact a consequence of Corollary 8.6.10 (see the proof of Theorem 12.3.1) but we prefer to give a proof independent of Holmgren's uniqueness theorem. Thus note that if $\theta \in \Gamma(P, N)$ it follows from Theorem 12.4.4 that we can make another shift of line of integration to prove that

$$\check{E}(\phi) = (2\pi)^{-n} \int \hat{\phi}(\xi + i\tau N + i\sigma\theta)/P(\xi + i\tau N + i\sigma\theta)\, d\xi, \qquad \tau < \tau_0,\ \sigma < 0.$$

When $\sigma \to -\infty$ we obtain as above that $\check{E}(\phi) = 0$ if $\langle x, \theta \rangle \geqq 0$ when $x \in \operatorname{supp} \phi$. Hence $\operatorname{supp} E \subset \{x; \langle x, \theta \rangle \geqq 0\}$, which proves that $\operatorname{supp} E \subset \Gamma^\circ(P, N)$.

To prove that no smaller convex cone contains the support of E, we shall use the following converse of Theorem 12.5.1.

Theorem 12.5.2. *Assume that a differential operator $P(D)$ has a fundamental solution with support in a closed cone K with vertex at 0 and no point $\neq 0$ in common with the half space $\{x; \langle x, N \rangle \geqq 0\}$. Then P is hyperbolic with respect to N.*

Admitting this result for a moment, we note that if the support of the fundamental solution constructed in Theorem 12.5.1 is a subset of a closed convex cone K with vertex at 0, then all proper planes of support of K must be non-characteristic. The set K° of all θ such that

$\langle x, \theta \rangle > 0$ if $0 \neq x \in K$ is thus a convex subset of the open set $\{\theta;\, P_m(\theta) \neq 0\}$, and it contains N. Hence K° is contained in the component $\Gamma(P, N)$, which proves that $K \supset \Gamma^\circ(P, N)$.

Proof of Theorem 12.5.2. Let H be the half space $\{x;\, \langle x, N \rangle \geqq 0\}$. If $f \in C_0^\infty(H)$, the convolution $u = E * f$ satisfies the equation $P(D)u = f$ and $\operatorname{supp} u \subset H$. If we prove that the plane $\langle x, N \rangle = 0$ is non-characteristic it follows from Theorem 12.3.1 that P is hyperbolic with respect to N. In view of Theorem 8.6.7 it is thus sufficient to show that $u = 0$ if $P(D)u = 0$ and $\operatorname{supp} u \subset H$. But this is obvious since $u = E * (P(D)u)$ (see the proof of Theorem 6.2.3).

Using Theorem 12.4.6 we shall now show that the fundamental solution of an operator $P(D)$ with lower order terms can be expressed in terms of that of the principal part $P_m(D)$ and its powers as suggested by the formal expansion

$$P^{-1} = (P_m - (P_m - P))^{-1} = \sum_0^\infty (P_m - P)^k P_m^{-k-1}.$$

Theorem 12.5.3. *Let $P(D)$ be hyperbolic with respect to N, with principal part P_m, and denote by E_k the fundamental solution of $P_m(D)^{k+1}$ which has support in $\Gamma^\circ(P, N)$. Then*

$$\sum_0^\infty (P_m(D) - P(D))^k E_k \tag{12.5.5}$$

converges in $B^{\mathrm{loc}}_{\infty, \tilde{P}_m}$ to the fundamental solution of $P(D)$ with support in $\Gamma^\circ(P, N)$.

Proof. By (i)′ in Theorem 12.4.6, or rather (12.4.4), we have

$$|P_m(\xi + i\tau N) - P(\xi + i\tau N)| / |P_m(\xi + i\tau N)| \leqq C/|\tau|, \quad |\tau| > 1,\ \xi \in \mathbb{R}^n.$$

When $\tau < -2C$ it follows that the formal expansion above converges,

$$P(\xi + i\tau N)^{-1} = \sum_0^\infty (P_m(\xi + i\tau N) - P(\xi + i\tau N))^k / P_m(\xi + i\tau N)^{-k-1},$$

when $\xi \in \mathbb{R}^n$. We have uniform convergence after multiplication by $\tilde{P}_m(\xi) \leqq C\,|P_m(\xi + i\tau N)|$. If we enter this expansion in (12.5.3), the theorem is proved.

Remark. It is clear that E_k is homogeneous of degree $-n + (k+1)m$ (Theorem 7.1.16) with Fourier transform $P_m(.\, - i0N)^{-k-1}$ (cf. Theorem 3.1.15). Since $P_m(D) - P(D)$ is of order $m-1$, the general term in (12.5.5) is a sum of terms of degree of homogeneity $\geqq -n + m + k$ and

$\leqq -n+(k+1)m$. Thus there are only finitely many terms of the same homogeneity. In Section 12.6 we shall use Theorem 12.5.3 to study the fundamental solution in $\Gamma^\circ(P, N)$.

Now that we have a fundamental solution available it is easy to solve the Cauchy problem for vanishing Cauchy data:

Theorem 12.5.4. *Let* $f \in \mathscr{D}'(\mathbb{R}^n)$ *and* $\operatorname{supp} f \subset H = \{x;\ \langle x, N\rangle \geqq 0\}$. *If* $P(D)$ *is hyperbolic with respect to* N, *the equation* $P(D)u = f$ *has one and only one solution* u *with support in* H, *and* $u = E * f$ *where* E *is the fundamental solution of Theorem 12.5.1. If* $f \in B_{p,k}^{\mathrm{loc}}(\mathbb{R}^n)$ *it follows that* $u \in B_{p,\tilde{P}k}^{\mathrm{loc}}(\mathbb{R}^n)$.

Proof. If $P(D)u = f$ we have (see the proof of Theorem 6.2.3)

$$E * f = E * (P(D)u) = (P(D)E) * u = u.$$

This proves the uniqueness, which is also a consequence of Corollary 8.6.9. On the other hand, if we define $u = E * f$ we obtain in the same way that $P(D)u = (P(D)E) * f = f$. The last statement in the theorem follows by a simple modification of Theorem 10.1.24 which may be proved by the reader. The proof is complete.

Using Theorem 10.1.25 we can easily pass to existence theorems of a classical form, involving the function space C^j instead of the spaces $B_{p,k}$. Since the spaces C^j are by no means suitable for giving precise statements concerning the regularity properties of the solutions, we shall not aim at giving even the very best results that could be obtained from Theorem 12.5.4.

Corollary 12.5.5. *Let* $f \in C^{j+r}(\mathbb{R}^n)$ *where* $r = [n/2]+1$, *and let* $\operatorname{supp} f \subset H$. *The solution with support in* H *of the equation* $P(D)u = f$ *given by Theorem 12.5.4 then belongs to* $C^j(\mathbb{R}^n)$.

Proof. Writting $k_r(\xi) = (1+|\xi|)^r$ we have $C_0^{j+r} \subset B_{2,k_{j+r}}$ by Parseval's formula. Hence $f \in B_{2,k_{j+r}}^{\mathrm{loc}}(\mathbb{R}^n)$. Noting that $\tilde{P}$ is bounded from below we obtain from Theorem 12.5.4 that $u \in B_{2,k_{j+r}}^{\mathrm{loc}}(\mathbb{R}^n)$. Since $1/k_r \in L^2$ it follows from Theorem 10.1.25 that $u \in C^j(\mathbb{R}^n)$ (see also Lemma 7.6.3, Corollary 7.9.4).

The proof of Theorem 12.1.2 will now give

Theorem 12.5.6. *Let* H *be the half space* $\{x;\ \langle x, N\rangle \geqq 0\}$ *and let* $P(D)$ *be an operator of order* m *which is hyperbolic with respect to* N. *Let*

$j \geqq m$ and set $r=[n/2]+1$. For all $f \in C^{j+r}(H)$ and $\phi_k \in C^{m-k+j+r}(\partial H)$, $0 \leqq k < m$, there exists one and only one solution $u \in C^j(H)$ of the Cauchy problem

$$P(D)u = f \quad \text{in } H \tag{12.5.6}$$

$$\langle D, N\rangle^k u = \phi_k \quad \text{in } \partial H,\ k=0, \dots, m-1. \tag{12.5.7}$$

At a point $x \in H$ the solution u is uniquely determined by the restrictions of f and of $\{\phi_k\}$ to the cone $\{x\} - \Gamma^\circ(P, N)$.

Proof. We may assume that $N=(1,0,\dots,0)$. First we determine $v \in C^{m+j+r}(H)$ so that when $x_1 = 0$

$$D^\alpha(f - P(D)\,v) = 0, \quad |\alpha| \leqq j+r; \quad D_1^k\, v = \phi_k \text{ when } 0 \leqq k < m. \tag{12.5.8}$$

It suffices to take $\alpha = (k, 0, \dots, 0)$, $0 \leqq k \leqq j+r$. Since $P(D)$ contains a term $P_m(N) D_1^m$ with non-zero coefficient, the conditions (12.5.8) allow us to compute successively

$$\phi_m \in C^{j+r}, \phi_{m+1} \in C^{j+r-1}, \dots, \phi_{m+j+r} \in C^0$$

so that (12.5.8) is equivalent to

$$D_1^k\, v = \phi_k \quad \text{when } x_1 = 0, \quad 0 \leqq k \leqq m+j+r. \tag{12.5.9}$$

By Corollary 1.3.4 we can now choose $v \in C^{m+j+r}(H)$ so that (12.5.9) is fulfilled.

Set $f_1 = f - P(D)v$ in H, $f_1 = 0$ in $\complement H$. Then $f_1 \in C^{j+r}$ by (12.5.8) so Corollary 12.5.5 shows that we can find $u_1 \in C^j(\mathbb{R}^n)$ with support in H so that $P(D)u_1 = f_1$. Then $D_1^k u_1 = 0$ when $x_1 = 0$ and $k \leqq j$, hence if $k \leqq m$, so $u = u_1 + v$ satisfies (12.5.6), (12.5.7).

Since the last statement in Theorem 12.5.6 follows immediately from Corollary 8.6.11 (or from the proof if one makes sure that the x_1 axis is in $\Gamma^\circ(P, N)$), the proof is now complete.

Corollary 12.5.7. *If $f \in C^\infty(H)$ and $\{\phi_k\} \in C^\infty(\partial H)$, the Cauchy problem* (12.5.6), (12.5.7) *has a unique solution $u \in C^\infty(H)$.*

Proof. The hypotheses of Theorem 12.5.6 are fulfilled for every j, and the solution u given in that theorem is independent of j.

Since the solution of the Cauchy problem obtained in Theorem 12.5.6 only depends on the values of f and of ϕ_k inside the cone $\{x\} - \Gamma^\circ(P, N)$ with vertex at x, it is clear that the Cauchy problem can also be studied locally. The plane $\langle x, N\rangle = 0$ may also be replaced by a curved "spacelike" surface, that is, a surface whose normals all belong to the cone $\Gamma(P, N)$. We leave the exact statements and proofs to the reader.

12.6. The Singularities of the Fundamental Solution

Let $P(D)$ be hyperbolic with respect to N, and denote the principal part by P_m. If $\xi \neq 0$ we denote by $P_{m,\xi}$ the lowest order homogeneous part in the Taylor expansion at 0 of

$$\eta \to P_m(\xi+\eta).$$

$P_{m,\xi}$ is also hyperbolic with respect to N (Lemma 8.7.2) and therefore also with respect to every $\theta \in \Gamma(P_{m,\xi}, N)$. From Theorem 10.2.12 we obtain a lower bound for the wave front set of the fundamental solution of $P(D)$:

Theorem 12.6.1. *If E is the fundamental solution of $P(D)$ with support in $\Gamma^\circ(P,N)$ and $0 \neq \xi \in \mathbb{R}^n$, then $\Gamma^\circ(P_{m,\xi}, N)$ is contained in the closed convex cone with vertex at 0 generated by $\{x; (x,\xi) \in WF(E)\}$.*

Proof. If $P_{m,\xi}$ is of degree μ, we have

$$P_{m,\xi}(\eta) = \lim_{\varepsilon \to 0} \varepsilon^{-\mu} P_m(\xi+\varepsilon\eta) = \lim_{\varepsilon \to 0} \varepsilon^{m-\mu} P_m(\xi/\varepsilon+\eta)$$

so $P_{m,\xi}$ is a localization of P_m at ∞ in the direction ξ (Definition 10.2.6). Since the homogeneous part P_j of P of degree j is weaker than P_m (Theorem 12.4.6) it follows that $\varepsilon^{m-\mu} P_j(\xi/\varepsilon+\eta)$ is bounded as $\varepsilon \to 0$. The limit must be

$$\sum_{j-|\alpha|=m-\mu} P_j^{(\alpha)}(\xi)\eta^\alpha/\alpha!$$

since negative powers of ε cannot occur. The degree is $\mu+j-m$ so $P_{m,\xi}$ is the principal part of the localization

$$P_\xi(\eta) = \lim_{\varepsilon \to 0} \varepsilon^{m-\mu} P(\xi/\varepsilon+\eta).$$

Now Theorem 10.2.12 shows that P_ξ has a fundamental solution E_ξ with $\operatorname{supp} E_\xi \times \{\xi\} \subset WF(E)$, hence $\operatorname{supp} E_\xi \subset \Gamma^\circ(P,N)$. By Theorems 12.5.2 and 12.5.1 it follows that P_ξ is hyperbolic with respect to N and that $\Gamma^\circ(P_{m,\xi}, N)$ is the closed convex cone with vertex at 0 generated by $\operatorname{supp} E_\xi$. This proves the theorem.

Theorem 12.6.2. *For the fundamental solution E of $P(D)$ with support in $\Gamma^\circ(P,N)$ we have*

$$WF_A(E) \subset \{(x,\xi);\ \xi \neq 0,\ x \in \Gamma^\circ(P_{m,\xi}, N)\}. \tag{12.6.1}$$

Note that $0 \in \Gamma^\circ(P_{m,\xi}, N)$ for every ξ so the right-hand side contains $(0,\xi)$ for every $\xi \neq 0$ as is inevitable since E is a fundamental solution (cf. Example 8.2.6). Theorem 12.6.2 follows from Theorems 8.7.5 and

8.4.18 if $P=P_m$. The general proof is similar to that of Theorem 8.3.7. We begin with a substitute for Lemma 8.3.6 providing a large zero free region.

Lemma 12.6.3. *Let $\xi_0 \in \mathbb{R}^n \setminus \{0\}$ and let Γ_1 be an open convex cone containing N such that $\bar{\Gamma}_1 \setminus \{0\} \subset \Gamma(P_{m,\xi_0}, N)$. Then one can find a conic neighborhood W of ξ_0 in $\mathbb{R}^n \setminus \{0\}$ and $\gamma > 0$ such that for large $|\xi|$*

$$|P(\xi + i\tau N + i\theta)| \geqq \gamma \tag{12.6.2}$$

if $\tau < \tau_0'$, $\xi \in W$, $\theta \in -\Gamma_1$, $|\theta| \leqq \gamma |\xi|$, $|\tau| \leqq \gamma |\xi|$.

Proof. As in the proof of Theorem 12.4.6 we introduce the homogenized polynomial

$$p(\xi, s) = s^m P(\xi/s)$$

where ξ is now taken near ξ_0 while $s>0$ is small. By (12.3.2) we have

$$p(\xi + i\tau N, s) \neq 0 \quad \text{if } \tau < \tau_0 s,\ s>0, \tag{12.6.3}$$

where we assume $\tau_0 < 0$. Let μ be the degree of P_{m,ξ_0}. It follows from Lemma 8.7.2 that

$$p(\xi_0 + iz\theta, 0) = (iz)^\mu P_{m,\xi_0}(\theta)(1+O(z)) \tag{12.6.4}$$

if $\theta \in -\bar{\Gamma}_1$, $|\theta|=1$.

As in the proof of Theorem 12.4.4 we now consider the equation

$$p(\xi + i\tau N + i\sigma\theta, s) = 0$$

where $\theta \in \bar{\Gamma}_1$ and $|\theta|=1$. We can choose $r>0$ and $\delta'>0$ so that there are precisely μ roots σ with $|\sigma|<r$ when $\max(|\tau|, |s|, |\xi-\xi_0|) \leqq \delta'$. If $s>0$ it follows from (12.6.3) that $\operatorname{Re}\sigma \neq 0$ provided that $\tau < \tau_0 s$. When $s=0$ and $\theta = N$ the roots are on the line $\operatorname{Re}\sigma = -\tau > 0$, so for reasons of continuity we obtain $\operatorname{Re}\sigma > 0$ for the μ roots if $\tau < \tau_0 s$, $s \geqq 0$. Hence we conclude as at the end of the proof of Lemma 8.7.4 that

$$|p(\xi + i\tau N + i\sigma\theta, s)| \geqq c|\sigma|^\mu$$

if $0 \geqq \sigma \geqq -r/2$, $\tau < \tau_0 s$, $\theta \in \Gamma_1$, $|\theta|=1$, $\max(|\tau|, |s|, |\xi - \xi_0|) \leqq \delta'$

With ξ/s, τ/s, σ/s as new variables we can write this result in the form

$$|P(\xi + i\tau N + i\sigma\theta)| \geqq c|\sigma|^\mu s^{\mu-m}$$

if $0 \geqq \sigma \geqq -r/2s$, $\tau < \tau_0$, $|s| < \delta'$, $|\tau| < \delta'/s$, $|\xi - \xi_0/s| < \delta'/s$, $\theta \in \Gamma_1$, $|\theta| = 1$.

If we choose $s = |\xi_0|/|\xi|$ and set $W = \{\xi;\ |\xi/|\xi| - \xi_0/|\xi_0|| < \delta'/|\xi_0|\}$ we obtain

$$|P(\xi + i\tau N + i\theta)| \geqq c_1 |\theta|^\mu |\xi|^{m-\mu}$$

if $\tau < \tau_0$, $\xi \in W$, $\theta \in -\Gamma_1$, $|\theta| < r|\xi|/2|\xi_0|$, $|\tau| < \delta'|\xi|/|\xi_0|$, $|\xi| > |\xi_0/\delta'|$.

If $\tau<\tau_0'=\tau_0-1$ we can replace τ by $\tau+1$ and θ by $\theta-N$. Since $|\theta-N|$ has a positive lower bound when $\theta\in-\Gamma_1$ we obtain (12.6.2). The lemma is proved.

Proof of Theorem 12.6.2. From (12.5.3) it follows that $E=(a-\Delta)^n E_1$,

$$E_1(x)=(2\pi)^{-n}\int e^{i\langle x,\xi+i\tau N\rangle}/(Q(\xi+i\tau N)P(\xi+i\tau N))\,d\xi \tag{12.6.5}$$

where $\tau<\tau_0$ is fixed, $Q(\xi)=(a+\langle\xi,\xi\rangle)^n$ and a is chosen so large that $Q(\xi+i\tau N)$ has no zero when ξ is real. The integrand is then $O(|\xi|^{-2n})$ at ∞ so the integral is absolutely convergent. As in the proof of Theorem 8.3.7 we shall deform the cycle of integration near ξ_0. Choose a non-negative C^∞ function χ which is homogeneous of degree 0 and equal to 1 in a neighborhood of ξ_0 while $\chi=0$ outside the neighborhood W in Lemma 12.6.3. With $\theta\in-\Gamma_1$, $|\theta|=\gamma$, and a large t we consider the chain

$$G_{\theta,t}\colon \mathbb{R}^n\ni\xi\to\xi+i\tau N+i\chi(\xi)\min(|\xi|,t)\theta,\qquad |\xi|>C,$$

where C is chosen so large that (12.6.2) is valid and there is a lower bound for Q also. By Stokes' formula we can shift the integration in (12.6.5) to $G_{\theta,t}$ for large $|\xi|$. If x is in the half space H_θ defined by $\langle x,\theta\rangle>0$ we may let $t\to\infty$ by dominated convergence and are left with integration over

$$G_\theta\colon \mathbb{R}^n\ni\xi\to\xi+i\tau N+i\chi(\xi)|\xi|\theta,\qquad |\xi|>C.$$

The integral when ξ is in a conic neighborhood of ξ_0 where $\chi(\xi)=1$ defines an analytic function E_0 in H_θ. Just as in the proof of Theorem 8.3.7 it follows in view of Theorem 8.4.8 that $WF_A(E_1-E_0)$ has no element with frequency ξ_0. Thus

$$\{x;\,(x,\xi_0)\in WF_A(E)\}\subset \complement H_\theta$$

for every $\theta\in-\Gamma(P_{m,\xi_0},N)$. The intersection of $\complement H_\theta$ for all such θ is by definition $\Gamma^\circ(P_{m,\xi_0},N)$ which completes the proof.

For a homogeneous hyperbolic polynomial P_m the Fourier transform of the fundamental solution E with support in $\Gamma^\circ(P_m,N)$ is $1/P_m(\xi-i0N)$, and that of $D^\alpha E$ is $\xi^\alpha/P_m(\xi-i0N)$, which is homogeneous of degree $|\alpha|-m$. To invert the Fourier transformation it is natural to exploit the homogeneity by integrating first with respect to the radial variable. To do so it is convenient to use (7.1.24) in order to avoid logarithmic terms, so we pause first to study this formula using the notion of wave front set.

Recall that u in (7.1.24) is a homogeneous distribution of degree $-n-k$ which has parity opposite to k. Hence $(x,\xi)\in WF(u)$ implies

$\langle x, \xi\rangle = 0$ (by Theorem 8.3.1) and $(-x, -\xi) \in WF(u)$ (by Theorem 8.2.4). The differential of $f(x, \xi) = \langle x, \xi\rangle$, $(x, \xi) \in \mathbb{R}^{2n} \setminus \{0\}$, is

$$f'(x, \xi)(y, \eta) = \langle y, \xi\rangle + \langle x, \eta\rangle = \langle (y, \eta), {}^t f'(x, \xi) 1\rangle$$

where ${}^t f'(x, \xi) 1 = (\xi, x)$. With σ_k defined by (7.1.23) we obtain

$$WF(\sigma_k(\langle x, \xi\rangle)) = \{(x, \xi; \eta, y);\ \langle x, \xi\rangle = 0,\ (\eta, y) = t(\xi, x) \text{ for some } t \neq 0\}, \tag{12.6.6}$$

by Theorem 8.2.4. (Equality must hold since σ_k is real and $0 \in \operatorname{sing\,supp} \sigma_k$.) In (12.6.6) η and y are vectors dual to x and ξ respectively. Since

$$WF(u(x) \otimes 1) = \{(x, \xi; \eta, 0);\ (x, \eta) \in WF(u)\} \tag{12.6.7}$$

by Theorem 8.2.9, the product $u(x)\sigma_k(\langle x, \xi\rangle)$ is defined when $x \neq 0$, for $y \neq 0$ in (12.6.6) then. Theorem 8.2.10 also gives

$$\begin{aligned} WF(u(x)\sigma_k(\langle x, \xi\rangle)) &\subset \{(x, \xi;\ t\xi + \eta, tx);\ \langle x, \xi\rangle = 0, t \neq 0, (x, \eta) \in WF(u)\} \\ &\cup WF(u \otimes 1) \cup WF(\sigma_k(\langle x, \xi\rangle)). \end{aligned} \tag{12.6.8}$$

Note that $(x, \xi; \eta, y) \in WF(u(x)\, \sigma_k(\langle x, \xi\rangle))$ implies $\langle x, \eta\rangle = 0$. The restriction to the unit sphere $|x| = 1$ is therefore well defined, and (7.1.24)′ is valid (cf. Theorem 5.2.1) in the sense of distribution theory,

$$\hat{u}(.) = \pi i^{-1-k} S_x(u(x)\sigma_k(\langle x, .\rangle)). \tag{12.6.9}$$

From Theorem 8.2.12 and (12.6.8) it follows that

$$WF(\hat{u}) \subset \{(\xi, tx);\ \xi \neq 0, tx \neq 0, (x, -t\xi) \in WF(u)\} \cup \{(0, x), x \neq 0\}.$$

Since $(x, \xi) \in WF(u) \Leftrightarrow (-x, -\xi) \in WF(u)$ this means that

$$WF(\hat{u}) \subset \{(\xi, x);\ (-x, \xi) \in WF(u)\} \cup T_0^*$$

as we already knew from Theorem 8.1.8.

Let us now examine if the argument in (12.6.9) can be fixed at a point $\xi \neq 0$. The restriction of $u(x)\sigma_k(\langle x, \xi\rangle)$ to $S^{n-1} \times \{\xi\}$ exists unless the wave front set contains an element of the form $(x, \xi; \eta, y)$ with $|x| = 1$ and η proportional to x. Since $\langle x, \eta\rangle = 0$ this means that η must be 0, and (12.6.8) shows that $(x, -t\xi) \in WF(u)$ then. If $\xi \neq 0$ and $(x, \xi) \notin WF(u)$ when $x \neq 0$ (and $\langle x, \xi\rangle = 0$) we conclude that $u(x)\sigma_k(\langle x, \xi\rangle)$ has a well defined restriction to $S^{n-1} \times \{\xi\}$. Hence (12.6.9) is valid pointwise in the open set

$$\{\xi \in \mathbb{R}^n;\ \xi \neq 0 \text{ and } (x, \xi) \notin WF(u) \text{ when } x \neq 0\} = (\mathbb{R}^n \setminus \{0\}) \setminus \operatorname{sing\,supp} \hat{u}.$$

All these facts are of course equally true with WF replaced by WF_A or WF_L.

We shall now return to the distribution $E_\alpha = D^\alpha E$ where E is the fundamental solution of $P_m(D)$ with support in $\Gamma^\circ(P_m, N)$. Then $\hat{E}_\alpha = \xi^\alpha / P_m(\xi - i0N)$ is homogeneous of degree $|\alpha| - m = -n - q$, where $q = m - n - |\alpha|$. The distribution

$$u(\xi) = \hat{E}_\alpha(\xi) - (-1)^q \hat{E}_\alpha(-\xi) = \xi^\alpha (1/P_m(\xi - i0N) - (-1)^n / P_m(\xi + i0N))$$

has the parity opposite to q. The support is in the zero set of P_m when n is even, and by Theorem 8.7.5 we always have

$$(\xi, x) \in WF(u) \Rightarrow x \in \Gamma^\circ(P_{m,\xi}, N) \cup \Gamma^\circ(P_{m,\xi}, -N).$$

The Fourier transform of u is $(2\pi)^n (E_\alpha(-x) - (-1)^q E_\alpha(x))$, so (12.6.9) gives in the distribution sense

$$\begin{aligned} (12.6.10) \qquad & E_\alpha(x) - (-1)^q E_\alpha(-x) \\ & = (2\pi)^{-n} \pi i^{-1-q} \int_{|\xi|=1} \sigma_q(-\langle x, \xi \rangle) \\ & \cdot \xi^\alpha (1/P_m(\xi - i0N) - (-1)^n / P_m(\xi + i0N)) \, d\omega(\xi) \end{aligned}$$

where $d\omega$ is the surface area on the unit sphere. When $x \neq 0$ the analytic singular support is contained in $W \cup (-W)$ where

$$(12.6.11) \qquad W = \bigcup_{\xi \neq 0} \Gamma^\circ(P_{m,\xi}, N)$$

is a closed cone because the set Γ° in Theorem 8.7.5 is closed. Since the second term on the left-hand side in (12.6.10) vanishes outside $-\Gamma^\circ(P_m, N)$ we obtain a formula for $E_\alpha(x)$ valid pointwise outside $W \cup \Gamma^\circ(P_m, -N)$.

(12.6.10) is in essence the Herglotz-Petrowsky formula. Usually it occurs in various equivalent forms. When n is even and P_m is of real principal type, the parenthesis on the right-hand side is a density on the surface $P_m = 0$, and if $q < 0$ then (12.6.10) can be written as an integral over its intersection with the unit sphere and the orthogonal plane of x. This is the original formula. One can also go in the opposite direction and rewrite (12.6.10) with contour integrals avoiding the singularities. Briefly this is done as follows.

The observations made after Definition 8.2.2 show that Theorem 8.1.6 can be strengthened to $f(. + iy) \to f_0$ in $\mathscr{D}'_{\Gamma^\circ}$ when $\Gamma \ni y \to 0$. Moreover, if the hypotheses of Theorem 8.1.6 are fulfilled and $x \to \theta(x) \in \Gamma$ is real analytic in some neighborhood of $\bar{X}$ then $f(x + i\varepsilon\theta(x)) \to f_0$ in $\mathscr{D}'_{\Gamma^\circ}$. Indeed, convergence in $\mathscr{D}'$ follows if we apply (3.1.20) to $f(z + i\varepsilon\theta(z))$, and if $\Gamma_1 \subset \Gamma \cup \{0\}$ then $f(z + i\varepsilon\theta(z))$ satisfies the hypotheses in Theorem 8.1.6 uniformly for small ε if Γ is replaced by Γ_1.

To be able to apply the preceding observation to (12.6.10) we observe that if $x\notin W\cup(-W)$ and $\xi\neq 0$ then $\langle x,\theta\rangle$ does not have a constant sign when $\theta\in\Gamma(P_{m,\xi},N)$. Hence we can choose $\theta\in\Gamma(P_{m,\xi},N)$ with $\langle x,\theta\rangle=0$. Since $\theta\in\Gamma(P_{m,\eta},N)$ for all η in a neighborhood of $\pm\xi$ (Lemma 8.7.4) and the cones $\Gamma(P_{m,\eta},N)$ are convex, we can use a partition of unity on S^{n-1} to combine finitely many choices of θ to a C^∞ function $\theta(\xi)$ in $\mathbb{R}^n\setminus 0$ with

$$\langle x,\theta(\xi)\rangle=0,\quad \theta(t\xi)=\theta(\xi),\quad t\neq 0,\quad \theta(\xi)\in\Gamma_{m,\xi}(N). \tag{12.6.12}$$

By regularization of θ restricted to $|\theta|<2$, for example by convolution with a Gaussian function, we can make θ real analytic.

Let

$$\omega(\zeta)=\zeta_1\,d\zeta_2\wedge\ldots\wedge d\zeta_n-\zeta_2\,d\zeta_1\wedge\ldots\wedge d\zeta_n+\ldots$$

be the Kronecker form. From (12.6.12), the discussion of boundary values of analytic functions above and the arguments which led to (12.6.10) we obtain if $x\notin W\cup\Gamma(P_m,-N)$

$$\text{(12.6.10)}'\quad E_\alpha(x)=\lim_{\varepsilon\to 0}(2\pi)^{-n}\pi i^{-1-q}\Big(\int_{\zeta=\xi-i\varepsilon\theta(\xi)}-(-1)^n\int_{\zeta=\xi+i\varepsilon\theta(\xi)}\Big)\cdot\sigma_q(-\langle x,\xi\rangle)\zeta^\alpha/P_m(\zeta)\,\omega(\zeta).$$

We could equally well have used the surface measure $d\omega(\xi)$ here but we shall now see that the Kronecker form makes the integral independent of ε so that no limit has to be taken.

a) If $q\geqq 0$ then

$$\sigma_q(-\langle x,\xi\rangle)=2^{-1}\operatorname{sgn}\langle -x,\xi\rangle\langle -x,\xi\rangle^q/q!.$$

Now $\langle x,\zeta\rangle^q\zeta^\alpha/P_m(\zeta)\omega(\zeta)$ is a closed form since the degree of homogeneity of the factor in front of $\omega(\zeta)$ is $-n$ (see the discussion following (3.2.25)). In (12.6.10)′ we have $\langle x,\xi\rangle=\langle x,\zeta\rangle$, and the form $\omega(\zeta)$ vanishes in the plane $\langle x,\zeta\rangle=0$. By Stokes' formula the integral is therefore independent of ε when ε is small enough. Let α_+ be the sum of the cycle

$$S^{n-1}\ni\xi\to\xi-i\varepsilon\theta(\xi) \tag{12.6.13}$$

and $(-1)^{n-1}$ times the cycle

$$\text{(12.6.13)}'\qquad S^{n-1}\ni\xi\to\xi+i\varepsilon\theta(\xi).$$

Then we have proved that for $q=m-n-|\alpha|\geqq 0$ and $x\notin W\cup\Gamma(P_m,-N)$

$$\text{(12.6.10)}''\quad D^\alpha E(x)=(2\pi)^{1-n}4^{-1}i\int_{\alpha_+}\langle ix,\zeta\rangle^q\operatorname{sgn}\langle x,\zeta\rangle\zeta^\alpha/P_m(\zeta)\omega(\zeta)/q!.$$

Note that $\langle x, \zeta \rangle$ is real on α_+ so $\operatorname{sgn}\langle x, \zeta \rangle$ is defined. We can omit this sign factor if we replace α_+ by the chain $\check{\alpha}_+$ which only differs by a change of orientation when $\langle x, \xi \rangle < 0$. Note that $\partial \check{\alpha}_+$ is then a cycle in the plane where $\langle x, \zeta \rangle = 0$.

b) If $q<0$ we have $\sigma_q(-\langle x, \xi \rangle) = \delta^{(k)}(-\langle x, \xi \rangle)$, $k=-1-q$. Assume for a moment that $x_1 \neq 0$ but $x_2 = \ldots = x_n = 0$. Then

$$\sigma_q(-\langle x, \xi \rangle) = (-x_1)^{-k} \partial_{\xi_1}^k \delta(-x_1 \xi_1) = (-x_1)^{-k} |x_1|^{-1} \delta^{(k)}(\xi_1).$$

In (12.6.10) we could have integrated over $\mathbb{R} \times S^{n-2}$ after replacing $d\omega(\xi)$ by the Kronecker form since the integrand is homogeneous of degree $-n$ and supported by the plane $\xi_1 = 0$. This means that we may integrate over $\xi = (\xi_1, \xi')$, $\xi' \in S^{n-2}$, in (12.6.10)'. Thus

$$E_\alpha(x) = \lim_{\varepsilon \to 0} (2\pi)^{-n} \pi i^{-1-q} |x_1|^{-1} \Big(\int_{\zeta = \xi - i\varepsilon\theta(\xi)} - (-1)^n \int_{\zeta = \xi + i\varepsilon\theta(\xi)} \Big)$$
$$\cdot (x_1^{-1} \partial/\partial \zeta_1)^k \zeta^\alpha / P_m(\zeta) (\zeta_2 d\zeta_3 \wedge \ldots \wedge d\zeta_n - \zeta_3 d\zeta_2 \wedge d\zeta_4 \ldots + \ldots)$$

where $\xi_1 = 0$ in the integral. (Note that $d\xi_2 \wedge \ldots \wedge d\xi_n = 0$.) By Stokes' formula the integral is independent of ε. The derivative can be interpreted as a residue which gives for $q = m-n-|\alpha| < 0$ and $x \notin W \cup \Gamma(P_m, -N)$

(12.6.10)''' $$D^\alpha E(x) = (2\pi)^{-n} (-1)^{q+1} 2^{-1} \int_{\alpha_-} \langle ix, \zeta \rangle^q \zeta^\alpha / P_m(\zeta) \omega(\zeta) (-q-1)!.$$

Here α_- is the sum of the cycle

(12.6.14) $$\{z \in \mathbb{C}, |z| = \delta\} \times \{\xi \in S^{n-1}; \langle x, \xi \rangle = 0\} \ni (z, \xi) \to zx + \xi - i\varepsilon\theta(\xi)$$

with ε sufficiently small and $\delta < \delta_\varepsilon$, and $(-1)^{n-1}$ times the cycle defined in the same way after a change of i to $-i$. The circle $|z| = \delta$ is given the positive orientation and the $n-2$ sphere is oriented as the boundary of the half sphere $\{\xi \in S^{n-1}; \langle x, \xi \rangle < 0\}$. This orientation takes care of the sign of x_1 introduced by the factor $|x_1|$ in the preceding formula, and the invariance of the formula makes it valid without any special assumption on the direction of x. Note that α_- is generated by a circle in the x direction with center at the boundary of $\check{\alpha}_+$ (or rather minus one half times the boundary if orientation and multiplicity are taken into account).

Summing up, we have proved

Theorem 12.6.4. *If P_m is hyperbolic with respect to N and E is the fundamental solution of $P_m(D)$ with support in $\Gamma(P_m, N)$, then* (12.6.10)'' *(resp.* (12.6.10)''') *is valid for* $x \notin W \cup \Gamma(P_m, -N)$ *and* $q = m - n - |\alpha| \geqq 0$ *(resp.* <0). *Here* W *is defined by* (12.6.11) *and* α_+, α_- *are defined as explained above.*

The following definition contains a sufficient condition for the vanishing of (12.6.10)'''

Definition 12.6.5. A point $x \notin W \cup \Gamma(P_m, -N)$ is said to satisfy the Petrowsky condition if the cycle

$$\{\xi \in S^{n-1};\ \langle x, \xi\rangle = 0\} \ni \xi \to \xi - i\varepsilon\theta(\xi)$$

is homologous to 0 in $\{\zeta \in \mathbb{C}^n;\ \langle x, \zeta\rangle = 0,\ P_m(\zeta) \neq 0\}$ when θ satisfies (12.6.12) and ε is sufficiently small.

It is clear that the definition does not depend on ε or the choice of x in a component of the complement of $W \cup \Gamma(P_m, -N)$.

Theorem 12.6.6. *Let the Petrowsky condition be satisfied in the component* V *of* $\complement(W \cup \Gamma(P_m, -N))$. *Then the fundamental solution of* $P_m(D)^k$ *with support in* $\Gamma(P_m, N)$ *coincides in* V *for every positive integer* k *with a homogeneous polynomial of degree* $mk - n$. *If* $P(D)$ *is hyperbolic with principal part* $P_m(D)$ *then the fundamental solution of* $P(D)^k$ *with support in* $\Gamma(P_m, N)$ *coincides in* V *with an entire function of exponential type.*

Proof. The cycle α_- is the same for P_m as for P_m^k. By assumption it is homologous to 0 in the set where $\langle x, \zeta\rangle \neq 0$, $P_m(\zeta) \neq 0$. If E_k is the fundamental solution of $P_m(D)^k$ with support in $\Gamma(P_m, N)$ it follows from (12.6.10)''' that $D^\alpha E_k(x) = 0$ in V when $|\alpha| > km - n$. By Taylor's formula and the analyticity and homogeneity of E_k in V it follows that E_k coincides in V with a homogeneous polynomial Q_k of degree $km - n$. By Theorem 12.5.3 the fundamental solution of $P(D) = P_m(D) - R(D)$ is

$$\sum_0^\infty R(D)^k E_{k+1}$$

in the sense of distribution theory. Since $R(D)$ is of degree $<m$ it is clear that $R(D)^k Q_{k+1}(z)$ is a polynomial in z containing only monomials z^α with $m + k - n \leqq |\alpha| \leqq m(k+1) - n$. Now we can estimate $D^\alpha E_k$, $|\alpha| = mk - n$, by applying (12.6.10)'' with $q = 0$ and P_m replaced by P_m^k. This gives

$$|D^\alpha Q_k| \leqq C^{k+1}, \qquad |\alpha| = mk - n.$$

The coefficient of $z^\alpha/\alpha!$ in $R(D)^k Q_{k+1}(z)$ is therefore at most C_1^k, hence

$$\sum_k |R(D)^k Q_{k+1}(z)| \leqq \sum |z^\alpha|/\alpha! \sum_{(|\alpha|+n)/m-1}^{|\alpha|+n-m} C_1^k.$$

The inner sum can be estimated by a constant times $(1+C_1)^{|\alpha|}$ which completes the proof.

Remark. In the second part it would have been sufficient to assume that E_k is a homogeneous polynomial in V for every k.

An example of the "Petrowsky lacunas" in Theorem 12.6.6 is the wave equation in an even number of dimensions (see (6.2.4)). The methods of algebraic geometry give a great deal of information on the converse of Theorem 12.6.6 but for that we must refer to the literature listed in the notes.

12.7. A Global Uniqueness Theorem

In this section we shall prove some rather precise results on the non-existence of solutions of the Cauchy problem in the non-hyperbolic case and on the support of the solution in the hyperbolic case.

Theorem 12.7.1. *Assume that no factor of the polynomial P has a principal part which is hyperbolic with respect to the vector $N \in \mathbb{R}^n$. If in a slab $X = \{x;\ a < \langle x, N\rangle < b\}$ we have a solution $u \in \mathscr{D}'(X)$ of the equation $P(D)u = 0$ such that* supp u *is bounded, it follows that $u = 0$ in X.*

Proof. It is sufficient to prove the theorem when P is *irreducible* since each irreducible factor of P may be considered separately. It is also sufficient to prove the theorem for solutions in C^∞. In fact, choose $\phi \in C_0^\infty(\mathbb{R}^n)$ such that $\int \phi\, dx = 1$ and $|\langle x, N\rangle| < 1$ if $x \in \operatorname{supp} \phi$. The convolution $u * \phi_\varepsilon$, where $\phi_\varepsilon(x) = \varepsilon^{-n}\phi(x/\varepsilon)$, is then defined and in C^∞ when $a + \varepsilon < \langle x, N\rangle < b - \varepsilon$, and $u * \phi_\varepsilon$ satisfies the hypotheses of the theorem there. If the theorem is proved for C^∞ functions it will thus follow that $u * \phi_\varepsilon = 0$, hence $u = \lim_{\varepsilon \to 0} u * \phi_\varepsilon = 0$ in X.

We can choose the coordinate system so that $X = \{x;\ -2 < x_n < 2\}$ and not all hyperplanes through the x_n axis are characteristic with respect to P. The partial Fourier-Laplace transform of u

$$\hat{u}_n(\zeta', x_n) = \int u(x) e^{-i\langle x', \zeta'\rangle}\, dx', \qquad -2 < x_n < 2,$$

where $x'=(x_1,\dots,x_{n-1})$ and $\zeta'=(\zeta_1,\dots,\zeta_{n-1})$, is then an entire analytic function of ζ' for fixed x_n. Since $P(D)u=0$ it follows that

(12.7.1) $$P(\zeta',D_n)\hat{u}_n(\zeta',x_n)=0, \quad -2<x_n<2.$$

If $P(\zeta',D_n)$ is independent of D_n we now conclude that $\hat{u}_n=0$, hence $u=0$ when $-2<x_n<2$. Thus we assume in what follows that the degree μ of $P(\zeta',D_n)$ with respect to D_n is positive. Since $\hat{u}_n$ satisfies the ordinary differential equation (12.7.1) it is a linear combination of exponential solutions. We shall isolate one of the components. To do so, we define a polynomial $R(\zeta',\sigma,\tau)$ of degree $m-1$ by the identity

(12.7.2) $$(\sigma-\tau)R(\zeta',\sigma,\tau)=P(\zeta',\sigma)-P(\zeta',\tau)$$

and set

(12.7.3) $$W(\zeta',\sigma,x_n)=R(\zeta',\sigma,D_n)\hat{u}_n(\zeta',x_n).$$

It follows from (12.7.1) and (12.7.2) that

$$(D_n-\sigma)W(\zeta',\sigma,x_n)=(P(\zeta',D_n)-P(\zeta',\sigma))\hat{u}_n=-P(\zeta',\sigma)\hat{u}_n.$$

Hence

(12.7.4) $$(D_n-\sigma)W(\zeta',\sigma,x_n)=0 \quad \text{if } P(\zeta',\sigma)=0.$$

Thus we have

(12.7.5) $$W(\zeta',\sigma,0)=e^{-i\sigma x_n}W(\zeta',\sigma,x_n); \qquad -2<x_n<2,\ P(\zeta',\sigma)=0.$$

Since R is a polynomial of degree $m-1$ and since $\hat{u}_n$ satisfies (7.3.3) as a function of ζ' for fixed x_n, we have

(12.7.6) $$|W(\zeta',\sigma,x_n)|\leqq C(1+|\sigma|)^{m-1}e^{M|\operatorname{Im}\zeta'|}$$

where M is an upper bound for $|x'|$ when $x\in\operatorname{supp}u$ (in X) and C is a constant which may depend on x_n. If we choose $x_n=\pm 1$ it thus follows that

(12.7.7) $$|W(\zeta',\sigma,0)|\leqq C(1+|\sigma|)^{m-1}\exp(M|\operatorname{Im}\zeta'|-|\operatorname{Im}\sigma|)$$

if $P(\zeta',\sigma)=0$.

Our aim now is to prove that $W(\zeta',\sigma,0)=0$ if $P(\zeta',\sigma)=0$. To do so we first observe that by Lemma A.1.2 in the appendix the irreducibility of P implies that

$$\Omega=\{\eta'\in\mathbb{C}^{n-1},\ P(\eta',\sigma) \text{ has } \mu \text{ different zeros}\}$$

is open and dense in $\mathbb{C}^{n-1}$, and it is easily seen that

$$\tilde{\Omega}=\{(\eta',\sigma);\ \eta'\in\Omega,\ P(\eta',\sigma)=0\}$$

is connected. That P_m is not hyperbolic with respect to $N=(0,1)$ means that either $P_m(0,1)=0$ or else the equation $P_m(\xi',\sigma)=0$ has a non-real root σ for some real ξ' and therefore a non-real root for every ξ' in an open subset of $\mathbb{R}^{n-1}$. In view of the homogeneity we obtain $P_m(\xi'/\sigma,1)=0$ and $1/\sigma$ is non-real. We have chosen the coordinates so that some plane containing the x_n axis is not characteristic, that is, $P_m(\xi',0)$ does not vanish identically. We can therefore choose $\xi'\in\mathbb{R}^{n-1}$ so that

$$P_m(\xi',0)\neq 0 \quad \text{but } P_m(c_0\xi',1)=0 \quad \text{either for } c_0=0 \text{ or for some } c_0 \text{ with } \operatorname{Im} c_0\neq 0. \tag{12.7.8}$$

With a fixed $\eta'\in\Omega$ we now consider the Riemann surface defined by

$$P(\tau\xi'+\eta',\sigma)=0. \tag{12.7.9}$$

As in the proof of Theorem 8.6.7 it follows from (12.7.8) that there exists a solution τ of (12.7.9) which for some integer p is an analytic function of $\sigma^{1/p}$ for large $|\sigma|$ and has an expansion

$$\tau(\sigma)=\sigma\sum_0^\infty c_j\sigma^{-j/p} \tag{12.7.10}$$

convergent when $|\sigma|^{1/p}\geqq R$, with leading term $c_0\sigma$ where c_0 is the constant in (12.7.8). The expansion (12.7.10) represents a part S of the Riemann surface (12.7.9).

Using (12.7.7) we shall now estimate $F(\sigma)=W(\tau(\sigma)\xi'+\eta',\sigma,0)$. Since $|\tau|/|\sigma|$ is bounded, we obtain for some constants A and B

$$|F(\sigma)|\leqq A e^{B|\sigma|}, \quad |\sigma|\geqq R^p. \tag{12.7.11}$$

A better estimate can be given on some rays. In fact, there exist arguments θ such that

$$M|\xi'||\operatorname{Im}(c_0 e^{i\theta})|<|\operatorname{Im} e^{i\theta}|/3=|\sin\theta|/3, \tag{12.7.12}$$

for since c_0 is either non-real or equal to 0, we can even make the left-hand side equal to 0 while the right hand side is $\neq 0$. The set of all θ satisfying (12.7.12) is open, and if θ satisfies (12.7.12) it is obvious that $\theta+k\pi$ also satisfies (12.7.12) for every integer k. When (12.7.12) is valid, we obtain using (12.7.7)

$$|F(re^{i\theta})|\leqq C_\theta e^{-r|\sin\theta|/3}, \quad r\geqq R^p. \tag{12.7.13}$$

Thus F is bounded on all rays with argument satisfying (12.7.12). Since S can be decomposed into angular domains with opening $<\pi$ by means of rays satisfying (12.7.12), it follows from (12.7.11) and the

Phragmén-Lindelöf theorem that F is bounded on S. Hence

$$F(\sigma)=\sum_0^\infty a_j \sigma^{-j/p}$$

where the series converges when $|\sigma|>R^p$. Now the function $F(\sigma)$ must behave asymptotically as a power of $\sigma^{-1/p}$ when $\sigma\to\infty$, if some $a_j\neq 0$, and this is impossible since (12.7.13) shows that F decreases exponentially along certain rays. Hence $F=0$ on S, thus $W(\tau\xi'+\eta',\sigma,0)=0$ in the connected component of S in the Riemann surface (12.7.9). It contains some point with $\tau=0$. In a neighborhood of $\eta'_0\in\Omega$ the zeros of $P(\eta',\sigma)=0$ are analytic functions $\sigma_j(\eta')$, $j=1,\dots,\mu$, and we have just seen that

$$\prod_1^\mu W(\eta',\sigma_j(\eta'),0)=0$$

for every η' near η'_0. But this proves that one of the factors vanishes identically, so $W(\eta',\sigma,0)=0$ in an open subset of the connected manifold $\tilde{\Omega}$. Hence this is true everywhere in $\tilde{\Omega}$. If $\eta'\in\Omega$ it follows from Lagrange's interpolation formula that the polynomials in σ

$$R(\eta',\sigma_j,\sigma)=P(\eta',\sigma)/(\sigma-\sigma_j), \qquad j=1,\dots,\mu,$$

form a basis for the polynomials in σ of degree $<\mu$. Since

$$R(\eta',\sigma_j(\eta'),D_n)\hat{u}_n(\eta',x_n)=W(\eta',\sigma_j(\eta'),x_n)=0,$$
$$|x_n|<2,\ \eta'\in\Omega,$$

by (12.7.3), it follows that $D_n^j\hat{u}_n(\eta',x_n)=0$ when $j<\mu$. In particular,

$$\hat{u}_n(\eta',x_n)=0 \qquad \text{if } |x_n|<2,\ \eta'\in\Omega.$$

Since Ω is dense it follows that $u=0$. The proof is complete.

We shall now give an application of Theorem 12.7.1.

Theorem 12.7.2. *Let X be an open subset of $\mathbb{R}^n$ and $u\in C^m(X)$ a solution of the equation $P(D)u=0$; let the plane $\Sigma=\{x;\ \langle x,N\rangle=0\}$ be non-characteristic with respect to the mth order operator $P(D)$, and assume that no irreducible factor of $P(D)$ has a principal part which is hyperbolic with respect to N. If the Cauchy data of u in $\Sigma\cap X$ have compact support, it follows then that $u=0$ in a neighborhood of $\Sigma\cap X$.*

Proof. Let $K'\subset X$ be a compact neighborhood in $\mathbb{R}^n$ of the support of the Cauchy data. Then it follows from Holmgren's uniqueness theorem (Theorem 8.6.5) that $u=0$ in a neighborhood of $\Sigma\cap\partial K'$. If we define $u_1=u$ in K' and $u_1=0$ elsewhere, it thus follows that $\operatorname{supp} u_1 \subset K'$ and that $u_1 \in C^m$, $P(D)u_1=0$ in a neighborhood of every point in Σ. If ε is a sufficiently small positive number, we have $P(D)u_1=0$ when $|\langle x, N\rangle|<\varepsilon$, so it follows from Theorem 12.7.1 that $u_1=0$ when $|\langle x, N\rangle|<\varepsilon$. This proves the theorem.

Remark. It follows from the proof that in the theorem above it would be sufficient to assume that the solution u exists in a neighborhood of $X \cap \Sigma$ in the half space $\langle x, N\rangle \geqq 0$.

The Cauchy data for which the Cauchy problem for a differential operator $P(D)$ satisfying the conditions in the theorem can be solved are thus *coherent* in the sense that they are uniquely determined in some sets if they are known in others. (There is of course no analytic relation which is satisfied by all "admissible" Cauchy data since the Cauchy problem can always be solved by Fourier transformation for data with Fourier transforms of compact support. These are dense in $\mathscr{S}$.)

We shall now prove that the conclusions of Theorem 12.7.2 are false if P has a factor with hyperbolic principal part. To do so we shall prove an existence theorem for the Cauchy problem with data in the (small) Gevrey class defined as follows:

Definition 12.7.3. If $\delta>0$ we denote by $\gamma^{(\delta)}$ or $\gamma^{(\delta)}(\mathbb{R}^n)$ the set of all $\phi \in C^\infty(\mathbb{R}^n)$ for which to every compact set K and every $\varepsilon>0$ there is a constant C_ε such that for every $\alpha \neq 0$.

$$|D^\alpha \phi(x)| \leqq C_\varepsilon \varepsilon^{|\alpha|}(|\alpha|!)^\delta, \quad x \in K. \tag{12.7.14}$$

We also set $\gamma_0^{(\delta)}=\gamma^{(\delta)} \cap C_0^\infty$.

By Stirling's formula we could of course replace $|\alpha|!$ by $|\alpha|^{|\alpha|}$ in the right-hand side. Since $\sum_1^\infty 1/k^\delta<\infty$ when $\delta>1$, it follows at once from Theorem 1.4.2 that $\gamma_0^{(\delta)}$ is so large when $\delta>1$ that one can find cutoff functions there; it is of course an algebra (Proposition 8.4.1).

The following lemma is a supplement to Theorem 7.3.1.

Lemma 12.7.4. *An entire function $\Phi(\zeta)$, $\zeta \in \mathbb{C}^n$, is the Fourier-Laplace transform of a function $\phi \in \gamma_0^{(\delta)}$ with support in the compact convex set K with supporting function H if and only if to every $B>0$ there exists a constant K_B such that*

$$|\Phi(\zeta)| \leqq K_B \exp(H(\operatorname{Im} \zeta)-B|\operatorname{Re} \zeta|^{1/\delta}), \quad \zeta \in \mathbb{C}^n. \tag{12.7.14$'$}$$

Proof. Since $\zeta^\alpha \Phi(\zeta)$ is the Fourier-Laplace transform of $D^\alpha \phi$ it follows from (7.3.1) and (12.7.14) with a new constant C_ε that

$$|\zeta|^k |\hat{\phi}(\zeta)| \leqq C_\varepsilon (n\varepsilon)^k k^{\delta k} e^{H(\operatorname{Im}\zeta)}.$$

This implies that

$$|\hat{\phi}(\zeta)| \leqq C_\varepsilon (n\varepsilon k^\delta / |\operatorname{Re}\zeta|)^k e^{H(\operatorname{Im}\zeta)}.$$

We now choose k as the largest integer $<(|\operatorname{Re}\zeta|/en\varepsilon)^{1/\delta}$ and obtain

$$|\hat{\phi}(\zeta)| \leqq C_\varepsilon e^{-k} e^{H(\operatorname{Im}\zeta)}.$$

Since $k > (|\operatorname{Re}\zeta|/en\varepsilon)^{1/\delta} - 1$, the estimate (12.7.14)′ follows with $B = (en\varepsilon)^{-1/\delta}$.

To prove the converse we note that by Theorem 7.3.1 every function satisfying (12.7.14)′ is the Fourier-Laplace transform of a function $\phi \in C_0^\infty$ with support in K. We have

$$|D^\alpha \phi| \leqq (2\pi)^{-n} \int |\xi^\alpha \Phi(\xi)| d\xi \leqq (2\pi)^{-n} K_B \int |\xi|^{|\alpha|} \exp(-B|\xi|^{1/\delta}) d\xi.$$

In polar coordinates the right-hand side becomes the integral defining the Γ-function and this proves (12.7.14) if $\varepsilon > B^{-\delta}$. The proof is complete.

We can now state an existence theorem which shows that Theorem 12.7.2 is not true for an operator with hyperbolic principal part. (Note that such an operator need not be hyperbolic in view of Theorem 12.4.6.)

Theorem 12.7.5. *Let the principal part of $P(D)$ be hyperbolic with respect to N and of order m. Then the Cauchy problem*

$$P(D)u = f \quad \text{in } \mathbb{R}^n, \tag{12.7.15}$$

$$\langle D, N\rangle^k u = \phi_k \quad \text{in } \Sigma, k = 0, \dots, m-1, \tag{12.7.16}$$

where $\Sigma = \{x \in \mathbb{R}^n, \langle x, N\rangle = 0\}$, has a solution $u \in \gamma^{(\delta)}(\mathbb{R}^n)$ for arbitrary $f \in \gamma^{(\delta)}(\mathbb{R}^n)$ and $\phi_k \in \gamma^{(\delta)}(\Sigma)$, provided that $1 < \delta \leqq m/(m-1)$.

In particular, this means that the admissible Cauchy data are not coherent if the principal part of $P(D)$ is hyperbolic. It is easy to deduce from the proof that $P(D)$ has a fundamental solution with support in $\Gamma^\circ(P_m, N)$ which belongs to the dual space of $\gamma_0^{(\delta)}$. In particular, there is a hyperfunction fundamental solution. Thus the conditions in Theorem 12.4.6 are caused by our insistence that the fundamental solution shall be in $\mathscr{D}'$.

We shall first prove the theorem when $f = 0$ and ϕ_k have compact support. If we choose the coordinate system so that $N = (0, \dots, 0, 1)$, it follows from Corollary 8.6.11 that a solution of (12.7.15), (12.7.16) must have compact support when x_n is bounded. As in Section 12.2 we can therefore study the Cauchy problem by taking the partial Fourier transform $\hat{u}_n(\xi', x_n)$ of u with respect to x'. Then we obtain a Cauchy problem for ordinary differential equations

$$P(\xi', D_n)\hat{u}_n(\xi', x_n) = 0; \tag{12.7.15$'$}$$

$$D_n^k \hat{u}_n(\xi', x_n) = \hat{\phi}_k(\xi'), \quad x_n = 0, \ k < m. \tag{12.7.16$'$}$$

To study this Cauchy problem we first estimate the characteristic roots.

Lemma 12.7.6. *If the principal part P_m of P is hyperbolic with respect to $N = (0, \dots, 0, 1)$, it follows that the roots of the equation $P(\xi', \tau) = 0$ satisfy the inequalities*

$$|\tau| \leqq C(|\xi'|+1), \quad |\operatorname{Im}\tau| \leqq C(|\xi'|+1)^{1-1/m}, \quad \xi' \in \mathbb{R}^{n-1}. \tag{12.7.17}$$

Proof. If we write

$$P(\xi', \tau) = \sum_0^m \tau^{m-j} a_j(\xi')$$

the polynomial a_j is of degree j at most and the constant a_0 is $\neq 0$ since $P_m(N) \neq 0$ and $N = (0, \dots, 0, 1)$. We may assume that $a_0 = 1$. If C_1 is a constant such that $|a_j(\xi')| \leqq (C_1(|\xi'|+1))^j$ for every j, we have

$$\left|\sum_1^m \tau^{m-j} a_j(\xi')\right| \leqq |\tau|^m \sum_1^m 2^{-j} < |\tau|^m \quad \text{if } |\tau| > 2C_1(|\xi'|+1).$$

This proves the first inequality in (12.7.17). (The assumption that ξ' is real was not used here.) To prove the second inequality we first note that for real ξ' we have

$$P_m(\xi', \tau) = \prod_1^m (\tau - \tau_j)$$

where the zeros τ_j are real. Hence (cf. (8.7.4))

$$|P_m(\xi', \tau)| \geqq |\operatorname{Im}\tau|^m, \quad \xi' \in \mathbb{R}^{n-1}.$$

Using the part of (12.7.17) which has already been proved, we obtain if $P(\xi', \tau) = 0$

$$|P_m(\xi', \tau)| = |P(\xi', \tau) - P_m(\xi', \tau)| \leqq C_2(|\xi'|+1)^{m-1}.$$

Hence it follows that $|\operatorname{Im}\tau|^m \leqq C_2(|\xi'|+1)^{m-1}$, which completes the proof.

To solve (12.7.15)$'$, (12.7.16)$'$ we must also recall an elementary fact concerning the case of an ordinary differential equation:

Lemma 12.7.7. *Let $p(D)=D^m+a_1D^{m-1}+\dots$ be an ordinary differential operator with constant coefficients. If $0\leqq k<m$ the Cauchy problem*

$$p(D)v_k=0 \quad \text{on } \mathbb{R}, \quad D^j v_k(0)=\delta_{jk} \quad \text{if } 0\leqq j<m, \tag{12.7.18}$$

where δ_{jk} is the Kronecker delta, has a unique solution. If $p(\tau)=0$ implies $|\tau|\leqq A$ and $|\operatorname{Im}\tau|\leqq B$, then

$$|D^j v_k(t)|\leqq 2^m(A+1)^{j+m-k}e^{(B+1)|t|}, \quad j=0,1,\dots \tag{12.7.19}$$

Proof. The uniqueness and existence are elementary. We claim that

$$v_k(t)=(2\pi i)^{-1}\int_{\partial\omega} e^{it\tau}q(\tau)/p(\tau)\,d\tau \tag{12.7.20}$$

where $\omega=\{\tau; |\tau|<A+1, |\operatorname{Im}\tau|<B+1\}$ contains the zeros of p and q is a polynomial of degree $<m$. That $p(D)v_k=0$ when v_k is defined by (12.7.20) follows at once from Cauchy's integral formula. The boundary conditions mean that the residue at ∞ of $\tau^j q(\tau)/p(\tau)$ shall be δ_{jk} when $j<m$, that is,

$$q(\tau)/p(\tau)=\tau^{-1-k}+O(\tau^{-m-1}), \quad \tau\to\infty.$$

This means that $q(\tau)$ is the polynomial part of $p(\tau)\tau^{-1-k}$. The coefficients of p are bounded by those of $(\tau+A)^m$ so those of q are bounded by those of the polynomial part of $(\tau+A)^m\tau^{-1-k}$. Thus

$$|q(\tau)|\leqq(2A+1)^m(A+1)^{-1-k}\leqq 2^m(A+1)^{m-1-k}, \quad \tau\in\partial\omega,$$

and $|p(\tau)|\geqq 1$ on $\partial\omega$ since the distance to the zeros of p is at least 1. Since the length of $\partial\omega$ is $\leqq 2\pi(A+1)$ and $|\tau|\leqq A+1$ on $\partial\omega$, we obtain (12.7.19) by differentiating (12.7.20) and an obvious estimate of the integral.

Proof of Theorem 12.7.5. First assume that $f=0$ and that all $\phi_k\in\gamma_0^{(\delta)}$. Let $F_k(\xi',x_n)$ be the solution of the Cauchy problem

$$P(\xi',D_n)F_k(\xi',x_n)=0; \quad D_n^jF_k(\xi',0)=\delta_{jk}, \ j<m,$$

where $k<m$. Then the solution of (12.7.15)′, (12.7.16)′ is

$$\hat u_n(\xi',x_n)=\sum_0^{m-1}\hat\phi_k(\xi')F_k(\xi',x_n). \tag{12.7.21}$$

We use Lemma 12.7.4 to estimate $\hat\phi_k(\xi')$ and Lemmas 12.7.6, 12.7.7 to estimate the derivatives of F_k. It follows that the function $\hat u_n$ defined by (12.7.21) has the estimate

$$|D_n^j\hat u_n(\xi',x_n)|\leqq K_BK^{j+1}(|\xi'|+1)^{j+m}$$
$$\cdot\exp((K|x_n|-B)(1+|\xi'|)^{1/\delta})$$

where B is arbitrary. (Here we have used that $1/\delta \geqq 1-1/m$ by hypothesis.) If we now define u as the inverse partial Fourier transform of $\hat{u}_n$ then

$$|D^\alpha u(x)| \leqq K_B(2\pi)^{-n} K^{|\alpha|+1} \int (|\xi'|+1)^{|\alpha|+m} \cdot \exp((K|x_n|-B)|\xi'|^{1/\delta}) d\xi'.$$

Since B can be chosen arbitrarily large, it follows as in the second half of the proof of Lemma 12.7.4 that $u \in \gamma^{(\delta)}$, and it is clear that u satisfies (12.7.15), with $f=0$, and (12.7.16). Thus the theorem is proved in this special case.

Next let $f=0$ and $\phi_k \in \gamma^{(\delta)}$. Choose $\psi_\nu \in \gamma_0^{(\delta)}$ so that $\psi_\nu(x)=1$ when $|x|<\nu$. Since $\psi_\nu \phi_k \in \gamma_0^{(\delta)}$, we have proved that the equation $P(D)u=0$ has a solution u_ν with the Cauchy data $\psi_\nu \phi_k$. From Corollary 8.6.11 it follows that $u_{\nu+1}=u_\nu$ for large ν on any compact set. Hence $u = \lim_{\nu\to\infty} u_\nu$ exists and belongs to $\gamma^{(\delta)}$. It is obvious that $P(D)u=0$ and that u has the Cauchy data (12.7.16).

Now assume that $f \in \gamma_0^{(\delta)}$. If E is a fundamental solution of $P(D)$, we have $v=E*f \in \gamma^{(\delta)}$. In fact, if K is a compact set in $\mathbb{R}^n$ and if $K' = K - \operatorname{supp} f = \{x-y; x \in K, y \in \operatorname{supp} f\}$, we can find C and k such that

$$|E(\chi)| \leqq C \sum_{|\beta| \leqq k} \sup |D^\beta \chi|, \quad \chi \in C_0^\infty(K'),$$

which gives

$$|D^\alpha(E*f)(x)| \leqq C \sum_{|\beta| \leqq k} \sup |D^{\alpha+\beta} f|, \quad x \in K.$$

With the notation $u_1 = u-v$ the Cauchy problem (12.7.15), (12.7.16) is now equivalent to

$$P(D)u_1=0, \quad \langle D, N\rangle^k u_1 = \phi_k - \langle D, N\rangle^k v \quad \text{in } \Sigma,\ k<m,$$

and we have just proved that this problem has a solution $u_1 \in \gamma^{(\delta)}$.

Finally, the existence of a solution for any f and ϕ_k in $\gamma^{(\delta)}$ follows if we replace f by $\psi_\nu f$ and let $\nu \to \infty$. The proof is complete.

Remark. The existence theory of Section 12.5 could also have been established with the methods used to prove Theorem 12.7.5.

Example. The Cauchy problem for the heat equation $\partial^2 u/\partial x^2 = \partial u/\partial t$ can be solved for arbitrary data in $\gamma^{(2)}$ on the t-axis. This case is classical and elementary, for the equation gives

$$\partial^{2k} u/\partial x^{2k} = \partial^k u/\partial t^k, \quad \partial^{2k-1} u/\partial x^{2k+1} = \partial^{k+1} u/\partial x \partial t^k.$$

The Taylor expansion with respect to x

$$u(x,t)=\sum_0^\infty \partial^k u(0,t)/\partial t^k x^{2k}/(2k)!$$
$$+\sum_0^\infty \partial^{k+1} u(0,t)/\partial t^k \partial x x^{2k+1}/(2k+1)!$$

is convergent since the data are in $\gamma^{(2)}$, and it solves the Cauchy problem.

We shall now give an application of Theorem 12.7.1 to the Cauchy problem with data of compact support.

Theorem 12.7.8. *Let P be an irreducible polynomial with principal part P_m which is hyperbolic with respect to N. Let K be a convex compact subset of the plane $\langle x, N\rangle = 0$ and y a point in this plane outside K. If $u \in C^m(H)$ where $H=\{x; \langle x, N\rangle \geqq 0\}$, is a solution of the equation $P(D)u=0$ with Cauchy data vanishing outside K, and if the support of u is contained in $\{y\}+\Gamma^\circ(P_m, N)$ except for a bounded set, it then follows that $u=0$ identically in H.*

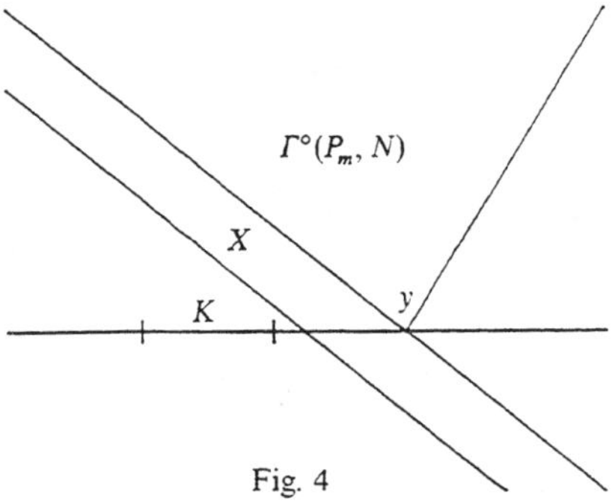

Fig. 4

Note that it follows from Theorem 12.5.6 that $\operatorname{supp} u \subset K + \Gamma^\circ(P_m, N)$. Combined with Theorem 12.7.8 this gives very precise information about the support of u.

Proof. Since K is convex and $y \notin K$, we can find $\xi \in \mathbb{R}^n$ such that

$$\langle x, \xi\rangle < \langle y, \xi\rangle - \varepsilon \quad \text{if } x \in K, \tag{12.7.22}$$

where $\varepsilon > 0$. Let τ_0 be the largest zero of the equation $P_m(\xi+\tau N)=0$, and set $\eta=\xi+\tau_0 N$. Then $P_m(\eta)=0$ and η is on the boundary of $\Gamma(P_m, N)$. Since $\langle x, N\rangle = \langle y, N\rangle = 0$ when $x \in K$, it follows from (12.7.22) that

$$\langle x, \eta\rangle < \langle y, \eta\rangle - \varepsilon \quad \text{if } x \in K. \tag{12.7.23}$$

Now set
$$X=\{x;\ \langle y,\eta\rangle-\varepsilon<\langle x,\eta\rangle<\langle y,\eta\rangle\}$$
and let v be the function which is equal to u in $X\cap H$ and is equal to 0 in $X\cap\complement H$. Since $X\cap K$ is empty, it is clear that $v\in C^m(X)$ and that $P(D)v=0$ in X. Now X does not meet the cone $\{y\}+\Gamma^\circ(P_m,N)$ for since η is in the closure of $\Gamma(P_m,N)$ we have $\langle x-y,\eta\rangle\geqq 0$ in this cone. Hence the support of v in X is bounded by hypothesis, so that $v=0$ identically in virtue of Theorem 12.7.1 which is applicable since $P_m(\eta)=0$.

Now consider the function u_1 which is equal to u when $x\in H$ and $\langle x,\eta\rangle\geqq\langle y,\eta\rangle$ and which is equal to 0 elsewhere in H. Since $u=v=0$ in $X\cap H$, it follows that $u_1\in C^m(H)$ and that $P(D)u_1=0$ in H. The Cauchy data of u_1 vanish, so we must have $u_1=0$. Since $u_1=u$ in $\{y\}+\Gamma^\circ(P_m,N)$ it follows that $u=0$ in $\{y\}+\Gamma^\circ(P_m,N)$. Hence the hypotheses in the theorem show that the support of u in H is bounded, and another application of Holmgren's uniqueness theorem now proves that $u=0$ identically in H. The proof is complete.

12.8. The Characteristic Cauchy Problem

Let $P(D)$ be a partial differential operator with constant coefficients in $\mathbb{R}^n$, let N be a real vector $\neq 0$ and set
$$H=\{x;\ \langle x,N\rangle\geqq 0\}.$$
We shall study the Cauchy problem for $P(D)$ in the half space H with vanishing data on the boundary ∂H when this is characteristic with respect to $P(D)$. Thus we want to find a solution of the equation

$$P(D)u=f \quad \text{in } \mathbb{R}^n \tag{12.8.1}$$

with support in H when f is given with support in H. By Theorem 8.6.7 we know that there is no uniqueness unless growth conditions are imposed which will not be done here. The question is whether an existence theorem is valid. The methods employed in Sections 10.5 and 10.6 can be adapted to the proof of an existence theorem for the Cauchy problem if there is a solution with $f=\delta$, that is, a fundamental solution (of finite order) with support in H. A sufficient condition for this is the *Petrowsky condition* that the coefficient of the highest power of τ in $P(\xi+\tau N)$ is independent of ξ and that, for some C

$$\operatorname{Im}\tau>C \quad \text{if } P(\xi+\tau N)=0,\ \xi\in\mathbb{R}^n. \tag{12.8.2}$$

Indeed, we can then define a fundamental solution E by (12.5.3) and use part of the proof of Theorem 12.5.1 to show that $\operatorname{supp}E\subset H$.

However, (12.8.2) is not a necessary condition for P to have a fundamental solution with support in H. In fact, if E is a fundamental solution of $P(D)$ then $Ee^{i\langle x,\theta\rangle}$ is a fundamental solution of $P(D-\theta)$ for every $\theta \in \mathbb{C}^n$. Hence the set of all $P(D)$ having a fundamental solution (of finite order) with support in H is invariant under translations. On the other hand, condition (12.8.2) is not translation invariant since it is satisfied by the Schrödinger operator $P(\xi)=\xi_2-\xi_1^2$, $N=(0,1)$ but not by $P(\xi+\theta)$ if $\theta=(i,0)$. However, condition (iii) in the following main result of this section shows that the lack of translation invariance is the only essential flaw of condition (12.8.2).

Theorem 12.8.1. *The following conditions on $P(D)$ and H are equivalent:*

(i) *$P(D)$ has a fundamental solution with support in H.*

(ii) *There exists a compact neighborhood K of 0 such that for some constants C and μ*

$$|u(0)| \leqq C \sum_{|\alpha|\leqq\mu} \sup_{-H} |D^\alpha P(D)u|, \quad u \in C_0^\infty(K). \tag{12.8.3}$$

(iii) *There exist constants A_1 and A_2 with $A_1>0$ such that for every solution $\tau(\zeta)$ of $P(\zeta+\tau N)=0$ which is analytic and single valued in a ball B with real center and radius A_1 we have*

$$\sup_B \operatorname{Im} \tau(\zeta) \geqq A_2. \tag{12.8.4}$$

(iv) *If $1\leqq p<\infty$ and $k\in\mathscr{K}$, the equation (12.8.1) has a solution $u\in B^{\text{loc}}_{p,\tilde{P}k}(\mathbb{R}^n)$ with support in H for every $f\in B^{\text{loc}}_{p,k}(\mathbb{R}^n)$ with support in H.*

(v) *The equation $P(D)u=f$ has a solution $u\in\mathscr{D}'_F(\mathbb{R}^n)$ with support in H for every $f\in\mathscr{D}'_F(\mathbb{R}^n)$ with support in H.*

The definitions used here are explained in Section 10.1.

Proof. The implications (iv) ⇒ (v) ⇒ (i) are obvious. If (i) is valid and $\chi\in C_0^\infty$ is equal to 1 in K, then

$$u(0)=\langle \check{E}, P(D)u\rangle = \langle \chi\check{E}, P(D)u\rangle, \quad u\in C_0^\infty(K).$$

Hence (12.8.3) follows from Theorem 2.3.10 since the intersection of $-H$ and a ball containing $\operatorname{supp}\chi$ is convex.

If $\phi\in C_0^\infty(K)$ and $\phi=1$ in a neighborhood of 0 then $K'=(-H)\cap\operatorname{supp} d\phi$ is a compact set $\subset(-H)\cap K\setminus\{0\}$, and (12.8.3) implies that with new constants

$$|u(0)| \leqq C \sum_{|\alpha|\leqq\mu} \sup_{K'} |D^\alpha u|, \quad \text{if } u\in C^\infty(\mathbb{R}^n),\ P(D)u=0. \tag{12.8.3)'}$$

In fact, we need only apply (12.8.3) to $v=\phi u$ noting that

$$P(D)v = \sum_{\alpha\neq 0} (P^{(\alpha)}(D)u) D^\alpha\phi/\alpha!$$

has support in K'. Condition (iii) will follow by application of (12.8.3)′ to superpositions of exponential solutions of the equation $P(D)u=0$. However, a few preliminary remarks are required before the proof.

First, it is sufficient to prove that (ii) ⇒ (iii) when P is irreducible. In fact, if $P=\prod P_j$ is a decomposition of P as a product of irreducible factors, it is obvious that the function τ in condition (iii) satisfies one of the equations $P_j(\zeta+\tau N)=0$, and (12.8.3)′ remains valid with $P(D)$ replaced by $P_j(D)$ since $P_j(D)u=0$ implies $P(D)u=0$. Choose the coordinates so that H is defined by the inequality $x_n \geqq 0$ and write

$$P(\xi)=\sum_0^m a_j(\xi')\xi_n^j$$

where $\xi'=(\xi_1,\dots,\xi_{n-1})$ and $a_m(\xi')$ is not identically 0. (Since ∂H may be characteristic the order of P may exceed m.) We may assume that $m>0$ for (iii) is void if $m=0$. Let $\Delta(\xi')$ be the discriminant of P as a polynomial in ξ_n, that is,

$$\Delta(\xi')=a_m^{2m-2}\prod_{i<k}(\tau_i-\tau_k)^2$$

where $\tau_1,\dots,\tau_m$ are the zeros of $P(\xi',\tau)$. Since P is irreducible we know that Δ is not identically 0, for P and $\partial P/\partial \xi_n$ would have a common factor then. Over the set in $\mathbb{C}^{n-1}$ where $a_m(\zeta')\Delta(\zeta')\neq 0$ the equation $P(\zeta)=0$ defines an m valued analytic function. More precisely, if B is a ball contained in the set

$$\{\zeta';\ \zeta'\in\mathbb{C}^{n-1},\ a_m(\zeta')\Delta(\zeta')\neq 0\}$$

there exist m analytic function $\tau_j(\zeta')$, $j=1,\dots,m$, defined in B such that $\tau_j(\zeta')\neq\tau_k(\zeta')$ for every $\zeta'\in B$ if $j\neq k$, and

$$P(\zeta)=a_m(\zeta')\prod_1^m(\zeta_n-\tau_j(\zeta')),\qquad \zeta'\in B.$$

The functions $\tau_j(\zeta')$ are uniquely determined apart from the labelling. This follows from the implicit function theorem (Lemma A.1.1) and analytic continuation along the radii from the center. The following lemma will permit us to find balls avoiding the zeros of a_m and of Δ.

Lemma 12.8.2. *There is a constant $\gamma=\gamma_{M,\nu}$ such that if $R\not\equiv 0$ is a polynomial in $\mathbb{C}^\nu$ of degree $\leqq M$ and $B\subset\mathbb{C}^\nu$ is a ball with radius t, it follows that there is a ball $B'\subset B$ with radius γt where R has no zeros. If B has real center we can take B' with real center.*

Proof. We may assume that B is the unit ball and that $\tilde{R}(0)=1$. Since the set of such R is compact and the largest radius of an open ball $B'\subset B$ where $R\neq 0$ is a positive lower semicontinuous function of R, hence has a positive lower bound, the lemma is proved.

Next we observe that if P is of degree $\leqq M$, then (cf. Lemma 12.7.6)

$$|a_m(\zeta')\zeta_n| \leqq C(1+|\zeta'|)^{M-m+1}, \quad \text{if } P(\zeta)=0. \tag{12.8.5}$$

In fact, if $w=a_m(\zeta')\zeta_n$ we have the equation

$$w^m + a_{m-1}w^{m-1} + a_{m-2}a_m w^{m-2} + \ldots + a_0 a_m^{m-1} = 0.$$

Since the degree of a_j is at most $M-j$, the degree of the coefficient of w^{m-k} is at most $M-(m-k)+(k-1)(M-m)=k(M-m+1)$. It follows that $w(1+|\zeta'|)^{-M+m-1}$ satisfies an equation with leading coefficient 1 and uniformly bounded coefficients. This implies (12.8.5).

Proof that (ii) $\Rightarrow$ (iii). Let U be the set of all (ξ_0', R, A) such that $\xi_0' \in \mathbb{R}^{n-1}$, $R>0$, $A>0$, and

(a) $a_m(\zeta')\Delta(\zeta') \neq 0$ when $|\zeta'-\xi_0'|<R+1$,

(b) $\operatorname{Im}\tau(\zeta') \leqq -A$ when $|\zeta'-\xi_0'|<R$ for one of the analytic solutions of $P(\zeta',\tau)=0$ defined then.

We shall first prove that there is a bound for $\min(A,R)$ when $(\xi_0', R, A) \in U$ and then use Lemma 12.8.2 to deduce (iii). By Lemma 11.1.4 there is a fixed positive lower bound for $|a_m(\zeta')|$ when $|\zeta'-\xi_0'|<R$, for some derivative of a_m is a constant. With $k=M-m+1$ we therefore obtain from (12.8.5)

$$|\tau(\zeta')| \leqq C(1+|\zeta'|)^k, \quad |\zeta'-\xi_0'|<R. \tag{12.8.5$'$}$$

Choose a fixed $\phi \in C_0^\infty(\{\xi;\, |\xi|<1/2\})$ with $\int \phi(\xi)\,d\xi = 1$ and set

$$u(x) = \int \exp i(\langle x', \xi'\rangle + x_n\tau(\xi'))\phi((\xi'-\xi_0')/R)\,d\xi'/R^n. \tag{12.8.6}$$

It is clear that $P(D)u=0$ and that $u(0)=1$. We shall apply (12.8.3)$'$ to u. To estimate the right-hand side we observe that $x_n \leqq 0$ in K' and that there is some $\delta>0$ such that either $|x'| \geqq \delta$ or $x_n \leqq -\delta$ if $x \in K'$.

1° Let $x_n \leqq 0$ and $|x'| \geqq \delta$. When $|\zeta'-\xi_0'|<R$ we have

$$\operatorname{Im}\tau(\zeta') \leqq 0, \quad |\tau(\zeta')| \leqq C(1+|\xi_0'|+R)^k.$$

Now

$$x'^\beta D^\alpha u(x) = \int e^{i\langle x',\xi'\rangle}(-D)_{\xi'}^\beta(\xi'^{\alpha'}\tau(\xi')^{\alpha_n}e^{ix_n\tau(\xi')}\phi((\xi'-\xi_0')/R))\,d\xi'/R^n,$$

and by Cauchy's inequality

$$|D_{\xi'}^\gamma(\xi'^{\alpha'}\tau(\xi')^{\alpha_n}e^{ix_n\tau(\xi')})| \leqq |\gamma|!\,C^{|\alpha|}(1+|\xi_0'|+R)^{k|\alpha|}(R/2)^{-|\gamma|},$$

if $|\xi'-\xi_0'|<R/2$. Hence we obtain for any N

$$|D^\alpha u(x)| \leqq C_N(1+|\xi_0'|+R)^{k\mu}R^{-N} \quad \text{if } |\alpha| \leqq \mu,\ |x'| \geqq \delta,\ x_n \leqq 0. \tag{12.8.7}$$

2° Next assume that $x_n \leqq -\delta < 0$. Then $|\exp(ix_n \tau(\xi'))| \leqq e^{-A\delta}$, and taking $\beta = 0$ in the preceding argument we obtain

$$|D^\alpha u(x)| \leqq C(1+|\xi'_0|+R)^{k\mu} e^{-A\delta} \quad \text{if } |\alpha| \leqq \mu,\ x_n \leqq -\delta. \tag{12.8.7$'$}$$

When (12.8.3)′ is combined with (12.8.7), (12.8.7)′ we obtain

$$1 \leqq C_N(1+|\xi'_0|+R)^{k\mu}(R^{-N}+e^{-A\delta}) \quad \text{if } (\xi'_0, R, A) \in U. \tag{12.8.8}$$

The function

$$f(t) = \sup\{s;\ (\xi'_0, s, s) \in U \text{ for some } \xi'_0 \in \mathbb{R}^{n-1} \text{ with } |\xi'_0| \leqq t\}$$

is finite. It follows from Theorems A.2.9 and A.2.2 that U is semi-algebraic, for b) in the definition of U means that one can choose τ_0 with $P(\xi'_0, \tau_0) = 0$ so that $\operatorname{Im}\tau \leqq -A$ if $(\tau, \zeta', \tau_0, \xi'_0)$ is in the set E in Theorem A.2.9 and $|\zeta' - \xi'_0| < R$. Hence Corollary A.2.6 shows that f is bounded unless f grows as a power of t when $t \to \infty$. By (12.8.8) we have

$$1 \leqq C_N(1+t+f(t))^k (f(t)^{-N} + e^{-\delta f(t)}) \tag{12.8.8$'$}$$

so f cannot grow as a power of t. Thus f is bounded, that is, for some s_0

$$(\xi'_0, s, s) \in U \Rightarrow s \leqq s_0.$$

Now if $\tau(\zeta')$ is a solution of $P(\zeta', \tau(\zeta')) = 0$ with $\operatorname{Im}\tau(\zeta') \leqq A_2 < 0$ in a ball with real center and radius A_1, it follows from Lemma 12.8.2 that we can find ξ'_0 so that $(\xi'_0, R, -A_2) \in U$, $R+1 = \gamma A_1$. If $\gamma A_1 = s_0 + 2$ it follows that $-A_2 \leqq s_0$, so (iii) is fulfilled if we choose $A_2 < -s_0$. This completes the proof of the implication (ii) ⇒ (iii).

The remaining implication (iii) ⇒ (iv) is the hardest part of Theorem 12.8.1. Again we may assume that P is irreducible. In fact, if $P = \prod P_j$ is a decomposition of P in irreducible factors, the condition (iii) is fulfilled by each P_j if it is fulfilled by P. In view of (10.4.3) the implication (iii) ⇒ (iv) follows for P if it is known for each P_j. We may of course choose coordinates as above. Condition (iii) then gives that for an analytic solution $\tau(\zeta')$ of the equation $P(\zeta', \tau(\zeta')) = 0$ in a ball B with radius A_1 and real center we have

$$\sup_B \operatorname{Im}\tau(\zeta') \geqq A_2 - A_1.$$

Then

$$P_1(\zeta) = P(\zeta', \zeta_n + i(A_2 - A_1) - i)$$

satisfies the same condition with $A_2 - A_1$ replaced by 1, and

$$P_1(D) = e^{-(1+A_1-A_2)x_n} P(D) e^{(1+A_1-A_2)x_n}$$

so (iv) is valid for $P(D)$ and for $P_1(D)$ at the same time. From now on we therefore assume condition (iii) in the form

$$\sup_B \operatorname{Im} \tau(\zeta') \geqq 1 \tag{12.8.9}$$

when $P(\zeta', \tau(\zeta'))=0$ and τ is analytic in a ball with radius A_1 and real center. For the sake of convenience we assume that $A_1 \geqq 1$.

To clarify what must be done we note that the equation $P(D)u=f$ means that

$$\check{u}(P(D)v)=\check{f}(v), \quad v \in C_0^\infty(\mathbb{R}^n).$$

This implies that $\check{f}(v)$ can be estimated in terms of the restriction of $P(D)v$ to the half space $-H$. Conversely, if this is proved we can obtain u by applying the Hahn-Banach theorem. Thus we must show that when $v \in C_0^\infty(\mathbb{R}^n)$ it is possible to estimate the restriction of v to $-H$ in terms of the restriction of $P(D)v$ to that half space. Writting $P(D)v=g$ we obtain by taking Fourier transforms with respect to the variables $x_1, \ldots, x_{n-1}$

$$P(\zeta', D_n)\hat{v}(\zeta', x_n)=\hat{g}(\zeta', x_n); \quad \zeta' \in \mathbb{C}^{n-1}, \ x_n \in \mathbb{R}.$$

There are two difficulties in solving this ordinary differential equation:

a) The leading coefficient $a_m(\xi')$ may have zeros.

b) The equation $P(\xi', \tau)=0$ may for large $\xi' \in \mathbb{R}^{n-1}$ have roots with large negative imaginary part, and (12.8.9) only guarantees that *one* can be moved to the upper half plane by a moderate change of ξ'.

The first problem is not hard to solve for already in Section 7.3 we have seen how to avoid the zeros of a polynomial. Problem b) is much more serious. It requires us to remove only one linear factor $D_n - \tau(\zeta')$ at a time with ζ' where $\operatorname{Im} \tau(\zeta') > 1$, and then use the maximum principle to derive conclusions for a general ζ'. It is this procedure which, so far at any rate, has made it impossible to use constructive arguments such as in Section 10.2.

As a preliminary step we shall prove some elementary facts concerning integration of ordinary differential equations. Let $k \in \mathscr{K}(\mathbb{R})$ so that for some constants C and N

$$k(\xi+\eta) \leqq k(\xi)(1+C|\eta|)^N; \quad \xi, \eta \in \mathbb{R}. \tag{12.8.10}$$

With $1 \leqq p \leqq \infty$ we shall use the norm

$$\|u\|_{p,k}=(2\pi)^{-1/p}\|k\hat{u}\|_{L^p}, \quad u \in \mathscr{S}(\mathbb{R}).$$

If $f \in \mathscr{S}$ and q is a polynomial in one variable with no real zero, the equation $q(D)u=f$ has a unique solution in $\mathscr{S}$, and

$$\|u\|_{p,|q|k}=\|q(D)u\|_{p,k}. \tag{12.8.11}$$

Now we are really only interested in restrictions to $\mathbb{R}_-$, so we introduce the quotient norms defined by

(12.8.12) $$\|u\|_{p,k}^- = \inf_v \|v\|_{p,k}; \quad v=u \text{ on } \mathbb{R}_-;\ v,u\in\mathscr{S}.$$

In order that $u=v$ on $\mathbb{R}_-$ shall be equivalent to $q(D)u=q(D)v$ on $\mathbb{R}_-$ for $u,v\in\mathscr{S}$, we require that any non-trivial solution of the homogeneous equation $q(D)w=0$ is exponentially increasing at $-\infty$, that is, that $q(\tau)=0$ implies $\operatorname{Im}\tau>0$. In that case we can pass to quotient norms in (12.8.11) and obtain

Lemma 12.8.3. *If $u\in\mathscr{S}(\mathbb{R})$ and $q(\tau)$ is a polynomial with all zeros in the half plane $\operatorname{Im}\tau>0$, we have*

(12.8.13) $$\|u\|_{p,|q|k}^- = \|q(D)u\|_{p,k}^-.$$

In particular, if $\lambda\in\mathbb{C}$ and $\operatorname{Im}\lambda>0$ we have

(12.8.14) $$\operatorname{Im}\lambda\,\|u\|_{p,k}^- \leqq \|(D-\lambda)u\|_{p,k}^-.$$

The last statement follows of course from the fact that $|\tau-\lambda| \geqq \operatorname{Im}\lambda$ when τ is real. For general q we have $\|q(D)u\|_{p,k}^- \leqq \|u\|_{p,|q|k}^-$.

To clarify the direction of the argument we now state an intermediate result on the way to the implication (iii) ⇒ (iv).

Proposition 12.8.4. *Let P be irreducible and satisfy* (12.8.9). *Then there exist constants C, κ, R such that if $v\in C_0^\infty$, $P(D)v=g$ and $|x'|\leqq M$ in* $\operatorname{supp} v$ *when $x_n\leqq 0$ it follows that for $1\leqq p\leqq\infty$ and $k\in\mathscr{K}(\mathbb{R})$*

(12.8.15) $$\sup_{\mathbb{R}^{n-1}} \|\hat v(\xi',x_n)\|_{p,k}^- \leqq C e^{\kappa M} \sup_{|\operatorname{Im}\zeta'|<R} \|\hat g(\zeta',x_n)\|_{p,k/\tilde P}^-.$$

Note that while k is a function of ξ_n only, the quotient $k/\tilde P$ also depends on ζ'. Later on we shall also allow k to depend on ξ'.

Lemma 12.8.5. *If $V(\zeta')=\|\hat v(\zeta',.)\|_{p,k}^-$ then $\log V(\zeta')$ is a plurisubharmonic function of $\zeta'\in\mathbb{C}^{n-1}$ and*

(12.8.16) $$V(\zeta')\leqq e^{M|\operatorname{Im}\zeta'|}V, \quad V=\sup_{\mathbb{R}^{n-1}} V(\xi').$$

Proof If $F(\zeta')$ is an analytic function with values in a Banach space then $\log\|F(\zeta')\|$ is plurisubharmonic since it is an upper semicontinuous function which is the supremum of $\log|\langle F(\zeta'),g\rangle|$ when g runs over the unit ball in the dual space. (See Section 4.1 for basic facts on plurisubharmonic functions.) For $x_n\leqq 0$ it follows from Theorem 7.3.1

that

$$e^{-M|\mathrm{Im}\,\zeta'|}\hat{v}(\zeta', x_n)$$

converges to 0 as $\zeta' \to \infty$, uniformly in x_n. This remains true after any number of differentiations with respect to x_n. Hence $V(\zeta')e^{-M|\mathrm{Im}\,\zeta'|} \to 0$ as $\zeta' \to \infty$. If $\xi', \eta' \in \mathbb{R}^{n-1}$ the difference

$$\log V(\xi' + w\eta') - \log V - M|\eta'|\,\mathrm{Im}\,w$$

is therefore a subharmonic function of $w \in \mathbb{C}$ which is $\leqq 0$ on the real axis and $\to -\infty$ as $w \to \infty$ in the upper half plane. Hence it is $\leqq 0$ when $\mathrm{Im}\,w > 0$. Taking $w = i$ we obtain (12.8.16).

(12.8.16) is a well known consequence of the Phragmén-Lindelöf theorem for entire function of exponential type. The following is essentially the Hadamard three circle theorem.

Lemma 12.8.6. *If w is a plurisubharmonic function, $w \leqq 0$ in the ball $\{\zeta \in \mathbb{C}^\nu, |\zeta| < R_0\}$, and $w \leqq -1$ in a smaller ball $\{\zeta; |\zeta - \zeta_0| < R_1\}$ then*

(12.8.17) $$w(\zeta) \leqq \delta(R_0, R_1)(|\zeta|/R_0 - 1), \quad |\zeta| < R_0,$$

where $\delta(R_0, R_1) > 0$.

Proof. If $\zeta_0 = 0$ then the maximum principle gives

$$w(\zeta) \leqq (\log|\zeta/R_0|)/\log(R_0/R_1), \quad R_1 < |\zeta| < R_0.$$

Since $\log t \leqq t - 1$ and $-1 \leqq t - 1$ when $t > 0$ it follows that

(12.8.18) $$w(\zeta) \leqq (|\zeta|/R_0 - 1)/\max(1, \log(R_0/R_1)).$$

If $|\zeta_0| < R_1/2$ we can apply this result with R_1 replaced by $R_1/2$. Assume now that $|\zeta_0| \geqq R_1/2$, thus $3R_1/2 \leqq R_0$. Then we have $w \leqq -1$ resp. $w \leqq 0$ in the balls with center $\zeta_0' = \zeta_0(1 - R_1/2|\zeta_0|)$ and radius $R_1/2$ resp. $3R_1/2$. Thus we obtain by (12.8.18)

$$w(\zeta_0' + \zeta) \leqq (2|\zeta|/3R_1 - 1)/\log 3, \quad |\zeta| < 3R_1/2.$$

In particular,

$$w(\zeta_0' + \zeta) \leqq -1/(9\log 3), \quad |\zeta| < 4R_1/3.$$

If we iterate this argument at most k times where k is the smallest integer such that $(4/3)^k R_1 > 2/3 R_0$, we have proved (12.8.17) with δ bounded below by a constant times a power of R_1/R_0.

Next we shall prove a few lemmas which show that algebraic functions are just as well behaved as polynomials when one keeps away from their branch points.

Lemma 12.8.7. *Let Z be a connected open set in $\mathbb{C}^{n-1}$ and $F_m(Z)$ the set of all analytic functions f in Z such that for some polynomial $R \not\equiv 0$ of degree $\leqq m$ in $\mathbb{C}^n$ the equation $R(\zeta', f(\zeta'))=0$ is valid in Z. Let Z_1 and Z_2 be non-empty open sets $\Subset Z$. Then there is a constant C such that*

$$\sup_{Z_2} |f| \leqq C \sup_{Z_1} |f|, \quad f \in F_m(Z). \tag{12.8.19}$$

Proof. It is sufficient to prove the statement when $Z_1 \Subset Z_2 \Subset Z$ are balls with center at the origin. The reduction to the case $n=2$ is then obvious. Without restriction we may assume that

$$\sup_{Z_1} |f| = 1,$$

for $F_m(Z)$ is closed under multiplication by complex numbers. We shall write z instead of ζ_1 and w instead of ζ_2. The equation satisfied by f may be written

$$\sum_{j+k \leqq m} a_{jk} f^j z^k = 0,$$

and we can normalize the coefficients so that $\sum |a_{jk}| = 1$.

For any choice (a_{jk}^0) of the coefficients we can find a disc Z' with $Z_2 \subset Z' \Subset Z$ and a closed line segment L joining a point in Z_1 to one in $\partial Z'$ such that the degree of $\sum a_{jk}^0 w^j z^k$ in w for each $z \in L \cup \partial Z'$ is equal to its largest possible degree. This requires only that we avoid the finitely many zeros of the leading coefficient. We can then find a number $W > 1$ such that

$$\sum a_{jk}^0 w^j z^k \neq 0 \quad \text{if } |w| = W \text{ and } z \in L \cup \partial Z'.$$

It follows that for all (a_{jk}) sufficiently close to (a_{jk}^0) we have

$$\sum a_{jk} w^j z^k \neq 0 \quad \text{if } |w| = W \text{ and } z \in L \cup \partial Z'.$$

If $\sum a_{jk} f(z)^j z^k = 0$ in Z and $|f(z)| \leqq 1$ when $z \in Z_1$, we conclude that $|f(z)| < W$ when $z \in L \cup \partial Z'$. In fact, this is a connected set which meets Z_1, so $|f(z)| < W$ at some point and $|f(z)| \neq W$ at every point in $L \cup \partial Z'$. Hence the maximum principle gives that $|f(z)| < W$ when $z \in Z_2$. Since the set of all (a_{jk}) with $\sum |a_{jk}| = 1$ is compact, the proof is complete.

Note that it follows from Lemma 12.8.7 that the set of all $f \in F_m(Z)$ with $\sup_{Z_1} |f| = 1$ is compact for the topology of uniform convergence on compact subsets of Z (cf. Theorem 4.4.2).

Lemma 12.8.8. *Let the hypotheses be as in Lemma 12.8.7 and let in addition Z_0 be a non-empty open set $\Subset Z_1$. Then there are constants C_z and $r > 0$ such that if $f \in F_m(Z)$ and $\sup_{Z_0} \operatorname{Im} f \geqq 0$ we have*

$$\sup_{Z_2} |D^\alpha f| \leqq C_\alpha \sup_{Z_1} \operatorname{Im} f, \quad \alpha \neq 0, \tag{12.8.19$'$}$$

$$\sup_{Z_1} \operatorname{Im} f \leqq C_0 \inf_B \operatorname{Im} f \tag{12.8.19$''$}$$

for some ball $B \subset Z_1$ of radius r.

Proof. We may assume that $\sup_{Z_0} \operatorname{Im} f = 0$, thus $\operatorname{Im} f(z_0) = 0$ for some $z_0 \in \bar{Z}_0$; we may also assume that $f(z_0) = 0$ since a real constant may be added to f. Furthermore we can normalize f so that $\sup_{Z_2} |f| = 1$, which makes the set M of all such f compact. There is a bound depending on α for the left-hand side of (12.8.19)$'$. Moreover, $\sup_{Z_1} \operatorname{Im} f > 0$ for if $\operatorname{Im} f \leqq 0$ in Z_1 then $\operatorname{Im} f = 0$ in a neighborhood of z_0 since $\operatorname{Im} f(z_0) = 0$, so f is a real constant in Z which must be equal to 0 since $\operatorname{Re} f(z_0) = 0$. Thus $\sup_{Z_1} \operatorname{Im} f$ is a positive continuous function of $f \in M$, hence has a positive lower bound which proves (12.8.19)$'$. It is also clear that (12.8.19)$''$ is valid for a fixed f, hence for all f in a neighborhood. By the compactnes we obtain the general validity of (12.8.19)$''$ for some $r > 0$.

Lemma 12.8.9. *Let $Z' \Subset Z$ and let Z be connected. If k is a positive integer there is a constant C such that*

$$\prod_1^k \sup_{Z'} |f_j| \leqq C \sup_{Z'} \left| \prod_1^k f_j \right|; \qquad f_1, \ldots, f_k \in F_m(Z). \tag{12.8.20}$$

Proof. The supremum in the right-hand side is a positive continuous function of $(f_1, \ldots, f_k)$ in the compact set defined by $\sup_{Z'} |f_j| = 1$ for every j. Hence it has a positive lower bound which proves the lemma.

As in the proof that (ii) ⇒ (iii) we shall write

$$P(\zeta) = a_m(\zeta') \prod_1^m (\zeta_n - \tau_j(\zeta')).$$

It is then possible to describe $\tilde{P}$ in terms of the zeros τ_j:

Lemma 12.8.10. *There is a constant C such that*

$$C^{-1} \tilde{P}(\eta) \leqq |a_m(\eta')| \prod_1^m (|\eta_n - \tau_j(\eta')| + 1 + \sup_{|\zeta' - \eta'| < 1} |\tau_j'(\zeta')|) \leqq C \tilde{P}(\eta) \tag{12.8.21}$$

provided that $\Delta(\zeta') a_m(\zeta') \neq 0$ when $|\zeta' - \eta'| < 2$.

Proof. By (10.4.1) and Lemma 11.1.4 we have

$$|a_m(\eta')| \leqq \sup_{|\zeta'-\eta'|<1} |a_m(\zeta')| \leqq C\tilde{a}_m(\eta') \leqq C'|a_m(\eta')|.$$

Since

$$\sup_{|\zeta-\eta|<1} |\zeta_n - \tau_j(\zeta')| \leqq |\eta_n - \tau_j(\eta')| + 1 + \sup_{|\zeta'-\eta'|<1} |\tau_j'(\zeta')| \tag{12.8.22}$$

the left-hand inequality (12.8.21) follows in view of (10.4.1). For an analytic function F in the unit disc in $\mathbb{C}$ we have

$$|F'(0)| \leqq \sup |F|.$$

Hence

$$\max(|\eta_n - \tau_j(\eta')| + 1, \sup_{2|\zeta'-\eta'|<1} |\tau_j'(\zeta')|/2) \leqq \sup_{|\zeta-\eta|<1} |\zeta_n - \tau_j(\zeta')|,$$

and if we now use (12.8.19), (12.8.20) the second the second inequality (12.8.21) follows.

Proof of Proposition 12.8.4. In addition to the notation $V(\zeta') = \|\hat{v}(\zeta',.)\|_{\bar{p},k}$ and $V = \sup_{\mathbb{R}^{n-1}} V(\xi')$ in Lemma 12.8.5 we set

$$G = \sup_{|\mathrm{Im}\,\zeta'|<R} \|\hat{g}(\zeta',.)\|_{\bar{p},k/\bar{P}}.$$

To estimate $V(\xi')$ we first note that by Lemma 12.8.2 we can find $\eta' \in \mathbb{R}^{n-1}$ with $|\eta'-\xi'| + A_1 + 3 < R = (A_1+3)/\gamma$ so that

$$\Delta(\zeta') a_m(\zeta') \neq 0 \quad \text{when } |\zeta'-\eta'| < A_1 + 3.$$

Application of Lemma 12.8.6 to a linear function of $\log V(\zeta')$ gives

$$\log V(\xi') \leqq (1-\delta) \sup_{|\zeta'-\xi'|<R} \log V(\zeta') + \delta \sup_{|\zeta'-\eta'|<A_1} \log V(\zeta').$$

By Lemma 12.8.5 we have $V(\zeta') \leqq V e^{M|\mathrm{Im}\,\zeta'|}$, hence

$$V(\xi') \leqq (Ve^{RM})^{1-\delta} \sup_{|\zeta'-\eta'|<A_1} V(\zeta')^{\delta}. \tag{12.8.23}$$

By hypothesis $\mathrm{Im}\,\tau_j(\zeta') \geqq 1$ for some ζ' with $|\zeta'-\eta'| \leqq A_1$. Hence

$$\mathrm{Im}\,\tau_j(\zeta') + 2A_1 T_j \geqq 1 \quad \text{when } |\zeta'-\eta'| < A_1 \quad \text{if } T_j = \sup_{|\zeta'-\eta'|<A_1} |\tau_j'(\zeta')|.$$

For the zeros of the polynomial in ζ_n defined by

$$q(\zeta', \zeta_n) = \prod_1^m (\zeta_n - \tau_j(\zeta') - 3iA_1 T_j)$$

we have therefore $\mathrm{Im}\,\zeta_n \geqq 1 + A_1 T_j$. The reason for introducing this polynomial is that Lemma 12.8.3 gives

$$\|\hat{v}(\zeta', x_n)\|_{\bar{p},k} = \|q(\zeta', D_n)\hat{v}(\zeta', x_n)\|_{\bar{p},k/|q|}, \quad |\zeta'-\eta'| < A_1. \tag{12.8.24}$$

For real ξ_n and $|\zeta'-\eta'|<A_1$ we have

$$1+A_1 T_j \leqq \operatorname{Im}\tau_j(\zeta')+3A_1 T_j \leqq |\xi_n-\tau_j(\zeta')-3iA_1 T_j|,$$

hence

$$|\xi_n-\tau_j(\eta')| \leqq |\xi_n-\tau_j(\zeta')-3iA_1 T_j|+5A_1 T_j \leqq 6|\xi_n-\tau_j(\zeta')-3iA_1 T_j|.$$

If we note that $\tilde{P}(\xi) \leqq C\tilde{P}(\eta',\xi_n)$ by (10.1.8) and recall (12.8.21), we obtain

$$\tilde{P}(\xi) \leqq C|a_m(\eta')|\,|q(\zeta',\xi_n)|, \quad |\zeta'-\eta'|<A_1. \tag{12.8.25}$$

We introduce the weight function

$$k_P(\xi_n)=|a_m(\eta')|\,k(\xi_n)/\tilde{P}(\xi',\xi_n),$$

which of course depends on ξ' too, and can then use (12.8.25) to write (12.8.24) in the form

$$V(\zeta') \leqq C\|q(\zeta',D_n)\hat{v}(\zeta',x_n)\|^{-}_{p,k_P}, \quad \text{if } |\zeta'-\eta'|<A_1. \tag{12.8.24$'$}$$

By (10.1.8) and the definition of G we have

$$|a_m(\eta')|^{-1}\|\hat{g}(\zeta',.)\|^{-}_{p,k_P} \leqq CG \quad \text{if } |\zeta'-\eta'|<A_1+1.$$

If we recall from the proof of Lemma 12.8.10 that $|a_m(\eta')/a_m(\zeta')|$ is bounded when $|\zeta'-\eta'|<A_1+1$, we obtain

$$\Big\|\prod_1^m (D_n-\tau_j(\zeta'))\hat{v}(\zeta',.)\Big\|^{-}_{p,k_P} \leqq C'G, \quad |\zeta'-\eta'|<A_1+1. \tag{12.8.26}$$

Set

$$g_j(\zeta',x_n)=\prod_1^j (D_n-\tau_\nu(\zeta'))\prod_{j+1}^m (D_n-\tau_\nu(\zeta')-3iA_1 T_\nu)\hat{v}(\zeta',x_n),$$

$$N_j=\sup_{|\zeta'-\eta'|<A_1+1}\|g_j\|^{-}_{p,k_P}.$$

Then (12.8.26) gives us control of N_m, and the right-hand side of (12.8.24)$'$ is bounded by CN_0. Now we know from Lemma 12.8.8 that the ball $\{\zeta';\,|\zeta'-\eta'|<A_1+1\}$ contains a ball B_j of fixed radius r such that

$$T_j \leqq C_0 \operatorname{Im}\tau_j(\zeta'), \quad \zeta'\in B_j.$$

By (12.8.14) it follows for $j=1,\dots,m$ that when $\zeta'\in B_j$

$$T_j\Big\|\prod_1^{j-1}(D_n-\tau_\nu(\zeta'))\prod_{j+1}^m (D_n-\tau_\nu(\zeta')-3iA_1 T_\nu)\hat{v}(\zeta',.)\Big\|^{-}_{p,k_P} \leqq C_0 N_j.$$

Hence

$$\|g_{j-1}(\zeta',.)\|^{-}_{p,k_P} \leqq (1+3A_1 C_0)N_j \quad \text{if } \zeta'\in B_j. \tag{12.8.27}$$

To apply Lemma 12.8.6 we also need some bound for $\|g_{j-1}(\zeta', x_n)\|^-_{p,k_P}$ when $|\zeta'-\eta'|<A_1+2$. Then we have

$$\prod_1^{j-1}|\xi_n-\tau_\nu(\zeta')|\prod_j^m|\xi_n-\tau_\nu(\zeta')-3iA_1T_\nu|$$
$$\leqq\prod_1^m(|\xi_n-\tau_j(\eta')|+CT_j)\leqq C'\tilde{P}(\xi)/|a_m(\eta')|$$

by the second part of (12.8.21) and (10.1.8), for

$$\sup_{|\zeta'-\eta'|<A_1+2}|\tau_j(\zeta')-\tau_j(\eta')|\leqq C''\sup_{|\zeta'-\eta'|<A_1}|\tau_j(\zeta')-\tau_j(\eta')|\leqq 2C''A_1T_j$$

by Lemma 12.8.7. Hence we obtain with a new constant C

$$\|g_{j-1}(\zeta', x_n)\|^-_{p,k_P}\leqq C\|\hat{v}(\zeta', x_n)\|^-_{p,k}\leqq CVe^{RM}, \quad |\zeta'-\eta'|<A_1+2. \tag{12.8.28}$$

If we now apply Lemma 12.8.6 to a suitable linear function of $\log\|g_{j-1}(\zeta', x_n)\|^-_{p,k_P}$ and use (12.8.27), (12.8.28), it follows that

$$N_{j-1}\leqq CN_j^\varepsilon(Ve^{RM})^{1-\varepsilon}, \quad 1\leqq j\leqq m; \quad \varepsilon=\delta(A_1+2,r)/(A_1+2).$$

Thus we obtain inductively for decreasing j

$$N_j\leqq C^{(1+\varepsilon+\ldots+\varepsilon^{m-j-1})}N_m^{\varepsilon^{m-j}}(Ve^{RM})^{1-\varepsilon^{m-j}}.$$

When $j=0$ we conclude using (12.8.24)′ and (12.8.26) that

$$V(\zeta')\leqq C_1G^{\varepsilon^m}(Ve^{RM})^{1-\varepsilon^m} \quad \text{if } |\zeta'-\eta'|<A_1.$$

Combining this estimate with (12.8.23) we obtain

$$V(\xi')\leqq C_2G^{\delta\varepsilon^m}(Ve^{RM})^{1-\delta\varepsilon^m}.$$

The supremum of the left hand side is V. After cancellation of a power of V we find that

$$V\leqq C_3Ge^{\kappa M}$$

where $\kappa=R(\delta^{-1}\varepsilon^{-m}-1)$. This completes the proof of Proposition 12.8.4.

If $k\in\mathscr{K}(\mathbb{R}^n)$ we shall write

$$\|u\|^-_{p,k}=\inf_v\|v\|_{p,k}; \quad u,v\in\mathscr{S}(\mathbb{R}^n), \ u=v \text{ in } -H.$$

Our next goal is to pass to such norms in (12.8.15).

Lemma 12.8.11. *If* $k\in\mathscr{K}(\mathbb{R}^n)$ *and* $k_{\xi'}(\xi_n)=k(\xi',\xi_n)$ *then*

$$\|u\|^-_{p,k}=((2\pi)^{1-n}\int(\|\hat{u}(\xi', x_n)\|^-_{p,k_{\xi'}})^p d\xi')^{1/p}, \quad u\in\mathscr{S}. \tag{12.8.29}$$

Proof. If $u, v \in \mathscr{S}$ and $u=v$ in $\mathbb{R}^n_-$ we have

$$\|v\|^p_{p,k} = (2\pi)^{1-n} \int (\|\hat{v}(\xi', x_n)\|_{p,k_{\xi'}})^p d\xi'$$
$$\geqq (2\pi)^{1-n} \int (\|\hat{u}(\xi', x_n)\|^-_{p,k_{\xi'}})^p d\xi'.$$

Thus we have at least inequality $\geqq$ in (12.8.29). When proving the opposite inequality we may assume that $\hat{u}(\xi', x_n)$ vanishes identically for large $|\xi'|$, for such functions are dense in $\mathscr{S}$. Let Φ be a positive continuous function. For each ξ' we can choose $w_{\xi'} \in \mathscr{S}(\mathbb{R})$ vanishing on $\mathbb{R}_-$ so that

$$\|\hat{u}(\xi', .) + w_{\xi'}\|_{p,k_{\xi'}} < \|\hat{u}(\xi', .)\|^-_{p,k_{\xi'}} + \Phi(\xi').$$

For reasons of continuity we have

$$\|\hat{u}(\eta', .) + w_{\xi'}\|_{p,k_{\eta'}} < \|\hat{u}(\eta', .)\|^-_{p,k_{\eta'}} + \Phi(\eta')$$

for all η' in a neighborhood of ξ' also. For large η' we can take $w=0$. Using a partition of unity in ξ' we can therefore find $W(\xi', x_n) \in \mathscr{S}(\mathbb{R}^n)$ vanishing when $x_n < 0$ and for large $|\xi'|$ so that for all ξ'

$$\|\hat{u}(\xi', .) + W(\xi', .)\|_{p,k_{\xi'}} < \|u(\xi', .)\|^-_{p,k_{\xi'}} + \Phi(\xi').$$

Taking $\hat{v} = \hat{u} + W$ we conclude that

$$\|u\|^-_{p,k} \leqq ((2\pi)^{1-n} \int (\|\hat{u}(\xi', .)\|^-_{p,k_{\xi'}})^p d\xi')^{1/p} + ((2\pi)^{1-n} \int \Phi(\xi')^p d\xi')^{1/p}.$$

Since Φ can be chosen so that the last integral is arbitrarily small, we obtain the remaining inequality $\leqq$ in (12.8.29).

Theorem 12.8.12. *Assume that P satisfies* (iii) *in Theorem 12.8.1 with $A_2 = A_1 + 1$ and $N = (0, \ldots, 0, 1)$. Then one can find κ and for arbitrary $p \in [1, \infty]$ and $k \in \mathscr{K}(\mathbb{R}^n)$ a constant C such that*

$$\|v\|^-_{p,k} \leqq C e^{\kappa M} \|P(D)v\|^-_{p,k/\tilde{P}} \tag{12.8.30}$$

when $v \in C_0^\infty$ and $|x'| \leqq M$ if $(x', x_n) \in \operatorname{supp} v$, $x_n \leqq 0$.

Proof. Assuming first that k is independent of ξ' we shall modify (12.8.15). Let $\chi \in C_0^\infty(\mathbb{R}^{n-1})$ be equal to 1 when $|x'| < 1$ and 0 when $|x'| > 2$. If $g = P(D)v$ and $\chi_M(x') = \chi(x'/M)$, we have $\chi_M g = g$ when $x_n < 0$, hence

$$\hat{g}(\zeta', x_n) = (2\pi)^{1-n} \int \hat{\chi}_M(\zeta' - \eta') \hat{g}(\eta', x_n) d\eta', \quad x_n < 0.$$

To estimate the right-hand side of (12.8.15) we note that (10.1.8) gives

$$\|\hat{g}(\eta', x_n)\|^-_{p,k/\tilde{P}_{\zeta'}} \leqq (1 + C|\zeta' - \eta'|)^N \|\hat{g}(\eta', x_n)\|^-_{p,k/\tilde{P}_{\eta'}}.$$

If $|\operatorname{Im} \theta'| < R$ then

$$(1 + C|\theta'|)^N |\hat{\chi}_M(\theta')| \leqq C'(1 + C|\theta'|)^N M^{n-1} e^{2MR} (1 + M|\theta'|)^{-N}$$
$$\leqq C' M^{n-1} e^{2MR} \quad \text{if } M > C.$$

Replacing the constant κ in (12.8.15) by $\kappa+2R+1$ we obtain

(12.8.15)′ $$\sup \|\hat{v}(\xi', x_n)\|_{\bar{p},k} \leqq C\, e^{\kappa M} \int \|g(\xi', x_n)\|_{\bar{p},k/\tilde{P}_{\xi'}} d\xi'.$$

Choose a fixed function $\phi \in C_0^\infty(\mathbb{R}^{n-1})$ with support in the unit ball and $\hat{\phi}(0)=1$. Set $\phi_{\eta'}(x')=e^{i\langle x',\eta'\rangle}\phi(x')$ so that $\hat{\phi}_{\eta'}(\xi')=\hat{\phi}(\xi'-\eta')$. Then we have $|x'|\leqq M+1$ if $(x',x_n)\in \operatorname{supp} v*\phi_{\eta'}$ and $x_n\leqq 0$. (Here $*$ stands for convolution in x'.) Hence it follows from (12.8.15)′ with $k=k_{\eta'}$ that

$$\sup_{\xi'} |\hat{\phi}(\xi'-\eta')|\, \|\hat{v}(\xi', x_n)\|_{\bar{p},k_{\eta'}} \leqq C e^{\kappa(M+1)} \int |\hat{\phi}(\xi'-\eta')|\, \|g(\xi', x_n)\|_{\bar{p},k_{\eta'}/\tilde{P}_{\xi'}} d\xi'.$$

In fact, the constant in (12.8.15)′ does not depend on k. Choosing $\xi'=\eta'$ in the left-hand side, we obtain by (10.1.1)

$$\|\hat{v}(\eta', x_n)\|_{\bar{p},k_{\eta'}} \leqq C' e^{\kappa M} \int |\hat{\phi}(\xi'-\eta')|(1+C|\xi'-\eta'|)^N \|g(\xi', x_n)\|_{\bar{p},k_{\xi'}/\tilde{P}_{\xi'}} d\xi'.$$

Since $\hat{\phi}(\xi')(1+C|\xi'|)^N \in L^1$, it follows in view of (12.8.29) that (12.8.30) is valid.

A local existence theorem follows at once from Theorem 12.8.12.

Theorem 12.8.13. *Let $1\leqq p\leqq\infty$, let $k\in\mathscr{K}(\mathbb{R}^n)$ and let $X \Subset \mathbb{R}^n$. Assume that P satisfies condition* (iii) *of Theorem 12.8.1. If $f\in B_{p,k}(\mathbb{R}^n)$ and $\operatorname{supp} f\subset H$, it follows that one find $u\in B_{p,\tilde{P}k}(\mathbb{R}^n)$ such that $\operatorname{supp} u\subset H$ and $P(D)u=f$ in X.*

Proof. It is no restriction to assume that f has compact support. Since we shall construct u with compact support it is permissible to assume that (iii) is fulfilled with $A_2=A_1+1$. The equation $P(D)u=f$ means that

$$\check{u}(P(D)v)=\check{f}(v), \quad v\in C_0^\infty(-X).$$

We have by Theorem 12.8.12 if $1/p+1/p'=1$

$$|\check{f}(v)|\leqq C\|v\|_{\bar{p}',1/k} \leqq C'\|P(D)v\|_{\bar{p}',1/\tilde{P}k}, \quad v\in C_0^\infty(-X).$$

The linear form

$$P(D)v \to \check{f}(v)$$

on $P(D)C_0^\infty(-X)$ can be extended by the Hahn-Banach theorem to a linear form $\check{U}$ on $C_0^\infty(\mathbb{R}^n)$ such that

$$|\check{U}(w)|\leqq C'\|w\|_{\bar{p}',1/\tilde{P}k}.$$

Thus U is a distribution with support in H such that $P(D)U=f$ in X. Furthermore, $U\in B_{p,\tilde{P}k}$ if $p>1$ (Theorem 10.1.14); in case $p=1$ the Fourier transform of U is a measure $d\mu$ with

$$\int \tilde{P}k|d\mu|<\infty.$$

Now take $\psi\in C_0^\infty$ so that $\psi=1$ in X, and set $u=\psi U$. Then we have $P(D)u=f$ in X, the support of u is a compact subset of H and the Fourier transform of u is a C^∞ function satisfying the same conditions as $\hat{U}$ (Theorem 10.1.15). Hence $u\in B_{p,\tilde{P}k}(\mathbb{R}^n)$ even when $p=1$.

In particular, it follows from Theorem 12.8.13 that one can find $E\in B_{\infty,\tilde{P}}(\mathbb{R}^n)$ so that $\operatorname{supp} E\subset H$ and $P(D)E-\delta$ vanishes in X. Thus E is almost a fundamental solution when X is chosen large. It allows us to prove an approximation theorem similar to Theorem 10.5.2 which will lead to global existence theorems.

Theorem 12.8.14. *Let $X_1\subset X_2$ be bounded open sets in $\mathbb{R}^n$ and let $P(D)$ be a differential operator which satisfies condition* (iii) *in Theorem 12.8.1. Denote by $\mathcal{N}_j$ the set of all solutions $u\in C^\infty(X_j)$ of the equation $P(D)u=0$ such that $\operatorname{supp} u\subset X_j\cap H$, and give $\mathcal{N}_j$ the topology induced by $C^\infty(X_j)$. If*

$$\mu\in\mathscr{E}'(\mathbb{R}^n),\qquad H^0\cap\operatorname{supp}\mu\subset\bar{X}_2,$$
$$H^0\cap\operatorname{supp} P(-D)\mu\Subset X_1\Rightarrow H^0\cap\operatorname{supp}\mu\Subset X_1, \tag{12.8.31}$$

it follows that the restrictions to X_1 of the elements in $\mathcal{N}_2$ form a dense subspace $\mathcal{N}_2'$ of $\mathcal{N}_1$. Here H^0 denotes the interior of H.

Proof. The statement will follow from the Hahn-Banach theorem if we prove that every $v\in\mathscr{E}'(X_1)$ which is orthogonal to $\mathcal{N}_2'$ is also orthogonal to $\mathcal{N}_1$. Choose $E\in\mathscr{E}'(H)$ so that $P(D)E-\delta=f$ vanishes in a large open set X to be specified later, and set $u=E*\phi$, $\phi\in C_0^\infty$. Then

$$P(D)u=\phi+\phi*f$$

and $\phi*f=0$ in X_2 if $f=0$ in $X_2-\operatorname{supp}\phi$. Set

$$Y=\{y;\,\bar{X}_2-\{y\}\subset X\}$$

which is an open set. We have $Y\supset\bar{X}_2$ if X is chosen so large that $X\supset\bar{X}_2-\bar{X}_2$ as we assume from now on. Then $P(D)u=0$ in X_2 if $\phi\in C_0^\infty(Y\setminus X_2)$, and $\operatorname{supp} u\subset H$ if $\operatorname{supp}\phi\subset H$. Hence

$$0=v(\phi*E)=\phi*E*\check{v}(0)=(\check{E}*v)(\phi),\qquad \phi\in C_0^\infty(H^0\cap Y\setminus\bar{X}_2),$$

that is, $\mu=\check{E}*v$ vanishes in $H^0\cap Y\setminus\bar{X}_2$. Furthermore,

$$P(-D)\,\mu=v+\check{f}*v$$

and $\check{f}*v=0$ in Y since if $y\in\operatorname{supp}\check{f}*v$ we have $y=x-z$, hence $z=x-y$ for some $x\in X_1$ and $z\notin X$.

If we choose $\chi\in C_0^\infty(Y)$ equal to 1 in a neighborhood of $\bar{X}_2$, it follows now that $\mu_1=\chi\mu\in\mathscr{E}'(Y)$ and $H^0\cap\operatorname{supp}\mu_1\subset\bar{X}_2$, $H^0\cap\operatorname{supp} P(-D)\,\mu_1$

$\subset X_1$. Hence $H^0 \cap \operatorname{supp} \mu_1 \Subset X_1$ by (12.8.31). We can thus choose $\chi_1 \in C_0^\infty(X_1)$ so that $\chi_1 = 1$ in a neighborhood of $H^0 \cap \operatorname{supp} \mu_1$. With $\mu_2 = \chi_1 \mu_1$ and $v_2 = P(-D)\mu_2$, we then have $v_2 \in \mathscr{E}'(X_1)$ and $v - v_2 \in \mathscr{E}'(X_1)$,

$$H^0 \cap \operatorname{supp}(v - v_2) = \emptyset.$$

If $u \in \mathscr{N}_1$ we obtain

$$v_2(u) = (P(-D)\mu_2)(u) = \mu_2(P(D)u) = 0,$$

and $(v - v_2)(u) = 0$ in virtue of Theorem 2.3.3. Thus $v(u) = 0$, which completes the proof.

We can use Holmgren's uniqueness theorem to construct sets X_j for which (12.8.31) is valid. Let μ be the order of P and let p be the principal part. If $\xi \in \mathbb{R}^n$ and $p(\xi) \neq 0$, the polynomial $p(N + \tau \xi)$ does not have more than μ zeros since it does not vanish identically. Hence we can choose $\varepsilon > 0$ so that $p(N + \tau \xi) \neq 0$ when $0 < \tau \leqq \varepsilon$. This means that (8.6.17) is true with $N_2 = N$ and $N_1 = N + \varepsilon \xi$. Choosing n vectors $\xi^1, \dots, \xi^n$ such that every $\xi \in \mathbb{R}^n$ is a linear combination with non-negative coefficients of $\xi^1, \dots, \xi^n$ and $-N$, we set $N^j = N + \varepsilon \xi^j$ with ε positive but so small that (8.6.17) is valid with $N_2 = N$ and $N_1 = N^j$ for every $j = 1, \dots, n$. Then the set

$$X = \{x \in \mathbb{R}^n;\ \langle x, N^j \rangle < 1, j = 1, \dots, n;\ \langle x, N \rangle > -1\}$$

is an open neighborhood of 0, and X is bounded since

$$\langle x, -N \rangle < 1, \qquad \langle x, \xi^j \rangle < 2/\varepsilon \qquad \text{if } x \in X$$

which implies that every linear form $\langle x, \xi \rangle$ is bounded from above when $x \in X$. From Corollary 8.6.10 it follows that when $0 < t < s$ we have

(12.8.32)
$$\begin{gathered} \mu \in \mathscr{D}'(\mathbb{R}^n), \qquad H^0 \cap \operatorname{supp} \mu \subset s\bar{X}, \\ H^0 \cap \operatorname{supp} P(-D)\mu \subset t\bar{X} \Rightarrow H^0 \cap \operatorname{supp} \mu \subset t\bar{X}. \end{gathered}$$

In fact, in the wedge $\langle x, N \rangle > 0$, $\langle x, N^j \rangle > t$ we have $P(-D)\mu = 0$, and since $\mu = 0$ in the part of the wedge where $\langle x, N^j \rangle > s$, Corollary 8.6.10 shows that $\mu = 0$ in the whole wedge.

We are now prepared for the proof of a theorem containing the remaining implication (iii) $\Rightarrow$ (iv) in Theorem 12.8.1.

Theorem 12.8.15. *Assume that $P(D)$ satisfies condition* (iii) *of Theorem 12.8.1. Let $f \in B^{\text{loc}}_{p_j, k_j}(\mathbb{R}^n)$, $j = 1, 2, \dots$ where $k_j \in \mathscr{K}$ and $1 \leqq p_j < \infty$, and let* $\operatorname{supp} f \subset H$. *Then there exists a solution u of the equation $P(D)u = f$ such that* $\operatorname{supp} u \subset H$ *and $u \in B^{\text{loc}}_{p_j, \tilde{P} k_j}(\mathbb{R}^n)$, $j = 1, 2, \dots$.*

Proof. We have made all the preparations required to repeat the proof of Theorem 10.6.7 so we shall only indicate the argument briefly. For $\nu=1,2,\dots$ we can find $u_\nu \in \bigcap B_{p_j,k_j}(\mathbb{R}^n)$ with $\operatorname{supp} u_\nu \subset H$ satisfying the equation $P(D)u_\nu=f$ in $X_\nu=\nu X$. In fact, if $\phi_\nu \in C_0^\infty(X_\nu)$ is equal to 1 in $X_{\nu-1}$ we can set $u_\nu=E*(\phi_{\nu+1}f)$ with $E\in B_{\infty,\tilde{P}}(\mathbb{R}^n)$, $\operatorname{supp} E\subset H$ and $P(D)E=\delta$ in a sufficiently large open set. The existence of E is guaranteed by Theorem 12.8.13. Then $P(D)(u_\nu-u_{\nu-1})=0$ in $X_{\nu-1}$. If $\chi\in C_0^\infty(X\cap H)$, $\int\chi\,dx=1$ and $\chi_\varepsilon(x)=\varepsilon^{-n}\chi(x/\varepsilon)$, then $v_\varepsilon=\chi_\varepsilon*(u_\nu-u_{\nu-1})\to u_\nu-u_{\nu-1}$ in $B_{p_j,k_j}(\mathbb{R}^n)$ as $\varepsilon\to 0$, and $P(D)v_\varepsilon=0$ in $X_{\nu-2}$ if $\varepsilon<1$. Hence it follows from Theorem 12.8.13 that we can find $w\in C^\infty(X_{\nu+1})$ with $\operatorname{supp} w\subset H$ and $P(D)w=0$ approximating v_ε arbitrarily closely in $X_{\nu-2}$. Replacing u_ν by $u_\nu-\phi_{\nu+1}w$ we can therefore successively achieve that

$$\|\phi_{\nu-2}(u_\nu-u_{\nu-1})\|_{p_j,k_j}<2^{-j},\qquad j\leqq\nu.$$

But this means that $u=\lim_{\nu\to\infty} u_\nu$ exists, $u\in\bigcap B^{\mathrm{loc}}_{p_j,k_j}(\mathbb{R}^n)$, $\operatorname{supp} u\subset H$, and $P(D)u=f$ in $\mathbb{R}^n$. The proof is complete.

Definition 12.8.16. We shall say that $P(D)$ is an evolution operator with respect to H if the equivalent conditions in Theorem 12.8.1 are fulfilled.

The notion of evolution operator agrees well with the classification of differential operators in Section 10.4:

Theorem 12.8.17. *Let $P(D)$ be an evolution operator with respect to H, let E be the set of all Q which are equally strong as P and let E_0 be the component of P in E. Then it follows that every $Q\in E_0$ is an evolution operator with respect to H.*

However, every $Q\in E$ need not be an evolution operator with respect to H. An example is $P(\xi)=\xi_1^2+i\xi_2$ and $Q(\xi)=\xi_1^2-i\xi_2$, corresponding to the heat operator with time x_2 and $-x_2$ respectively.

Proof of Theorem 12.8.17. We can assume that H is defined by $x_n\geqq 0$. If X is a bounded open set containing 0, we have by Theorem 12.8.12

$$\|v\|^-_{1,1}\leqq C\|P(D)v\|^-_{1,1/\tilde{P}},\qquad v\in C_0^\infty(X).$$

(The hypothesis $A_2=A_1+1$ in Theorem 12.8.12 is immediately removed by a complex translation of P in the ξ_n direction.) The proof of Theorem 12.8.12 shows that C only depends on A_1 and A_2 in addition to the dimension n and the degree μ of P. For the polynomial $P_\varepsilon(\xi)=P(\xi/\varepsilon)$ condition (iii) of Theorem 12.8.1 is fulfilled with A_1,A_2 replaced by $\varepsilon A_1, \varepsilon A_2$, hence with $A_1=1$ and $A_2=-1$ if ε is small enough. Thus

(12.8.32)
$$\|v\|^-_{1,1} \leqq C_{n\mu} \|P_\varepsilon(D) v\|^-_{1,1/\tilde{P}_\varepsilon}, \quad v \in C_0^\infty(X),$$

if $\varepsilon < \varepsilon(P)$. Now assume that $Q \prec P$,

(12.8.33)
$$\tilde{Q}(\xi) \leqq M \tilde{P}(\xi), \quad \xi \in \mathbb{R}^n.$$

By Theorem 10.4.3 it follows that

$$\tilde{Q}(\xi, t) \leqq C'_n M \tilde{P}(\xi, t), \quad \xi \in \mathbb{R}^n,\ t > 1,$$

that is,

(12.8.33)′
$$\tilde{Q}_\varepsilon(\xi) \leqq C'_n M \tilde{P}_\varepsilon(\xi), \quad \xi \in \mathbb{R}^n,\ 0 < \varepsilon < 1.$$

From (12.8.32) we therefore obtain

$$\|v\|^-_{1,1} \leqq C_{n\mu} \|(P_\varepsilon(D) + Q_\varepsilon(D)) v\|^-_{1,1/\tilde{P}_\varepsilon} + M C'_n C_{n\mu} \|v\|^-_{1,1}, \quad v \in C_0^\infty(X).$$

Thus

$$|v(0)| \leqq \|v\|^-_{1,1} \leqq 2 C_{n\mu} \|(P_\varepsilon(D) + Q_\varepsilon(D)) v\|^-_{1,1/\tilde{P}_\varepsilon}, \quad v \in C_0^\infty(X),$$

provided that $M < 1/(2 C'_n C_{n\mu})$. This implies that $P_\varepsilon + Q_\varepsilon$ satisfies condition (ii) in Theorem 12.8.1, so $P_\varepsilon + Q_\varepsilon$ and therefore $P+Q$ is an evolution operator.

We have now proved that if (12.8.33) is valid and M is smaller than a constant depending only on n and μ, then $P+Q$ is an evolution operator with respect to H if P is. The set of all $R \in E$ such that R is an evolution operator with respect to H is therefore both open and closed, hence a union of components of E. This proves the theorem.

In the two dimensional case it is easy to describe the evolution operators with respect to the half space $x_2 \geqq 0$. To do so we observe that the zeros of $P(\xi_1, \tau)$ have Puiseux series expansions at ∞

$$\tau(\xi_1) = \sum_{-\infty}^{k} c_j \xi_1^{j/p}, \quad c_k \neq 0.$$

Since

$$\operatorname{Im} \tau > -C(|\xi_1|^{(k-p)/p} + 1)$$

by condition (iii) in Theorem 12.8.1, and the argument of ξ_1 can be chosen as $\nu\pi$ for any integer ν, it is easily seen that one of the following cases must occur:

(a) $k \leqq 0$, that is, τ is bounded at infinity.

(b) k is a positive even multiple of p and $\operatorname{Im} c_k > 0$.

(c) k is a positive multiple of p, $\operatorname{Im} c_k = 0$ and $c_j = 0$ when $k-p < j < k$.

Now if $\tau=\tau_j(\xi_1)$ is a zero of $P(\xi_1,\tau)$ then $\operatorname{Im}\tau_j$ grows as $|\xi_1|^{k/p}$ in case b), and $|\tau_j'|$ grows as $|\xi_1|^{(k-p)/p}$ in case (c). Hence it follows from Lemma 12.8.10 that P is equally strong as

$$p(\xi)=a\,\xi_1^{\mu}\prod_1^m(\xi_2-c_i\,\xi_1^{\mu_i}). \tag{12.8.34}$$

Here $a\,\xi_1^{\mu}$ is the leading term of the coefficient of ξ_2^m in P, and $c_i\,\xi_1^{\mu_i}$ are the leading terms in the Puiseux series expansions for τ_i. Thus c_i is real unless μ_i is even and $\operatorname{Im} c_i>0$. It is not hard to show that $P-p$ is actually dominated by p, so P and p are in the same component of the set of operators equally strong as p. Thus the evolution operators are precisely the operators (12.8.34) and their components in the set of equally strong operators. Besides the tangential operator D_1 we have hyperbolic factors D_2-cD_1 with real c, factors $D_2-cD_1^{\nu}$ with c real and $\nu>1$ of Schrödinger equation type, and finally factors $D_2-cD_1^{\nu}$ with ν even and $\operatorname{Im} c>0$ of heat equation type. In the two dimensional case the proof of Theorem 12.8.1 can of course be shortened a great deal since we have such good control of the zeros.

12.9. Mixed Problems

When solving the Cauchy problem for say the wave equation in $\mathbb{R}^{3+1}$ we assumed Cauchy data known in all of $\mathbb{R}^3$ at time $t=0$. If Cauchy data are only given in a subset X of $\mathbb{R}^3$ the Cauchy problem has a unique solution at (x,t) only when all light rays arriving at (x,t) must have started in X at time 0. To go beyond this set it is physically plausible that one needs a boundary condition on $\{(x,t);\ x\in\partial X,\ t\geqq 0\}$ which controls the radiation flowing into the cylinder $X\times\mathbb{R}_+$. This gives what is called a mixed problem. In the spirit of this volume we shall only consider the case where X is a half space and leave the much more delicate problem when ∂X is curved to Volume III.

Thus assume given a partial differential operator $P(D)$ with constant coefficients in $\mathbb{R}^n$ and two half spaces

$$H=\{x\in\mathbb{R}^n;\ \langle x,N\rangle\geqq 0\},\qquad H'=\{x\in\mathbb{R}^n;\ \langle x,\theta\rangle\geqq 0\},$$

where N and θ are linearly independent. The mixed problem we shall study is to find u such that

$$P(D)u=f \quad \text{in } H\cap H' \tag{12.9.1}$$

$$D^{\alpha}(u-\phi)=0 \quad \text{in } H'\cap\partial H \quad \text{when } |\alpha|<m, \tag{12.9.2}$$

$$B_j(D)u=\phi_j,\quad j=1,\dots,\mu \quad \text{on } H\cap\partial H'. \tag{12.9.3}$$

First we shall determine necessary conditions in order that there shall exist a unique solution $u \in C^\infty(H \cap H')$ for arbitrary $f, \phi \in C^\infty(H \cap H')$ and $\phi_j \in C^\infty(H \cap \partial H')$ vanishing of infinite order at $\partial H \cap \partial H'$. We shall then show that these conditions imply the existence of a unique solution for quite general data.

The reader who has just been exposed to the difficulties presented by the characteristic Cauchy problem should appreciate that we assume ∂H non-characteristic, that is, if P_m is the principal part,

$$P_m(N) \neq 0.$$

It is then necessary to require that $P(D)$ be hyperbolic with respect to N. In fact, let $f \in C_0^\infty(H)$, choose $t \in (H' \setminus \partial H') \cap \partial H$ and consider the mixed problem (12.9.1)–(12.9.3) with $\phi = \phi_j = 0$ and f replaced by $x \to f(x - at)$ with a so large that this is in $C_0^\infty(H \cap H')$. If a solution u_a exists then $P(D) u_a(. + at) = f$ in $(H' - \{at\}) \cap H$ and u_a has vanishing Cauchy data on ∂H. Because of the uniqueness of the Cauchy problem it follows that the equation $P(D)u = f$ has a solution in $C^\infty(H)$ with vanishing Cauchy data, so $P(D)$ *is hyperbolic* by Theorem 12.3.1. This will be assumed in what follows. Thus we have by Theorem 12.4.4

$$P(\zeta) \neq 0 \quad \text{if } \operatorname{Im} \zeta \in \{\tau_0 N\} - \Gamma(P, N). \tag{12.9.4}$$

Proposition 12.9.1. *Let m_+, m_-, m_0 be the number of positive, negative and vanishing zeros of $P_m(\theta - \tau N)$ as a polynomial in τ. Then the limits*

$$q_{m_0}(\xi) = \lim_{\varepsilon \to 0} P_m(\theta + \varepsilon \xi)\, \varepsilon^{-m_0}, \tag{12.9.5}$$
$$q(\xi) = \lim_{\varepsilon \to 0} \varepsilon^{m - m_0} P(\xi + \theta/\varepsilon)$$

exist, $q(\zeta + t\theta) = q(\zeta)$ for every t, q_{m_0} is the principal part of q and

$$q(\zeta) \neq 0 \quad \text{if } \operatorname{Im} \zeta \in \{\tau_0 N\} - \Gamma(P, N) + \mathbb{R}\theta. \tag{12.9.6}$$

The equation

$$P(\zeta + w\theta) = 0, \quad \operatorname{Im} \zeta \in \{\tau_0 N\} - \Gamma(P, N), \tag{12.9.7}$$

is of degree $m_+ + m_-$ with respect to w. It has m_+ (resp. m_-) *roots with* $\operatorname{Im} w > 0$ (resp. $\operatorname{Im} w < 0$), *and the leading coefficient is $q(\zeta)$.*

Proof. The polynomial $P_{m.\theta}(\xi)$ which is the homogeneous part of lowest degree of $\xi \to P_m(\theta + \xi)$ is hyperbolic with respect to N (Lemma 8.7.2), hence different from 0 at N. Thus $P_m(\theta + \varepsilon N)$ vanishes at $\varepsilon = 0$ precisely of the order of the degree of $P_{m.\theta}$ which is therefore equal to m_0. Thus the limit q_{m_0} exists and is not identically 0. From the proof of Theorem 12.6.1 it follows that the limit q also exists and is a localization of P at infinity in the direction θ. It is clear that $q(\xi+t\theta)=q(\xi)$ (cf. Theorem 10.2.8) so (12.9.6) follows from (12.9.4) and Hurwitz' theorem.

We have proved now that (12.9.7) is of degree $m_+ + m_-$ with respect to w, and that the leading coefficient is $q(\zeta)$. By (12.9.4) there are no real roots. To determine the number of roots in each half plane we take $\zeta=-i\tau N$ and set $w=\tau W$. Since for $\tau\to+\infty$

$$P(-i\tau N+\tau W\theta)(i/\tau)^m \to P_m(N+iW\theta)$$

which has m_+ (m_-) zeros with $\operatorname{Im} W>0$ ($\operatorname{Im} W<0$), the proposition is proved.

The Cauchy problem was solved by a Fourier-Laplace transformation in all variables (see (12.5.3)). To solve the mixed problem we shall have to take Fourier-Laplace transforms only along the boundary plane $\partial H'$. This leaves us with an ordinary differential equation with boundary conditions and we digress now to discuss such problems.

Thus let $p(D)$, $D=-id/dt$, be a differential operator with constant coefficients on $\mathbb{R}$. We assume that p has leading coefficient one and no real zeros, and write

$$p(\tau)=p_+(\tau)\,p_-(\tau)$$

where p_+ and p_- have leading coefficient 1, degree m_+ and m_- respectively, and all zeros in the half planes $\operatorname{Im}\tau>0$ ($\operatorname{Im}\tau<0$). The solutions of the equation $p(D)u=0$ can uniquely be written $u=u_+ + u_-$ where $p_+(D)u_+=0$, $p_-(D)u_-=0$. Here u_+ is exponentially decreasing on $\mathbb{R}_+$ and exponentially increasing on $\mathbb{R}_-$; the opposite is true for u_-.

Let us now consider a boundary problem

$$(12.9.8)\qquad p(D)u=f \text{ on } \mathbb{R}_+,\qquad b_j(D)u(0)=\phi_j,\quad j=1,\dots,\mu$$

where $f\in\mathscr{S}(\mathbb{R})$, $\phi_j\in\mathbb{C}$, and u is required to be in $\mathscr{S}(\overline{\mathbb{R}}_+)$. A solution u_0 of the differential equation on $\mathbb{R}$ is given by $\hat u_0=\hat f/p$. Writing $u=u_0+v$ we are left with the problem

$$(12.9.8)'\qquad p(D)v=0 \text{ on } \mathbb{R}_+,\qquad b_j(D)v(0)=\psi_j,\quad j=1,\dots,\mu,$$

where $\psi_j=\phi_j-b_j(D)u_0(0)$. The equation $p(D)v=0$ is equivalent to $p_+(D)v=0$ when v is bounded on $\mathbb{R}_+$. We can divide b_j by p_+,

$$b_j(D)=p_+(D)\,q_j(D)+r_j(D),\qquad j=1,\dots,\mu$$

where the degree of r_j is less than m_+. Then (12.9.8)′ is equivalent to

$$p_+(D)v=0, \quad r_j(D)v(0)=\psi_j, \quad j=1,\dots,\mu. \tag{12.9.8$''$}$$

The boundary conditions are μ linear equations in the Cauchy data $v(0),\dots,D^{m_+-1}v(0)$ so there exists a unique solution if and only if $r_1,\dots,r_\mu$ are linearly independent and $\mu=m_+$. This means that $b_1,\dots,b_\mu$ are linearly independent modulo p_+ and that $\mu=m_+$. An analytic condition for this is the non-vanishing of the Lopatinski determinant

$$L(b_1,\dots,b_{m_+};p_+)=\det((2\pi i)^{-1}\int b_j(\tau)\tau^{k-1}/p_+(\tau)\,d\tau)_{j,k=1}^{m_+} \tag{12.9.9}$$

where the integral is taken along the boundary of a disc containing all zeros of $p_+(\tau)$. Indeed,

$$(f,g)\to(2\pi i)^{-1}\int f(\tau)g(\tau)/p_+(\tau)\,d\tau=\langle f,g\rangle_{p_+}$$

is a bilinear form on the vector space of polynomials in τ modulo $p_+(\tau)$ for it vanishes when f or g is divisible by p_+. It is non-degenerate since $\langle f,g\rangle_{p_+}=0$ for all g means precisely that there are no negative powers of τ in the Laurent expansion of $f(\tau)/p_+(\tau)$ at ∞. The classes $\operatorname{mod} p_+(\tau)$ of $\tau^{k-1}, k=1,\dots,m_+$, are a basis so it follows that $b_1,\dots,b_{m_+}$ are linearly independent $\operatorname{mod} p_+$ if and only (12.9.9) is different from 0. The Lopatinski determinant does not change if $p_+(\tau)$ and all $b_j(\tau)$ are replaced by $p_+(\tau-a)$ and $b_j(\tau-a)$ for some $a\in\mathbb{R}$. In fact, this just causes τ^{k-1} to be replaced by $(\tau+a)^{k-1}$, and

$$(\tau+a)^{j-1}=\sum_{k\leq j} c_{jk}(a)\tau^{k-1}$$

where $\det c_{jk}(a)=1$ since $c_{jk}(a)-\delta_{jk}$ is strictly triangular. Note also that $\langle f,g\rangle_{p_+}$ is a polynomial in the coefficients of f, g, p_+ since at infinity

$$1/p_+(\tau)=\tau^{-m_+}\sum_0^\infty(1-p_+(\tau)/\tau^{m_+})^k.$$

If $\mu=m_+$ and the Lopatinski determinant is not 0 we can find the solution $u\in\mathscr{S}(\bar{\mathbb{R}}_+)$ of (12.9.8) as follows. Let $E_-(y)$ be the fundamental solution of $p_-(D)$ with support in $\mathbb{R}_-$. It is exponentially decreasing. Since the equation $p_-(D)v=0$ has no solution in $\mathscr{S}(\bar{\mathbb{R}}_+)$ we must have

$$p_+(D)u=f*E_-=\int_{-\infty}^0 f(.-t)E_-(t)\,dt=g \tag{12.9.10}$$

where the last equality is a definition. We shall now prove that

$$LD^j u(0)=\sum_1^{m_+} t_{jk}\phi_k+\sum s_{jk}D^k g(0), \quad j=0,1,\dots \tag{12.9.11}$$

where L is the Lopatinski determinant, t_{jk} and s_{jk} are polynomials in the coefficients of $p_+, b_1, \dots, b_{m_+}$ and

$$m_+ + k \leqq \max(j, \deg b_1, \dots, \deg b_{m_+})$$

in the second sum. The identity is equivalent to

$$L\tau^j = \sum_1^{m_+} t_{jk} b_k(\tau) + \sum s_{jk} \tau^k p_+(\tau).$$

Thus t_{jk} are obtained from the equations

$$(12.9.12) \qquad L\langle \tau^j, \tau^{i-1}\rangle_{p_+} = \sum_1^{m_+} t_{jk} \langle b_k(\tau), \tau^{i-1}\rangle_{p_+}, \qquad i = 1, \dots, m_+,$$

thus by multiplying the vector $\langle \tau^j, \tau^{i-1}\rangle_{p_+}$, $i=1, \dots, m_+$, by the cofactor matrix of the Lopatinski matrix $\langle b_k(\tau), \tau^{i-1}\rangle_{p_+}$. The polynomial $\sum s_{jk} \tau^k$ is then obtained by division with $p_+(\tau)$.

Returning now to the situation in Proposition 12.9.1 we write

$$(12.9.13) \quad P(\zeta + w\theta) = q(\zeta) P_+(w, \zeta) P_-(w, \zeta), \quad \operatorname{Im} \zeta \in \{\tau_0 N\} - \Gamma(P, N)$$

where P_+ and P_- are polynomials in w of degree m_+ and m_- with zeros in the upper and lower half planes respectively and leading coefficients 1. By a translation of P we can always make $\tau_0 = 1$. Then $|q(\zeta)|$ has a positive lower bound if $\operatorname{Im} \zeta \in -\Gamma(P, N)$, and we obtain using (10.4.3)

$$(12.9.14) \quad \sum |D_w^j P_+(0, \zeta)| \sum |D_w^j P_-(0, \zeta)| \leqq C \tilde{P}(\zeta), \qquad \operatorname{Im} \zeta \in -\Gamma(P, N).$$

Since the sums are bounded from below by $m_+!$ and $m_-!$ this gives a polynomial bound for all coefficients of P_+ and P_-. If $\mu = m_+$ we can form the Lopatinski determinant

$$(12.9.15) \qquad L(\zeta) = L(B_1(\zeta + w\theta), \dots, \quad B_{m_+}(\zeta + w\theta); P_+(w, \zeta)),$$
$$\operatorname{Im} \zeta \in -\Gamma(P, N)$$

where ζ is regarded as a parameter. By the translation invariance of the Lopatinski determinant $L(\zeta) = L(\zeta + t\theta)$, $t \in \mathbb{R}$, which means that $L(\zeta)$ can be regarded as an analytic function in the image of $\mathbb{R}^n - i\Gamma(P, N)$ in $\mathbb{C}^n / \mathbb{C}\theta$. Also q is analytic there and nowhere equal to 0.

When studying (12.9.1)–(12.9.3) with $\phi = 0$ it is convenient to extend u and f by 0 in $H' \cap \complement H$. More generally we shall consider distribution solutions u of (12.9.1) in the interior of H' which vanish in $\complement H$ and can be regarded as C^∞ functions of $\langle x, \theta\rangle \in \bar{\mathbb{R}}_+$ with values in $\mathscr{D}'(\mathbb{R}^{n-1})$ satisfying (12.9.3). This assumes that we choose coordinates $x_1 = \langle x, \theta\rangle$, $x_2, \dots, x_n$ but the choice of $x_2, \dots, x_n$ is irrelevant. To simplify we shall usually assume $x_n = \langle x, N\rangle$ in the proofs. By the support of a solution u in this class we shall mean the closure of the support of u as a distribution in the interior of H'.

Theorem 12.9.2. *Assume that the rank of the Lopatinski matrix*

$$L_{jk}(\zeta)=(2\pi i)^{-1}\int \lambda^{k-1}B_j(\zeta+\lambda\theta)/P_+(\lambda,\zeta)\,d\lambda,$$
$$k=1,\ldots,m_+,\qquad j=1,\ldots,\mu,$$

is less than m_+ *for every* ζ *with* $\operatorname{Im}\zeta\in\{\tau_0 N\}-\Gamma(P,N)$. *Then one can find a* C^∞ *function* u *of* $\langle x,\theta\rangle\in\bar{\mathbb{R}}_+$ *with values in* $\mathscr{D}'(\mathbb{R}^{n-1})$ *such that*

$$0\in\operatorname{supp}u\subset H'\cap\Gamma^\circ(P,N),\qquad P(D)u=0 \text{ in } H',$$
$$B_j(D)u=0 \text{ on } \partial H',\ j=1,\ldots,\mu.$$

Proof. Introducing additional B_j's and changing the labelling if necessary we may assume that the rank of the Lopatinski matrix with $1\leqq j<m_+$ is m_+-1 for general ζ. The identity

$$L(B_1(\zeta+w\theta),\ldots,B_{m_+-1}(\zeta+w\theta),B(w);P_+(w,\zeta))$$
$$=(2\pi i)^{-1}\int B(w)H(w,\zeta)/P_+(w,\zeta)\,dw$$

for every polynomial $B(w)$ defines uniquely a polynomial $H(w,\zeta)$ in w of degree m_+-1 with coefficients which are analytic when $\operatorname{Im}\zeta\in\{\tau_0N\}-\Gamma(P,N)$, and not all 0 where the Lopatinski matrix has rank m_+-1. Since

$$H(w+a,\zeta)=H(w,\zeta+a\theta),\qquad a\in\mathbb{R},$$

it follows that the coefficients do not vanish identically when $\langle\zeta,\theta\rangle=0$ either. In what follows we assume $\tau_0=1$.

We choose coordinates above so that $N=(0,\ldots,0,1)$ and $\theta=(1,0,\ldots,0)$ and we write $\zeta'=(\zeta_2,\ldots,\zeta_n)$. Then $H(w,(0,\zeta'))$ has analytic coefficients when $(0,\operatorname{Im}\zeta')\in-\Gamma(P,N)$, and they are not identically 0. Set

$$U(x_1,\zeta')=(2\pi i)^{-1}\int H(w,(0,\zeta'))\,e^{iwx_1}/P_+(w,(0,\zeta'))\,dw$$

with the integral taken around the boundary of an open set in the half plane $\operatorname{Im}w>-1$ containing a disc of radius 1 about each zero of P_+. This is an analytic function of ζ', and by (12.9.14) we have for every j

$$|D_{x_1}^j U(x_1,\zeta')|\leqq C_j e^{x_1}(1+|\zeta'|)^{M_j}.$$

It is clear that $P(D_1,\zeta')U(x_1,\zeta')=0$. Hence U is the Fourier-Laplace transform with respect to x' of a distribution u in H' with $P(D)u=0$ and $\operatorname{supp}u\subset H'\cap H$, which is a C^∞ function of x_1 with distribution values (cf. Theorem 7.4.3). To estimate the support it is best to study the boundary values on $\partial H'$ first. For any differential operator $B(D)$ with constant coefficients the boundary value of $B(D)u$ when $x_1=0$ is the inverse Fourier-Laplace transform of

$$\begin{aligned}F(\zeta')&=(2\pi i)^{-1}\int H(w,(0,\zeta'))\,B(w,\zeta')/P_+(w,(0,\zeta'))\,dw\\&=L(B_1(\zeta+w\theta),\dots,B_{m_+-1}(\zeta+w\theta),B(\zeta+w\theta);P_+(w,\zeta))\end{aligned}$$

where $\zeta=(0,\zeta')$. However, the Lopatinski determinant is independent of ζ_1 which proves that F is analytic in

$$\Omega=\{\zeta';(\lambda,\operatorname{Im}\zeta')\in-\Gamma(P,N)\text{ for some }\lambda\}.$$

When $B=B_j$ for some j we have $F=0$ so $B_j(D)u=0$ on $\partial H'$. For any B there is a polynomial bound for F. Since the dual cone of $\Gamma(P,N)+\mathbb{R}\theta$ is $\Gamma^\circ(P,N)\cap\partial H'$ it follows that

$$\operatorname{supp}B(D)u|_{x_1=0}\subset\Gamma^\circ(P,N)\cap\partial H'.$$

Extend u now to $\mathbb{R}^n$ so that $u=0$ in $\complement H'$. Then

$$P(D)u=f,\qquad\operatorname{supp}f\subset\Gamma^\circ(P,N)\cap\partial H'$$

and $\operatorname{supp}u\subset H$. Thus $u=E*f$ where E is the fundamental solution of $P(D)$ with support in $\Gamma^\circ(P,N)$ which proves that $\operatorname{supp}u\subset\Gamma^\circ(P,N)$. If $0\notin\operatorname{supp}u$ it follows from Theorem 7.4.3 that for some $C,M,c>0$

$$|F(\zeta')|\leqq C(1+|\zeta'|)^M e^{c\operatorname{Im}\zeta_n},\qquad\zeta'\in\Omega.$$

Now F is algebraic, so a standard application of the Tarski-Seidenberg Theorem A.2.2, which is left for the reader to do in detail, gives that

$$\sup_{\xi'}|F(\xi'+i\tau N')|(1+|\xi'+i\tau N'|)^{-M}$$

cannot decrease faster than a power of $|\tau|^{-1}$ as $\tau\to-\infty$ unless $F\equiv0$. Since $H\neq0$ this does not happen for every B. Hence $0\in\operatorname{supp}u$ and the theorem is proved.

By convolving u with a C_0^∞ density in $\partial H'$ with support in $H\cap\partial H'$ we obtain a non-trivial C^∞ solution of (12.9.1)–(12.9.3) with all the data f,ϕ,ϕ_j equal to 0. Thus uniqueness of (12.9.1)–(12.9.3) implies that the rank of the Lopatinski matrix is m_+ for general ζ. This is not yet sufficient for uniqueness; the following theorem will suggest a condition which must be added.

Theorem 12.9.3. *Let $\mu=m_+$ and let $u\in C^\infty(H')$ be a solution of* (12.9.1) *and* (12.9.3) *with $f=0$, $\phi_j=0$, such that* $\operatorname{supp}u\subset H\cap H'$. *Then $\mathscr{L}*u=0$ in H' if $\mathscr{L}$ is the distribution with support in $\partial H'\cap\Gamma^\circ(P,N)$ and $\hat{\mathscr{L}}(\zeta)=L(\zeta)$ defined by* (12.9.15).

We recall that $L(\zeta)$ is independent of ζ_1 with our standard coordinates. Thus $\mathscr{L}=\mathscr{L}_\theta\otimes\delta(\langle x,\theta\rangle)$ where $\mathscr{L}_\theta$ is a distribution in the plane $\partial H'$, so the convolution acts only for fixed $\langle x,\theta\rangle$.

Proof of Theorem 12.9.3. Starting from (12.9.11) we shall first derive a representation formula for $\mathscr{L}_\theta*(B(D)v)|_{\partial H'}$ when $v\in C_0^\infty(\mathbb{R}^n)$. The theorem will follow when it is applied to the product of u and a cutoff function. (Here $B(D)$ is an arbitrary differential operator with constant coefficients.)

If we apply (12.9.11) to $P_-(w,(0,\zeta'))$ and set $P(D)v=f$, $v\in C_0^\infty$, we obtain using (12.9.10) and an explicit formula for E_-

$$\begin{aligned}&L(\zeta')B(D_1,\zeta')\hat{v}(0,\zeta')\\&\quad=\sum T_k(\zeta')B_k(D_1,\zeta')\hat{v}(0,\zeta')\\&\quad\quad-\sum\int_0^\infty q(\zeta')^{-1}S_k(\zeta')D_1^k\hat{f}(x_1,\zeta')\,dx_1(2\pi)^{-1}\int e^{-ix_1w}/P_-(w,\zeta')\,dw.\end{aligned}$$

Here the integral is taken over the boundary of an open set containing the zeros of P_-. The coefficients $T_k(\zeta')$ and $S_k(\zeta')/q(\zeta')$ are analytic and polynomially bounded in $\{\zeta';(0,\operatorname{Im}\zeta')\in-\Gamma(P,N)\}$. Hence it follows that

$$\mathscr{L}_\theta*B(D)v|_{\partial H'}=\sum\mathscr{T}_k*B_k(D)v|_{\partial H'}+\sum(\mathscr{S}_k*D_1^kP(D)v)|_{\partial H'}\tag{12.9.16}$$

where $\hat{\mathscr{T}}_k(\zeta')=T_k(\zeta')$ and $\mathscr{S}_k(x_1,.)=0$, $x_1\geqq0$,

$$\hat{\mathscr{S}}_k(x_1,\zeta')=-(2\pi)^{-1}q(\zeta')^{-1}S_k(\zeta')\int e^{ix_1w}/P_-(w,\zeta')\,dw,\qquad x_1<0.$$

These distributions are supported by the cone $\Gamma^\circ(P,N)+\mathbb{R}(1,0,\dots,0)$.

If u satisfies the homogeneous mixed problem as assumed in the theorem then $u(x)=0$ for $x_n<\varepsilon x_1$ by Holmgren's uniqueness theorem, if ε is the smallest positive solution of $P_m(\varepsilon,0,\dots,-1)=0$. If $y\in\operatorname{supp}u$ and

$$y\in\{(0,x')\}-(\Gamma^\circ(P,N)+\mathbb{R}\theta)$$

it follows that $0\leqq\varepsilon y_1\leqq y_n\leqq x_n$, $|y'-x'|\leqq C(x_n-y_n)$. This means that y belongs to a compact set K. Let $\chi\in C_0^\infty(\mathbb{R}^n)$ be equal to 1 in a neighborhood V of K and apply (12.9.16) to the product of χ and a C^∞ extension of u to $\mathbb{R}^n$. Then $B_k(D)v=0$ and $P(D)v=0$ in $V\cap H'$ so it follows that $\mathscr{L}_\theta*B(D)u=0$ at $(0,x')$. Thus $D_1^j\mathscr{L}_\theta*u=0$ in $\partial H'$ for every j, for we can take $B(D)=D_1^j$. Set $U=\mathscr{L}*u$ in H' and $U=0$ in $\complement H'$. Then we have $P(D)U=0$ in $\mathbb{R}^n$ since $U\in C^\infty$ and this is true in H'. Now $U=0$ in $\complement H$ so it follows that $U=0$, which completes the proof.

One inference which can immediately be drawn from Theorem 12.9.3 is that if $\mu = m_+ + 1$ and the Lopatinski determinant of $B_1, \ldots, B_{m_+}$ is not identically 0 then the mixed problem (12.9.1)–(12.9.3) cannot be solved for arbitrary data. In fact, let $f = \phi = 0$, $\phi_j = 0$ for $j \leqq m_+$ and $\phi_{m_+ + 1} \in C_0^\infty$. Then the equation $B_{m_+ + 1}(D)u|_{\partial H'} = \phi_{m_+ + 1}$ gives by Theorem 12.9.3

$$\mathscr{L}_\theta * \phi_{m_+ + 1} = 0.$$

As in the proof of Theorem 12.9.2 it follows that $0 \in \operatorname{supp} \mathscr{L}_\theta$. If $\mathscr{L}_\theta^T \in \mathscr{E}'$ is equal to $\mathscr{L}_\theta$ when $x_n \leqq T$, say, it follows that $x_n \geqq T$ in $\operatorname{supp} \mathscr{L}_\theta^T * \phi_{m_+ + 1}$. Hence $x_n \geqq T$ in $\operatorname{supp} \phi_{m_+ + 1}$ by the theorem of supports. Since T is arbitrary this proves that $\phi_{m_+ + 1} = 0$; the mixed problem does not have a solution for any non-trivial $\phi_{m_+ + 1}$ of compact support. Summing up, we conclude from Theorems 12.9.2 and 12.9.3 that a *necessary condition for existence and uniqueness of solutions of the mixed problem* (12.9.1)–(12.9.3) *is that* $\mu = m_+$ *and that the Lopatinski determinant is not identically* 0. This will be assumed from now on. We shall prove that everything is then quite analogous to the Cauchy problem for differential operators. After defining a principal symbol of L we shall first prove that there is uniqueness if and only if this principal symbol is different from 0 at N, and then that existence is equivalent to a condition which is perfectly analogous to hyperbolicity.

Lemma 12.9.4. *Let $g_+(w)$ and $g_-(w)$ be polynomials in one variable w of degree m_+ and m_-, with leading coefficient* 1 *and no zeros in common. For every polynomial $r(w)$ of degree $< m_+ + m_-$ with sufficiently small coefficients there exist uniquely determined $r_+(w)$ and $r_-(w)$ of degree $< m_+$ and m_- such that the coefficients are small and*

$$g_+(w)\, g_-(w) + r(w) = (g_+(w) + r_+(w))(g_-(w) + r_-(w)).$$

The coefficients of r_+ and r_- are analytic functions of those of r.

Proof. This follows at once from the implicit function theorem (Theorem A.1.1 with several unknowns), for the linearized equation

$$r(w) = g_+(w)\, r_-(w) + g_-(w)\, r_+(w)$$

has a unique solution given by the decomposition of $r(w)/g_+(w)\, g_-(w)$ in partial fractions.

Proposition 12.9.5. *Assume that the Lopatinski determinant $L(\zeta')$ which is analytic for $\operatorname{Im} \zeta' \in -\Gamma' = \{\theta';\ (\lambda, \theta') \in -\Gamma(P, N)$ for some $\lambda\}$ is not identically* 0. *Then there is an integer h such that $\varepsilon^h L(\zeta'/\varepsilon)$ is analytic in*

(ζ', ε) in a neighborhood of $(\mathbb{R}^{n-1} - i\Gamma') \times \{0\}$ and the limit $L_0(\zeta')$ when $\varepsilon \to 0$ is not identically 0.

Proof. Since

$$\int b_j(w)\, w^{k-1}/p_+(w)\, dw = \varepsilon^{m_+ - k} \int b_j(w/\varepsilon)\, w^{k-1}/\varepsilon^{m_+} p_+(w/\varepsilon)\, dw$$

we have if $\operatorname{Im} \zeta \in -\Gamma(P, N)$

$$\begin{aligned} L(\zeta/\varepsilon) &= L(B_1(\zeta/\varepsilon + w\theta), \dots; P_+(w, \zeta/\varepsilon)) \\ &= \varepsilon^{\Sigma(m_+ - k)} L(B_1(\zeta/\varepsilon + w\theta/\varepsilon), \dots; \varepsilon^{m_+} P_+(w/\varepsilon, \zeta/\varepsilon)). \end{aligned}$$

From the equation

$$P(\zeta/\varepsilon + w\theta) = q(\zeta/\varepsilon)\, P_+(w, \zeta/\varepsilon)\, P_-(w, \zeta/\varepsilon)$$

it follows that

$$\varepsilon^m P(\zeta/\varepsilon + w\theta/\varepsilon) = \varepsilon^{m_0} q(\zeta/\varepsilon)\, \varepsilon^{m_+} P_+(w/\varepsilon, \zeta/\varepsilon)\, \varepsilon^{m_-} P_-(w/\varepsilon, \zeta/\varepsilon)$$

where $\varepsilon^{m_+} P_+(w/\varepsilon, \zeta/\varepsilon)$ and $\varepsilon^{m_-} P_-(w/\varepsilon, \zeta/\varepsilon)$ have leading coefficients 1 and all zeros in the half planes $\operatorname{Im} w > 0$ resp. $\operatorname{Im} w < 0$. The left-hand side is a polynomial in ε reducing to

$$P_m(\zeta + w\theta) = q_{m_0}(\zeta)\, P_{m_+}(w, \zeta)\, P_{m_-}(w, \zeta)$$

when $\varepsilon = 0$. If we apply Lemma 12.9.4 after division by $\varepsilon^{m_0} q(\zeta/\varepsilon) = q_{m_0}(\zeta) + O(\varepsilon)$ it follows that $\varepsilon^{m_+} P_+(w/\varepsilon, \zeta/\varepsilon)$ is analytic in ε and ζ and reduces to $P_{m_+}(w, \zeta)$ when $\varepsilon = 0$. The proposition is an immediate consequence if h is chosen minimal so that $\varepsilon^h L(\zeta'/\varepsilon)$ is analytic at $\varepsilon = 0$.

Since $L_0(t\zeta') = t^h L_0(\zeta')$ for $t > 0$ and $\{w \in \mathbb{C}; \operatorname{Im} w\zeta' \in -\Gamma'\}$ is convex for every $\zeta' \neq 0$, we can extend L_0 uniquely to an analytic function in

$$\bigcup_{w \neq 0} w(\mathbb{R}^{n-1} - i\Gamma')$$

which is homogeneous of degree h. It will be called the *principal symbol* of the mixed problem (12.9.1)–(12.9.3). Without reference to our special coordinates it is invariantly defined in

$$\bigcup_{w} w(\mathbb{R}^n - i\Gamma(P, N))/\mathbb{C}\theta.$$

We shall now prove an analogue of Theorem 8.6.7.

Theorem 12.9.6. *If the principal symbol of the mixed problem* (12.9.1)–(12.9.3) (*with $\mu = m_+$*) *vanishes at $N/\mathbb{R}\theta$ then one can find a solution $u \in C^\infty(H \cap H')$ of the homogeneous mixed problem such that $H \cap \partial H' \subset \operatorname{supp} u$.*

Proof. As in the proof of Theorem 8.6.7 we first look for a solution of the equation

$$L(sN'+t\xi')=0$$

where t/s is small and $\operatorname{Im} s$ large negative. Here $\xi'\in\mathbb{R}^{n-1}$ will be chosen later. Writing $t=ws$ and $s=1/\varepsilon$ we obtain the equation

$$\varepsilon^h L((N'+w\xi')/\varepsilon)=0$$

which is analytic at $\varepsilon=w=0$ and reduces to $L_0(N'+w\xi')=0$ when $\varepsilon=0$. Choose ξ' so that $L_0(N'+w\xi')$ is not identically 0. Then we obtain a Puiseux series expansion

$$t(s)=s\sum_1^\infty c_j(s^{-1/p})^j$$

converging for $|s^{1/p}|>M$, say, such that $L(sN'+t(s)\xi')=0$ when s is in an angular neighborhood of the negative imaginary axis.

Our construction can now proceed as a combination of those in Theorem 12.9.2 and in Theorem 8.6.7. Since the rank of the Lopatinski matrix is $<m_+$ when $\zeta=(0,sN'+t(s)\xi')$ we can find a polynomial $H(w,s)$ in w of degree $<m_+$ whose coefficients are analytic functions of $s^{-1/p}$ at 0 and not all identically 0, such that

$$\int B_j(w,sN'+t(s)\xi')\,H(w,s)/P_+(w,(0,sN'+t(s)\xi'))\,dw=0,$$
$$j=1,\dots,m_+.$$

The coefficients are just suitable minors of the Lopatinski matrix. Thus

$$U(x_1,s)=(2\pi i)^{-1}\int H(w,s)\,e^{iwx_1}/P_+(w,(0,sN'+t(s)\xi'))\,dw$$

satisfies the equation $P(D_1,sN'+t(s)\xi')\,U(x_1,s)=0$ and the boundary conditions $B_j(D_1,sN'+t(s)\xi')\,U(0,s)=0$ when $C(|t(s)|+1)<-\operatorname{Im} s$ for a constant C so large that this implies $\operatorname{Im}(sN'+t(s)\xi')\in-\Gamma'$. We have a polynomial bound for $U(x_1,s)$. Let $1-1/p<\rho_0<\rho<1$ and let γ_τ be the curve

$$\mathbb{R}\ni\sigma\to\sigma-i(\sigma^2+1)^{\rho_0/2}-i\tau.$$

$U(x_1,s)$ is defined on γ_τ if τ is large enough, and

$$u(x)=\int_{\gamma_\tau} e^{i\langle x',sN'+t(s)\xi'\rangle-(is)^\rho}\,U(x_1,s)\,ds$$

is independent of τ. (We take $-\pi<\arg s<0$.) It follows that $u\in C^\infty(H')$, $P(D)u=0$ in H', $B_j(D)u=0$ on $\partial H'$, and when $\tau\to\infty$ we obtain $u=0$ when $\langle x,N\rangle<0$. From the proof of Proposition 12.9.5 it follows that $D_1^j U(0,s)$ is a meromorphic function of $s^{1/p}$ at ∞ and not identically 0 for some j. The proof of Theorem 8.6.7′ then gives with no change that $D_1^j u(0,x')$ is analytic when $\langle x',N'\rangle>0$. Hence the support is $H\cap\partial H'$ for some j, which completes the proof.

Theorem 12.9.6 has a converse which we state as a uniqueness theorem:

Theorem 12.9.7. *If the principal symbol of the mixed problem* (12.9.1)–(12.9.3) (*with* $\mu=m_+$) *does not vanish at* $N/\mathbb{R}\theta$ *then a solution* $u\in C^\infty(H\cap H')$ *of the mixed problem is uniquely determined by the data* f, ϕ, ϕ_j.

Proof. If the data are 0 we have $\mathscr{L}*u=0$ in H' if u is defined as 0 in $H'\cap\complement H$. By Proposition 12.9.5 we know that $\varepsilon^h L(-i\zeta'/\varepsilon)$ is analytic and never 0 for ζ' in a neighborhood of N' and small ε. Taking ε purely imaginary we conclude that L^{-1} has a polynomially bounded extension to a complex conic neighborhood of N' at ∞. Hence Theorem 8.6.15 shows that

$$(x',N')\notin WF_A(u(x_1,.)),\qquad x'\in\mathbb{R}^{n-1},$$

for every fixed x_1. Now Holmgren's uniqueness theorem gives $u=0$ (cf. Corollary 8.6.9), for $u=0$ when $\langle x',N'\rangle<0$.

From Theorem 8.6.8 we can of course obtain much more precise uniqueness theorems but we leave this to study the existence of solutions of the mixed problem. First we prove an analogue of Theorem 12.3.1.

Theorem 12.9.8. *If the mixed problem* (12.9.1)–(12.9.3) ($\mu=m_+$) *has a unique solution* $u\in C^\infty(H\cap H')$ *for arbitrary* C^∞ *data vanishing at* ∂H *then there exists a constant* $\gamma_0\leqq\tau_0$ *such that*

$$L(\xi'+i\gamma N')\neq 0 \quad \textit{if } \xi'\in\mathbb{R}^{n-1} \textit{ and } \gamma<\gamma_0. \tag{12.9.17}$$

Proof. By Banach's theorem the solution depends continuously on the data. For a fixed point $y\in\partial H'\cap(H\setminus\partial H)$ we can therefore find a compact set $K\subset H\cap H'$ and constants C, k such that

$$\sum_{|\alpha|\leqq m}|D^\alpha u(y)|\leqq C\Big(\sum_{|\alpha|\leqq k}\sup_K|D^\alpha f|+\sum_j\sum_{|\alpha|\leqq k}\sup_{K\cap\partial H'}|D^\alpha\phi_j|\Big) \tag{12.9.18}$$

if $\phi=0$ and f, ϕ_j vanish of infinite order on ∂H. If $U\in C^\infty(H')$ and

$$P(D)U=0 \ \text{ in } H',\quad B_j(D)U=0 \ \text{ on } \partial H',\quad j=1,\dots,m_+,$$

we can apply (12.9.18) to $u=\chi(\langle .,N\rangle)U$ with $\chi\in C^\infty(\mathbb{R})$ equal to 0 on $(-\infty,0)$ and equal to 1 on (c,∞) where $c=\langle y,N\rangle/2$. Then

$$P(D)u=0 \quad \text{when } \langle x,N\rangle>c;$$

$$B_j(D)u=0 \quad \text{on } \partial H' \text{ when } \langle x,N\rangle>c.$$

Writing $K'=\{x;\, x\in K, \langle x,N\rangle\leqq c\}$ we obtain with new constants C', k'

$$\sum_{|\alpha|\leqq m}|D^\alpha U(y)|\leqq C'\sum_{|\alpha|\leqq k'}\sup_{K'}|D^\alpha U|. \tag{12.9.18)'$$

If $L(\zeta')=0$, $\zeta'=\xi'+i\gamma N'$, we can choose a polynomial $H(w)$ of degree $<m_+$ such that $\tilde{H}(0)=1$ and

$$U_1(x_1)=(2\pi i)^{-1}\int H(w)e^{iwx_1}/P_+(w,(0,\zeta'))dw$$

satisfies not only the equation $P(D_1,\zeta')U_1(x_1)=0$ but also the boundary conditions $B_j(D_1,\zeta')U_1(0)=0$. This is equivalent to the equations

$$(2\pi i)^{-1}\int H(w)B_j(w,\zeta')/P_+(w,(0,\zeta'))dw=0$$

which is a system of m_+ equations for the coefficients of $H(w)$ with determinant $L(\zeta')=0$. Now we apply (12.9.18)′ to $U(x)=U_1(x_1)e^{i\langle x',\zeta'\rangle}$ and obtain

$$\sum_{j\leqq m}|D_1^j U_1(0)|\leqq C(1+|\zeta'|)^M e^{c\gamma} \tag{12.9.18)''$$

for some C and M. Here $D_1^j U_1(0)$ is the residue at infinity of $w^j H(w)/P_+(w,(0,\zeta'))$, hence $H(w)$ is the polynomial part of

$$\sum_{j<m_+} D_1^j U_1(0)P_+(w,(0,\zeta'))/w^{j+1}.$$

In view of (12.9.14) it follows that

$$1=\tilde{H}(0)\leqq C\sum|D_1^j U_1(0)|\tilde{P}((0,\zeta')).$$

Using (12.9.18)″ we now obtain

$$1\leqq C'(1+|\zeta'|)^{M'}e^{c\gamma}, \qquad \gamma<\tau_0,\; L(\zeta')=0.$$

This is quite analogous to (12.3.5) so (12.9.17) follows from the Tarski-Seidenberg theorem by repeating the end of the proof of Theorem 12.3.1. We leave the details for the reader.

Definition 12.9.9. The mixed problem (12.9.1)–(12.9.3) is called hyperbolic if $\mu=m_+$, the principal symbol L_0 does not vanish at $N/\mathbb{R}\theta$ and (12.9.17) is valid for some γ_0.

As in Theorem 12.4.4 we can extend the zero free region of L:

Theorem 12.9.10. *If the mixed problem is hyperbolic then the component* Σ *of* $N'=N/\mathbb{R}\theta$ *in* $\{\xi'\in\Gamma(P,N)/\mathbb{R}\theta,\ L_0(\xi')\neq 0\}$ *is an open convex cone,*

$$L_0(\zeta')\neq 0, \quad \operatorname{Im}\zeta'\in -\Sigma, \tag{12.9.19}$$

$$L(\zeta')\neq 0, \quad \operatorname{Im}\zeta'\in \gamma_0 N'-\Sigma. \tag{12.9.20}$$

Proof. If $\xi'\in\mathbb{R}^{n-1}$ it follows from Proposition 12.9.5 that

$$L_0(\xi'+zN')=\lim_{\varepsilon\to+0}\varepsilon^h L((\xi'+zN')/\varepsilon), \qquad \operatorname{Im} z<0.$$

Since $L((\xi'+zN')/\varepsilon)\neq 0$ by (12.9.17) when $\operatorname{Im} z<\varepsilon\gamma_0$ and the limit $L_0(\xi'+zN')=z^h L_0(\xi'/z+N')$ is not identically 0, it follows from Hurwitz' theorem that $L_0(\xi'+zN')\neq 0$ when $\operatorname{Im} z<0$. In view of the homogeneity of L_0 we conclude that

$$F_{\xi'}(z)=L_0(\xi'+zN')$$

is analytic and $\neq 0$ when $\operatorname{Im} z\neq 0$. If $\xi'\in\Sigma$ then

$$\operatorname{Im}(\xi'+zN')/i=-\xi'-\operatorname{Re} z N'\in -\Gamma(P,N)/\mathbb{R}\theta, \qquad \operatorname{Re} z\geqq 0,$$

so $F_{\xi'}(z)$ is analytic when $\operatorname{Re} z\geqq 0$, and $F_{\xi'}(z)\neq 0$ when $\operatorname{Re} z=0$, because $F_{\xi'}(0)\neq 0$. Since

$$L_0(zN'+\xi')=z^h L_0(N'+\xi'/z)$$

and $L_0(N')\neq 0$ we can find a constant C such that $F_{\xi'}(z)\neq 0$ if $|z|>C|\xi'|$. Now $F_{N'}(z)=(z+1)^h L_0(N')$ has no zero in the half plane $\operatorname{Re} z\geqq 0$, so this remains true for reasons of continuity for every $F_{\xi'}(z)$, $\xi'\in\Sigma$. In particular $L_0(\lambda N'+\mu\xi')\neq 0$ if $\lambda,\mu\geqq 0$, $\lambda+\mu>0$, so Σ is star shaped with respect to N'.

Next we prove (12.9.20). Let $\xi'\in\mathbb{R}^{n-1}$ and $\eta'\in\Sigma$. If $\gamma<\gamma_0$ then

$$G_\gamma(z)=L(\xi'+i\gamma N'+iz\eta')$$

is an analytic function of z when $\operatorname{Re} z\leqq 0$, and there is no zero with $\operatorname{Re} z=0$. Since

$$(iz)^{-h}L(\xi'+i\gamma N'+iz\eta')\to L_0(\eta')\neq 0, \qquad z\to\infty,$$

uniformly in γ when γ is in a compact set, there is no zero of G_γ far away so the number of zeros with $\operatorname{Re} z\leqq 0$ is independent of γ. If γ is a very large negative number we have for some constant C

$$L(\xi'+i\gamma N'+iz\eta')\neq 0 \quad \text{if } \operatorname{Re} z\leqq 0,\ |z/\gamma|>C,$$

for $(iz)^{-h}L(\xi'+i\gamma N'+iz\eta')$ will be close to $L_0(\eta')$. On the other hand,

$$(i\gamma)^{-h}L(\xi'+i\gamma N'+i\gamma w\eta')\to L_0(N'+w\eta'), \qquad \gamma\to-\infty,$$

uniformly when $\operatorname{Re} w \geqq 0$ and $|w| \leqq C$. We proved above that the limit has no zeros in the half plane $\operatorname{Re} w \geqq 0$. Hence

$$L(\xi' + i\gamma N' + iz\eta') \neq 0$$

if $\operatorname{Re} z \leqq 0$ and γ is a sufficiently large negative number. This remains true for all $\gamma < \gamma_0$ and proves (12.9.20). Going back to the argument at the beginning of the proof with N' replaced by η' we deduce (12.9.19). We also conclude that Σ is starshaped with respect to every $\eta' \in \Sigma$, that is, Σ is convex. The proof is complete.

Lemma 12.9.11. *It follows from* (12.9.20) *that for some C and M*

$$(1 + |\zeta'|)^{-M} \leqq C|L(\zeta')|, \qquad \operatorname{Im} \zeta' \in (\gamma_0 - 1)N' - \bar{\Sigma}.$$

Proof. By (12.9.20) we have

$$f(t) = \inf\{|L(\zeta')|;\ |\zeta'| \leqq t,\ \operatorname{Im} \zeta' \in (\gamma_0 - 1)N' - \bar{\Sigma}\} > 0, \qquad t \in \mathbb{R}_+,$$

for $N' + \bar{\Sigma} \subset \Sigma$ and the set of ζ' considered is compact. From Corollary A.2.4 and Theorem A.2.5 we obtain that $f(t) = At^a(1 + o(1))$, $t \to \infty$, where $A \neq 0$ and a is a rational number. Thus $(1+t)^{-M} \leqq Cf(t)$ if $M \geqq -a$ and C is large enough, which proves the lemma.

We are now ready to construct "fundamental solutions" of the mixed problem:

Theorem 12.9.12. *If the mixed problem* (12.9.1)–(12.9.3) *is hyperbolic, then one can find distributions e_k in H', $k = 1, \dots, m_+$ which are C^∞ functions of $\langle x, \theta\rangle$ with values in $\mathscr{D}'(\mathbb{R}^{n-1})$ such that*

$$\operatorname{supp} e_k \subset \Sigma^\circ + \Gamma^\circ(P, N),\ P(D)e_k = 0 \quad \text{in } H', \tag{12.9.21}$$

$$B_j(D)e_{k|\partial H'} = \delta_{jk}\delta(x').$$

Here Σ° is the dual cone $\subset \partial H'$ of the component Σ of $N/\mathbb{R}\theta$ in $\{\xi' \in \Gamma(P, N)/\mathbb{R}\theta,\ L_0(\xi') \neq 0\}$, where L_0 is the principal symbol of the mixed problem. The notation δ_{jk} stands for the Kronecker delta.

Proof. Let $H_k(w, \zeta')$, $k = 1, \dots, m_+$, be a polynomial of degree $< m_+$ in w and set when $\operatorname{Im} \zeta' \in (\gamma_0 - 1)N' - \Sigma$

$$E_k(x_1, \zeta') = (2\pi i)^{-1} \int H_k(w, \zeta') e^{iwx_1}/P_+(w, (0, \zeta'))\, dw.$$

Then we have $P(D_1, \zeta')E_k(x_1, \zeta') = 0$, and

$$B_j(D_1, \zeta')E_k(0, \zeta') = (2\pi i)^{-1} \int B_j(w, \zeta')\, H_k(w, \zeta')/P_+(w, (0, \zeta'))\, dw$$

$$= \sum_{r=1}^{m_+} H_{rk}(\zeta')(2\pi i)^{-1} \int w^{r-1} B_j(w, \zeta')/P_+(w, (0, \zeta'))\, dw$$

where we have written

$$H_k(w,\zeta')=\sum_1^{m_+} H_{rk}(\zeta')w^{r-1}.$$

Thus $B_j(D_1,\zeta')E_k(0,\zeta')$ is the product of the Lopatinsky matrix and the matrix (H_{rk}). We obtain $B_j(D_1,\zeta')E_k(0,\zeta')=\delta_{jk}$, the Kronecker delta, by choosing for (H_{rk}) the cofactor matrix of the Lopatinski matrix, divided by L. It follows from Lemma 12.9.11 that the coefficients of $H_k(w,\zeta')$ are analytic and polynomially bounded when $\operatorname{Im}\zeta'\in(\gamma_0-1)\cdot N'-\Sigma$. Hence Theorem 7.4.3 shows that $E_k(x_1,\zeta')$ is the Fourier-Laplace transform with respect to x' of a C^∞ function $e_k(x_1,.)$ of x_1 with values in $\mathscr{D}'(\mathbb{R}^{n-1})$ supported by Σ°. As in the proof of Theorem 12.9.2 we can improve the information on the support by defining $e_k(x_1,.)=0$ when $x_1<0$. Then $\operatorname{supp} e_k\subset H$ and $\operatorname{supp} P(D)e_k\subset\{0\}\times\Sigma^\circ$. Since $e_k=E(P,N)*(P(D)e_k)$ this completes the proof of (12.9.21).

Corollary 12.9.13. *If the mixed problem* (12.9.1)–(12.9.3) *is hyperbolic then it has a unique solution* $u\in C^\infty(H\cap H')$ *for all* C^∞ *data* f, ϕ, ϕ_j *which are compatible at* $\partial H\cap\partial H'$ *in the sense that a formal power series solution exists at every point there. The solution is determined at* $x\in H\cap H'$ *by the data* ϕ_j *in* $R_x=\partial H'\cap(\{x\}-\Gamma^\circ(P,N)-\Sigma^\circ)$ *and the data* f, ϕ *in* $(\{x\}\cup R_x)-\Gamma^\circ(P,N)$.

Proof. We can extend f and ϕ to C^∞ functions F and Φ in $\mathbb{R}^n$ and find a solution u_0 of the Cauchy problem (12.9.1), (12.9.2) in H with these data. (See Theorem 12.5.6.) With $u=u_0+v$ the mixed problem is then reduced to

$$P(D)v=0 \quad\text{in } H\cap H', \qquad D^\alpha v=0 \quad\text{in } H'\cap\partial H,\ |\alpha|<m,$$
$$B_j(D)v=\phi_j-B_j(D)u_0=\psi_j, \qquad j=1,\dots,m_+, \quad\text{on } H\cap\partial H'.$$

All derivatives of v must then vanish on $\partial H\cap\partial H'$ for a formal power series solution so the compatibility conditions mean that ψ_j vanishes of infinite order at $\partial H'\cap\partial H$ since they are preserved by the reduction. We extend ψ_j to a C^∞ function on $\partial H'$ equal to 0 in $\partial H'\cap\complement H$. Now a solution of the mixed problem is given by

$$v=\sum e_k*\psi_k$$

with the convolution taken in the x' variables only.

By Theorem 12.9.7 the solution constructed is the only solution of the mixed problem. We have $v=0$ at $x\in H\cap H'$ if $\psi_j=0$ in R_x, which follows if $\phi_j=0$ in R_x and the data F, Φ of the Cauchy problem vanish in $R_x-\Gamma^\circ(P,N)$. For a suitable choice of the extensions F, Φ this is true if f and ϕ vanish in a neighborhood of $R_x-\Gamma^\circ(P,N)$.

Since u_0 vanishes at x if F and Φ vanish in $\{x\}-\Gamma^\circ(P,N)$, we have proved that u vanishes at x if the data vanish in a neighborhood of the listed sets. The statement in the corollary follows if we apply this result to $u(x-\varepsilon N)$ and let $\varepsilon \to 0$.

The existence proof here is of course applicable also when the data are distributions. The fundamental solution $E(P,N)$ is a C^∞ function of x_1 with values in $\mathscr{D}'(\mathbb{R}^{n-1})$ for $x_1 \geqq 0$ and for $x_1 \leqq 0$. Let f be a C^∞ function of $x_1 \geqq 0$ with values in $\mathscr{D}'(\mathbb{R}^{n-1})$, supported by the half space H, and let $\phi_j \in \mathscr{D}'(\partial H')$, $\operatorname{supp} \phi_j \subset H \cap \partial H'$. Then the proof of Corollary 12.9.13 shows that there is a unique solution of (12.9.1), (12.9.3) which is a C^∞ function of $x_1 \geqq 0$ with values in $\mathscr{D}'(\mathbb{R}^n)$ and support in H. (It is easier to pose zero Cauchy data in the case of distribution solutions.)

Example 12.9.14. For the wave operator $P(D)=D_1^2+\ldots+D_{n-1}^2-D_n^2$ ($n \geqq 4$) we shall determine when the oblique derivative boundary operator $B(D)=\sum_1^n b_j D_j$ gives a hyperbolic mixed problem ($b_1=1$). We have

$$L(\zeta')=\sum_1^n b_j \zeta_j, \quad \zeta_1=(\zeta_n^2-\zeta_2^2-\ldots-\zeta_{n-1}^2)^{\frac{1}{2}}$$

where $\operatorname{Im}\zeta_1>0$ and $\operatorname{Im}\zeta'$ is in the backward light cone, and

$$L_0(\xi')=\sum_2^n b_j \xi_j-(\xi_n^2-\xi_2^2-\ldots-\xi_{n-1}^2)^{\frac{1}{2}}$$

when ξ' is in the forward light cone. The non-characteristic condition is therefore $b_n \neq 1$. If we set $\xi''=(\xi_2,\ldots,\xi_{n-1})$ and

$$W=\left\{\sum_2^{n-1} b_j \xi_j;\ |\xi''|=1\right\}$$

which is either an ellipse (possibly degenerate) or the closed interior of an ellipse, the other condition for hyperbolicity is that

$$F(z)=(z^2-1)^{\frac{1}{2}}+b_n z$$

takes no value in W when $\operatorname{Im} z<0$. Here $F(z)=z(b_n-(1-z^{-2})^{\frac{1}{2}})$, $|z|>1$, so F is analytic and odd outside $(-1,1)$; thus F must not take any value in W when $\operatorname{Im} z \neq 0$.

1) If $\operatorname{Re} b_n>0$ then 0 is in the range of $F(z)$ when $\operatorname{Im} z<0$ unless $0<b_n<1$. Then we must have $W \subset \mathbb{R}$, that is, all b_j are real.

2) If $b_n=0$ then $\{F(z), \operatorname{Im} z<0\}$ is the upper half plane apart from $[0,i]$. Thus $W \subset \mathbb{R}$, that is, all b_j are real, or $W \subset [-i,i]$, that is, all b_j are purely imaginary and $\sum_2^{n-1} |b_j|^2 \leqq 1$.

3) If $\operatorname{Re} b_n = 0$ and $\operatorname{Im} b_n \neq 0$ then all b_j must be purely imaginary and

$$|b_2|^2 + \ldots + |b_{n-1}|^2 \leqq |b_n|^2 + 1,$$

which also covers the second case in 2) when $\operatorname{Im} b_n$ becomes 0.

4) If $\operatorname{Re} b_n < 0$ then the range of F when $\operatorname{Im} z \neq 0$ is the exterior of the ellipse $E = \{b_n x + iy;\ (x, y) \in S^1\}$ apart from the curves $\pm F([1, \infty])$. It follows that W is in the closed interior of E unless b_n is real and $W \subset \mathbb{R}$.

Summing up, we have two main cases.

a) All b_j are real and $b_n < 1$. The principal symbol vanishes at a point ξ' in the forward light cone if

$$\xi_1 = (\xi_n^2 - \xi_2^2 - \ldots - \xi_{n-1}^2)^{\frac{1}{2}}$$

satisfies $\xi_1 = \sum_2^n b_j \xi_j$, that is, $\sum_2^n b_j \xi_j > 0$ and

$$\xi_n^2 = \xi_2^2 + \ldots + \xi_{n-1}^2 + \Big(\sum_2^n b_j \xi_j\Big)^2.$$

Thus

$$\Sigma = \left\{\xi';\ \xi_n > \left(\xi_2^2 + \ldots + \xi_{n-1}^2 + \left(\max\left(0, \sum_2^n b_j \xi_j\right)\right)^2\right)^{\frac{1}{2}}\right\}.$$

The propagation speed described by Σ° is increased in all directions if $(b_2, \ldots, b_n)$ is in the interior of the forward light cone and in no direction if it is in the backward light cone. When it is twice the speed of light Fig. 5 shows the outer boundary of the support of the solution of the mixed problem for the wave equation with $f = \delta_{2,0,0}$ and $\phi = 0$, for fixed integer x_3.

b) A calculation which we leave for the reader shows that if we write $(b_2, \ldots, b_{n-1}, -b_n) = F + iE$ with F and E real, then $1 + [E, E] \geqq 0$, F is in the forward light cone and with $[\ ,\ ]$ denoting the Lorentz scalar product (as in Section 12.1)

$$[E, F]^2 \leqq [F, F](1 + [E, E]).$$

This sums up the remaining cases and the propagation speed is unchanged in view of the Lorentz invariance of the conditions.

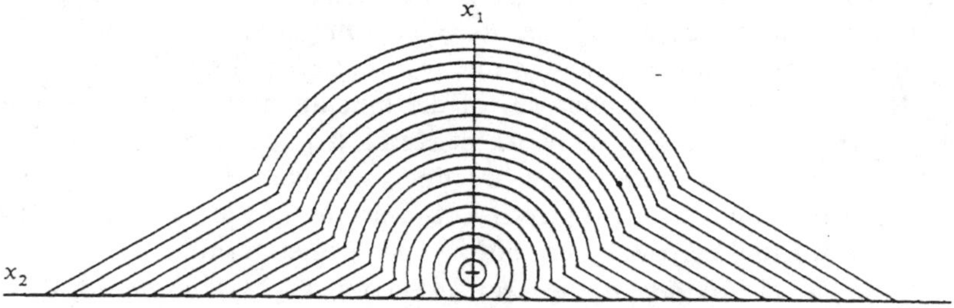

Fig. 5

Notes

Section 12.1 is devoted to the explicit formulas of M. Riesz [1, 4] for the wave equation in $\mathbb{R}^{3+1}$. The formula (12.1.13) here differs from that of Riesz by containing the Gaussian curvature rather than a mean curvature related to the embedding of Σ in $\mathbb{R}^4$. The two formulas are related by a degenerate form of the Gauss equations. The asymptotic expansion of the solution of the oscillatory Cauchy problem discussed in Section 12.2 has a long tradition in optics at least for the leading terms. (See for example Sommerfeld [1] on the phase shift at the caustics.) The asymptotic behavior at simple caustics goes back to Airy [1] and was made precise by Ludwig [2]. For the case of more general caustics the reader should consult Duistermaat [1] and Guillemin-Sternberg [1].

It was already emphasized by Hadamard [1] that in the C^∞ case as opposed to the analytic case of Cauchy-Kovalevsky the solution of the Cauchy problem need not depend continuously on the data. (In the terminology of Hadamard, the Cauchy problem need not be correctly posed.) In view of the closed graph theorem this means that solutions need not exist for arbitrary C^∞ data. Petrowsky [2] essentially characterized the operators for which the Cauchy problem can always be solved in $\mathscr{S}$ or $\mathscr{S}'$. (His condition is the inequality (12.3.5) here.) By the Holmgren uniqueness theorem this implies existence of solutions of the Cauchy problem in $\mathscr{D}'$ and C^∞ also. Gårding [1] showed that (12.3.1), (12.3.2) is necessary and sufficient for uniqueness and existence of solutions of the Cauchy problem in C^∞. The passage from (12.3.5) to (12.3.1) required a rather complicated algebraic discussion which has now become simple and standardized thanks to the Tarski-Seidenberg theorem. (The weaker Petrowsky condition (12.3.5) remains interesting though for we shall see in Chapter XVI that it is the appropriate condition for convolution operators.) Gårding also proved the precise properties of hyperbolic polynomials given in Section 8.7 and the beginning of Section 12.4. The description of the lower order terms in Theorem 12.4.6 was conjectured by Gårding [1] and proved by Svensson [1] (see also Chaillou [1] and Munster [1, 2]). In the two dimensional case it is due to E.E. Levi [1] and Lax [1].

The study of the singularities of the fundamental solution in Section 12.6 has its roots in the work of Herglotz [1] and Petrowsky [4]. The general results given here are due to Atiyah-Bott-Gårding [1]. The reader should turn to Atiyah-Bott-Gårding [2] for a study of the necessity of the Petrowsky condition and further results.

The global uniqueness theorem in Section 12.7 is due to John [3] in the non-characteristic case and to Brodda [1] in the characteristic case. The existence theorem which follows is proved along lines which go back to Cauchy [1]; there is now a vast literature on such existence theorems in Gevrey classes for operators with variable coefficients also.

The work of Petrowsky [2] was not restricted to the non-characteristic Cauchy problem. However, a complete study of the characteristic Cauchy problem was first given in Hörmander [21]. The passage from local to global results relies on an idea of Malgrange similar to those used in Section 10.6. Complete results do not exist on the more general problem of characterizing operators having a fundamental solution with support in a convex cone with an edge of arbitrary codimension. The best results are due to Enqvist [1]. The difficulty of the problem is underlined by a result of his stating that even if a fundamental solution exists with support in Γ_j for every j and Γ_j decreases, there need not be any with support in the intersection of all Γ_j. Uniqueness theorems for the characteristic Cauchy problem with global conditions can be found in Täcklind [1] and Gelfand-Šilov [1].

General results on the constant coefficient mixed problem were first stated by Hersh [1, 2]. Sakamoto [1] observed that there are "some rough discussions" and completed the argument under some additional hypotheses. In particular she assumed that the plane carrying the mixed data is non-characteristic, and no compelling reasons were given for the number of boundary conditions chosen. These points were supplemented by Shibata [1]. In addition to these results we have added here a fairly complete discussion of uniqueness, completely similar to Holmgren's uniqueness theorem. It should be observed that the papers by Hersh also contain a study of some mixed problems with infinite propagation speed. The concluding example is largely taken from Gårding [6].

Chapter XIII. Differential Operators of Constant Strength

Summary

In this chapter we shall study differential operators which in the spaces $B_{p,k}$ can be considered as bounded perturbations of differential operators with constant coefficients. This requires that the constant coefficient operators obtained by "freezing" the argument in the coefficients at a point x_0 have a strength independent of x_0. By means of a simple perturbation argument most of the results which we have proved for differential operators with constant coefficients can be extended locally to differential operators having constant strength in this sense.

After a discussion of local existence theorems in Sections 13.2 and 13.3 and of hypoellipticity in Section 13.4 we turn in Section 13.5 to global existence questions. It turns out that solution exist globally if and only if the adjoint operator and its localizations at infinity are injective. When the coefficients are real analytic this is always true by Holmgren's uniqueness theorem (see Section 8.6). However, no such uniqueness theorem is valid when the coefficients are just in C^∞. In fact, we construct in Section 13.6 a number of examples of non-uniqueness for the Cauchy problem including an elliptic equation which has a non-trivial solution of compact support.

13.1. Definitions and Basic Properties

We wish to consider differential operators which in $B_{p,k}$ are bounded perturbations of differential operators with constant coefficients. In view of Theorem 10.3.6 this leads to the following definition.

Definition 13.1.1. A differential operator $P(x, D)$ defined for $x \in X$ is said to have constant strength in X if for arbitrary fixed $x, y \in X$ the differential operators $P(x, D)$ and $P(y, D)$ with constant coefficients are equally strong, that is,

$$\tilde{P}(x,\xi)/\tilde{P}(y,\xi)\leqq C_{x,y}, \qquad \xi\in\mathbb{R}^n.$$

The following lemma puts this condition in a form which is usually more convenient.

Lemma 13.1.2. *Let $P(x,D)$ have constant strength. With a fixed x^0 set $P_0(D)=P(x^0,D)$ and let $P_0,\dots,P_r$ be a basis in the finite dimensional vector space of operators with constant coefficients weaker than P_0. Then we have*

$$P(x,D)=P_0(D)+\sum_0^r c_j(x)\,P_j(D) \tag{13.1.1}$$

where the coefficients c_j are uniquely determined, vanish at x^0, and have the same differentiability and continuity properties as the coefficients of $P(x,D)$.

The proof is trivial.

We next show that an operator $P(x,D)$ and its adjoint ${}^tP(x,D)$ are simultaneously of constant strength. We recall that the adjoint is defined by

$$\int v\,{}^tP(x,D)u\,dx=\int (P(x,D)v)\,u\,dx$$

when v or u has compact support.

Theorem 13.1.3. *Let $P(x,D)$ be differential operator of order m with coefficients in $C^\infty(X)$ and constant strength in X. Then it follows that ${}^tP(x,D)$ is also of constant strength and is as strong as $P(x,-D)$ for every x.*

Proof. Using the representation (13.1.1) we obtain

$$\begin{aligned}{}^tP(x,D)&=P_0(-D)+\sum_0^r P_j(-D)c_j\\ &=P_0(-D)+\sum_\alpha\sum_{j=0}^r ((-D)^\alpha c_j)P_j^{(\alpha)}(-D)/\alpha!.\end{aligned}$$

Since $c_j(x^0)=0$ we have in particular

$${}^tP(x^0,D)=P_0(-D)+\sum_{\alpha\neq 0}\sum_{j=0}^r (-D)^\alpha c_j(x^0)P_j^{(\alpha)}(-D)/\alpha!.$$

Now we have when $\alpha\neq 0$

$$P_j^{(\alpha)}(-D)\ll P_j(-D)\prec P_0(-D),$$

so it follows from Corollary 10.4.8 that ${}^tP(x^0, D)$ is equally strong as $P(x^0, -D)$. Since this is true for every $x^0 \in X$, it is clear that ${}^tP(x, D)$ is of constant strength.

Multiplication of operators of constant strength also gives an operator of constant strength:

Theorem 13.1.4. *Let $P(x, D)$ and $Q(x, D)$ be of constant strength in X and assume that the coefficients of $P(x, D)$ are in $C^m(X)$ where m is the order of $Q(x, D)$. Then the operator $R(x, D) = Q(x, D)\,P(x, D)$ is of constant strength. Moreover, $R(x, \xi)$ is for every x as strong as $Q(x, \xi)P(x, \xi)$.*

Proof. Using (13.1.1) and Leibniz' formula we obtain

$$R(x, D) = Q(x, D)P(x, D) \tag{13.1.2}$$
$$= Q(x, D)P_0(D) + \sum_{\alpha} \sum_{j=0}^{r} D^\alpha c_j(x) Q^{(\alpha)}(x, D) P_j(D)/\alpha!.$$

Since

$$Q_0^{(\alpha)}(D) P_j(D) \ll Q_0(D) P_j(D) \prec Q_0(D) P_0(D), \quad \text{if } \alpha \neq 0,$$

it follows from Corollary 10.4.8 that $R(x^0, D)$ is as strong as $Q_0(D)P_0(D)$. This completes the proof.

Remark. It is important to note that the class of differential operators of constant strength is tied to the vector space structure in $\mathbb{R}^n$ and therefore not invariant for non-linear coordinate transformations.

13.2. Existence Theorems when the Coefficients are Merely Continuous

The following result extends Theorem 10.3.7.

Theorem 13.2.1. *Let $P(x, D)$ have continuous coefficients and be of constant strength in a neighborhood of $x^0 \in \mathbb{R}^n$. If X is a sufficiently small open neighborhood of x^0, we can find a linear operator E in $L^2(X)$ such that*

$$P(x, D)Ef = f, \quad f \in L^2(X), \tag{13.2.1}$$

$$EP(x, D)u = u, \quad u \in C_0^\infty(X), \tag{13.2.2}$$

(13.2.3) *$Q(D)E$ is a bounded operator in $L^2(X)$ if $Q(D) \prec P(x^0, D)$.*

Remarks. 1. The interpretation of (13.2.1) requires some comment since the multiplication of arbitrary distributions by continuous functions is not defined. The meaning of (13.2.1) is that, with $P(x,D)$ written in the form (13.1.1), we have

$$P_0(D)Ef+\sum_0^r c_j P_j(D)Ef=f,$$

and this has a sense since $P_j(D)Ef\in L^2(X)$ in virtue of (13.2.3).

2. According to Schwartz' kernel theorem (Theorem 5.2.1) the bounded linear operator E in $L^2(X)$ has a kernel $E(x,y)$ which is a distribution in $X\times X$. The identities (13.2.1) and (13.2.2) show that E is a two-sided fundamental solution of $P(x,D)$, that is, if the coefficients are smooth enough

$$P(x,D_x)\,E(x,y)={}^tP(y,D_y)\,E(x,y)=\delta(x-y).$$

Proof of Theorem 13.2.1. Let X_1 be a bounded neighborhood of x^0 such that $P(x,D)$ has constant strength and continuous coefficients in $\bar{X}_1$. If $X\subset X_1$, we can apply Theorem 10.3.7 with $P=P_0$ and obtain a bounded linear mapping E_0 in $L^2(X)$ such that

$$P_0(D)E_0f=f,\quad f\in L^2(X), \tag{13.2.4}$$

$$E_0P_0(D)u=u,\quad u\in C_0^\infty(X), \tag{13.2.5}$$

$$N(P_j(D)E_0)<C,\quad j=0,\dots,r. \tag{13.2.6}$$

Here N denotes the operator norm in $L^2(X)$ and C can be chosen independent of X when $X\subset X_1$. The operators P_j are those introduced in Lemma 13.1.2. We wish to find a solution of the equation

$$P(x,D)u=P_0(D)u+\sum_0^r c_j(x)P_j(D)u=f\in L^2(X) \tag{13.2.7}$$

which is if the form $u=E_0g$, $g\in L^2(X)$. This means that we want to solve the following equation for g

$$g+\sum_0^r c_jP_j(D)E_0g=f. \tag{13.2.8}$$

If A denotes the operator

$$g\to\sum_0^r c_jP_j(D)E_0g,$$

it follows from (13.2.6) that

$$N(A)\leqq C\sum_0^r \sup_X|c_j|.$$

Since the coefficients c_j are continuous and $c_j(x^0)=0$, we thus have $N(A)<1/2$ if X is a sufficiently small neighborhood of x^0. Hence the operator $I+A$ has an inverse with norm at most equal to 2. (I denotes the identity operator.) This means that the Equation (13.2.8) has the unique solution $g=(I+A)^{-1}f$. If we set $E=E_0(I+A)^{-1}$, we have $Ef=E_0 g$ where g satisfies (13.2.8). This proves (13.2.1). From (13.2.6) and the fact that the operators $P_j(D)$, $j=0,\dots,r$, form a basis for the operators weaker than $P_0(D)$, it follows immediately that (13.2.3) is valid. Finally, if $f=P(x,D)u$, $u\in C_0^\infty(X)$, the unique solution of (13.2.8) is $g=P_0(D)u$. Hence $Ef=E_0 g=u$, which proves (13.2.2).

13.3. Existence Theorems when the Coefficients are in C^∞

The aim of this section is to show that Theorem 13.2.1 remains valid if L^2 is replaced by an arbitrary space $B_{p,k}$. In order to obtain such results with X independent of the choice of $k\in\mathscr{K}$, we shall replace k by another function $k_\delta\in\mathscr{K}$ chosen according to Theorem 10.1.5. Thus we have $B_{p,k}=B_{p,k_\delta}$, and the following lemma shows the advantage of using the weight functions k_δ instead of k.

Lemma 13.3.1. *Let $k\in\mathscr{K}$ and let k_δ be defined as in Theorem 10.1.5. For every $\phi\in\mathscr{S}$ there exists a positive number δ_0 such that*

$$\|\phi u\|_{p,k_\delta}\leqq 2\|\phi\|_{1,1}\|u\|_{p,k_\delta} \tag{13.3.1}$$

if $0<\delta<\delta_0$; $u\in B_{p,k}=B_{p,k_\delta}$.

Note that the norm on ϕ in the right-hand side *does not depend on* k. The number δ_0 is of course dependent on k but that will have no importance in the application below.

Proof of Lemma 13.3.1. In virtue of Theorem 10.1.15 we only have to show that

$$\|\phi\|_{1,M_{k_\delta}}\leqq 2\|\phi\|_{1,1}$$

when $\delta<\delta_0$. But this is obvious since it follows from (10.1.10), (10.1.11) that

$$\|\phi\|_{1,M_{k_\delta}}=(2\pi)^{-n}\int|\hat\phi(\xi)|M_{k_\delta}(\xi)d\xi\to(2\pi)^{-n}\int|\hat\phi(\xi)|d\xi=\|\phi\|_{1,1}$$

when $\delta\to 0$.

When using Lemma 13.3.1 we need the following additional information.

Lemma 13.3.2. *Let $\psi\in C_0^\infty(\mathbb{R}^n)$ and set $\psi_\varepsilon(x)=\psi((x-x^0)/\varepsilon)$. If $h\in C^\infty$ in a neighborhood of x^0 and $h(x^0)=0$, it follows that $\|\psi_\varepsilon h\|_{1,1}=O(\varepsilon)$ when $\varepsilon\to 0$.*

Proof. By Theorem 1.1.9 we can choose $h_j\in C_0^\infty$ so that

$$h(x)=\sum(x_j-x_j^0)h_j(x)$$

in a neighborhood of 0. Using Theorem 10.1.15 we obtain for small ε

$$\begin{aligned}\|\psi_\varepsilon h\|_{1,1}&\leqq\sum\|h_j\|_{1,1}\,\|(x_j-x_j^0)\,\psi_\varepsilon\|_{1,1}\\&=\varepsilon\sum\|h_j\|_{1,1}\,\|x_j\psi\|_{1,1}\end{aligned}$$

which proves the lemma.

We can now prove the main result of this section.

Theorem 13.3.3. *Let $P(x,D)$ have C^∞ coefficients and be of constant strength in a neighborhood of $x^0\in\mathbb{R}^n$. If X is a sufficiently small open neighborhood of x^0, there exists a linear mapping E of $\mathscr{E}'(\mathbb{R}^n)$ into $\mathscr{E}'(\mathbb{R}^n)$ with the following properties*

$$P(x,D)Ef=f \quad \text{in } X \text{ if } f\in\mathscr{E}'(\mathbb{R}^n), \tag{13.3.2}$$

$$EP(x,D)u=u \quad \text{in } X \text{ if } u\in\mathscr{E}'(X), \tag{13.3.3}$$

$$\|Ef\|_{p,\tilde{P}_0k}\leqq C_k\|f\|_{p,k} \tag{13.3.4}$$

if $f\in\mathscr{E}'(\mathbb{R}^n)\cap B_{p,k}$ and $k\in\mathscr{K}$.

Here C_k is independent of f, and (13.3.4) *means in particular that*

$$Ef\in B_{p,\tilde{P}_0k} \quad \text{if } f\in\mathscr{E}'(\mathbb{R}^n)\cap B_{p,k}.$$

Proof. Writing $X_\varepsilon=\{x;\,|x-x^0|<\varepsilon\}$, we choose $\varepsilon_0>0$ so that $P(x,D)$ has C^∞ coefficients and constant strength in X_{ε_0}. Let χ be a function in $C_0^\infty(\mathbb{R}^n)$ such that $\chi=1$ in a neighborhood of the ball $\{x;\,|x|\leqq 2\varepsilon_0\}$ and set $F_0=\chi E_0$ where $E_0\in B^{\text{loc}}_{\infty,\tilde{P}_0}(\mathbb{R}^n)$ is a fundamental solution of $P_0(D)$. If $\operatorname{supp} f\subset X_{\varepsilon_0}$, we then have $F_0*f=E_0*f$ in X_{ε_0}, hence

$$P_0(D)(F_0*f)=F_0*(P_0(D)f)=f \tag{13.3.5}$$

in X_{ε_0} if $f\in\mathscr{E}'(X_{\varepsilon_0})$. Also note that

$$F_0\in B_{\infty,\tilde{P}_0}. \tag{13.3.6}$$

Choose $\psi\in C_0^\infty(\mathbb{R}^n)$ so that $\psi(x)=1$ when $|x|\leqq 1$ and $\operatorname{supp}\psi$ is contained in the ball $\{x;\,|x|\leqq 2\}$. Writing $\psi_\varepsilon(x)=\psi((x-x^0)/\varepsilon)$, we shall prove below that there exists a positive number ε_1 such that the equation

$$g+\sum_0^r\psi_\varepsilon c_jP_j(D)(F_0*g)=\psi_\varepsilon f \tag{13.3.7}$$

has a unique solution $g\in\mathscr{E}'(\mathbb{R}^n)$ for an arbitrary $f\in\mathscr{E}'(\mathbb{R}^n)$, if $0<\varepsilon<\varepsilon_1<\varepsilon_0/2$. Accepting this result for a moment we define

$$Ef=F_0*g. \tag{13.3.8}$$

Since $\operatorname{supp}\psi_\varepsilon\subset X_{\varepsilon_0}$, we have $g\in\mathscr{E}'(X_{\varepsilon_0})$. Hence it follows from (13.3.5) and (13.3.7) that

$$\begin{aligned}P(x,D)Ef=P(x,D)(F_0*g)&=P_0(D)(F_0*g)+\sum_0^r c_jP_j(D)(F_0*g)\\&=g+\sum_0^r\psi_\varepsilon c_jP_j(D)(F_0*g)=\psi_\varepsilon f=f\quad\text{in } X_\varepsilon,\end{aligned}$$

which proves (13.3.2). If $u\in\mathscr{E}'(X_\varepsilon)$, the equation (13.3.7) with $f=P(x,D)u$ is satisfied by $g=P_0(D)u$, for then it follows from (13.3.5) that $F_0*g=u$ in $X_{2\varepsilon}$, and both sides of (13.3.7) have their supports contained in $X_{2\varepsilon}$. Hence $EP(x,D)u=F_0*(P_0(D)u)=u$ in X_ε.

To prove the existence and uniqueness of solutions of the Equation (13.3.7) and to prove (13.3.4), we shall study the mapping A_ε defined by

$$g\to\sum_0^r\psi_\varepsilon c_jP_j(D)(F_0*g),\qquad g\in\mathscr{D}'(\mathbb{R}^n). \tag{13.3.9}$$

We shall estimate the norm of A_ε in B_{p,k_δ} when $k\in\mathscr{K}$ and k_δ is defined as in Theorem 10.1.5. First note that (13.3.6) implies that there is a constant C such that $|P_j(\xi)\hat F_0(\xi)|\leqq C$, $j=0,\dots,r$. If δ is sufficiently small, it thus follows from Lemma 13.3.1 and Theorem 10.1.12 that

$$\begin{aligned}\|A_\varepsilon g\|_{p,k_\delta}&\leqq 2\sum_0^r\|\psi_\varepsilon c_j\|_{1,1}\|P_j(D)F_0*g\|_{p,k_\delta}\\&\leqq 2C\sum_0^r\|\psi_\varepsilon c_j\|_{1,1}\|g\|_{p,k_\delta},\qquad g\in B_{p,k_\delta}.\end{aligned} \tag{13.3.10}$$

Now we see from Lemma 13.3.2 that we can choose ε_1 so that $0<\varepsilon_1<\varepsilon_0/2$ and

$$\sum_0^r\|\psi_\varepsilon c_j\|_{1,1}<1/4C\qquad\text{if } 0<\varepsilon<\varepsilon_1. \tag{13.3.11}$$

Note that this choice of ε_1 is independent of the function k. From (13.3.10) it then follows that

$$\|A_\varepsilon g\|_{p,k_\delta}\leqq 2^{-1}\|g\|_{p,k_\delta},\qquad g\in B_{p,k_\delta}. \tag{13.3.12}$$

The equation (13.3.7) can be written in the form $g+A_\varepsilon g=\psi_\varepsilon f$. Since $B_{p,k_\delta}=B_{p,k}$, we obtain immediately from (13.3.12) that the equation (13.3.7) has one and only one solution $g\in B_{p,k}$ when $f\in B_{p,k}$, and $g\in\mathscr{E}'(\mathbb{R}^n)$ since ψ_ε has compact support. Noting that every finite set of distributions in $\mathscr{E}'(\mathbb{R}^n)$ is contained in $B_{p,k}$ for some k, we conclude that the equation (13.3.7) has one and only one solution $g\in\mathscr{E}'(\mathbb{R}^n)$ for every $f\in\mathscr{E}'(\mathbb{R}^n)$. From (13.3.12) we get the estimate $\|g\|_{p,k_\delta}\leqq 2\|\psi_\varepsilon f\|_{p,k_\delta}$, hence

$$\|Ef\|_{p,\tilde{P}_0k_\delta}\leqq\|F_0\|_{\infty,\tilde{P}_0}\|g\|_{p,k_\delta}\leqq 4\|F_0\|_{\infty,\tilde{P}_0}\|f\|_{p,k_\delta}\|\psi_\varepsilon\|_{1,1}.$$

This proves (13.3.4), for $\|f\|_{p,k_\delta}$ and $\|f\|_{p,k}$ are equivalent norms.

The following corollary extends Theorem 10.3.2.

Corollary 13.3.4. *Let $P(x,D)$ and X be as described in Theorem 13.3.3. If $u\in\mathscr{E}'(X)$ and $P(x,D)u\in B_{p,k}$, it follows that $u\in B_{p,\tilde{P}_0k}$.*

In the following corollary it is essential that we could choose the same X for every k in Theorem 13.3.3.

Corollary 13.3.5. *Let $P(x,D)$ and X be as described in Theorem 13.3.3. The equation $P(x,D)u=f$ then has a solution $u\in C^\infty(X)$ for every $f\in C^\infty(\mathbb{R}^n)$.*

By Schwartz' kernel theorem (Theorem 5.2.1) the mapping E in Theorem 13.3.3 has a kernel $e(x,y)\in\mathscr{D}'(\mathbb{R}^n\times\mathbb{R}^n)$. From (13.3.2), (13.3.3) it follows that in $X\times X$

$$P(x,D_x)e(x,y)={}^tP(y,D_y)e(x,y)=\delta(x-y).$$

The singularities of $e(x,y)$ are very similar to those of the kernel of a convolution operator (cf. (8.2.15)):

Theorem 13.3.6. *From* (13.3.4) *it follows that*

$$(x,y,\xi,\eta)\in WF(e)\Rightarrow\xi+\eta=0 \tag{13.3.13}$$

if e is the kernel of E.

Proof. By (13.3.4) and Theorem 10.1.14 we have

$$|\langle Ef,g\rangle|\leqq C_k\|f\|_{p,k}\|g\|_{p',1/\tilde{P}_0k}\quad\text{if } f,g\in C_0^\infty(\mathbb{R}^n)\text{ and } k\in\mathscr{K}.$$

If we replace $f(y)$ by $f(y)\,e^{i\langle y,\eta\rangle}$ and $g(x)$ by $g(x)\,e^{i\langle x,\xi\rangle}$ and write $u=(g\otimes f)\,e$, it follows that

$$|\hat{u}(\xi,-\eta)|\leqq C_k\|k\hat{f}(.-\eta)\|_{L^p}\|\hat{g}(.-\xi)/\tilde{P}_0k\|_{L^{p'}}. \tag{13.3.14}$$

Since

$$|k(\theta)\hat{f}(\theta-\eta)|\leqq k(\eta)(1+C|\theta-\eta|)^N|\hat{f}(\theta-\eta)|$$

and $\hat{f} \in \mathscr{S}$, we can estimate the L^p norm by a constant times $k(\eta)$. Using a similar estimate for the other norm in (13.3.14) we obtain

$$|\hat{u}(\xi, -\eta)| \leqq C'_k k(\eta)/k(\xi). \tag{13.3.15}$$

We shall now show that (13.3.15) implies that $\hat{u}(\xi, -\eta)$ is rapidly decreasing if $(\xi, \eta) \to \infty$ in the complement of a conic neighborhood W of the diagonal in $\mathbb{R}^n \times \mathbb{R}^n$.

We argue by contradiction. Assume that there is a sequence $(\xi_j, \eta_j) \notin W$ tending to ∞ such that for some fixed N

$$|\hat{u}(\xi_j, -\eta_j)| > (1+|\xi_j|+|\eta_j|)^{-N}, \quad j=1, 2, \ldots. \tag{13.3.16}$$

Since $k \in \mathscr{K}$ implies $1/k \in \mathscr{K}$, the condition (13.3.15) is symmetric in ξ and η so we may assume that $|\eta_j| \leqq |\xi_j|$. We can also pass to a subsequence to make sure that the sequence is so lacunary that for all j

$$|\xi_{j+1}| \geqq |\xi_j|^2 \geqq 2|\xi_j|. \tag{13.3.17}$$

Now we introduce

$$k(\xi) = \max_i (1+|\xi-\xi_i|)^{-1}(1+|\xi_i|)$$

which is obviously a function in $\mathscr{K}$. We have $k(\xi_j) \geqq (1+|\xi_j|)$. Note that $k(\xi) \leqq 2$ unless $|\xi-\xi_i| < |\xi_i|/2$ for some i, thus

$$|\xi_i|/2 < |\xi| < 3|\xi_i|/2.$$

Since $|\eta_j| \leqq |\xi_j| \leqq |\xi_{j+1}/2|$ it follows that

$$k(\eta_j) \leqq \max(2, (1+|\xi_{j-1}|), (1+|\xi_j|)/(1+|\xi_j - \eta_j|)).$$

By hypothesis $(\xi_j, \eta_j) \notin W$ so $|\xi_j - \eta_j| > c|\xi_j|$ for a fixed $c > 0$. Hence

$$k(\eta_j) \leqq C_1(1+|\xi_{j-1}|) \leqq 2\, C_1(1+|\xi_j|)^{\frac{1}{2}}$$

by (13.3.17), which implies $k(\eta_j)/k(\xi_j) < 2\, C_1(1+|\xi_j|)^{-\frac{1}{2}}$. If we now apply (13.3.15) with k replaced by k^{2N+1} we have a contradiction with (13.3.16) which completes the proof.

We know from Theorems 12.4.6 and 12.8.17 that $P(x, D)$ is hyperbolic (resp. an evolution operator) for every fixed $x \in X$ if X is connected and this is true for $x = x^0$. The Cauchy problem can then be solved as in the constant coefficient case:

Theorem 13.3.7. *Assume that the hypotheses in Theorem 13.3.3 are fulfilled and in addition that $P(x^0, D)$ is hyperbolic with respect to N* (resp. *an evolution operator with respect to the half space H*

$=\{x;\ \langle x, N\rangle \geqq 0\})$. *Then it follows that*

(13.3.18) $\quad \operatorname{supp} Ef \subset \Gamma^\circ(P,N)+\operatorname{supp} f \quad$ resp. $\operatorname{supp} Ef \subset H+\operatorname{supp} f$

if $f\in\mathscr{E}'$ *and* E *is the mapping constructed in Theorem 13.3.3.*

Proof. This follows by inspecting the proof of Theorem 13.3.3. First the convolution with F_0 has the required properties if we choose a fundamental solution E_0 with support in $\Gamma^\circ(P,N)$ resp. H. (Actually we must use Theorem 12.8.13 and the comment after it.) It follows that A_ε also has the property (13.3.18), and so has the inverse of $(I+A_\varepsilon)$ since it can be expressed by the Neumann series. This completes the proof.

13.4. Hypoellipticity

The results of Section 11.1 are now easily extended to operators of constant strength:

Theorem 13.4.1. *Let* $P(x,D)$ *be an operator of constant strength with coefficients in* $C^\infty(X)$. *Assume that the operator* $P_0(D)=P(x^0,D)$ *is hypoelliptic for some* $x^0\in X$. *If* $u\in\mathscr{D}'(X)$ *and* $P(x,D)u\in B^{\text{loc}}_{p,k}(X)$, *it follows that* $u\in B^{\text{loc}}_{p,\tilde{P}_0 k}(X)$. *In particular,* $P(x,D)u\in C^\infty(X)$ *implies that* $u\in C^\infty(X)$.

Note that by Theorem 11.1.9 the hypotheses in the theorem are in fact fulfilled for every $x^0\in X$.

Proof of Theorem 13.4.1. In view of the previous remark and Theorem 10.1.20 it is sufficient to prove the statement in a neighborhood of x^0. To do so we first decompose P according to Lemma 13.1.2. If $d(\xi)$ is the distance from $\xi\in\mathbb{R}^n$ to the zeros of $P_0(\zeta)$, we have $d+1\in\mathscr{K}$ in virtue of (11.4.10), and Theorem 10.4.3 gives

$$\sum_\alpha |P_j^{(\alpha)}(\xi)|^2 (d(\xi)+1)^{2|\alpha|} \leqq C\sum_\alpha |P_0^{(\alpha)}(\xi)|^2 (d(\xi)+1)^{2|\alpha|} \leqq C'\tilde{P}_0(\xi)^2,$$

where the last estimate follows from Lemma 11.1.4, and C, C' are constants. With the notation

$$\tilde{P}_j'(\xi)^2=\sum_{\alpha\neq 0}|P_j^{(\alpha)}(\xi)|^2$$

we thus have with a constant C

$$\tilde{P}_j'(\xi)\leqq C\tilde{P}_0(\xi)/(d(\xi)+1). \tag{13.4.1}$$

Let Y be a neighborhood of x^0 such that $Y \Subset X$ and Theorem 13.3.3 applies in Y. Then we have $u \in B_{p,k'}^{\text{loc}}(Y)$ for some $k' \in \mathscr{K}$. Repeating the proof of Theorem 11.1.7 we shall prove that

$$u \in B_{p,k_\nu}^{\text{loc}}(Y), \qquad \nu = 0, 1, \ldots \tag{13.4.2}$$

where $k_\nu = \inf(\tilde{P}_0 k, (d+1)^\nu k')$. Since it follows as in the proof of Theorem 11.1.8 that $B_{p,k_\nu}^{\text{loc}} = B_{p,\tilde{P}_0 k}^{\text{loc}}$ for large ν, this will prove the theorem.

(13.4.2) is trivial when $\nu = 0$ since $k_0 \leqq k'$. Assuming that (13.4.2) is proved for one value of ν we shall now prove that ν can be replaced by $\nu+1$. Thus take $\phi \in C_0^\infty(Y)$ and set $C_j = c_j$ when $j \neq 0$, $C_0 = c_0 + 1$, where c_j are the coefficients in (13.1.1). Leibniz' formula then gives

$$P(x,D)(\phi u) = \phi P(x,D) u + \sum_{j=0}^{r} \sum_{\alpha \neq 0} (D^\alpha \phi)\, C_j P_j^{(\alpha)}(D) u / \alpha!.$$

The terms in the sum are in

$$B_{p,k_\nu/\tilde{P}_j'} \subset B_{p,k_{\nu+1}/\tilde{P}_0} \qquad \text{for } k_{\nu+1} \tilde{P}_j' / \tilde{P}_0 \leqq C k_\nu$$

by (13.4.1). Since $\phi P(x,D) u \in B_{p,k}$ and $k_{\nu+1}/\tilde{P}_0 \leqq k$, it follows that $P(x,D)(\phi u) \in B_{p,k_{\nu+1}/\tilde{P}_0}$. We now only have to apply Corollary 13.3.4 to conclude that $\phi u \in B_{p,k_{\nu+1}}$. Hence (13.4.2) is valid with ν replaced by $\nu+1$. The proof is complete.

It follows immediately from Theorem 10.4.9 that elliptic operators satisfy the hypotheses of Theorem 13.4.1. For elliptic operators Corollary 8.3.2 already gave a stronger result than the last statement in Theorem 13.4.1. It is valid in general:

Theorem 13.4.2. *If $P(x,D)$ satisfies the hypotheses of Theorem 13.4.1 then*

$$WF(u) = WF(P(x,D)u), \qquad u \in \mathscr{D}'(X), \tag{13.4.3}$$

$$\operatorname{sing\,supp} u = \operatorname{sing\,supp} P(x,D) u, \qquad u \in \mathscr{D}'(X). \tag{13.4.4}$$

Proof. (13.4.4) is a consequence of (13.4.3) and was also proved in Theorem 13.4.1. To prove (13.4.3) we first choose an arbitrary $x^0 \in X$ and then an open neighborhood $Y \Subset X$ of x^0 where Theorem 13.3.3 is applicable. We may assume that $u \in \mathscr{E}'(Y)$. If $(x^0, \xi^0) \notin WF(P(x,D)u)$, we can choose $\phi \in C_0^\infty(Y)$ equal to 1 in a neighborhood of x^0 so that $Y \times \{\xi^0\}$ does not meet $WF(\phi P(x,D) u$. Hence $Y \times \{\xi^0\}$ does not meet $WF(E(\phi P(x,D) u))$ by Theorem 13.3.6, and $E((1-\phi) P(x,D) u)$ is in C^∞ in the open neighborhood of x^0 where $\phi = 1$, by Theorem 13.4.1. Since $u = E(P(x,D) u)$ in Y, it follows that $(x^0, \xi^0) \notin WF(u)$. This proves that

$$WF(u) \subset WF(P(x,D)u)$$

and the opposite inclusion is always true.

The following terminology is clearly consistent with Definition 11.1.2.

Definition 13.4.3. A differential operator $P(x, D)$ with coefficients in $C^\infty(X)$ is called hypoelliptic if (13.4.4) is valid; it is called micro-hypoelliptic if even (13.4.3) is fulfilled.

The following theorem combined with Theorem 13.4.2 shows that these notions coincide for operators of constant strength.

Theorem 13.4.4. *If $P(x, D)$ has constant strength in X and coefficients in $C^\infty(X)$, then $P(x, D)$ is hypoelliptic if and only if the constant coefficient operator $P(x^0, D)$ is hypoelliptic for every fixed $x^0 \in X$.*

Proof. (13.4.4) remains valid when X is replaced by an open set $Y \Subset X$ where Theorem 13.3.3 is applicable. If $x^0 \in Y$ then

$$u = E\delta_{x^0} \in B^{\mathrm{loc}}_{\infty, \tilde{P}_0}(Y), \qquad P(x, D)u = \delta_{x^0}.$$

If (13.4.4) is valid it follows that $u \in C^\infty(Y \setminus \{x^0\})$. Let Q_0 be a localization of P_0 at ∞. As in the proof of Theorem 10.2.12 we can choose a sequence $\eta_j \to \infty$ such that

$$P_0(\xi + \eta_j)/\tilde{P}_0(\eta_j) \to Q_0(\xi),$$

and since P_k is weaker than P_0 the limits

$$Q_k(\xi) = \lim_{j \to \infty} P_k(\xi + \eta_j)/\tilde{P}_0(\eta_j)$$

also exist after we pass to a suitable subsequence. From the proof of Theorem 10.2.12 we know in addition since $u \in B^{\mathrm{loc}}_{\infty, \tilde{P}_0}$ that the limit

$$v = \lim u\tilde{P}_0(\eta_j) e^{-i\langle \cdot - x^0, \eta_j \rangle} \in B^{\mathrm{loc}}_{\infty, \tilde{Q}_0}$$

exists if we pass to a suitable subsequence, and $v = 0$ in $Y \setminus \{x^0\}$ since $u \in C^\infty$ there. From the equation $P(x, D)u = \delta_{x^0}$ it follows that

$$(Q_0(D) + \sum c_j(x) Q_j(D))v = \delta_{x^0}$$

so v is not equal to 0. If $\phi \in C_0^\infty(Y)$ is equal to 1 near x^0 then $w = \phi v$ is supported by x^0 and not 0, so $\hat{w} = e^{i\langle x^0, \xi \rangle}\psi(\xi)$ where ψ is a non-zero polynomial such that $\psi\tilde{Q}_0$ is bounded. Hence ψ and Q_0 are both bounded, so all localizations of P_0 at infinity are constant. In view of condition (iv) in Theorem 11.1.1 this means precisely that P_0 is hypoelliptic.

However, we wish to emphasize that there are large classes of hypoelliptic operators which are not of constant strength, and that

Theorem 13.4.4 is not applicable to them. For example, the Kolmogorov Equation (7.6.13), corresponding to $-D_1^2+x_1 iD_2-iD_3$, is easily seen to be hypoelliptic by using the fundamental solution constructed in Section 7.6. However, for fixed x_1 it operates only along a two dimensional plane so it is not hypoelliptic when the coefficients are frozen.

13.5. Global Existence Theorems

In this section we shall examine when it is possible to remove the hypothesis in Theorem 13.3.3 that X is small. We must then consider the localizations at infinity of an operator $P(x,D)$ of constant strength. These occurred implicitly already in the proof of Theorem 13.4.4; modifying Definition 10.2.6 we shall say that $Q(x,D)$ is a localization at infinity if Q is not identically 0 and

$$Q(x,\xi)=\lim_{j\to\infty} a_j P(x,\xi+\xi_j)$$

for some sequence $\xi_j\in\mathbb{R}^n$ tending to ∞ and some sequence $a_j>0$. Note that

$$\tilde{Q}(x,0)=\lim_{j\to\infty} a_j\tilde{P}(x,\xi_j).$$

Writing $P_0(\xi)=P(x^0,\xi)$ for some fixed x^0 we have in particular $a_j\tilde{P}_0(\xi_j)\to\tilde{Q}(x^0,0)$, and $\tilde{Q}(x^0,0)>0$ since otherwise

$$\lim a_j\tilde{P}(x,\xi_j)=\lim a_j\tilde{P}_0(\xi_j)\tilde{P}(x,\xi_j)/\tilde{P}(x^0,\xi_j)$$

would be 0 for every x. Thus we only change Q by a positive factor if we take $a_j=1/\tilde{P}_0(\xi_j)$, thus

$$Q(x,\xi)=\lim_{j\to\infty} P(x,\xi+\xi_j)/\tilde{P}_0(\xi_j). \tag{13.5.1}$$

This makes Q normalized by $\tilde{Q}(x^0,0)=1$. It is clear that Q is also of constant strength. From (13.1.1) we obtain at least for a subsequence

$$Q(x,D)=Q_0(D)+\sum c_j(x)\,Q_j(D),$$
$$Q_j(\xi)=\lim_{k\to\infty} P_j(\xi+\xi_k)/\tilde{P}_0(\xi_k)$$

so the coefficients of $Q(x,D)$ are also in C^∞.

In the following theorem we denote by $B_{p,k}(\bar{X})$ the set of restrictions to X of elements in $B_{p,k}$, that is, $B_{p,k}(\bar{X})=B_{p,k}/N_{p,k}(X)$ where

$$N_{p,k}(X)=\{u\in B_{p,k};\ u=0 \text{ in } X\}.$$

If $p<\infty$ and $k'(\xi)=1/k(-\xi)$ then Theorem 10.1.14 identifies the dual of $B_{p,k}$ with $B_{p',k'}$ so the dual of $B_{p,k}(\bar{X})$ is the annihilator $B^0_{p',k'}(\bar{X})$ of $N_{p,k}(X)$ in $B_{p',k'}$. Its elements have support in $\bar{X}$; on the other hand

$$B^0_{p',k'}(\bar{X})\supset\mathscr{E}'(X)\cap B_{p',k'}. \tag{13.5.2}$$

In fact, it follows from Theorems 10.1.16 and 10.1.17 that every $u\in N_{p,k}(X)$ is the limit in $B_{p,k}$ of a sequence $u_j\in\mathscr{S}$ vanishing on any compact subset of X for large j.

Theorem 13.5.1. *Let $P(x,D)$ be an operator with C^∞ coefficients and constant strength in the open set $X\subset\mathbb{R}^n$. Let $p\neq\infty$ and assume that $P(x,D)B_{p,k\tilde{P}_0}(\bar{Y})$ has finite codimension in $B_{p,k}(\bar{Y})$ for all $k\in\mathscr{K}$ and all $Y\Subset X$. Then ${}^tQ(x,D)$ is injective on $\mathscr{E}'(X)$ if Q is any localization of P at infinity. If $P(x,D)B_{p,k\tilde{P}_0}(\bar{Y})=B_{p,k}(\bar{Y})$ for all $k\in\mathscr{K}$ and $Y\Subset X$ then ${}^tP$ is also injective on $\mathscr{E}'(X)$.*

Proof. We start with the last statement. If the continuous operator $P(x,D)$ from $B_{p,k_P}(\bar{Y})$ to $B_{p,k}(\bar{Y})$ is surjective, $k_P=k\tilde{P}_0$, then the adjoint ${}^tP$: $B^0_{p',k'}(\bar{Y})\to B^0_{p',k'_P}(\bar{Y})$ is injective. Thus (13.5.2) shows that ${}^tPu=0$ implies $u=0$ if $u\in\mathscr{E}'(Y)$ and $u\in B_{p',k'}$. This is true for every $u\in\mathscr{E}'(X)$ if we choose $Y\supset\operatorname{supp}u$ and $k(\xi)=(1+|\xi|)^s$ with sufficiently large s.

Now assume only that $P(x,D)B_{p,k_P}(\bar{Y})$ has finite codimension N in $B_{p,k}(\bar{Y})$. Since $\mathscr{S}$ is dense in $B_{p,k}$ it follows that $C^\infty(\bar{Y})$ is dense in $B_{p,k}(\bar{Y})$. We can therefore choose $\phi_1,\dots,\phi_N\in C^\infty(\bar{Y})$ so that they span a supplementary space of $P(x,D)B_{p,k_P}(\bar{Y})$ in $B_{p,k}(\bar{Y})$. Then

$$T(v,a_1,\dots,a_N)=P(x,D)v+\sum_{j=1}^N a_j\phi_j$$

is a surjective map from $B_{p,k_P}(\bar{Y})\oplus\mathbb{C}^N$ to $B_{p,k}(\bar{Y})$. Hence the adjoint ${}^tT$ is an injective map with closed range, so the closed graph theorem shows that it has a continuous inverse. Thus

$$\|v\|_{p',k'}\leqq C\Big(\|{}^tPv\|_{p',k'_P}+\sum_{j=1}^N|\langle v,\phi_j\rangle|\Big),\quad v\in B^0_{p',k'}(\bar{Y}). \tag{13.5.3}$$

If Q is a localization of P at infinity given by (13.5.1), we have

$${}^tQ(x,\xi)=\lim{}^tP(x,\xi-\xi_j)/\tilde{P}_0(\xi_j).$$

Now apply (13.5.3) to $v(x)=w(x)e^{-i\langle x,\xi_j\rangle}$, $w\in C_0^\infty(Y)$, and let $j\to\infty$. Then the sum in (13.5.3) tends to 0. If

$$\lim_{j\to\infty}k(\xi+\xi_j)=k_0(\xi)>0 \tag{13.5.4}$$

exists then $\hat{w}(\xi)/k(-\xi+\xi_j)\to\hat{w}(\xi)/k_0(-\xi)$ in $L^{p'}$ norm because

$$1/k(-\xi+\xi_j)\leqq(1+C|\xi|)^N/k(\xi_j).$$

Hence the left-hand side of (13.5.3) converges to $\|w\|_{p',k_0'}$. We have

$$\tilde{P}_0(\xi_j)k_P'(\xi-\xi_j)=\tilde{P}_0(\xi_j)/(k(\xi_j-\xi)\tilde{P}_0(\xi_j-\xi))\to 1/(k_0(-\xi)\tilde{Q}_0(-\xi))$$

so (13.5.3) implies

$$\|w\|_{p',k_0'}\leqq C\,\|{}^tQ\,w\|_{p',k_{0Q}'},\qquad w\in C_0^\infty(Y). \tag{13.5.3$'$}$$

For any s we can choose k so that $k_0(\xi)=(1+|\xi|)^s$, at least if the sequence ξ_j is so thin that $|\xi_j-\xi_k|>j$ if $j\neq k$, as we may assume. It suffices to prove this when $s=1$ and then we just have to take

$$k(\xi)=\min(1+|\xi-\xi_k|).$$

The triangle inequality gives $k(\xi+\eta)\leqq(1+|\eta|)\,k(\xi)$, so $k\in\mathscr{K}$. We have $k(\xi+\xi_j)=1+|\xi|$ if $|\xi|<j/2$, for $|\xi+\xi_j-\xi_k|\geqq|\xi_j-\xi_k|-j/2>j/2$ then, $k\neq j$. Hence (13.5.3)′ is valid with $k_0(\xi)=(1+|\xi|)^s$. The injectivity of ${}^tQ$ now follows as in the beginning of the proof if we extend the validity of (13.5.3)′ to all $w\in\mathscr{E}'(Y)\cap B_{p',k_0'}$; if $p'=\infty$ we require that $k_0'(\xi)\,\hat{w}(\xi)\to 0$ as $\xi\to\infty$. Then the regularization $w*\phi_\varepsilon$ in Theorem 10.1.17 converges to w in $B_{p',k_0'}$ norm as $\varepsilon\to 0$. Since (13.5.3)′ is valid for $w*\phi_\varepsilon$ when ε is small, we obtain (13.5.3)′ for w when $\varepsilon\to 0$. The proof is complete.

If $P(x,D)$ has constant strength and analytic coefficients then ${}^tP(x,D)$ is injective on $\mathscr{E}'$. In fact, assume that $0\neq u\in\mathscr{E}'$ and that ${}^tP(x,D)u=0$. If $\xi\in\mathbb{R}^n\setminus 0$ and $x\in\operatorname{supp}u$ is chosen so that $\langle x,\xi\rangle$ is maximal, it follows from Theorem 8.6.5 that (x,ξ) is in the characteristic set of ${}^tP$. Hence it follows from Theorem 13.1.3 that the principal part of P_0 vanishes at ξ for every ξ which is impossible. Since the localizations of P at infinity also have constant strength, the conclusions in Theorem 13.5.1 are therefore valid. However, we shall see in Section 13.6 that even elliptic equations with C^∞ coefficients may have solutions of compact support. This makes the following converse of Theorem 13.5.1 interesting.

Theorem 13.5.2. *Let $P(x,D)$ be an operator with C^∞ coefficients and constant strength in the open set $X\subset\mathbb{R}^n$. Assume that ${}^tQ$ is injective on $\mathscr{E}'(X)$ if Q is any localization of P at ∞. Then*

$$R(\bar{Y})=\{w\in\mathscr{E}'(\bar{Y});\ {}^tPw=0\} \tag{13.5.5}$$

is a finite dimensional subspace of $C_0^\infty(\bar{Y})$ if Y is any open set $\Subset X$, and for every $f\in B_{p,k}(\bar{Y})$ with $\langle f,\phi\rangle=0$ for all $\phi\in R(\bar{Y})$ the equation $P(x,D)u=f$ has a solution $u\in B_{p,k\tilde{P}_0}(\bar{Y})$.

Note that $\langle f, \phi\rangle$ is defined since Theorem 2.3.3 shows that $\langle F, \phi\rangle = 0$ for every $F \in \mathscr{D}'(\mathbb{R}^n)$ vanishing in Y.

For the proof of Theorem 13.5.2 we need two lemmas. The first of them shows that the Fourier transforms of arbitrary non-smooth distributions of compact support can be localized at infinity; the second expresses $B_{p,k}$ in terms of spaces $B_{\infty,k'}$.

Lemma 13.5.3. *For every $u \in \mathscr{E}'(\mathbb{R}^n) \setminus C_0^\infty$ one can choose a sequence $\xi_j \to \infty$ in $\mathbb{R}^n$ and constants $t_j \in \mathbb{C}$ such that $t_j \exp(-i\langle ., \xi_j\rangle) u$ has a limit $u_0 \neq 0$ in $\mathscr{E}'$ as $j \to \infty$ with* $\operatorname{supp} u_0 \subset \operatorname{sing\,supp} u$.

Proof. We must choose ξ_j and t_j so that $t_j \hat{u}(\xi + \xi_j) \to \hat{u}_0$ in $\mathscr{S}'$. To do so we first observe that for some positive constants C and M

$$|\hat{u}(\xi)| \leqq C(1+|\xi|)^M, \qquad \xi \in \mathbb{R}^N.$$

Since $u \notin C_0^\infty$ we can also choose $\eta_j \to \infty$ in $\mathbb{R}^n$ so that for some other constants

$$|\hat{u}(\eta_j)| \geqq c(1+|\eta_j|)^\mu.$$

We shall choose ξ_j fairly close to η_j so that $\hat{u}(\xi_j)$ is not accidentally small. To do so we set

$$k(\xi) = k_N(\xi) = (1+|\xi|)^{-N}$$

where $N > M$ so that $|\hat{u}(\xi)| k(\xi - \eta_j) \to 0$ at ∞. Let ξ_j be a point where

$$|\hat{u}(\xi)| k(\xi - \eta_j)$$

attains its maximum. Then

$$\begin{aligned} c(1+|\eta_j|)^\mu \leqq |\hat{u}(\eta_j)| \leqq |\hat{u}(\xi_j)| k(\xi_j - \eta_j) &\leqq C(1+|\xi_j|)^M k(\xi_j - \eta_j) \\ &\leqq C(1+|\eta_j|)^M (1+|\xi_j - \eta_j|)^{M-N}, \end{aligned}$$

hence

$$(1+|\xi_j - \eta_j|)^{N-M} \leqq C(1+|\eta_j|)^{M-\mu}.$$

If $N > 2M - \mu$ it follows that $|\xi_j - \eta_j|/|\eta_j| \to 0$ as $j \to \infty$. In particular, $\xi_j \to \infty$. Since

$$|\hat{u}(\xi)| k(\xi - \eta_j) \leqq |\hat{u}(\xi_j)| k(\xi_j - \eta_j)$$

we have

$$|\hat{u}(\xi)| \leqq |\hat{u}(\xi_j)| (1+|\xi - \xi_j|)^N,$$

that is,

$$|\hat{u}(\xi + \xi_j)| / |\hat{u}(\xi_j)| \leqq (1+|\xi|)^N, \qquad \xi \in \mathbb{R}^n.$$

Now a sequence of distributions with support in $\bar{Y}$ which is bounded in B_{∞,k_N} is precompact in $B_{\infty,k_{N+1}}$ (Theorem 10.1.10). Passing to a subsequence we may therefore assume that $u e^{-i\langle ., \xi_j\rangle} / \hat{u}(\xi_j)$ has a limit

u_0 in $B_{\infty,k_{N+1}}$. We have $u_0 \neq 0$ since $\hat{u}_0(0)=1$. The estimates above give

$$|\hat{u}(\xi_j)| \geqq C_1(2+|\xi_j|)^{-N}, \qquad C_1 = \sup_{|\xi|<1} |\hat{u}(\xi)| > 0.$$

If $\phi \in C_0^\infty$ and $\phi u \in C_0^\infty$ it follows that

$$\langle u e^{-i\langle \cdot, \xi_j\rangle}/\hat{u}(\xi_j), \phi\rangle = \widehat{(u\phi)}(\xi_j)/\hat{u}(\xi_j) \to 0 \quad \text{as } j\to\infty,$$

which proves that $\operatorname{supp} u_0 \subset \operatorname{sing\,supp} u$. The proof is complete.

Remark. We do not really need the last part of the lemma but have included it to remind the reader of the similar proof of Theorem 10.2.12. If θ is in the set $\Sigma(u)$ defined as in Section 8.1, that is, $(x,\theta)\in WF(u)$ for some x, then we can choose η_j in any given conic neighborhood of θ. The constructed sequence ξ_j will also belong to any conic neighborhood of θ when j is large.

Lemma 13.5.4. *Let $k\in\mathscr{K}$ and define $k_\eta\in\mathscr{K}$ for $\eta\in\mathbb{R}^n$ by*

$$k_\eta(\xi) = (1+|\xi-\eta|)^{-M} k(\xi), \qquad \xi\in\mathbb{R}^n, \tag{13.5.6}$$

where $M>n$ and

$$k(\xi) \leqq C(1+|\xi-\theta|)^M k(\theta); \qquad \xi,\theta\in\mathbb{R}^n. \tag{13.5.7}$$

If K is a compact set in $\mathbb{R}^n$ and $1\leqq p<\infty$ it follows that $\|u\|_{p,k}$ is equivalent to the norm

$$|||u|||_{p,k} = \left(\int \|u\|^p_{\infty,k_\eta}\, d\eta\right)^{1/p}$$

for $u\in\mathscr{E}'(K)\cap B_{p,k}$.

Proof. Since $|\hat{u}(\eta)k(\eta)| \leqq \|u\|_{\infty,k_\eta}$ we have $\|u\|_{p,k} \leqq |||u|||_{p,k}$. To prove an estimate in the opposite direction we choose $\chi\in C_0^\infty$ with $\chi=1$ in a neighborhood of K. Then $\hat{u}=(2\pi)^{-n}\hat{u}*\hat{\chi}$ if $u\in\mathscr{E}'(K)$. Since

$$k_\eta(\xi) \leqq C(1+|\xi-\theta|)^{2M} k(\theta)(1+|\theta-\eta|)^{-M}$$

it follows in view of Hölder's inequality that

$$|k_\eta(\xi)\hat{u}(\xi)| \leqq C\, \|\hat{\chi}(1+|.|)^{2M}\|_{p'} \left(\int |\hat{u}(\theta)k(\theta)(1+|\theta-\eta|)^{-M}|^p d\theta\right)^{1/p}.$$

Hence

$$\|u\|^p_{\infty,k_\eta} \leqq C_1 \int |\hat{u}(\theta)k(\theta)(1+|\theta-\eta|)^{-M}|^p d\theta.$$

Integrating with respect to η we now obtain $|||u|||^p_{p,k} \leqq C_2 \|u\|^p_{p,k}$ which completes the proof.

Proof of Theorem 13.5.2. If $w \in R(\bar{Y}) \setminus C_0^\infty$ we can by Lemma 13.5.3 choose $\xi_j \to \infty$ in $\mathbb{R}^n$ and $t_j \in \mathbb{C}$ so that

$$w_j = t_j w e^{-i\langle .,\xi_j\rangle} \to w_0 \neq 0.$$

Now the equation ${}^tPw=0$ gives ${}^tP(w_j e^{i\langle .,\xi_j\rangle})=0$, hence

$${}^tP(x, D+\xi_j)w_j = 0.$$

Passing to a subsequence we may assume that the limit

$$Q(x,\xi) = \lim_{j\to\infty} P(x, \xi-\xi_j)/\tilde{P}_0(-\xi_j)$$

exists. As in the proof of Theorem 13.5.1 we have

$${}^tP(x, \xi+\xi_j)/\tilde{P}_0(-\xi_j) \to {}^tQ(x,\xi).$$

Hence ${}^tQ(x,D)w_0=0$ which by hypothesis implies $w_0=0$. This is a contradiction proving that $R(\bar{Y}) \subset C_0^\infty(\bar{Y})$. Now $R(\bar{Y})$ is a Banach space with the maximum norm for example, and by the closed graph theorem the C^1 norm is an equivalent norm in $R(\bar{Y})$. Thus the unit ball is compact, so $R(\bar{Y})$ is finite dimensional.

To prove the theorem it suffices to show that with the notation used in the proof of Theorem 13.5.1 we have the estimate

$$\|v\|_{p',k'} \leq C\|{}^tPv\|_{p',k_P'} \quad \text{if } v \in C_0^\infty(\bar{Y}) \text{ and } v \perp R(\bar{Y}). \tag{13.5.8}$$

In fact, then we obtain

$$|\langle f, v\rangle| \leq C\|f\|_{p,k}\|{}^tPv\|_{p',k_P'} \quad \text{if } v \in C_0^\infty(\bar{Y}) \text{ and } v \perp R(\bar{Y}). \tag{13.5.9}$$

Since f is orthogonal to $R(\bar{Y})$ this inequality is in fact valid for all $v \in C_0^\infty(\bar{Y})$ for v can be written $v = v_1 + v_2$ with $v_1 \perp R(\bar{Y})$ and $v_2 \in R(\bar{Y})$. The estimate (13.5.9) is equivalent to the same estimate with v replaced by v_1. The linear form

$${}^tPv \to \langle f, v\rangle, \quad v \in C_0^\infty(\bar{Y}),$$

is thus continuous on a subspace of $B_{p',k_P'}$. By the Hahn-Banach theorem it can be extended to a continuous linear form u on $B_{p',k_P'}$ such that

$$\langle u, {}^tPv\rangle = \langle f, v\rangle, \quad v \in C_0^\infty(Y).$$

This means that $P(x,D)u=f$ in Y and $u \in B_{p,k_P}(\bar{Y})$ if $p \neq 1$; as in the proof of Theorem 12.8.13 the conclusion remains valid when $p=1$.

First we shall prove (13.5.8) when $p'=\infty$. If (13.5.8) were not true then, we could find $v_j \in C_0^\infty(\bar{Y})$ orthogonal to $R(\bar{Y})$ so that

$$\|v_j\|_{\infty,k'} = 1, \quad \|{}^tPv_j\|_{\infty,k_P'} < 1/j. \tag{13.5.10}$$

Choose ξ_j so that $|\hat{v}_j(\xi_j)|\,k'(\xi_j)\to 1$ as $j\to\infty$ and define w_j by

$$\hat{w}_j(\xi)=\hat{v}_j(\xi+\xi_j)/\hat{v}_j(\xi_j).$$

Then

$$|\hat{w}_j(\xi)|\leqq k'(\xi_j)/(k'(\xi+\xi_j)\,|\hat{v}_j(\xi_j)|\,k'(\xi_j))<2C(1+|\xi|)^M \tag{13.5.11}$$

for large j so there is a subsequence which has a limit w in $\mathscr{E}'$ with $\hat{w}(0)=1$. (See the proof of Lemma 13.5.3.) As at the beginning of the proof we may assume that

$${}^tP(x,\xi+\xi_j)/\tilde{P}_0(-\xi_j)\to {}^tQ(x,\xi).$$

Since $w_j=v_je^{-i\langle .,\xi_j\rangle}/\hat{v}_j(\xi)$ we obtain

$${}^tQ(x,D)w=\lim e^{-i\langle .,\xi_j\rangle}\,{}^tP v_j/(\tilde{P}_0(-\xi_j)\hat{v}_j(\xi_j)).$$

If we now use the second part of (13.5.10) we find that the Fourier transform of the right-hand side at ξ is bounded by

$$1/|j\,k'_P(\xi+\xi_j)\tilde{P}_0(-\xi_j)\hat{v}_j(\xi_j)|=j^{-1}(k'_P(\xi_j)/k'_P(\xi+\xi_j))/|k'(\xi_j)\hat{v}_j(\xi_j)|$$

which tends to 0 as $j\to\infty$ uniformly on compact sets since $k'_P\in\mathscr{K}$. Thus ${}^tQ(x,D)\,w=0$ so by hypothesis Q is not a localization at ∞. This means that ξ_j has a finite limit point ξ_0, and therefore that

$${}^tQ(x,\xi)={}^tP(x,\xi+\xi_0)/\tilde{P}_0(-\xi_0),$$

Hence

$${}^tP(x,D)(we^{i\langle .,\xi_0\rangle})=0.$$

$$v_0=we^{i\langle .,\xi_0\rangle}=\lim v_j/\hat{v}_j(\xi_j)\in\mathscr{E}'(\bar{Y})$$

satisfies the equation ${}^tP(x,D)v_0=0$, that is, $v_0\in R(\bar{Y})$. On the other hand, v_0 is orthogonal to $R(\bar{Y})$ since this is true for every v_j, so $v_0=0$ which is a contradiction because $\hat{v}_0(\xi_0)=1$. This proves (13.5.8) with $p'=\infty$.

The proof also gives that

$$\|v\|_{\infty,k'}\leqq C_1\,\|{}^tPv\|_{\infty,k'_P}\quad\text{if } v\in C_0^\infty(\bar{Y}) \text{ and } v\perp R(\bar{Y}), \tag{13.5.8$'$}$$

for every $k\in\mathscr{K}$ such that (13.5.7) is valid with fixed C and M. In fact, we would otherwise obtain (13.5.10) for some $k_j\in\mathscr{K}$ satisfying (13.5.7). The estimate (13.5.11) follows as before and the rest of the proof is unchanged. In particular, we can apply (13.5.8)$'$ to

$$k_\eta(\xi)=(1+|\xi+\eta|)^M k(\xi).$$

Then $k'_\eta(\xi)=(1+|\xi-\eta|)^{-M}k'(\xi)$. Since (13.5.8)$'$ gives

$$\|v\|_{\infty,k'_\eta}\leqq C_1\,\|{}^tPv\|_{\infty,(k_\eta,P)'} \tag{13.5.12}$$

if $v \in C_0^\infty(\bar{Y})$ and $v \perp R(\bar{Y})$, where C_1 does not depend on η, we obtain (13.5.8) if we take M large enough, raise (13.5.12) to the power p', and integrate using Lemma 13.5.4. The theorem is proved.

We have already observed that the hypotheses of Theorem 13.5.2 are fulfilled and that $R(\bar{Y})=0$ if P has analytic coefficients (and constant strength). Another case where the hypotheses are fulfilled is when the localizations of P at infinity are all hyperbolic. This follows from Theorem 13.3.7. In particular, Theorem 13.5.2 is therefore applicable to operators of real principal type with variable coefficients only in the lower order terms. However, we shall show in Section 13.6 that the hypotheses of Theorem 13.5.2 are not fulfilled in general and that $R(\bar{Y})$ may have positive dimension even in the elliptic case.

13.6. Non-uniqueness for the Cauchy Problem

If the boundary of the half space

$$H_N=\{x \in \mathbb{R}^n; \langle x, N\rangle \geqq 0\}$$

is characteristic with respect to $P(D)$ we know from Theorem 8.6.7 that the equation $P(D)u=0$ has a solution with support equal to H_N. This is quite obvious if $P(D)$ is homogeneous, for then we have $P(D)u=0$ when u is any function of $\langle x, N\rangle$. In this section we shall construct examples which show that the uniqueness theorems of Section 8.6 for the non-characteristic Cauchy problem break down completely for operators with C^∞ coefficients. This may seem rather surprising at first sight but might be suspected because of the vital role played by analytic continuation in Section 8.6.

In this section we shall use the notation

$$\tilde{P}_N(\xi)=\Big(\sum_j |D_N^j P(\xi)|^2\Big)^{\frac{1}{2}}, \quad D_N=\langle D, N\rangle, \tag{13.6.1}$$

which means that we consider the restriction of P to lines with direction N. As usual we denote by P_m the homogeneous part of P of order m.

Theorem 13.6.1. *Let $P(D)$ and $Q(D)$ be two differential operators with constant coefficients of order $\leqq m$ such that $P_m(N) \neq 0$ and*

$$\sup_{\mathbb{R}^n} \tilde{Q}_N(\xi)/\tilde{P}_N(\xi)=\infty. \tag{13.6.2}$$

Then the equation

$$(P(D)+a(x)\,Q(D))\,u=0 \tag{13.6.3}$$

has a solution $u\in C^\infty(\mathbb{R}^n)$ *with* $\operatorname{supp} u=H_N$ *for some* $a\in C^\infty(\mathbb{R}^n)$ *with* $\operatorname{supp} a\subset H_N$.

The hypothesis that the degree of Q is at most m could be dropped here but this would lengthen the proof. Before giving the rather long proof we note some consequences.

Corollary 13.6.2. *Let $P(D)$ and $Q(D)$ be two differential operators with constant coefficients such that the order of Q is at most equal to the order of P but Q is not weaker than P. For any $N\in\mathbb{R}^n\setminus 0$ one can then find $u\in C^\infty(\mathbb{R}^n)$ with $\operatorname{supp} u=H_N$ and $a\in C^\infty(\mathbb{R}^n)$ with $\operatorname{supp} a\subset H_N$ so that* (13.6.3) *is fulfilled.*

Proof. Let m be the order of P. If $P_m(N)=0$ we can take $a=0$ by Theorem 8.6.7. If $P_m(N)\neq 0$ the statement follows from Theorem 13.6.1, for if $\tilde{Q}_N(\xi)\leqq C\tilde{P}_N(\xi)$ then $|Q(\xi)|\leqq C'\tilde{P}(\xi)$. This implies that Q is weaker than P (Theorem 10.4.3) which is a contradiction.

Theorem 13.6.1 also gives examples of constant strength:

Example 13.6.3. Let $P(\xi)=\xi_1^2+\xi_2^2-\xi_3^2$ and $N=(0,1,0)$. Then $\tilde{P}_N(\xi)\leqq |\xi_1^2-\xi_3^2|+2$ when $\xi_2=0$. If Q is linear and $|Q(\xi)|/\tilde{P}_N(\xi)$ is bounded, it follows that $Q(\xi_1,0,\pm\xi_1)=0$ so Q is a linear function of ξ_2 only. If $Q(\xi)$ is a linear function which is not independent of (ξ_1,ξ_3) it follows from Theorem 13.6.1 that we can find a with support in the half space $x_2\geqq 0$ such that (13.6.3) has a solution with support equal to that half space.

The following is just a more general version of this example:

Corollary 13.6.4. *If $P(D)$ is homogeneous of degree m and $k\geqq 0$, then the conclusions of Theorem 13.6.1 are valid for some homogeneous Q of order $m-k$ unless*

$$\xi\in\mathbb{R}^n,\quad \langle D,N\rangle^j P(\xi)=0,\quad j\leqq k \Rightarrow \xi=0.$$

Example 13.6.5. There exist functions u and a in $C^\infty(\mathbb{R}^2)$ such that

$$D_2 u+aD_1 u=0$$

and $\operatorname{supp} u=\{x;\,x_2\geqq 0\}\supset\operatorname{supp} a$.

Here we could replace D_2 by any differential operator which does not contain D_1, in particular any positive power of D_2. This example can be given a more spectacular form:

Corollary 13.6.6. *Let $Q(x,D)$ be a differential operator of order m with coefficients in $C^\infty(X)$, X open in $\mathbb{R}^n$, and let $\psi \in C^\infty(X)$ be real valued and satisfy the characteristic equation*

$$Q_m(x, \operatorname{grad} \psi) = 0. \tag{13.6.4}$$

Let $\phi \in C^\infty(X)$ be real valued and assume that $\operatorname{grad}\phi$ and $\operatorname{grad}\psi$ are linearly independent at a point $x^0 \in X$. Then there is an open set Y with $x^0 \in Y \subset X$ and a differential operator $P(x,D)$ with coefficients in $C^\infty(Y)$ whose principal part $P_m(x,D)$ is equal to $Q_m(x,D)$ when $\phi(x) \geqq \phi(x^0)$ such that the equation $P(x,D)u=0$ has a solution $u \in C^\infty(Y)$ with support equal to

$$\{x; x \in Y, \phi(x) \leqq \phi(x^0)\}. \tag{13.6.5}$$

Proof. The statement is invariant for a change of coordinates so we may assume that $x^0=0$ and that

$$\psi(x) = x_1, \qquad \phi(x) = -x_2.$$

Then it follows from (13.6.4) that we can write

$$Q_m(x,D) = \sum_2^n q_j(x,D) D_j$$

where q_j is of order $m-1$. Now let $u(x)=u(x_1,x_2)$ and $a(x)=a(x_1,x_2)$ be chosen according to Example 13.6.5. Then

$$Q_m(x,D)u = q_2(x,D) D_2 u = -q_2(x,D) a D_1 u.$$

Hence the operator

$$P(x,D) = Q_m(x,D) + q_2(x,D) a D_1$$

has the required properties.

The preceding corollary is always applicable if Q_m has constant coefficients and is not elliptic.

Corollary 13.6.7. *Assume in addition to the hypotheses of Corollary 13.6.6 that*

$$\sum_1^n Q_m^{(j)}(x, \operatorname{grad}\psi)\, \partial\phi/\partial x_j = 0. \tag{13.6.6}$$

(This means that ϕ is constant on the bicharacteristics generating the level surfaces of ψ.) Then there is an open set Y with $x^0 \in Y \subset X$ and a differential operator $P(x,D)$ with coefficients in $C^\infty(Y)$ and principal

part equal to $Q_m(x,D)$ *in* Y, *such that the equation* $P(x,D)u=0$ *has a solution* $u\in C^\infty(Y)$ *with support equal to the set* (13.6.5).

Proof. We may again assume that $\psi(x)=x_1$, $\phi(x)=-x_2$. The meaning of (13.6.4) and (13.6.6) is then that the coefficients of D_1^m and $D_1^{m-1}D_2$ in $Q_m(x,D)$ must vanish. Hence we can write

$$Q_m(x,D)=q_2(x,D)\,D_2^2+\sum_3^n q_j(x,D)\,D_j$$

where q_2 is of order $m-2$ and q_j is of order $m-1$ when $j>2$. We now apply the observation made after Example 13.6.5 to choose C^∞ functions a and u of (x_1,x_2) vanishing for $x_2\leqq 0$ such that the support of u is exactly the half space $x_2\geqq 0$ and

$$D_2^2u+aD_1u=0.$$

Then it follows that

$$Q_m(x,D)\,u=-q_2(x,D)\,aD_1\,u.$$

Hence the operator $P(x,D)$ defined by

$$P(x,D)=Q_m(x,D)+q_2(x,D)\,aD_1$$

has all the required properties.

Corollary 13.6.7 is applicable if Q_m has constant coefficients and the surface $\phi(x)=\phi(x^0)$ is a cylinder with bicharacteristic generators conjugate to a characteristic plane different from the tangent plane to the surface $\phi(x)=\phi(x^0)$ at x^0. In fact, this follows if we replace ϕ by a function whose level surfaces in a neighborhood of x^0 only differ from the surface $\phi(x)=\phi(x^0)$ by a translation, and let ψ be a linear function which is constant in the characteristic plane. We note that the surface $\phi(x)=\phi(x^0)$ can be chosen convex in all directions except that of the bicharacteristic generator.

After all these examples of applications of Theorem 13.6.1 we now turn to its proof. The first point is to express the hypothesis in a form which is more useful in the proof.

Lemma 13.6.8. *Let* P *and* Q *satisfy the hypotheses of Theorem 13.6.1. Then it is possible to find a Laurent series*

$$\xi(t)=\sum_{-\infty}^{\mu} c_j t^j,\qquad c_j\in\mathbb{R}^n,$$

converging for large t, *and a positive integer* κ *such that when* $t\to\infty$

$$P(\xi(t)+z\,t^\kappa N)\,t^{-g(P)}\to c(P)\,z^{d(P)},\qquad (13.6.7)$$
$$Q(\xi(t)+z\,t^\kappa N)\,t^{-g(Q)}\to c(Q)\,z^{d(Q)}$$

for some constants $c(P)$, $c(Q) \neq 0$ *and integers* $0 < g(P) < g(Q)$, $d(P) > d(Q) \geqq 0$.

Proof. $\tilde{Q}_N(\eta)/\tilde{P}_N(\eta)$ is continuous since $\tilde{P}_N(\eta) \geqq m!\,|P_m(N)| > 0$. If we apply Theorem A.2.8 to

$$E = \{(R, S, \eta);\ \tilde{Q}_N(\eta)^2 = S\tilde{P}_N(\eta)^2, |\eta|^2 = R^2\}$$

it follows that the maximum of $\tilde{Q}_N(\eta)/\tilde{P}_N(\eta)$ when $\eta \in \mathbb{R}^n$ and $|\eta| = R$ is attained for $\eta = \eta(R)$ where $\eta(R)$ is algebraic for large R. From the Puiseux series expansion of $\eta(R)$ it follows that $\xi(t) = \eta(t^r)$ has a Laurent series expansion of the desired form for some positive integer r.

By Taylor's formula we have

$$P(\xi(t) + z\, t^\kappa N) = \sum D_N^j P(\xi(t))\, (i\, z\, t^\kappa)^j / j!. \tag{13.6.8}$$

Here $D_N^j P(\xi(t))$ is asymptotic to a constant times t^{μ_j} for some integer μ_j, or else $D_N^j P(\xi(t))$ is identically 0 in which case we define $\mu_j = -\infty$. The sum grows as t raised to the power

$$g_P(\kappa) = \max_{j \geqq 0} (\kappa j + \mu_j(P)), \qquad \kappa \geqq 0,$$

which is a finite, convex, increasing, piecewise linear function. $\tilde{P}_N(\xi(t))\, t^{-g_P(0)}$ has a limit $\neq 0$ as $t \to \infty$, so $g_Q(0) > g_P(0)$ by (13.6.2) if g_Q is defined similarly. On the other hand, $g_P(\kappa) \geqq \kappa m$ since $P_m(N) \neq 0$ so $g_Q(\kappa) \leqq g_P(\kappa)$ when κ is large. It is therefore possible to find a rational number $\kappa > 0$ such that

$$g_P(\kappa) < g_Q(\kappa)$$

and g_P, g_Q are linear in a neighborhood of κ, with slopes d_P and d_Q such that $d_P > d_Q$. In fact, g_P has to increase faster than g_Q some time before catching up. Now we have

$$\kappa j + \mu_j(P) \leqq g_P(\kappa), \qquad j \geqq 0,$$

with equality for just one value j_0 of j, so $g_P(\sigma) = \sigma j_0 + \mu_{j_0}(P)$ in a neighborhood of κ. Hence $j_0 = d_P$ and it follows from (13.6.8) that the first part of (13.6.7) is valid with $g_P(\kappa) = g(P)$ and $d(P) = d_P = j_0$. The second part follows in the same way. Replacing t by an integral power of t we can make κ an integer, which completes the proof.

Proof of Theorem 13.6.1. We shall define a by the differential Equation (13.6.3). The problem is to construct u so that $a = -P(D)\,u/Q(D)\,u$ is smooth. Set $b_\nu = 1/\nu$, $t_\nu = 2^\nu$ and

$$u_\nu(x) = \exp i(\langle x, \xi(t_\nu)\rangle + \phi_\nu(\langle x, N\rangle)), \tag{13.6.9}$$
$$\langle x, N\rangle \in I_\nu = (b_{\nu+1}, b_{\nu-1}).$$

Here ϕ_ν will be chosen later so that u_ν satisfies a differential equation of the desired form. With some large ν_0 we shall then set

$$u(x)=\sum_{\nu_0}^{\infty}\chi_\nu(\langle x,N\rangle)\,u_\nu(x), \qquad x\in\mathbb{R}^n, \tag{13.6.10}$$

where $\chi_\nu\in C_0^\infty(I_\nu)$ is 1 in a neighborhood of $(b'_\nu, b'_{\nu-1})$; $b'_\nu=(b_\nu+b_{\nu+1})/2$. (The first term will be slightly modified.) Then $u(x)=u_\nu(x)+u_{\nu+1}(x)$ when $\langle x,N\rangle$ is close to b'_ν, When $\langle x,N\rangle$ is at some distance to the left (resp. right) of b'_ν, the function u_ν (resp. $u_{\nu+1}$) will then be cut off. To make $a=-P(D)u/Q(D)u$ smooth we have to make sure that $u_{\nu+1}$ (resp. u_ν) is much larger than u_ν (resp. $u_{\nu+1}$) then. When $\langle x,N\rangle$ is near b'_ν the two terms must therefore be of the same size, and we shall make both satisfy the same differential equation $P(D)u+aQ(D)u=0$ with a constant a then. This can be done by choosing ϕ_ν and $\phi_{\nu+1}$ as linear functions with suitable slopes there. Since (13.6.7) gives information on the ratio of P and Q at $\xi(t)+z\,t^\kappa N$, it is natural to choose the slope of ϕ_ν of the order of magnitude t_ν^κ.

After this outline we begin the detailed construction. Set

$$\phi'_\nu(s)=t_\nu^\kappa\,\psi_\nu(s).$$

We shall determine ψ_ν so that near b'_ν and $b'_{\nu-1}$, respectively, ψ_ν is equal to constants σ_ν^- and σ_ν^+ satisfying

$$(P/Q)(\xi(t_\nu)+t_\nu^\kappa\,\sigma_\nu^-\,N)=(P/Q)(\xi(t_{\nu+1})+t_{\nu+1}^\kappa\,\sigma_{\nu+1}^+\,N). \tag{13.6.11}$$

With the notation of Lemma 13.6.8 the left-hand side is asymptotically equal to

$$(c(P)/c(Q))\,t_\nu^{g(P)-g(Q)}(\sigma_\nu^-)^{d(P)-d(Q)}$$

so the Equation (13.6.11) is closely approximated by

$$2^{g(Q)-g(P)}\,(\sigma_\nu^-)^{d(P)-d(Q)}=(\sigma_{\nu+1}^+)^{d(P)-d(Q)}.$$

If we choose $\sigma_\nu^-=i$, it follows that (13.6.11) for large ν has a solution $\sigma_{\nu+1}^+$ with

$$\sigma_{\nu+1}^+\to 2^\gamma i, \qquad \nu\to\infty. \tag{13.6.12}$$

Here $\gamma=(g(Q)-g(P))/(d(P)-d(Q))>0$ so $2^\gamma>1$. This means that the slope $-t_{\nu+1}^\kappa\operatorname{Im}\sigma_{\nu+1}^+$ of $-\operatorname{Im}\phi_{\nu+1}$ as a function of $\langle x,N\rangle$ will be far smaller than the slope $-t_\nu^\kappa\operatorname{Im}\sigma_\nu^-$ of $-\operatorname{Im}\phi_\nu$, so the ratio $|u_{\nu+1}/u_\nu|$ will decrease fast when $\langle x,N\rangle$ increases in a neighborhood of b'_ν.

Choose $B\geqq\operatorname{Im}\sigma_{\nu+1}^+$ for every ν, which implies that $B>1$. For every $M>0$ we can choose $\psi_\nu\in C^\infty(I_\nu)$ so that for large ν the following conditions are fulfilled for some constants $C_k>0$ and some compact

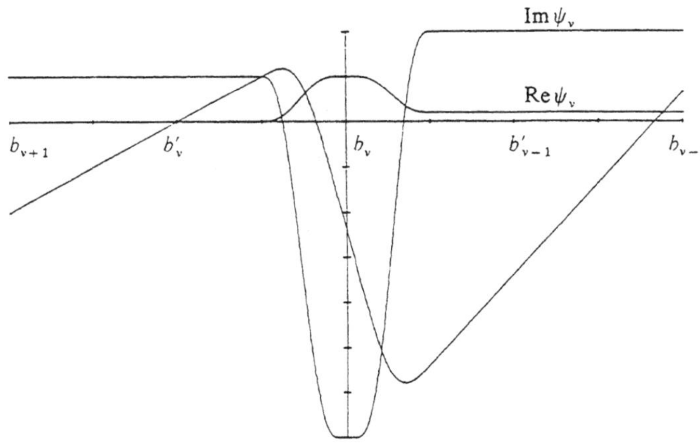

Fig. 6. Re ψ_ν, Im ψ_ν and $8\int \operatorname{Im}\psi_\nu ds/|I_\nu|$ when $\sigma_{\nu+1}=0.2+2i$

set $K\subset\{z\in\mathbb{C}, z\neq 0, \operatorname{Im} z\leqq B\}$

$$\psi_\nu(s)=\sigma_\nu^- \quad \text{when } |s-b'_\nu|<1/4\nu^2, \tag{13.6.13}$$
$$\psi_\nu(s)=\sigma_\nu^+ \quad \text{when } |s-b'_{\nu-1}|<1/4(\nu-1)^2;$$
$$\psi_\nu(s)\in K \text{ when } s\in I_\nu;\ |\psi_\nu^{(k)}(s)|\leqq C_k\nu^{2k} \text{ if } s\in I_\nu,\ k=0,1,\dots;$$
$$\int_{b'_\nu}^{b'_{\nu-1}} \operatorname{Im}\psi_\nu(s)\,ds<-M/\nu^2.$$

In fact, it suffices to piece together by a partition of unity the prescribed values σ_ν^- and σ_ν^+ near b'_ν and $b'_{\nu-1}$ and a value $1-iA$ near b_ν, where A is chosen large. Since the length of I_ν is $2/(\nu^2-1)$ it follows from (1.4.2) that the derivatives of order k of the functions in the partition of unity (with three terms) can be made $O(\nu^{2k})$.

Having chosen ψ_ν we determine ϕ_ν successively so that

$$\phi'_\nu=t_\nu^\kappa\psi_\nu,\qquad \phi_\nu(b'_\nu)=\phi_{\nu+1}(b'_\nu). \tag{13.6.14}$$

In view of (13.6.12) we have for some positive C_1, C_2 when ν is large

$$C_1 2^{\kappa\nu}\leqq \operatorname{Im}(\phi_{\nu+1}(s)-\phi_\nu(s))/(s-b'_\nu) \tag{13.6.15}$$
$$\leqq C_2 2^{\kappa\nu},\ |s-b'_\nu|<1/4\nu^2.$$

This gives the information about the relative sizes of u_ν and $u_{\nu+1}$ needed to prove that a is smooth. To prove that u itself is smooth we must also have an estimate of u_ν. From the last part of (13.6.13) it follows that

$$\operatorname{Im}(\phi_\nu(b'_{\nu-1})-\phi_\nu(b'_\nu))<-M2^{\kappa\nu}/\nu^2,$$

that is,

$$\operatorname{Im}\phi_{\nu+1}(b'_\nu) > \operatorname{Im}\phi_\nu(b'_{\nu-1}) + M 2^{\kappa\nu}/\nu^2.$$

Hence

$$\operatorname{Im}\phi_\nu(b'_{\nu-1}) > M 2^{\kappa(\nu-1)}/\nu^2$$

if we make sure that this is true at the beginning of the construction. Since $\operatorname{Im}\phi'_\nu \leqq 2^{\kappa\nu} B$, it follows that

$$\operatorname{Im}\phi_\nu(s) \geqq 2^{\kappa\nu}/\nu^2, \quad b'_\nu \leqq s \leqq b'_{\nu-1} \tag{13.6.16}$$

provided that

$$M 2^{\kappa(\nu-1)}/\nu^2 - 2^{\kappa\nu} B(b'_{\nu-1} - b'_\nu) \geqq 2^{\kappa\nu}/\nu^2,$$

which is true for large ν if $M > (B+1)2^\kappa$. We fix M now so that this condition is fulfilled. In view of (13.6.15) the estimate (13.6.16) is also valid with ν replaced by $\nu+1$ or $\nu-1$ in the right-hand side when $b'_\nu - 1/4\nu^2 < s < b'_\nu$ or $b'_{\nu-1} < s < b'_{\nu-1} + 1/4(\nu-1)^2$.

Now choose $\chi_\nu \in C_0^\infty(b'_\nu - 1/4\nu^2, b'_{\nu-1} + 1/4(\nu-1)^2)$ so that $\chi_\nu = 1$ in $(b'_\nu - 1/8\nu^2, b'_{\nu-1} + 1/8(\nu-1)^2)$ and

$$|\chi_\nu^{(k)}(s)| \leqq C_k \nu^{2k}. \tag{13.6.17}$$

This is again possible by (1.4.2). With some ν_0 so large that the preceding estimates and some later ones are valid for $\nu \geqq \nu_0$ we define u by (13.6.10). (The first term will be modified later on.) In view of (13.6.13), (13.6.16) and (13.6.17) derivatives of order k of the νth term can be estimated by

$$C'_k \exp(-2^{\kappa(\nu-1)}/\nu^2)(t_\nu^\kappa + |\xi(t_\nu)|)^k,$$

which converges to 0 very rapidly when $\nu \to \infty$. Since only two terms in (13.6.10) are simultaneously different from 0, it follows that $u \in C^\infty$.

We set $a = 0$ when $\langle x, N\rangle \leqq 0$ and $a = -P(D)u/Q(D)u$ when $\langle x, N\rangle > 0$. In the slab where

$$|\langle x, N\rangle - b'_\nu| < 1/8\nu^2$$

the construction has been made so that $u = u_\nu + u_{\nu+1}$ and $-a$ is the constant (13.6.11) which tends to 0 when $\nu \to \infty$ in view of (13.6.7) since $g(P) < g(Q)$. The derivatives of a are all 0 in this set. Now let

$$b'_\nu + 1/8\nu^2 < \langle x, N\rangle < b'_\nu + 1/4\nu^2.$$

In this slab u_ν is a pure exponential and is much larger than $u_{\nu+1}$. We therefore write $u = u_\nu(1 + R_\nu)$,

$$\begin{aligned} R_\nu(x) = \chi_{\nu+1}(\langle x, N\rangle) \exp i(&\langle x, \xi(t_{\nu+1}) - \xi(t_\nu)\rangle \\ &+ \phi_{\nu+1}(\langle x, N\rangle) - \phi_\nu(\langle x, N\rangle)). \end{aligned}$$

If $\sigma \geqq \kappa$ and $\xi(t)=O(t^\sigma)$ it follows from (13.6.15), (13.6.13) and (13.6.17) that

(13.6.18) $$|D^\alpha R_\nu(x)| \leqq C_\alpha 2^{|\alpha|\sigma\nu} \exp(-c 2^{\kappa\nu}/\nu^2)$$

for some positive constants C_α and c. Since u_ν is a pure exponential, we have

$$\begin{aligned}(Q(D)u)/u_\nu &= Q(\xi(t_\nu)+\sigma_\nu^- t_\nu^\kappa N + D)(1+R_\nu)\\ &= Q(\xi(t_\nu)+\sigma_\nu^- t_\nu^\kappa N)(1+S_\nu)\end{aligned}$$

where for some constant C

(13.6.18)′ $$|D^\alpha S_\nu(x)| \leqq C_\alpha 2^{|\alpha|\sigma\nu} \exp(C\nu - c 2^{\kappa\nu}/\nu^2).$$

In particular, $|S_\nu|<1/2$ when ν is large. Replacing Q by P we also obtain

$$(P(D)u)/u_\nu = P(\xi(t_\nu)+\sigma_\nu^- t_\nu^\kappa N)(1+T_\nu)$$

where T_ν also satisfies estimates of the form (13.6.18)′. Thus

$$a = -(P/Q)(\xi(t_\nu)+\sigma_\nu^- t_\nu^\kappa N)(1+T_\nu)/(1+S_\nu),$$

so it follows from (13.6.7) and (13.6.18)′ that a and all its derivatives have bounds converging to 0 as $\nu\to\infty$. The same argument is applicable when

$$b'_{\nu-1} - 1/4(\nu-1)^2 < \langle x, N\rangle < b'_{\nu-1} - 1/8(\nu-1)^2.$$

It remains to consider the set where

(13.6.19) $$b'_\nu + 1/4\nu^2 < \langle x, N\rangle < b'_{\nu-1} - 1/4(\nu-1)^2.$$

There we have $u=u_\nu$, and (13.6.7) gives

$$Q(D)u = c(Q)\, t_\nu^{g(Q)} f_{Q\nu}(\langle x, N\rangle)\, u,$$

where

$$\begin{aligned} f_{Q\nu}(s) &= e^{-i\phi_\nu(s)} q_\nu(D_s/t_\nu^\kappa)\, e^{i\phi_\nu(s)},\\ q_\nu(z) &= Q(\xi(t_\nu)+z\, t_\nu^\kappa N)/(t_\nu^{g(Q)} c(Q)).\end{aligned}$$

Here $q_\nu(z)-z^{d(Q)}$ is a polynomial with coefficients which are $O(t_\nu^{-1})$. Recalling that $\phi'_\nu/t_\nu^\kappa=\psi_\nu$, we obtain

$$f_{Q\nu}(s) = \psi_\nu(s)^{d(Q)}(1+r_{Q\nu}(s))$$

where $r_{Q\nu}(s)$ is a polynomial in ψ_ν, ψ_ν^{-1} and the derivatives of ψ_ν, with coefficients $O(1/t_\nu)$. Since t_ν grows exponentially it follows from (13.6.13) that $r_{Q\nu}(s)$ and every one of its derivatives $\to 0$ as $\nu\to\infty$. If $r_{P\nu}$ is defined in the same way, we have in the set defined by (13.6.19)

$$a = -P(D)u/Q(D)u = -(c(P)/c(Q))\, t_\nu^{g(P)-g(Q)}\, \psi_\nu^{d(P)-d(Q)}(1+r_{P\nu})/(1+r_{Q\nu})$$

where the argument is x on the left and $s=\langle x,N\rangle$ on the right. Since $g(P)<g(Q)$ it follows that the right-hand side and all its derivatives tend to 0 exponentially when $\nu\to\infty$. In view of Corollary 1.1.2 it follows that $a\in C^\infty$ when $\langle x,N\rangle<b'_{\nu_0-1}$ if ν_0 is chosen large enough.

Now change the definitions of χ_{ν_0} and ϕ_{ν_0} when $\langle x,N\rangle>b'_{\nu_0-1}$ so that $\chi_{\nu_0}=1$ and ϕ_{ν_0} is linear in $(b'_{\nu_0-1}-1/4(\nu_0-1)^2,\infty)$. It is then clear that $u\in C^\infty(\mathbb{R}^n)$, and $a\in C^\infty(\mathbb{R}^n)$ since a is constant when $\langle x,N\rangle>b'_{\nu_0-1}-1/4(\nu_0-1)^2$. The proof of Theorem 13.6.1 is now complete.

Remark. The proof shows that $\sup|a|$ is as small as we please if ν_0 is chosen large. Thus the perturbation of P in (13.6.3) can be made small in the whole space. For any compact interval $I\subset\mathbb{R}$ we can construct u and a satisfying (13.6.3) and

$$\operatorname{supp} u=\operatorname{supp} a=\{x;\ \langle x,N\rangle\in I\}$$

by making the same construction as above near the two boundaries of the slab.

If P is hyperbolic with respect to N and Q is weaker than P, it follows from Theorem 13.3.7 that u must be equal to 0 identically if $(P+aQ)u=0$ and $\operatorname{supp} u\subset H_N$, provided that $\sup|a|$ is small enough. Corollary 13.6.2 can therefore not be improved when P is hyperbolic. On the other hand, Theorem 13.6.1 fails completely when P is elliptic. Much greater care must then be exercised in the construction of examples of non-uniqueness for the Cauchy problem. In the proof of Theorem 13.6.1 the functions (13.6.9) were chosen rather arbitrarily. We just had to make sure that $\operatorname{Im}\phi_\nu$ and ϕ'_ν were of the right order of magnitude and that $\operatorname{Im}\phi_\nu-\operatorname{Im}\phi_{\nu+1}$ changed sign from $-$ to $+$ at the point b'_ν where a switch was made from $u_{\nu+1}$ to u_ν. The exponential increase of t_ν took care of the errors introduced. However, in the case of complex characteristics we cannot use such large frequencies. Instead the analogue of the function ϕ_ν has to be chosen so that it fits an initial analytic perturbation of the operator $P(D)$. To do so we shall use asymptotic expansions close to those of geometrical optics. (Cf. Section 12.2.)

Theorem 13.6.9. *Let $P(D)$ and $Q(D)$ be differential operators with constant coefficients and the order of $P(D)$ at least as large as that of $Q(D)$. Assume that there exists a sequence $\zeta_\nu\in Z_N$,*

$$Z_N=\{\xi+isN;\ \xi\in\mathbb{R}^n, s\leqq 0\},$$

and sequences T_ν, K_ν *of positive numbers such that*

$$P(\zeta_\nu + T_\nu z N)/P(\zeta_\nu) \to 1, \tag{13.6.20}$$
$$Q(\zeta_\nu + T_\nu z N)/Q(\zeta_\nu) \to 1, \qquad z \in \mathbb{C},$$

$$P(\zeta_\nu)/Q(\zeta_\nu) \to 0, \tag{13.6.21}$$

$$K_\nu(P(\zeta_\nu + T_\nu z N)/P(\zeta_\nu) - Q(\zeta_\nu + T_\nu z N)/Q(\zeta_\nu)) \to r(z), \qquad z \in \mathbb{C}, \tag{13.6.22}$$

$$T_\nu/K_\nu \to \infty, \qquad T_\nu K_\nu/(1+|\operatorname{Im} \zeta_\nu|) \to \infty. \tag{13.6.23}$$

Here r *is a polynomial in one variable and we assume that*

$$\operatorname{Im} r'(z) < 0 \quad \textit{for some } z. \tag{13.6.24}$$

Then one can find $a \in C^\infty(\mathbb{R}^n)$ *vanishing of a given, arbitrarily high order when* $\langle x, N\rangle = 0$, *such that* (13.6.3) *has a solution* $u \in C^\infty(\mathbb{R}^n)$ *with* $\operatorname{supp} u = H_N$.

Note that (13.6.23) implies that

$$T_\nu^2/(1+|\operatorname{Im} \zeta_\nu|) \to \infty, \qquad \text{thus } T_\nu \to \infty. \tag{13.6.23$'$}$$

If the conclusion of Theorem 13.6.9 is not already contained in Theorem 13.6.1 we have

$$\tilde{Q}_N(\xi) \leqq C \tilde{P}_N(\xi), \qquad \xi \in \mathbb{R}^n.$$

This will be assumed from now on. By Theorem 10.4.3 and Lemma 10.4.2 this implies that

$$\sum_j |\langle D, N\rangle^j Q(\zeta)|\, T^j \leqq C' \sum_j |\langle D, N\rangle^j P(\zeta)|\, T^j,$$
$$\zeta \in Z_N, \qquad T \geqq |\operatorname{Im} \zeta| + 1.$$

Thus the conditions (13.6.20), (13.6.21) require that $\operatorname{Im} \zeta_\nu \to \infty$ and that

$$T_\nu/|\operatorname{Im} \zeta_\nu| \to 0, \tag{13.6.25}$$

so T_ν will have to be somewhere between $|\operatorname{Im} \zeta_\nu|^{\frac{1}{2}}$ and $|\operatorname{Im} \zeta_\nu|$.

The condition (13.6.20) just means that T_ν is much smaller than the distance from ζ_ν to the zeros of P and Q in the direction N. The condition (13.6.24) is of course fulfilled unless $r'(z)$ is a constant r_1 with $\operatorname{Im} r_1 \geqq 0$. Since $r(0)=0$ this means that $r(z) = r_1 z$. Thus (13.6.24) is equivalent to

$$r(z) \text{ is not of the form } r_1 z \text{ with } \operatorname{Im} r_1 \geqq 0. \tag{13.6.24$'$}$$

Choose z_0 so that $\operatorname{Im} r'(z_0) < 0$, and replace ζ_ν by $\zeta_\nu + T_\nu z_0 N$. Then the conditions (13.6.20), (13.6.21) are still fulfilled, and $\zeta_\nu + T_\nu z_0 N \in Z_N$ by

(13.6.25) for large ν. Since

$$\begin{aligned}&P(\zeta_\nu+T_\nu zN)/P(\zeta_\nu+T_\nu z_0N)-Q(\zeta_\nu+T_\nu zN)/Q(\zeta_\nu+T_\nu z_0N)\\&\quad=(Q(\zeta_\nu)/Q(\zeta_\nu+T_\nu z_0N))((P(\zeta_\nu+T_\nu zN)/P(\zeta_\nu))(1+M_\nu)\\&\quad-Q(\zeta_\nu+T_\nu zN)/Q(\zeta_\nu))\end{aligned}$$

where

$$K_\nu M_\nu=K_\nu(P(\zeta_\nu)\,Q(\zeta_\nu+T_\nu z_0N)/(P(\zeta_\nu+T_\nu z_0N)\,Q(\zeta_\nu))-1)\to -r(z_0),$$

it follows that (13.6.22) remains valid with r replaced by $r(z+z_0)-r(z_0)$. Hence we may always assume that $\operatorname{Im} r'(0)<0$. If T_ν is now replaced by $\varepsilon_\nu T_\nu$ and K_ν is replaced by K_ν/ε_ν where ε_ν is a sequence converging to 0 so slowly that (13.6.23) remains valid, then r is replaced by $z\,r'(0)$. One can therefore assume in the proof of Theorem 13.6.9 that

$$r(z)=r_1 z \quad \text{where } \operatorname{Im} r_1<0. \tag{13.6.24)$''$ \text{(}$$

Before giving the very long proof of Theorem 13.6.9 we shall discuss some examples.

Example 13.6.10. Let $P(\xi)=(\xi_1^2+\xi_2^2+\xi_3^2)^a-\xi_1^{2a}/2$ and $Q(\xi)=\xi_2^b$ where $a>1$ and $b\leqq 2a$. Choose $\zeta_\nu=(\xi_{\nu 1},\xi_{\nu 2},-i\nu)$ where $\xi_{\nu 1},\xi_{\nu 2}$ are real and $\xi_{\nu 1}^2+\xi_{\nu 2}^2=\nu^2$. Then we obtain for $N=(0,0,1)$

$$P(\zeta_\nu+T_\nu zN)=(T_\nu^2z^2-2i\nu T_\nu z)^a-\xi_{\nu 1}^{2a}/2,\qquad Q(\zeta_\nu+T_\nu zN)=\xi_{\nu 2}^b.$$

Thus (13.6.20), (13.6.21) are fulfilled if

$$\xi_{\nu 1}^{2a}/\xi_{\nu 2}^b\to 0,\qquad \nu T_\nu/\xi_{\nu 1}^2\to 0.$$

With $K_\nu=\xi_{\nu 1}^{2a}(\nu T_\nu)^{-a}$ we obtain $r(z)=-2(-2iz)^a$. Condition (13.6.23) becomes

$$(\nu T_\nu)^a T_\nu/\xi_{\nu 1}^{2a}\to\infty,\qquad \xi_{\nu 1}^{2a}T_\nu(\nu T_\nu)^{-a}/\nu\to\infty.$$

Summing up, we have the conditions

$$\begin{aligned}&\xi_{\nu 1}^{2a}/\nu^b\to 0, &&\nu T_\nu/\xi_{\nu 1}^2\to 0,\\&T_\nu^{1+a}\xi_{\nu 1}^{-2a}\nu^a\to\infty, &&T_\nu^{a-1}\xi_{\nu 1}^{-2a}\nu^{1+a}\to 0.\end{aligned}$$

Since $|\xi_{\nu 1}|<\nu$, the last condition implies the second one. The last conditions are satisfied for an appropriate choice of T_ν if and only if

$$(\xi_{\nu 1}^{2a}\nu^{-1-a})^{a+1}(\xi_{\nu 1}^{-2a}\nu^a)^{a-1}\to\infty,$$

so our remaining conditions are

$$\xi_{\nu 1}\nu^{-b/2a}\to 0,\qquad \xi_{\nu 1}\nu^{-(3a+1)/4a}\to\infty.$$

These are compatible if and only if $(3a+1)/2<b$, that is,

$$b>2a-(a-1)/2.$$

When $a=2$ we can take $b=4$ so Q is of the same order as P. Starting with $a=4$ we can choose Q of lower order than P. The case $a=2$, $b=4$ is also contained in the following

Corollary 13.6.11. *Let P a homogeneous polynomial, and assume that $P(\zeta)=0$ for some $\zeta\neq 0$ with $\operatorname{Im}\zeta$ proportional to N. Unless P has a polynomial factorization*

$$P=P_1^j P_2;\qquad D_N P_1(\zeta)\neq 0,\ P_2(\zeta)\neq 0,$$

one can for every homogeneous Q with $\deg Q=\deg P$, $Q(\zeta)\neq 0$ and $Q(\xi+zN)\equiv Q(\xi)$ find $u\in C^\infty(\mathbb{R}^n)$ and $a\in C^\infty(\mathbb{R}^n)$ such that (13.6.3) *is fulfilled,* $\operatorname{supp} u=H_N$ *and a vanishes of any prescribed order when $\langle x,N\rangle=0$.*

Proof. We choose the coordinates so that $N=(0,\dots,0,1)$ and set $\zeta=(\xi_0',\lambda_0)$, $\xi_0'\in\mathbb{R}^{n-1}$. In view of Corollary 13.6.4 we may assume that $\operatorname{Im}\lambda_0\neq 0$, hence that $\operatorname{Im}\lambda_0<0$, for ζ may be replaced by $-\zeta$. Since $Q(\zeta)=Q(\xi_0')\neq 0$ we have $\xi_0'\neq 0$. Let λ_0 be a zero of multiplicity μ of $P(\xi_0',\lambda)$ as a polynomial in λ. Then the equation $P(\xi',\lambda)=0$ has μ roots, and the equation $\partial P(\xi',\lambda)/\partial\lambda=0$ has $\mu-1$ roots close to λ_0 if ξ' is close to ξ_0'. If $P=0$ at all such zeros of $\partial P/\partial\lambda$, then there can only be one, for a zero of P is a zero of $\partial P/\partial\lambda$ of multiplicity decreased by one. If we consider the decomposition of P in a product of irreducible polynomials, it follows easily that $P=P_1^\mu P_2$ with $P_2\neq 0$ and $D_n P_1\neq 0$ at (ξ_0',λ_0). In what follows we exclude this case.

Thus we have $\partial P(\xi',\lambda)/\partial\lambda=0$ but $P(\xi',\lambda)\neq 0$ for a sequence (ξ_ν',λ_ν) in Z_N converging to (ξ_0',λ_0). With $P^{(j)}(\xi',\lambda)=\partial^j P(\xi',\lambda)/\partial\lambda^j$ we have

$$P(\xi_\nu',\lambda_\nu+S_\nu z)=\sum_{j\neq 1} P^{(j)}(\xi_\nu',\lambda_\nu)(S_\nu z)^j/j!.$$

If S_ν is sufficiently small, then

$$P(\xi_\nu',\lambda_\nu+S_\nu z)/P(\xi_\nu',\lambda_\nu)-1=O(S_\nu \nu^{-1}(|z|^2+\dots+|z|^m)).$$

Now we set $\zeta_\nu=\rho_\nu(\xi_\nu',\lambda_\nu)$, $T_\nu=\rho_\nu S_\nu$ where $\rho_\nu\to+\infty$. Then

$$P(\zeta_\nu+T_\nu zN)=\rho_\nu^m P(\xi_\nu',\lambda_\nu+S_\nu z).$$

Since $P(\xi_\nu',\lambda_\nu)\to 0$ but $Q(\xi_\nu')\to Q(\xi_0')\neq 0$ it is now clear that (13.6.20) and (13.6.21) are valid. We obtain (13.6.22) for some sequence K_ν with $K_\nu\geqq \nu/S_\nu$ and some r of degree $\geqq 2$. The conditions (13.6.23) can be written

$$\rho_\nu S_\nu/K_\nu\to\infty,\qquad S_\nu K_\nu\to\infty.$$

The first is valid if ρ_ν is sufficiently rapidly increasing, and the second follows since $S_\nu K_\nu \geqq \nu$. The proof is complete.

Example 13.6.12. Let $P(\xi)=(\xi_1-i\xi_2)^a-\xi_1^{b-1}$, $Q(\xi)=\xi_1^b$, $N=(0,1)$. With $\zeta_\nu=(\nu,-i\nu)$ we have

$$P(\zeta_\nu+T_\nu zN)=(-iT_\nu z)^a-\nu^{b-1}, \qquad Q(\zeta_\nu+T_\nu zN)=\nu^b.$$

Hence (13.6.20) and (13.6.21) are fulfilled if $T_\nu^a \nu^{1-b}\to 0$. Then we obtain (13.6.22) with $K_\nu=\nu^{b-1}T_\nu^{-a}$ and $r(z)=-(-iz)^a$, so (13.6.24) is valid if $a>1$. All the required conditions are then

$$T_\nu \nu^{(1-b)/a}\to 0, \qquad T_\nu \nu^{-(b-1)/(a+1)}\to\infty, \qquad T_\nu \nu^{-(b-2)/(a-1)}\to 0.$$

These are compatible if and only if $(b-1)/(a+1)<(b-2)/(a-1)$, that is, $a<2b-3$ or

$$b>a-(a-3)/2.$$

Note that the multiplicity a has to be two units higher than in Example 13.6.10 if the order of Q shall be (strictly) smaller than that of P. There is of course no difficulty in extending the example to the general exceptional case in Corollary 13.6.11. When we have a factorization with $j>3$ an example of non-uniqueness is obtained which also contains a lower order term corresponding to D_1^{b-1} in the example above.

We shall now pass to the proof of Theorem 13.6.9. The first step is to rephrase the hypothesis by means of the Tarski-Seidenberg theorem.

Lemma 13.6.13. *Assume that the hypotheses of Theorem 13.6.9 are fulfilled but not* (13.6.2). *Then there exist rational functions* $\zeta(\varepsilon)$, $T(\varepsilon)$, $K(\varepsilon)$, $b(\varepsilon)$, $c(\varepsilon)$ *such that* $\zeta(\varepsilon)\in Z_N$, $T(\varepsilon)>0$, $K(\varepsilon)>0$, *for small* $\varepsilon>0$, *and for* $\varepsilon\to 0$

(13.6.20)′
$$c(\varepsilon)P(\zeta(\varepsilon)+T(\varepsilon)zN)-1=O(\varepsilon),$$
$$b(\varepsilon)Q(\zeta(\varepsilon)+T(\varepsilon)zN)-1=O(\varepsilon),$$

(13.6.21)′
$$b(\varepsilon)/c(\varepsilon)=O(\varepsilon),$$

(13.6.22)′
$$K(\varepsilon)(c(\varepsilon)P(\zeta(\varepsilon)+T(\varepsilon)zN)-b(\varepsilon)Q(\zeta(\varepsilon)+T(\varepsilon)zN))-r(z)=O(\varepsilon),$$

(13.6.23)′
$$K(\varepsilon)/T(\varepsilon)=O(\varepsilon),$$
$$(1+|\operatorname{Im}\zeta(\varepsilon)|)/(T(\varepsilon)K(\varepsilon))=O(\varepsilon),$$

(13.6.25)′
$$(1+T(\varepsilon))/|\operatorname{Im}\zeta(\varepsilon)|=O(\varepsilon).$$

One can choose $c(\varepsilon)$ *and* $b(\varepsilon)$ *so that* $b(\varepsilon)/c(\varepsilon)$ *is analytic on* $\mathbb{R}$.

Proof. Let E be the set of all $(1/\varepsilon, \zeta, T, K, c, b)$ with $\zeta \in Z_N$, $T>0$, $K>0$, $0<\varepsilon \leqq 1$, $c \in \mathbb{C}$, $b \in \mathbb{C}$, such that

$$|b|^2 \leqq |\varepsilon c|^2, \quad K \leqq \varepsilon T, \quad 1+T^2 \leqq \varepsilon^2 |\mathrm{Im}\,\zeta|^2, \quad 1+|\mathrm{Im}\,\zeta|^2 \leqq (\varepsilon T K)^2,$$

and the coefficients of the polynomials in z

$$cP(\zeta+TzN)-1, \quad bQ(\zeta+TzN)-1,$$
$$K(cP(\zeta+TzN)-bQ(\zeta+TzN))-r(z)$$

have modulus $\leqq \varepsilon$. This is a semi-algebraic set, and by hypothesis it contains points with $1/\varepsilon$ arbitrarily large. Hence it follows from Theorem A.2.8 that there are Puiseux series $\zeta(\varepsilon), \dots, b(\varepsilon)$ converging for small ε such that $(1/\varepsilon, \dots, b(\varepsilon)) \in E$. Choose an integer k so large that $\zeta(\varepsilon^k), \dots, b(\varepsilon^k)$ become Laurent series. Sufficiently high partial sums of these series will then have all the required properties except that $b(\varepsilon)/c(\varepsilon)$ may have a finite number of poles on $\mathbb{R} \setminus 0$. However, if ν is sufficiently large we can replace $c(\varepsilon)$ by $c(\varepsilon+h\varepsilon^\nu)$. If h is purely imaginary it is clear that $c(\varepsilon+h\varepsilon^\nu)$ cannot have a real zero $\varepsilon \neq 0$ except for finitely many values of h so this permits us to choose b/c analytic.

To avoid an interruption of the proof later on we give a version of the expansions of geometrical optics which will be needed. (See also Sections 7.7 and 12.2.)

Lemma 13.6.14. *Let $I \subset \mathbb{R}$ be a compact interval and let*

$$G_\delta(S, D_S) = \sum_0^m g_j(S,\delta)\, D_S^j, \qquad D_S = -i\, d/dS,$$

be an ordinary differential operator with C^∞ coefficients when $S \in I$ and $\delta \in \mathbb{R}$ is small. Assume that there exist positive integers m_0 and m_1 such that

$$\delta^{m_1} G_\delta(S, \delta^{-m_0} z) = H_\delta(S, z)$$

also has C^∞ coefficients and $H_0 \not\equiv 0$. Assume further that

$$H_0(S,z)=0, \qquad \partial H_0(S,z)/\partial z \neq 0 \qquad \text{when } z=\phi_0'(S), \tag{13.6.26}$$

where $\phi_0 \colon I \to \mathbb{C}$ is a C^∞ function. Then there exist C^∞ functions $\phi(S,\delta)$ and $W(S,\delta)$ when $S \in I$ and $|\delta|$ is small, such that

$$\phi(S,0)=\phi_0(S), \qquad W(S,0) \neq 0; \; S \in I; \tag{13.6.27}$$

$$\begin{aligned} &\exp(-i\phi(S,\delta)\,\delta^{-m_0})\, G_\delta(S, D_S)\,(W(S,\delta) \\ &\quad \cdot \exp(i\phi(S,\delta)\,\delta^{-m_0})) = R(S,\delta) \end{aligned} \tag{13.6.28}$$

where $R(S,\delta)$ is a C^∞ function vanishing of infinite order when $\delta=0$. If Γ_1 and Γ_2 are C^∞ curves in the (S,δ) plane intersecting $I \times \{0\}$ transver-

sally at different points, and if $g_m(S,\delta)$ does not vanish of infinite order on any one of them when $\delta=0$, then W can be chosen so that R vanishes of infinite order on Γ_1 and Γ_2 also.

Proof. Since $H_\delta(S,z)$ is a polynomial in z, thus analytic in z, it follows from the implicit function theorem and (13.6.26) that the equation $H_\delta(S,\psi(S,\delta))=0$ has a unique solution $\psi\in C^\infty$ when $S\in I$ and δ is small, such that $\psi(S,0)=\phi_0'(S)$. Choose ϕ with $\partial\phi(S,\delta)/\partial S=\psi(S,\delta)$, $\phi(S,0)=\phi_0(S)$.

The equation (13.6.28) can be written

$$\delta^{m_1} G_\delta(S, D_S+\phi_S'(S,\delta)\,\delta^{-m_0})\, W(S,\delta)=\delta^{m_1} R(S,\delta)$$

or

$$\delta^{-m_0}(H_\delta(S,\delta^{m_0} D_S+\phi_S'(S,\delta))-H_\delta(S,\phi_S'(S,\delta)))\, W(S,\delta) = \delta^{m_1-m_0} R(S,\delta). \tag{13.6.29}$$

When $\delta=0$ the left-hand side reduces to $LW(S,0)$ where

$$L=\partial H_0(S,z)/\partial z\, D_S+B, \qquad z=\phi_0'(S),$$

for some $B\in C^\infty$. By hypothesis the coefficient of D_S has no zero in I. Introducing the formal Taylor expansion

$$W(S,\delta)\sim\sum_0^\infty W^{(j)}(S,0)\,\delta^j/j!$$

in (13.6.29), we find that (13.6.29) is valid for some R vanishing of infinite order when $\delta=0$ if and only if a sequence of equations are satisfied, of the form

$$LW(S,0)=0,\dots, LW^{(j)}(S,0)+E_j=0,\dots$$

where E_j is determined by $W,\dots,W^{(j-1)}$. These can be solved successively, and $W(S,0)$ can be chosen with no zero in I. By Theorem 1.2.6 there exists a C^∞ function $W(S,\delta)$ with these derivatives when $\delta=0$, which proves the first part of the lemma.

To prove the last assertion we have to find a function $V\in C^\infty$ vanishing of infinite order when $\delta=0$ such that (13.6.28) is valid with W replaced by V apart from an error vanishing of infinite order on the curves Γ_j and $I\times\{0\}$. The difference $W-V$ will then have the required properties. The condition on V is that

$$\delta^{m_1} G_\delta(S, D_S+\phi_S'(S,\delta)\,\delta^{-m_0})\, V(S,\delta)-R(S,\delta)\,\delta^{m_1}$$

shall vanish of infinite order on the curves Γ_j. The differential operator has C^∞ coefficients and the coefficient of the highest derivative is $\delta^{m_1} g_m(S,\delta)$, which vanishes at most as a power of δ on Γ_j when $\delta=0$.

Now we require that derivatives of order $<m$ of V with respect to S shall vanish on Γ_j. Using the equation we can then (see Section 12.1) compute $D_S^m V$ on Γ_j, which yields a function vanishing of infinite order when $\delta \to 0$. Repeating the argument we find that V has the desired properties if the derivatives with respect to S on Γ_j are certain C^∞ functions vanishing of infinite order when $\delta \to 0$. We extend them so that they vanish when $\delta < 0$. By another application of Theorem 1.2.6 we can then find V so that V vanishes when $\delta \leqq 0$ and V has the required derivatives with respect to S on the transversal curves Γ_j. This completes the proof.

Remark. It is obvious that the lemma remains valid if the coefficients g_j are singular when $\delta = 0$ but $\delta^N g_j \in C^\infty$ for some integer $N > 0$. We may also replace δ^{m_0} by a C^∞ function of δ vanishing precisely of order m_0 when $\delta = 0$.

Proof of Theorem 13.6.9. With the notation of Lemma 13.6.13 we set

$$A(\varepsilon) = b(\varepsilon)/c(\varepsilon).$$

A is a rational function vanishing at 0 and with no poles on $\mathbb{R}$. With a sufficiently large integer ρ we shall take the coefficient a in (13.6.3) as $-A(\langle x, N\rangle^\rho)$ apart from a term vanishing in $\complement H_N$. For small $\delta > 0$ we set in analogy to (13.6.9)

$$u_\delta(x) = v_\delta(\langle x, N\rangle) \exp i\langle x, \zeta(\delta^\rho)\rangle.$$

The differential equation $(P(D) - A(\langle x, N\rangle^\rho)\, Q(D))\, u_\delta = 0$ can then be written

$$(P(\zeta(\delta^\rho) + D_s N) - A(s^\rho)\, Q(\zeta(\delta^\rho) + D_s N))\, v_\delta(s) = 0, \tag{13.6.30}$$

and we shall solve it approximately using Lemma 13.6.14. To be able to use (13.6.20)′, (13.6.22)′ we multiply by $c(\delta^\rho)$ and obtain the equivalent equation

$$\begin{aligned}((c(\delta^\rho)\, P(\zeta(\delta^\rho) + D_s N) - b(\delta^\rho)\, Q(\zeta(\delta^\rho) + D_s N))\\ + (1 - A(s^\rho)/A(\delta^\rho))\, b(\delta^\rho)\, Q(\zeta(\delta^\rho) + D_s N))\, v_\delta(s) = 0.\end{aligned} \tag{13.6.30′}$$

The difference $1 - A(s^\rho)/A(\delta^\rho)$ vanishes when $s = \delta$, and the first order term in the Taylor expansion at $s = \delta$ is

$$-A'(\delta^\rho)/A(\delta^\rho)\, \rho\, \delta^{\rho-1}(s - \delta) = -\rho j_0(1 + O(\delta^\rho))(s - \delta)/\delta,$$

if A has a zero of order j_0 at 0. To balance the two terms in (13.6.30)′ we therefore want $(s - \delta)/\delta$ and $1/K(\delta^\rho)$ to be of the same order of magnitude. Since $K(\varepsilon)$ may be replaced by the leading term in the

Laurent expansion at 0, we may assume that $K(\varepsilon)=\varepsilon^{-\kappa}$ where κ is a positive integer. Thus we wish $s-\delta$ to be of the order of magnitude $\delta^{1+\kappa\rho}$. To be able to apply Lemma 13.6.14 in a fixed interval we must now introduce a new variable S through

$$s=\delta+S\,\delta^{\kappa\rho+1}.$$

With the notation $v_\delta(s)=V_\delta(S)$ the equation (13.6.30)′ becomes

$$(K(\delta^\rho)(c(\delta^\rho)P-b(\delta^\rho)Q)(\zeta(\delta^\rho)+\delta^{-\kappa\rho-1}D_S N) + C(S,\delta)\,b(\delta^\rho)\,Q(\zeta(\delta^\rho)+\delta^{-\kappa\rho-1}D_S N))\,V_\delta(S)=0. \tag{13.6.30$''$}$$

Here

$$C(S,\delta)=K(\delta^\rho)(1-A(s^\rho)/A(\delta^\rho))\to-\rho j_0 S,\qquad \delta\to 0.$$

It is clear that $C(S,\delta)$ is analytic for small δ.

The coefficients of (13.6.30)″ become smooth after multiplication by some power of δ. If D_S is replaced by $\delta^{\kappa\rho+1}\,T(\delta^\rho)\,z$ we obtain a polynomial converging to

$$r(z)-\rho j_0 S \tag{13.6.31}$$

when $\delta\to 0$. Note that if $\rho>1$, as we assume from now on, then

$$\delta^{\kappa\rho+1}\,T(\delta^\rho)=\delta\,T(\delta^\rho)/K(\delta^\rho)\to\infty,\qquad \delta\to 0, \tag{13.6.32}$$

by the first part of (13.6.23)′. This allows us to apply Lemma 13.6.14 and the remark following its proof. In doing so we may assume that r is given by (13.6.24)″. Then the polynomial (13.6.31) has the unique zero $z(S)=\rho j_0 S/r_1$. Note that

$$\operatorname{Im} dz/dS=\rho j_0 \operatorname{Im} 1/r_1>0.$$

With $I=[-2,2]$ we now choose ϕ and W by applying Lemma 13.6.14 with $\phi_0(S)=j_0 S^2/2r_1$. For small δ we then obtain

$$\partial^2 \operatorname{Im}\phi(S,\delta)/\partial S^2\geqq c_0>0,\qquad S\in I. \tag{13.6.33}$$

The choice of the curves Γ_j in Lemma 13.6.14 is left open for the moment. Returning to the original variables we define

$$u_\delta(x)=\exp i(\langle x,\zeta(\delta^\rho)\rangle+\phi(S,\delta)\,\delta^{\kappa\rho+1}\,T(\delta^\rho))\,W(S,\delta),$$
$$r_\delta(x)=R(S,\delta)/(K(\delta^\rho)\,c(\delta^\rho)\,W(S,\delta))=R_1(S,\delta),$$

where $S=(\langle x,N\rangle-\delta)\,\delta^{-\kappa\rho-1}\in I$. Here R_1 is also a C^∞ function vanishing of infinite order when $\delta=0$ and on the curves Γ_j. We have

$$(P(D)-A(\langle x,N\rangle^\rho)\,Q(D))\,u_\delta=r_\delta u_\delta \tag{13.6.34}$$

if $(\langle x,N\rangle-\delta)\,\delta^{-\kappa\rho-1}\in I$. Furthermore,

$$Q(D)\,u_\delta=M_\delta u_\delta \tag{13.6.35}$$

where

$$M_\delta(x)=m(S,\delta)/b(\delta^\rho); \qquad m\in C^\infty \quad\text{and}\quad m(S,0)=1.$$

Following the proof of Theorem 13.6.1 we shall now piece together u by means of the functions u_δ.

For $\nu=1,2,\dots$ we put $\delta_\nu=\delta_1\,\nu^{-\gamma}$ where $\delta_1>0$ and $\gamma>0$. For the interval

$$I_\nu=\{s\in\mathbb{R};\,(s-\delta_\nu)\,\delta_\nu^{-\kappa\rho-1}\in I\}$$

the length $|I_\nu|$ is $4(\delta_1\,\nu^{-\gamma})^{\kappa\rho+1}$, and the distance $\delta_\nu-\delta_{\nu+1}$ between the centers of I_ν and $I_{\nu+1}$ is asymptotically $\delta_1\,\gamma\,\nu^{-\gamma-1}$. Choose γ so that

$$\gamma(\kappa\rho+1)=\gamma+1, \qquad \text{that is,}\quad \gamma=1/\kappa\rho,$$

and choose δ_1 so that

$$\delta_1\,\gamma=2\delta_1^{\kappa\rho+1}.$$

For large ν the end points of I_ν are then close to the centers of $I_{\nu+1}$ and of $I_{\nu-1}$. We shall switch from one u_{δ_ν} to the next when $\langle x,N\rangle$ is near the center of the interval where they are both defined. The center of the left half of I_ν is

$$B_\nu=\delta_\nu-\delta_\nu^{\kappa\rho+1}.$$

For large ν the center of the right half of I_ν is close to $B_{\nu-1}$. To confirm this we set $B_{\nu-1}=\delta_\nu+S\,\delta_\nu^{\kappa\rho+1}$ and obtain

$$S=(\delta_{\nu-1}-\delta_\nu)/\delta_\nu^{\kappa\rho+1}-(\delta_{\nu-1}/\delta_\nu)^{\kappa\rho+1}=f(\delta_\nu)$$

where f is a convergent power series in $1/\nu$, hence in $\delta_\nu^{\kappa\rho}$, with $f(0)=1$. We choose the curves Γ_j in Lemma 13.6.14 as $S=-1$ and $S=f(\delta)$. This guarantees that the right-hand side of (13.6.34) vanishes of infinite order where the switch over occurs.

When $\langle x,N\rangle\in I_\nu$ we set

$$U_\nu(x)=C_\nu\,u_{\delta_\nu}(x)$$

where $C_\nu>0$ is determined successively so that with the notation in (13.6.35) we have

(13.6.36) $\quad |M_{\delta_\nu}(x)\,U_\nu(x)|=|M_{\delta_{\nu-1}}(x)\,U_{\nu-1}(x)| \qquad$ when $\langle x,N\rangle=B_{\nu-1}$.

Note that the two sides are constant in this hyperplane. Choose $\chi\in C_0^\infty(-3/2,\,3/2)$ equal to 1 in $(-5/4,\,5/4)$, and set

$$u(x)=\sum_{\nu_0}^{\infty}U_\nu(x)\,\chi_\nu(x), \qquad \chi_\nu(x)=\chi((\langle x,N\rangle-\delta_\nu)\,\delta_\nu^{-\kappa\rho-1}).$$

When ν_0 is large and the first term is appropriately modified we shall see that $u\in C^\infty(\mathbb{R}^n)$ and that (13.6.3) is satisfied with a C^∞ function a such that $a(x)+A(\langle x,N\rangle^\rho)$ vanishes outside H_N.

The first step is to study $F_\nu = |M_{\delta_\nu}(x)\, U_\nu(x)|$ as a function of $s = \langle x, N\rangle$. Apart from a constant term $\log F_\nu(s)$ is equal to

$$\log |m(S, \delta_\nu)\, W(S, \delta_\nu)| + s\,|\mathrm{Im}\, \zeta(\delta_\nu^\varrho)| - \mathrm{Im}\, \phi(S, \delta_\nu)\, \delta_\nu^{\kappa\varrho+1}\, T(\delta_\nu^\varrho).$$

Here $S = (s - \delta_\nu)\, \delta_\nu^{-\kappa\varrho-1}$. By (13.6.25)′ we may assume that

$$T(\delta_\nu^\varrho)/|\mathrm{Im}\, \zeta(\delta_\nu^\varrho)| = O(\delta_\nu^\varrho),$$

and in view of (13.6.32) this implies that $d(\log F_\nu(s))/ds$ is asymptotically equal to $|\mathrm{Im}\, \zeta(\delta_\nu^\varrho)|$. Moreover, since

$$K(\varepsilon)/|\mathrm{Im}\, \zeta(\varepsilon)| = (K(\varepsilon)/T(\varepsilon))\,(T(\varepsilon)/|\mathrm{Im}\, \zeta(\varepsilon)|) = O(\varepsilon^2),$$

we have

$$\delta_\nu^{\kappa\varrho+1}\, |\mathrm{Im}\, \zeta(\delta_\nu^\varrho)| \geqq c_1\, \delta_\nu^{\kappa\varrho+1-(\kappa+2)\varrho} = c_1\, \delta_\nu^{1-2\varrho} \geqq c_2\, \nu^\gamma.$$

For large ν it follows that

$$F_\nu(B_\nu)/F_\nu(B_{\nu-1}) < C \exp(-c_3\, \nu^\gamma).$$

Since $F_\nu(B_\nu) = F_{\nu+1}(B_\nu)$ by (13.6.36) we conclude that

$$F_\nu(B_{\nu-1}) < C_1 \exp(-c_4\, \nu^{\gamma+1}).$$

Hence

$$F_\nu(s) < C_1 \exp(-c_4\, \nu^{\gamma+1}), \qquad B_\nu \leqq s \leqq B_{\nu-1},$$

and (13.6.37) below shows that the same estimate is valid in $\operatorname{supp} \chi_\nu$. It follows immediately that all derivatives of u have bounds converging to 0 when $\langle x, N\rangle \to 0$, so $u \in C^\infty(\mathbb{R}^n)$.

To show that $a = -P(D)\,u/Q(D)\,u$ is in C^∞ when $\langle x, N\rangle < B_{\nu_0-1}$ we first prove for some new constants c_1, c_2 that

(13.6.37) $$c_1\, T(\delta_\nu^\varrho) \leqq (s - B_\nu)^{-1} \log(F_\nu(s)/F_{\nu+1}(s)) \leqq c_2\, T(\delta_\nu^\varrho), \qquad s \in I_\nu \cap I_{\nu+1}.$$

Since $F_\nu(s)/F_{\nu+1}(s) = 1$ when $s = B_\nu$ we only have to examine the derivative of the logarithm. Set $S_j = (s - \delta_j)\, \delta_j^{-\kappa\varrho-1}$, $j = \nu, \nu+1$. Then

$$S_{\nu+1} - S_\nu - 2 \to 0$$

uniformly when $\nu \to \infty$, so $S_{\nu+1} > S_\nu + 1$ if ν is large. As we saw above, the derivative of $\log(F_\nu(s)/F_{\nu+1}(s))$ with respect to s is

(13.6.38) $$O(\delta_\nu^{-\kappa\varrho-1}) + |\mathrm{Im}\, \zeta(\delta_\nu^\varrho)| - |\mathrm{Im}\, \zeta(\delta_{\nu+1}^\varrho)| + T(\delta_{\nu+1}^\varrho)\, \mathrm{Im}\, \phi'(S_{\nu+1}, \delta_{\nu+1}) - T(\delta_\nu^\varrho)\, \mathrm{Im}\, \phi'(S_\nu, \delta_\nu).$$

Here $\delta_\nu^{-\kappa\varrho-1}/T(\delta_\nu^\varrho) \to 0$ by (13.6.32). Since

$$T(\varepsilon)\, \varepsilon^{-\kappa}/|\mathrm{Im}\, \zeta(\varepsilon)| \to \infty, \qquad \varepsilon \to 0,$$

by the second part of (13.6.23)′ and since $\mathrm{Im}\, \zeta(\varepsilon)$ has a pole when $\varepsilon = 0$, we have

$$T(\varepsilon)\, \varepsilon^{-\kappa-1}/|\mathrm{Im}\, \zeta'(\varepsilon)| \to \infty, \qquad \varepsilon \to 0.$$

Now $\delta_\nu^{-(\kappa+1)\rho}(\delta_\nu^\rho-\delta_{\nu+1}^\rho)=\delta_\nu^{-\kappa\rho}(1-(\delta_{\nu+1}/\delta_\nu)^\rho)\to\rho\,\gamma\,\delta_1^{-1/\gamma}$ so it follows that

$$|\operatorname{Im}\zeta(\delta_{\nu+1}^\rho)|-|\operatorname{Im}\zeta(\delta_\nu^\rho)|=o(T(\delta_\nu^\rho)),\qquad \nu\to\infty.$$

The last two terms in (13.6.38) are therefore dominating. If we recall (13.6.33) and that $S_{\nu+1}-S_\nu>1$ we may conclude that (13.6.38) is bounded from above and below by constants times $T(\delta_\nu^\rho)$, which proves (13.6.37).

Let us now study a in the neighborhood of B_ν where $u=U_\nu+U_{\nu+1}$. There we have

$$\begin{aligned}-a&=(P(D)\,U_\nu+P(D)\,U_{\nu+1})/(Q(D)\,U_\nu+Q(D)\,U_{\nu+1})\\&=A(\langle x,N\rangle^\rho)+(r_{\delta_\nu}U_\nu+r_{\delta_{\nu+1}}U_{\nu+1})/(M_{\delta_\nu}U_\nu+M_{\delta_{\nu-1}}U_{\nu+1}).\end{aligned}$$

U_ν dominates when $S=(s-B_\nu)\,\delta_\nu^{-\kappa\rho-1}>0$, so then we divide by $M_{\delta_\nu}U_\nu$. In view of (13.6.37) we have for some constant $c>0$

$$|1+M_{\delta_{\nu+1}}U_{\nu+1}/M_{\delta_\nu}U_\nu|\geqq c\min(S\,\delta_\nu^{\kappa\rho+1}\,T(\delta_\nu^\rho),1),$$

for $1-e^{-t}\geqq(1-e^{-1})\min(1,t)$, $t>0$. On the other hand, r_{δ_ν} and $r_{\delta_{\nu+1}}$ can be estimated by any desired power of δ_ν and S. It follows that

$$a(x)+A(\langle x,N\rangle^\rho)$$

in the set now considered can be estimated by any power of $1/\nu$. Since the derivatives of r_{δ_ν} have estimates similar to those we have used for r_{δ_ν}, the same is true for all the derivatives.

In the part of the left half of I_ν where $\chi_{\nu+1}$ is cutting off $U_{\nu+1}$, we know by (13.6.37) that $F_{\nu+1}(s)/F_\nu(s)$ is exponentially small, so similar estimates are immediately obtained there. The argument is even simpler than the corresponding point in the proof of Theorem 13.6.1 so we omit the details.

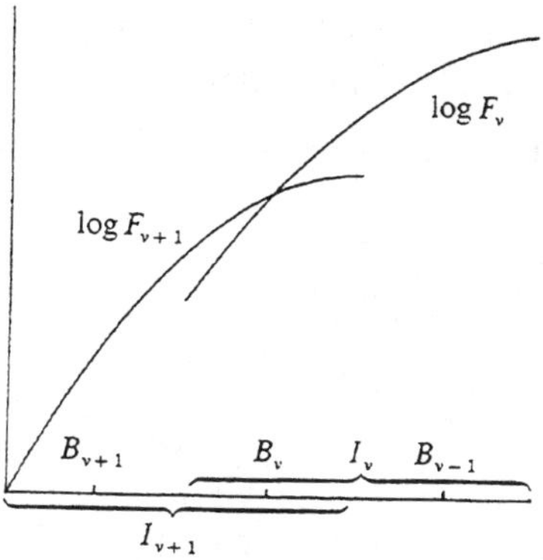

Fig. 7

In the middle of I_ν where $u=U_\nu$ we have

$$-a=A(\langle x,N\rangle^\rho)+r_{\delta_\nu}/M_{\delta_\nu},$$

and all derivatives of the second term have bounds converging to 0 as $\nu\to\infty$. This proves the smoothness of a for $\langle x,N\rangle \leqq B_{\nu_0-1}$. We can continue a as a function of $\langle x,N\rangle$ for $\langle x,N\rangle > B_{\nu_0-1}$ and define u there by just solving an ordinary differential equation. This completes the proof.

By modifying the preceding construction we shall now give an example which shows that the injectivity assumption in Theorem 13.5.2 is not automatically fulfilled.

Theorem 13.6.15. *There exists a fourth order elliptic differential operator P in $\mathbb{R}^3$ with C^∞ coefficients such that the equation $Pu=0$ has a non-trivial solution $u\in C_0^\infty(\mathbb{R}^3)$.*

Proof. Let $P(\xi)=(\xi_1^2+\xi_2^2+\xi_3^2)^2-\xi_1^4/2$, $Q(\xi)=\xi_2^4$, $N=(0,0,1)$ as in Example 13.6.10 with $a=2$, $b=4$. Choose $\zeta(\varepsilon)$, $c(\varepsilon)$, $b(\varepsilon)$, $K(\varepsilon)$, $T(\varepsilon)$ according to Lemma 13.6.13 with $r(z)=r_1 z$. Then we have for $\alpha\neq 0$

$$K(\varepsilon)\,c(\varepsilon)\,P^{(\alpha)}(\zeta(\varepsilon)+T(\varepsilon)\,zN)=O(\varepsilon), \tag{13.6.39}$$

$$K(\varepsilon)\,b(\varepsilon)\,Q^{(\alpha)}(\zeta(\varepsilon)+T(\varepsilon)\,zN)=O(\varepsilon). \tag{13.6.40}$$

In fact, $\partial P/\partial\xi_3=4\xi_3(\xi_1^2+\xi_2^2+\xi_3^2)$, $\partial^2 P/\partial\xi_3^2=4(\xi_1^2+\xi_2^2+3\xi_3^2)$, so it follows from (13.6.20)′, (13.6.22)′ that

$$\begin{aligned} &c(\varepsilon)\,P(\zeta(\varepsilon))\to 1, \qquad b(\varepsilon)\,Q(\zeta(\varepsilon))\to 1,\\ &T(\varepsilon)\,K(\varepsilon)\,c(\varepsilon)\,\zeta_3(\varepsilon)\,(\zeta_1(\varepsilon)^2+\zeta_2(\varepsilon)^2+\zeta_3(\varepsilon)^2)\to r_1/4,\\ &T(\varepsilon)^2\,K(\varepsilon)\,c(\varepsilon)\,(\zeta_1(\varepsilon)^2+\zeta_2(\varepsilon)^2+3\zeta_3(\varepsilon)^2)\to 0.\end{aligned} \tag{13.6.41}$$

By (13.6.21)′ we have $P(\zeta(\varepsilon))=o(|\zeta(\varepsilon)|^4)$, hence $|\zeta(\varepsilon)|=O(\operatorname{Im}\zeta_3(\varepsilon))$ since P is elliptic. From (13.6.23)′ it follows that $|\zeta(\varepsilon)|=o(T(\varepsilon)\,K(\varepsilon))$, so

$$|\zeta(\varepsilon)|^2(\zeta_1(\varepsilon)^2+\zeta_2(\varepsilon)^2+\zeta_3(\varepsilon)^2)=o(|\zeta(\varepsilon)|^4)$$

by (13.6.41). Thus

$$\zeta_1(\varepsilon)^4=2(\zeta_1(\varepsilon)^2+\zeta_2(\varepsilon)^2+\zeta_3(\varepsilon)^2)^2-2P(\zeta(\varepsilon))=o(|\zeta(\varepsilon)|^4)$$

which proves that $\zeta_1(\varepsilon)=o(|\zeta_3(\varepsilon)|)$, $|\zeta_2(\varepsilon)|/|\zeta_3(\varepsilon)|\to 1$ just as in the discussion of Example 13.6.10. Now (13.6.23)′ implies $\zeta(\varepsilon)=o(T(\varepsilon)^2)$ which gives $K(\varepsilon)\,c(\varepsilon)|\zeta(\varepsilon)|^3\to 0$ by (13.6.41). Since $T(\varepsilon)=o(|\zeta(\varepsilon)|)$ and $b(\varepsilon)=o(c(\varepsilon))$ we obtain (13.6.39) and (13.6.40).

It follows from (13.6.39), (13.6.40) that in the proof of Theorem 13.6.9 we may replace $\zeta(\delta_\nu^\rho)$ by the vector ζ_ν where the first two (real) components are replaced by the nearest integer. This does not affect the crucial discussion of the size of F_ν either. (Lemma 13.6.14 is applied to equations depending on ν in a non-essential way.) Hence we obtain functions u and a in Theorem 13.6.9 which are periodic with period 2π in x_1 and in x_2. If ν_0 is chosen large enough then $P(\xi) + \operatorname{Re} a(x)\, Q(\xi)$ is elliptic, and

$$u(x) = e^{i(x_1\xi_1 + x_2\xi_2)}\, v(x_3), \qquad x_3 > 1,$$

$v(1)=1$ say. Here ξ_1 and ξ_2 are real. Now

$$(P(D) + \overline{a(2N-x)}\, Q(D))\, \overline{u(2N-x)} = 0,$$
$$\overline{u(2N-x)} = e^{i(x_1\xi_1 + x_2\xi_2)}\, \overline{v(2-x_3)}, \qquad x_3 < 1.$$

Let $\psi \in C^\infty$ be equal to 1 on $(-\infty, 1)$ and then decrease to be 0 in $(1+\varepsilon, \infty)$ where ε is so small that $\operatorname{Re} v(x_3) > 1/2$ when $|x_3 - 1| < \varepsilon$. Set

$$U(x) = \psi(x_3)\, u(x) + (1 - \psi(x_3))\, \overline{u(2N-x)}$$

and define A in the same way. Then

$$(P(D) + A(x)\, Q(D))\, U(x) = F(x)$$

where $|x_3 - 1| \leqq \varepsilon$ in $\operatorname{supp} F$. Hence $F = BU$ with $B \in C^\infty$ so U satisfies the elliptic differential equation

$$(P(D) + A(x)\, Q(D) - B(x))\, U = 0.$$

We have $0 \leqq x_3 \leqq 2$ in the support.

Now consider x as toroidal coordinates, that is, consider the map

$$x \to ((x_3+1)\sin x_1, (4 + (x_3+1)\cos x_1)\cos x_2,$$
$$(4 + (x_3+1)\cos x_1)\sin x_2) = y$$

which in $\mathbb{R}^3$ modulo 2π in the x_1 and x_2 variables is a diffeomorphism in a neighborhood of $\operatorname{supp} u$. Let $U(x) = v(y)$. Then we obtain a differential equation of order four

$$R(y, D)\, v(y) = 0$$

where R has C^∞ coefficients and $\operatorname{Re} R_4(y, \xi)$ is elliptic in a neighborhood of $\operatorname{supp} v$. Let $0 \leqq \chi \in C_0^\infty$ have support in this neighborhood and be 1 in a neighborhood of $\operatorname{supp} v$. Then

$$\chi(y)\, R(y, D) + (1 - \chi(y))\, \Delta^2$$

is elliptic in $\mathbb{R}^3$ and annihilates the C_0^∞ function v. The proof is complete.

Example 13.6.16. Let $P(x, D)$ be the operator in Theorem 13.6.15 and set

$$P_0(x,y,D_x,D_y)=D_y P(x,D_x)+Q(x,D_x)$$

where $x\in\mathbb{R}^3$, $y\in\mathbb{R}$ and Q is of order $\leqq 4$. Then P_0 has constant strength, the same as $D_y\Delta_x^2$ (Corollary 10.4.8), and $P(x,D_x)$ is a localization at infinity. It annihilates the product of the function u in Theorem 13.6.15 by any function of y so Theorem 13.5.2 is not applicable for this operator.

Notes

There are classical methods to study an elliptic operator by regarding it locally as a perturbation of a constant coefficient operator. They are known under several names such as E.E. Levi's parametrix method and Korn's approximation. The precise discussion of regularity properties in Chapter X allows one to apply these methods to arbitrary operators of constant strength. This was first carried out by Peetre [1] and in a somewhat more precise way in the predecessor of this book. The results on hypoellipticity in Section 13.4 are due to Malgrange [2] and Hörmander [4], apart from Theorem 13.4.4, which is due to Taylor [1], and the refinements involving wave front sets. Section 13.5 is taken from Gudmundsdottir [1].

The non-uniqueness of the Cauchy problem for operators with C^∞ coefficients caused a great surprise in the 1950's. The first examples due to Pliš [2] and de Giorgi [1] contained already the basic construction technique which is still used. They were considerably generalized by Cohen [1] and Hörmander [33] to yield Theorem 13.6.1 here. An essential new idea was introduced by Pliš [1] in his proof of what is Theorem 13.6.15 here. The general version of his construction in Theorem 13.6.9 is taken from Hörmander [33] where additional references can be found. However, they do not cover the important recent development for which we refer to Alinhac [1, 2], Alinhac-Zuily [1] and Zuily [1].

Chapter XIV. Scattering Theory

Summary

A partial differential operator $P_0(D)$ in $\mathbb{R}^n$ with constant real coefficients defines a self-adjoint operator H_0 in $L^2(\mathbb{R}^n)$, the domain consisting of all $u \in L^2(\mathbb{R}^n)$ with $P_0(D)u \in L^2(\mathbb{R}^n)$. In fact, if F is the Fourier transformation $u \to \hat{u}$ then FH_0F^{-1} is the multiplication operator $P_0(\xi)$ defined for all $U \in L^2$ with $P_0(\xi)\,U(\xi) \in L^2$. This is obviously a self-adjoint operator with absolutely continuous spectrum. The spectral projection E_λ is given by multiplication by the characteristic function of $\{\xi; P_0(\xi) < \lambda\}$. (Note that if $\Lambda \subset \mathbb{R}$ is of measure 0 then $P_0^{-1}\Lambda$ is of measure 0 since P_0 can locally be taken as a coordinate outside the null set defined by $dP_0 = 0$. We assume of course that P_0 is not a constant.)

The purpose of scattering theory is to examine how the spectral decomposition of H_0 is changed if H_0 is perturbed to an operator H which is fairly close to H_0 at infinity. To find the spectral measure dE_λ of H it is useful to note that in view of Example 3.1.13 the resolvent

$$R(z) = (H - z)^{-1} = \int (\lambda - z)^{-1}\, dE_\lambda$$

determines dE_λ formally,

$$dE_\lambda = (2\pi i)^{-1}(R(\lambda + i0) - R(\lambda - i0)).$$

It is thus natural to study the behavior of the resolvent near the real axis. In the constant coefficient case above we obtain if λ is not a critical value of P_0, that is, $dP_0 \neq 0$ when $P_0 = \lambda$, that $(R(\lambda + i0) - R(\lambda - i0))/2\pi i$ is convolution by $F^{-1}\delta(P_0 - \lambda)$. Now we know from Theorem 7.1.26 that if $v = F^{-1}(\hat{u}\,\delta(P_0 - \lambda))$, $\hat{u} \in C_0^\infty$, then

$$\sup_{R>1} R^{-1} \int_{|x|<R} |v(x)|^2\, dx < \infty, \tag{14.1}$$

and Theorem 7.1.28 shows that this is the best possible estimate for the growth of v at infinity. This suggests that we should introduce the subspace of $L^2_{loc}(\mathbb{R}^n)$ satisfying (14.1), which is the dual of the subspace B of L^2 defined by

$$(14.2) \qquad \Big(\int_{|x|<1} |v(x)|^2\,dx\Big)^{\frac{1}{2}} + \sum_0^\infty \Big(2^j \int_{1<|x/2^j|<2} |v(x)|^2\,dx\Big)^{\frac{1}{2}} < \infty.$$

These spaces and some more general versions of them required for technical reasons are studied in Section 14.1. In Sections 14.2 and 14.3 we then show that the resolvent of $P_0(D)$ is indeed continuous in these spaces provided that P_0 has simple zeros in a certain sense.

In Section 14.4 we begin the study of perturbations

$$P(x,D) = P_0(x,D) + V(x,D)$$

assuming that $V(x,D)$ is of short range which roughly speaking means that V as a differential operator is strictly weaker than $P_0(D)$ and decreases at infinity somewhat faster than $1/|x|$. The precise definition we give involves compactness in spaces closely related to B and B^*. We also assume that $V(x,D)$ is symmetric with domain $\mathscr{S}$ and show that $P(x,D)$ with domain $\mathscr{S}$ has a self-adjoint closure H. As in quantum mechanics we then prove that the wave operators

$$(14.3) \qquad W_\pm = \lim_{t\to\pm\infty} e^{itH} e^{-itH_0}$$

exist as strong limits. They are isometric operators intertwining H and H_0,

$$(14.4) \qquad HW_\pm = W_\pm H_0.$$

In Section 14.6 it is proved that W_+ and W_- have the same range and that apart from the finitly many critical values of P_0 the spectrum of H is discrete in the orthogonal complement. Thus the scattering operator $S = W_-^* W_+$ is unitary and H is unitarily equivalent to the direct sum of H_0 and an operator with a complete set of eigenfunctions.

The key to the proof of the preceding statements is the study of the boundary values of the resolvent of H made in Section 14.5. These are proved to exist except at the critical values of P_0 and the eigenvalues of H. The main point is the proof that certain eigenfunctions of $P(x,D)$ in B^* are in L^2 and in fact rapidly decreasing. This leads at the same time to compactness properties which imply the discreteness of the point spectrum. It is then fairly easy to show in Section 14.6 that the range of W_+ (or W_-) is the orthogonal complement of the space spanned by the eigenfunctions.

The scattering operator S commutes with H_0 so FSF^{-1} commutes with multiplication by $P_0(\xi)$. In Section 14.6 we show that it induces a unitary operator S_λ in $L^2(\delta(P_0-\lambda))$ when λ is not a critical value of P_0, and that $S_\lambda - I$ is compact in $L^2(|P_0'|^\kappa\,\delta(P_0-\lambda))$ for every $\kappa\in[-1,1]$. This is the so called scattering matrix for the energy λ. In Section 14.7, finally, we prove that in the second order case there are no eigenvalues inside the continuous spectrum.

14.1. Some Function Spaces

In the summary of this chapter we have seen that it is natural to introduce the subspace B of $L^2(\mathbb{R}^n)$ defined by

$$\|v\|_B=\sum_1^\infty (R_j \int_{X_j} |v|^2\,dx)^{\frac{1}{2}}<\infty, \tag{14.1.1}$$

where

$$R_0=0,\quad R_j=2^{j-1}\quad \text{when } j>0,\quad X_j=\{x;\, R_{j-1}<|x|<R_j\}. \tag{14.1.2}$$

This is a Banach space with the norm defined by (14.1.1), and $C_0^\infty(\mathbb{R}^n)$ is a dense subspace. Regarding B as a l^1 space with Hilbert space values we conclude that the dual space B^* is the corresponding l^∞ space, that is, the set of all $u\in L^2_{\mathrm{loc}}(\mathbb{R}^n)$ with

$$\|u\|_{B^*}=\sup_{j>0}(R_j^{-1} \int_{X_j} |u|^2\,dx)^{\frac{1}{2}}<\infty. \tag{14.1.3}$$

Note that since R_j is in geometric progression we have

$$\|u\|_{B^*}^2 \leqq \sup_{R>1} \int_{|x|<R} |u|^2\,dx/R \leqq 4\,\|u\|_{B^*}^2.$$

Thus B^* is defined by (14.1) as stated in the summary. The following result is an immediate but important corollary of Theorem 7.1.26.

Theorem 14.1.1. *Let M be a C^1 hypersurface in $\mathbb{R}^n$ and K a compact subset of M. Then the restriction to K of the Fourier transform*

$$\mathscr{S}\ni v\to \hat{v}|_K\in L^2_K(dS)$$

extends by continuity to a surjective map T from B to $L^2_K(dS)$.

Proof. If $\hat{u}=\hat{u}_0\,dS$ has support in K, then Theorem 7.1.26 gives

$$|(\hat{u}_0,\hat{v}|_K)|=(2\pi)^n\,|(u,v)|\leqq(2\pi)^n\,\|u\|_{B^*}\|v\|_B\leqq C\,\|\hat{u}_0\|_{L^2}\|v\|_B$$

which proves the continuity. Since the adjoint of T

$$L^2_K(dS)\ni u_0\to F(u_0\,dS)\in B^*$$

is injective with closed range by Theorem 7.1.28, with $\phi\geqq 0$ and $\phi(0)>0$, it follows that T is surjective.

Remark. Note that B is not a reflexive space. In fact, the closure $\mathring{B}^*$ of $\mathscr{S}$ in B^* is obviously equal to the closure of L^2_{comp} in B^* and is defined by

$$\int_{|x|<R} |u|^2 \, dx/R \to 0, \qquad R \to \infty. \tag{14.1.3$'$}$$

The dual space of $\mathring{B}^*$ is equal to B so the dual space of B^* has B as a subspace of infinite codimension.

The norm in B is a majorant for a mixed L^1, L^2 norm:

Theorem 14.1.2. *Let $x=(x_1, x')$ where $x'=(x_2, \dots, x_n)$. If $u \in B$ and u is considered as a function of x_1 with values in $L^2(\mathbb{R}^{n-1})$ then*

$$\int_{-\infty}^{\infty} \|u(x_1)\|_{L^2} \, dx_1 \leqq \sqrt{2} \, \|u\|_B. \tag{14.1.4}$$

Proof. Let $u_j = u$ in X_j and $u_j = 0$ elsewhere. Then we have by Cauchy-Schwarz' inequality

$$\int_{-\infty}^{\infty} \|u_j(x_1)\|_{L^2} \, dx_1 \leqq (2R_j)^{\frac{1}{2}} \left(\int_{-\infty}^{\infty} \|u_j(x_1)\|^2_{L^2} \, dx_1 \right)^{\frac{1}{2}}$$
$$= (2R_j)^{\frac{1}{2}} \Big(\int_{X_j} |u(x)|^2 \, dx \Big)^{\frac{1}{2}}.$$

Hence we obtain (14.1.4) if we add and use (14.1.1).

By duality we obtain from Theorem 14.1.2 that $L^\infty(\mathbb{R}, L^2(\mathbb{R}^{n-1})) \subset B^*$ and that

$$\|u\|_{B^*} \leqq \sqrt{2} \sup_{x_1} \|u(x_1)\|_{L^2}, \qquad u \in L^\infty(\mathbb{R}, L^2(\mathbb{R}^{n-1})). \tag{14.1.4$'$}$$

In the study of the resolvent we shall need a number of properties of B and B^* which in particular allow us to localize the Fourier transforms of these spaces. We shall also need such results for more general spaces similar to B and B^* defined as follows. Let $c_1, c_2, \dots$ be a sequence of positive numbers such that for some constant M

$$c_j/M \leqq c_{j+1} \leqq M c_j, \qquad j = 1, 2, \dots \tag{14.1.5}$$

We then define B_c to be the space of all $v \in L^2_{\text{loc}}(\mathbb{R}^n)$ such that

$$\|v\|_{B_c} = \sum_1^{\infty} c_j \Big(\int_{X_j} |v|^2 \, dx \Big)^{\frac{1}{2}} < \infty. \tag{14.1.1$'$}$$

Thus B corresponds to $c_j = R_j^{\frac{1}{2}}$. The dual space B_c^* is the set of all $u \in L^2_{\text{loc}}(\mathbb{R}^n)$ with

$$\|u\|_{B_c^*} = \sup_{j>0} c_j^{-1} \Big(\int_{X_j} |u|^2 \, dx \Big)^{\frac{1}{2}} < \infty. \tag{14.1.3$''$}$$

The following lemma will allow us to express the B_c norm in terms of the norm in the Hilbert space $L^2_s, s\in\mathbb{R}$, defined as the set of all $u\in L^2_{\text{loc}}$ with

$$\|u\|_s^2=\int(1+|x|^2)^s|u(x)|^2\,dx<\infty. \tag{14.1.6}$$

Lemma 14.1.3. *Let N be the smallest integer with $2^N>M$. Then there is a constant C_M such that for all c satisfying* (14.1.5) *and* $k=1,2,\dots$

$$\|v\|_{B_c}\leqq C_M\,c_k(R_k^N\|v\|_{-N}+R_k^{-N}\|v\|_N),\qquad v\in L^2_{\text{loc}}(\mathbb{R}^n). \tag{14.1.7}$$

Proof. Since

$$R_j^2/4<1+R_{j-1}^2<1+|x|^2<1+R_j^2\leqq 2R_j^2,\qquad x\in X_j, \tag{14.1.8}$$

we have

$$\|v\|_{-N}\geqq(2R_j)^{-N}\|v\|_{L^2(X_j)},\qquad \|v\|_N\geqq(R_j/2)^N\|v\|_{L^2(X_j)}.$$

Hence

$$2^{-N}\|v\|_{B_c}\leqq\sum_{j\leqq k}R_j^N c_j\|v\|_{-N}+\sum_{j\geqq k}R_j^{-N}c_j\|v\|_N.$$

Here

$$R_{j+1}^N c_{j+1}/(R_j^N c_j)=2^N c_{j+1}/c_j>2^N/M,$$
$$R_{j+1}^{-N} c_{j+1}/(R_j^{-N} c_j)=2^{-N} c_{j+1}/c_j<2^{-N}M.$$

Hence we obtain (14.1.7) with $C_M=2^N/(1-M/2^N)$ if we replace the two series by geometric series with the same term for $j=k$.

The reason why (14.1.7) gives precise information on the B_c norm is that the two sides are essentially equivalent when the support of v is in X_j. This will be exploited in the proof of the following interpolation theorem.

Theorem 14.1.4. *Let N be the smallest integer $>M$. Then there is a constant C'_M such that if T is a bounded linear operator $L^2_{-N}\to L^2_{-N}$ which restricts to a bounded operator $L^2_N\to L^2_N$ and both have norm $\leqq A$, it follows that T maps B_c to B_c with norm $\leqq C'_M A$ if c satisfies* (14.1.5).

Proof. As in the proof of Theorem 14.1.2 we write $u=\sum_1^\infty u_j$ with $\operatorname{supp} u_j\subset\bar X_j$. Then (14.1.8) gives

$$\|u\|_{B_c}=\sum c_k\|u_k\|_{L^2}\geqq 2^{-N-1}\sum c_k(R_k^N\|u_k\|_{-N}+R_k^{-N}\|u_k\|_N).$$

Using (14.1.7) we obtain

$$\|Tu\|_{B_c}\leqq\sum\|Tu_k\|_{B_c}\leqq C_M\sum c_k(R_k^N\|Tu_k\|_{-N}+R_k^{-N}\|Tu_k\|_N)$$
$$\leqq AC_M\sum c_k(R_k^N\|u_k\|_{-N}+R_k^{-N}\|u_k\|_N)\leqq AC_M2^{N+1}\|u\|_{B_c}.$$

This proves the theorem with $C'_M=2^{N+1}C_M$.

Corollary 14.1.5. *Let $r\in C^N(\mathbb{R}^n)$ and assume that $D^\alpha r$ is bounded when $|\alpha|\leqq N$. Then the operator $r(D)=F^{-1}rF$ is bounded in B_c and*

$$\|r(D)u\|_{B_c}\leqq C''_M \sum_{|\alpha|\leqq N} \sup|D^\alpha r|\,\|u\|_{B_c}, \quad u\in B_c, \tag{14.1.9}$$

provided that c satisfies (14.1.5).

Proof. Since $\|u\|_N$ is equivalent to

$$\sum_{|\alpha|\leqq N} \|D^\alpha \hat{u}\|_{L^2}$$

by Parseval's formula (compare (7.9.2)) it is clear that $r(D)$ is continuous in this norm, with norm bounded by a constant times

$$R=\sum_{|\alpha|\leqq N} \sup|D^\alpha r|.$$

Now L^2_{-N} is the antidual of L^2_N which makes $\bar{r}(D)$ in L^2_N the adjoint of $r(D)$ in L^2_{-N}. Hence we have the same norm in L^2_{-N} which proves the corollary.

Corollary 14.1.6. *Let X_1 and X_2 be open sets in $\mathbb{R}^n$ and ψ a C^{N+1} diffeomorphism $X_1\to X_2$. Choose $\chi\in C_0^N(X_1)$ and set*

$$Tu=F^{-1}(\chi(\hat{u}\circ\psi)).$$

Then it follows that T is bounded in B_c when c satisfies (14.1.5), *with a norm which can be estimated in terms of the maximum of the derivatives of χ of order $\leqq N$ and of ψ,ψ^{-1} of order $\leqq N+1$.*

Proof. The adjoint of T is

$$T^*v=F^{-1}(\bar{\chi}\circ\psi^{-1}|D\psi^{-1}/D\eta|\,\hat{v}\circ\psi^{-1})$$

so it is of the same form as T. As in the proof of Corollary 14.1.5 it is therefore clear that T and T^* are bounded in L^2_N, and this proves the corollary.

The following result will allow us to localize Fourier transforms of elements in B_c. The proof depends on a slight modification of that of Theorem 14.1.4.

Theorem 14.1.7. *Let $\chi\in C_0^\infty(\mathbb{R}^n)$ and set $\chi(D-\eta)u=F^{-1}\chi(.-\eta)Fu$, $u\in\mathscr{S}'$. Then we have*

$$\int\|\chi(D-\eta)u\|^2_{B_c}\,d\eta\leqq C_{M,\chi}\|u\|^2_{B_c}, \quad u\in B_c, \tag{14.1.10}$$

for all c satisfying (14.1.5).

Proof. Let us first prove that for fixed N

(14.1.11) $$\int \|\chi(D-\eta)u\|_N^2\, d\eta \leqq C^2 \|u\|_N^2, \qquad u\in L_N^2,$$

(14.1.12) $$\int \|\chi(D-\eta)u\|_{-N}^2\, d\eta \leqq C^2 \|u\|_{-N}^2, \qquad u\in L_{-N}^2.$$

As in the proof of Corollary 14.1.5 it is clear that (14.1.11) is equivalent to

$$\iint \sum_{|\alpha|\leqq N} |D^\alpha(\chi(\xi-\eta)\hat u(\xi))|^2\, d\xi\, d\eta \leqq C' \int \sum_{|\alpha|\leqq N} |D^\alpha \hat u(\xi)|^2\, d\xi.$$

This estimate follows if we integrate first with respect to η in the left-hand side and observe that $\int |D^\beta \chi(\xi-\eta)|^2\, d\eta$ is independent of ξ. To prove (14.1.12) we compute the adjoint of the map T defined by

$$\mathscr{S}(\mathbb{R}^n) \ni u \to \chi(D-\eta)\, u(x) \in \mathscr{S}(\mathbb{R}^{2n}).$$

If $v\in\mathscr{S}(\mathbb{R}^{2n})$ and $\hat v(\xi,\eta)$ is the Fourier transform of $v(x,\eta)$ with respect to x then

$$\begin{aligned}(Tu,v) &= \iint \chi(D-\eta)\, u(x)\, \overline{v(x,\eta)}\, dx\, d\eta\\ &= (2\pi)^{-n} \iint \chi(\xi-\eta)\, \hat u(\xi)\, \overline{\hat v(\xi,\eta)}\, d\xi\, d\eta\end{aligned}$$

so the adjoint T^* is defined by

$$FT^*v(\xi) = \int \overline{\chi(\xi-\eta)}\, \hat v(\xi,\eta)\, d\eta.$$

Thus

$$\begin{aligned}\|T^*v\|_N^2 &\leqq C \sum_{|\alpha|\leqq N} \int |\int D_\xi^\alpha(\overline{\chi(\xi-\eta)}\, \hat v(\xi,\eta))\, d\eta|^2\, d\xi\\ &\leqq C' \sum_{|\alpha|\leqq N} \iint |D_\xi^\alpha \hat v(\xi,\eta)|^2\, d\eta\, d\xi\\ &\leqq C'' \iint (1+|x|^2)^N |v(x,\eta)|^2\, dx\, d\eta,\end{aligned}$$

which proves the estimate (14.1.12) of T.

The proof of Theorem 14.1.7 is now completed by decomposing u as in the proof of Theorem 14.1.4. (What is involved is actually a vector valued version of that result.) If we set $u_k^\eta = \chi(D-\eta)\, u_k$ then Lemma 14.1.3, (14.1.11), (14.1.12) and (14.1.8) give

$$\begin{aligned}\int \|u_k^\eta\|_{B_c}^2\, d\eta &\leqq 2(C_M c_k)^2 \int (R_k^{2N} \|u_k^\eta\|_{-N}^2 + R_k^{-2N} \|u_k^\eta\|_N^2)\, d\eta\\ &\leqq 2(C_M c_k)^2\, C^2 (R_k^{2N} \|u_k\|_{-N}^2 + R_k^{-2N} \|u_k\|_N^2)\\ &\leqq 4^{N+1} (C_M c_k)^2\, C^2 \|u\|_{L^2(X_k)}^2.\end{aligned}$$

Since $\chi(D-\eta)\, u = \sum u_k^\eta$ we obtain using Minkowski's inequality

$$\left(\int \|\chi(D-\eta)\, u\|_{B_c}^2\, d\eta\right)^{\frac12} \leqq \sum_k \left(\int \|u_k^\eta\|_{B_c}^2\, d\eta\right)^{\frac12} \leqq 2^{N+1}\, C C_M \|u\|_{B_c},$$

which completes the proof.

If $\|\chi\|_{L^2}=1$ it follows from (14.1.10) that

$$\begin{aligned}|(u,v)| &= |\int (\chi(D-\eta)u, \chi(D-\eta)v)\,d\eta| \\ &\leqq \int \|\chi(D-\eta)u\|_{B_c^*} \|\chi(D-\eta)v\|_{B_c} d\eta \\ &\leqq C_{M,\chi}^{\frac{1}{2}} (\int \|\chi(D-\eta)u\|_{B_c^*}^2 d\eta)^{\frac{1}{2}} \|v\|_{B_c},\end{aligned}$$

provided that $u \in L^2_{\mathrm{loc}} \cap \mathscr{S}'$ and that $\hat{v} \in C_0^\infty$. Hence

$$\|u\|_{B_c^*}^2 \leqq C_{M,\chi} \int \|\chi(D-\eta)u\|_{B_c^*}^2 d\eta, \quad u \in L^2_{\mathrm{loc}} \cap \mathscr{S}'. \tag{14.1.10$'$}$$

In our applications of the preceding results it will be more convenient to state them in terms of the spaces B and B^* rather than having to recall the spaces B_c and B_c^*.

Corollary 14.1.8. *Let $\mu \in C^1(\bar{\mathbb{R}}_+, \mathbb{R}_+)$ and assume that*

$$(1+t)|\mu'(t)| \leqq N\mu(t), \quad t \geqq 0. \tag{14.1.13}$$

If $\chi \in C_0^\infty$ and $\|\chi\|_{L^2}=1$ it follows that with $\tilde{\mu}(x)=\mu(|x|)$

$$\int \|\tilde{\mu}(\chi(D-\eta)u)\|_B^2 d\eta \leqq C_{N,\chi} \|\tilde{\mu}u\|_B^2, \quad u \in B, \tag{14.1.14}$$

$$\|\tilde{\mu}u\|_{B^*}^2 \leqq C_{N,\chi} \int \|\tilde{\mu}(\chi(D-\eta)u)\|_{B^*}^2 d\eta, \quad u \in L^2_{\mathrm{loc}} \cap \mathscr{S}'. \tag{14.1.15}$$

Proof. From (14.1.13) it follows that

$$((1+s)/(1+t))^N \leqq \mu(s)/\mu(t) \leqq ((1+t)/(1+s))^N, \quad 0<s \leqq t. \tag{14.1.16}$$

Hence $c_j = R_j^{\frac{1}{2}} \mu(R_j)$ satisfies (14.1.5) with $M=2^{N+\frac{1}{2}}$, and $\|u\|_{B_c}$ is equivalent to $\|\tilde{\mu}u\|_B$, so the estimate (14.1.14) is equivalent to (14.1.10). If instead we take $c_j^{-1} = R_j^{-\frac{1}{2}} \mu(R_j)$ then (14.1.15) follows from (14.1.10)$'$.

14.2. Division by Functions with Simple Zeros

Let p be a real valued function in $C^2(X)$ where X is an open set in $\mathbb{R}^n$. If the zeros of p are simple, the limits

$$(p \pm i0)^{-1} = \lim_{\varepsilon \to +0} (p \pm i\varepsilon)^{-1}$$

exist in $\mathscr{D}'^1(X)$. This follows from the remark after the proof of Theorem 6.1.2, for $(t \pm i\varepsilon)^{-1} \to (t \pm i0)^{-1}$ in $\mathscr{D}'^1(\mathbb{R})$ since $\log(t \pm i\varepsilon) \to \log(t \pm i0)$ in L^1_{loc}. Like the Dirac measure δ_0 the distributions $(t \pm i0)^{-1}$ on $\mathbb{R}$ are homogeneous of degree -1, so it is natural to expect that $(p \pm i0)^{-1}$ has properties similar to those of the simple layer $\delta_0(p) = dS/|p'|$ discussed in Theorem 7.1.26. When proving that this is true we consider first a neighborhood of a point $\xi^0 \in X$ where $\partial p/\partial \xi_1 > 0$. The inverse function theorem applied to the map

$$\xi \to (p(\xi), \xi'), \qquad \xi' = (\xi_2, \ldots, \xi_n),$$

shows that ξ^0 has a neighborhood $X_0 \subset X$ mapped diffeomorphically onto $X_1 \times X'$ where X_1 is a neighborhood of $p(\xi^0)$ in $\mathbb{R}$ and X' is a neighborhood of $\xi^{0\prime}$ in $\mathbb{R}^{n-1}$. We denote the inverse map by

$$(\lambda, \xi') \to (\Xi(\xi', \lambda), \xi').$$

Thus $p(\Xi(\xi', \lambda), \xi') = \lambda$.

Lemma 14.2.1. *Let $p \in C^{N+2}(X_0)$, choose $\chi \in C_0^\infty(X_0)$ and define u_z by*

$$\hat{u}_z = (p - z - i0)^{-1}(\chi \hat{f}), \qquad \operatorname{Im} z \geqq 0, \tag{14.2.1}$$

where $f \in \mathscr{S}$. Then we have for $\operatorname{Im} z \geqq 0$, with C independent of f,

$$\|u_z(x_1, .)\|_{L^2} \leqq C \int \|f(t, .)\|_{L^2} (H(x_1 - t) + (1 + |t - x_1|)^{-N})\, dt, \tag{14.2.2}$$

where H is the Heaviside function. When $x_1 \to +\infty$ we have

$$\|u_\lambda(x_1, .) - e^{ix_1 \Xi(D', \lambda)} u_\infty(x', \lambda)\|_{L^2} \to 0, \tag{14.2.3}$$

where $u_\infty(x', \lambda)$ is 0 when $\lambda \notin X_1$ and is defined for $\lambda \in X_1$ by

$$\hat{u}_\infty(\xi', \lambda) = i(\chi \hat{f}/(\partial p/\partial \xi_1))(\Xi(\xi', \lambda), \xi'). \tag{14.2.4}$$

Proof. Since u_z is an analytic function of z when $\operatorname{Im} z > 0$ and $\|u_z(x_1, .)\|_{L^2} \to 0$ as $z \to \infty$, it suffices in view of the maximum principle to prove (14.2.2) when $z = \lambda \in \mathbb{R}$. If $\lambda \in X_1$ we can write

$$\begin{aligned} \chi(\xi)(p(\xi) - \lambda - i\varepsilon)^{-1} &= (\xi_1 - \Xi(\xi', \lambda) - i\varepsilon q(\xi))^{-1} g(\xi), \\ q(\xi) &= (\xi_1 - \Xi(\xi', \lambda))/(p(\xi) - \lambda) > 0, \\ g(\xi) &= \chi(\xi) q(\xi) \in C^{N+1}. \end{aligned}$$

Hence it follows from Lemma 6.2.2 that

$$\hat{u}_\lambda(\xi) = (\xi_1 - \Xi(\xi', \lambda) - i0)^{-1} g(\xi) \hat{f}(\xi).$$

To determine the inverse Fourier transform with respect to ξ_1 we note that

$$\begin{aligned} g(\xi)/(\xi_1 - \Xi - i0) &= g(\Xi, \xi')/(\xi_1 - \Xi - i0) + G(\xi), \\ G(\xi) &= (g(\xi) - g(\Xi, \xi'))/(\xi_1 - \Xi). \end{aligned}$$

The inverse Fourier transform of the first term is the bounded function

$$i g(\Xi, \xi') e^{ix_1 \Xi} H(x_1).$$

The inverse Fourier transform of $g(\xi)/(\xi_1-\Xi-i0)$ is bounded for $|x_1|<1$ since $\operatorname{supp} g$ is compact. We have $G\in C^N$ and can write G as the sum of a function in C_0^N and one with all ξ_1 derivatives integrable Hence the inverse Fourier transform $K(x_1,\xi')$ of G has the bound

$$|K(x_1,\xi')|\leqq C(1+|x_1|)^{-N}.$$

Moreover, $(1+|x_1|)^N K(x_1,\xi')\to 0$ as $x_1\to\infty$ by the Riemann-Lebesgue lemma. If $U(x_1,\xi')$ and $F(x_1,\xi')$ are the Fourier transforms of u_λ and of f with respect to x' then

$$U(x_1,\xi')=\int F(t,\xi')(ig(\Xi,\xi')\,e^{i(x_1-t)\Xi(\xi',\lambda)}H(x_1-t)+K(x_1-t,\xi'))\,dt.$$

In view of Parseval's formula the estimate (14.2.2) follows for $z=\lambda$. If $p(\xi)\neq\lambda$ in $\operatorname{supp}\chi$ we have $(p-\lambda)^{-1}\chi\in C_0^{N+2}$, so the estimate (14.2.2) is then true with H replaced by 0 and N replaced by $N+2$, the constant C converging to 0 when $\lambda\to\infty$. This completes the proof of (14.2.2). Since

$$\int F(t,\xi')\,i\,g(\Xi,\xi')\,e^{-it\Xi(\xi',\lambda)}H(x_1-t)\,dt\to i\hat f(\Xi,\xi')\,g(\Xi,\xi'),\qquad x_1\to\infty,$$

we have also proved (14.2.3) and the lemma.

Already when $N=0$ it follows from (14.2.2) and (14.1.4), (14.1.4') that

$$\|u_z\|_{B^*}\leqq C'\|f\|_B. \tag{14.2.5}$$

Although weaker than (14.2.2) this estimate is more useful because of its invariant form. If $\phi\in C_0^0(\mathbb{R}^n)$ we also obtain from (14.2.3) that

$$\int|u_\lambda(x)|^2\phi(x/R)\,dx/R\to(2\pi)^{1-n}\int_{p(\xi)=\lambda}|\chi\hat f|^2\Big(\int_{s>0}\phi(sp')\,ds\Big)\,dS/|p'| \tag{14.2.6}$$

where dS is the surface measure on the surface $p(\xi)=\lambda$. In fact, $(e^{ix_1\Xi(D',\lambda)}u_\infty)(-x',\lambda)$ is the Fourier transform of

$$e^{ix_1\Xi}(2\pi)^{1-n}(i\chi\hat f/(\partial p/\partial\xi_1))(\Xi,\xi')$$

by (14.2.4), so Corollary 7.1.30 and (14.2.3) give

$$(2\pi)^{1-n}\int_{x_1>0}|u_\lambda(x)|^2\,\phi(x/R)\,dx/R$$
$$\to(2\pi)^{2-2n}\iint_{t>0}|\chi\hat f/(\partial p/\partial\xi_1))(\Xi,\xi')|^2\,\phi(t,-t\partial\Xi/\partial\xi')\,dt\,d\xi'.$$

By (14.2.2) the integral when $x_1<0$ converges to 0. Since $p(\Xi,\xi')=\lambda$ we have

$$(\partial p/\partial\xi_1)(\Xi,\xi')\,\partial\Xi/\partial\xi'+\partial p/\partial\xi'=0,$$

so (14.2.6) follows if we set $t=s\partial p/\partial\xi_1$ noting that $d\xi'/(\partial p/\partial\xi_1)=dS/|p'|$.

Theorem 14.2.2. *Let $p\in C^2(X)$ where X is an open set in $\mathbb{R}^n$, and assume that p is real valued and that $p'\neq 0$ in X. If $f\in B$ and $\chi\in C_0^\infty(X)$, then*

$$z\to u_z=F^{-1}((p-z)^{-1}\chi\hat{f}),\qquad \operatorname{Im} z\neq 0,$$

can be uniquely extended to $\mathbb{C}^\pm=\{z;\ \pm\operatorname{Im} z\geqq 0\}$ as a weak continuous function with values in B^*, and*

$$\|u_z\|_{B^*}\leqq C\|f\|_B;\qquad f\in B,\ z\in\mathbb{C}^\pm. \tag{14.2.7}$$

For every $\phi\in C_0^0(\mathbb{R}^n)$ we have when $R\to\infty$

$$\int|u_{\lambda\pm i0}(x)|^2\,\phi(x/R)\,dx/R \to(2\pi)^{1-n}\int_{p=\lambda}|\chi\hat{f}|^2\Big(\int_{\pm s>0}\phi(sp')\,ds\Big)\,dS/|p'|, \tag{14.2.8}$$

$$\int u_{\lambda+i0}(x)\,\overline{u_{\lambda-i0}(x)}\,\phi(x/R)\,dx/R\to 0. \tag{14.2.9}$$

Proof. If χ has sufficiently small support and $f\in\mathscr{S}$, the estimate (14.2.7) for $z\in\mathbb{C}^+$ follows from the proof of (14.2.5) above after a suitable change of labels and signs for the coordinates. We can always write $\chi=\sum\chi_j$ where the sum is finite and each term has small support, so (14.2.7) follows for $z\in\mathbb{C}^+$ and $f\in\mathscr{S}$. Since $\mathbb{C}^+\ni z\to u_z$ is continuous with values in $\mathscr{S}'$ and $\mathscr{S}$ is dense in B, it follows that $\mathbb{C}^+\ni z\to u_z$ is weak* continuous and satisfies (14.2.7) for every $f\in\mathscr{S}$. Hence there is a unique weak* continuous extension of u_z to $\mathbb{C}^+$ satisfying (14.2.7) for every $f\in B$.

In view of (14.2.7) it suffices to prove (14.2.8) and (14.2.9) when $\phi\in C_0^\infty$ and $f\in\mathscr{S}$. We have already proved (14.2.8) when χ has sufficiently small support. If $\operatorname{supp}\chi_j\cup\operatorname{supp}\chi_k$ is sufficiently small we obtain by polarization of this result, writing $u_z^j=F^{-1}((p-z)^{-1}\chi_j\hat{f})$,

$$\int u^j_{\lambda+i0}\,\overline{u^k_{\lambda+i0}}\,\phi(./R)\,dx/R \to(2\pi)^{1-n}\int_{p=\lambda}\chi_j\overline{\chi_k}|\hat{f}|^2\Big(\int_0^\infty\phi(sp')\,ds\Big)\,dS/|p'|. \tag{14.2.8$'$}$$

This is also true if $\operatorname{supp}\chi_j$ and $\operatorname{supp}\chi_k$ are disjoint. In fact, the Fourier transform of $u^j_{\lambda+i0}\,\phi(./R)$ is equal to

$$(2\pi)^{-n}((p-\lambda-i0)^{-1}\chi_j\hat{f})*R^n\hat{\phi}(R.)$$

and it tends to 0 in $C^\infty(\complement\operatorname{supp}\chi_j)$ since $\hat{\phi}\in\mathscr{S}$. Hence the scalar product with $(p-\lambda-i0)^{-1}\chi_k\hat{f}$ converges to 0. Adding up the estimates (14.2.8)$'$ we have proved (14.2.8) with the plus sign, and replacing p by $-p$ we obtain (14.2.7) and (14.2.8) with the opposite sign. Now (14.2.9) is obvious if χ has small support for under the hypotheses in Lemma 14.2.1 we have

$$u_{\lambda \pm i0}(x_1, .) \to 0 \quad \text{in } L^2(\mathbb{R}^{n-1}) \text{ when } x_1 \to \mp\infty.$$

(Note that by the Riemann-Lebesgue lemma we can replace the term $(1+|t-x_1|)^{-N}$ in (14.2.2) by $o((1+|t-x_1|)^{-N})$.) The extension to arbitrary supports is done exactly as in the proof of (14.2.8), and it completes the proof.

Corollary 14.2.3. *With the notation in Theorem 14.2.2 the following conditions are equivalent:*

(i) $u_{\lambda+i0} = u_{\lambda-i0}$

(ii)$_+$ $u_{\lambda+i0} \in \mathring{B}^*$

(ii)$_-$ $u_{\lambda-i0} \in \mathring{B}^*$

(iii) $\chi \hat{f} = 0$ *when* $p = \lambda$.

Proof. (i) is equivalent to (iii) since

$$u_{\lambda+i0} - u_{\lambda-i0} = 2\pi i F^{-1}(\chi\,\delta(p-\lambda)\,\hat{f}). \tag{14.2.10}$$

(ii)$_+$ and (ii)$_-$ are also equivalent to (iii) by (14.2.8) with $\phi \geqq 0$ and $\phi(0) > 0$.

Remark. Using (14.2.10) and Theorem 14.1.1 it is easy to see that (7.1.29) follows from (14.2.8) and (14.2.9) if $k=1$. We leave this as an exercise.

We shall now prove that $u(x)$ is almost as small as $|x|\,f(x)$ at infinity when condition (iii) in Corollary 14.2.3 is fulfilled.

Theorem 14.2.4. *Let $p \in C^{2+N}(X)$ where X is an open set in $\mathbb{R}^n$, and assume that p is real valued, $p' \neq 0$ in X. If $f \in B$, $\chi \in C_0^\infty(X)$, and $\chi \hat{f} = 0$ when $p = \lambda$ then*

$$u = F^{-1}((p-\lambda \mp i0)\,\chi \hat{f}) \tag{14.2.11}$$

is independent of the sign, and

$$\|\tilde{\mu} u\|_{B^*} \leqq C_N \|\tilde{\mu} f\|_B \tag{14.2.12}$$

if $\tilde{\mu}(x) = \mu(|x|)$ where μ is any increasing function $\in C^1(\mathbb{R})$ such that

$$(1+t)\,\mu'(t) \leqq N\mu(t), \qquad t > 0. \tag{14.2.13}$$

Proof. It is sufficient to prove (14.2.12) when the hypotheses of Lemma 14.2.1 are satisfied. From (14.2.13) it follows in view of (14.1.16) that

$$\mu(s) \leqq \mu(t)\,(1+|t-s|)^N.$$

Hence

$$\mu(|x_1|)(1+|t-x_1|)^{-N} \leqq \mu(|t|).$$

Since $\mu(|x_1|) \leqq \mu(|t|)$ if $t \leqq x_1 \leqq 0$ it follows from (14.2.2) that

$$\mu(|x_1|)\,\|u(x_1,.)\|_{L^2} \leqq C \int \mu(|t|)\,\|f(t,.)\|_{L^2}\,dt \quad \text{if } x_1 \leqq 0.$$

So far we have just used (14.2.11) with the upper sign. Using the lower sign is equivalent to changing the sign of x_1 so the preceding estimate is also valid for $x_1 > 0$. In view of (14.1.4), (14.1.4)′ it follows that

$$\|\mu(|x_1|)\,u\|_{B^*} \leqq 2C\,\|\tilde{\mu} f\|_B. \tag{14.2.14}$$

Now the hypotheses on p are still fulfilled if we change the coordinates so that x_1 is replaced by $x_1+\varepsilon x_j$ for sufficiently small ε. We have

$$|x| \leqq C \max(|x_1|, |x_1+\varepsilon x_2|, \ldots, |x_1+\varepsilon x_n|)$$

so (14.1.16) gives

$$\mu(|x|) \leqq C^N \max(\mu(|x_1|), \mu(|x_1+\varepsilon x_2|), \ldots, \mu(|x_1+\varepsilon x_n|)).$$

The estimate (14.2.12) is therefore a consequence of (14.2.14).

14.3. The Resolvent of the Unperturbed Operator

Let $P_0(\xi)$ be a real valued polynomial in $\xi \in \mathbb{R}^n$ such that $\Lambda(P_0)=\{0\}$ (cf. (10.2.8)). The set $Z(P_0)$ of critical values of P_0

$$Z(P_0)=\{\lambda;\, \lambda=P_0(\xi) \text{ and } dP_0(\xi)=0 \text{ for some } \xi\in\mathbb{R}^n\}$$

is a finite set. In fact, it is a null set by the Morse-Sard theorem and a semi-algebraic set by the Tarski-Seidenberg theorem (Theorem A.2.2). A complete proof follows also from Theorem A.2.8 which shows that if $Z(P_0)$ contains an interval I then there is a semi-algebraic curve $I \ni \lambda \to \xi(\lambda)$ such that $\lambda = P_0(\xi(\lambda))$ and $dP_0(\xi(\lambda))=0$, which is clearly impossible. When $\lambda \notin Z(P_0)$ the level set

$$M_\lambda=\{\xi;\, P_0(\xi)=\lambda\}$$

is a C^∞ submanifold of $\mathbb{R}^n$. To examine its behavior at infinity we note that since $\tilde{P}_0(\xi)\to\infty$ as $\xi\to\infty$ (Proposition 10.2.9) the localizations of $P_0(\xi)-\lambda$ at infinity (see Definition 10.2.6) do not depend on λ. We want them to have only simple zeros. Since the set $L(P_0)$ of localizations at infinity is invariant under translations, apart from the normalization, this means that we require that $Q(0)=0$ implies $dQ(0)$

$\neq 0$ when $Q \in L(P_0)$. Now $L(P_0)$ is compact, so our condition is that

$$1 \leqq C \sum_{|\alpha| \leqq 1} |Q^{(\alpha)}(0)|, \qquad Q \in L(P_0),$$

or equivalently

$$\tilde{P}_0(\xi) \leqq C' \sum_{|\alpha| \leqq 1} |P_0^{(\alpha)}(\xi)| \qquad \text{when } |\xi| \text{ is large.}$$

In a slightly different form, this is the condition in the following

Definition 14.3.1. P_0 will be called simply characteristic if

$$\tilde{P}_0(\xi) \leqq C\Big(\sum_{|\alpha| \leqq 1} |P_0^{(\alpha)}(\xi)| + 1\Big), \qquad \xi \in \mathbb{R}^n. \tag{14.3.1}$$

Examples are all hypoelliptic operators and all operators of real principal type (Definitions 11.1.2 and 8.3.5). More generally, (14.3.1) is always valid if the localizations of P_0 at infinity are of first order, that is,

$$P_0^{(\alpha)}(\xi)/\tilde{P}_0(\xi) \to 0 \quad \text{when } \xi \to \infty \text{ if } |\alpha| > 1.$$

When $\operatorname{Im} z \neq 0$ the notation

$$R_0(z) = (P_0(D) - zI)^{-1}$$

is usually employed for the L^2 resolvent. Thus

$$R_0(z) f = F^{-1}((P_0(.) - z)^{-1} \hat{f}) \tag{14.3.2}$$

when $f \in L^2$. If $\hat{f} \in C_0^\infty$ we can extend (14.3.2) to a continuous function with values in $\mathscr{S}'$ defined in $\mathbb{C}^+ \setminus Z(P_0)$ or $\mathbb{C}^- \setminus Z(P_0)$. Such functions f form a dense subset of $\mathscr{S}$, hence a dense subset of B. The notation $R_0(\lambda \pm i0)$ will still be used for the limits of (14.3.2) if $\hat{f} \in C_0^\infty$.

Theorem 14.3.2. *Assume that P_0 is simply characteristic and let K be a compact subset of $\mathbb{C}^+$ (or $\mathbb{C}^-$) containing no critical value of P_0. Then*

$$\|Q(D) R_0(z) f\|_{B^*} \leqq C \sup \tilde{Q}/\tilde{P}_0 \, \|f\|_B; \qquad \hat{f} \in C_0^\infty, \quad z \in K, \tag{14.3.3}$$

if $Q(D)$ is weaker than $P_0(D)$.

Proof. It suffices to prove that if 0 is not a critical value then (14.3.3) is valid for $z \in K_\delta = \{z \in \mathbb{C}^+; |\operatorname{Re} z| < \delta + \operatorname{Im} z\}$, if δ is small enough. The polynomials $P_{0\eta}(\xi) = P_0(\xi + \eta)/\tilde{P}_0(\eta)$ and their limits as $\eta \to \infty$ form a compact set of polynomials either satisfying the hypotheses on p in Theorem 14.2.2 with $X = \{\xi; |\xi| < \rho\}$, say, or else uniformly bounded from below in X. If $\chi_0 \in C_0^\infty(X)$ it follows from the uniformity of the proof of (14.2.7) that one can find C and ε such that for all η

$$\|F^{-1}(P_{0\eta} - z/\tilde{P}_0(\eta))^{-1} \chi_0 \hat{f}\|_{B^*} \leqq C \|f\|_B$$

if $f\in\mathscr{S}$ and $z/\tilde{P}_0(\eta)\in K_\varepsilon$. Let $\delta=\varepsilon\min\tilde{P}_0$, and choose χ_0 and $\chi\in C_0^\infty(X)$ so that $\chi_0=1$ in $\operatorname{supp}\chi$ and $\|\chi\|_{L^2}=1$. Replacing $\hat{f}(\xi)$ in this estimate by $\chi(\xi)\,Q(\xi+\eta)\,\hat{f}(\xi+\eta)$ we obtain if $z\in K_\delta$

$$\begin{aligned}\|\chi(D-\eta)\,Q(D)\,R_0(z)\,f\|_{B^*}&\leqq C\,\|Q(D)\,\chi(D-\eta)\,f\|_B/\tilde{P}_0(\eta)\\ &\leqq C'\,\|\chi(D-\eta)\,f\|_B\,\tilde{Q}(\eta)/\tilde{P}_0(\eta).\end{aligned}$$

Here we have used the notation in Theorem 14.1.7. The last estimate follows from Corollary 14.1.5, for

$$r(\xi)=Q(\xi)\,\chi_0(\xi-\eta)/\tilde{Q}(\eta)$$

and its derivatives have bounds independent of Q and η. If we square and integrate with respect to η, the estimate (14.3.3) follows in view of (14.1.10) and (14.1.10)'.

It follows from (14.3.3) that when $\hat{f}\in C_0^\infty$ the map

$$\mathbb{C}^\pm\smallsetminus Z(P_0)\ni z\to Q(D)\,R_0(z)\,f$$

is not only continuous with values in $\mathscr{S}'$ but weak* continuous with values in B^*. Since C_0^∞ is dense in $\mathscr{S}$ and $\mathscr{S}$ is dense in B it follows that $Q(D)\,R_0(z)$ extends to a continuous map from B to B^* which is weak* continuous as a function of z. We shall use the same notation for the extension. Note that if $z=\lambda\in\mathbb{R}\smallsetminus Z(P_0)$ we must indicate if λ is considered as an element of $\mathbb{C}^+$ or of $\mathbb{C}^-$, for we have

$$R_0(\lambda+i0)\,f-R_0(\lambda-i0)\,f=F^{-1}(2\pi i\,\delta_\lambda(P_0)\,\hat{f}) \tag{14.3.4}$$

by (14.2.10). This can be used to discuss the homogeneous equation $P_0(D)\,u=0$.

Theorem 14.3.3. *If* $u\in B^*$, $\lambda\in K\Subset\mathbb{R}\smallsetminus Z(P_0)$ *and* $(P_0(D)-\lambda)\,u=0$ *then* $\hat{u}=v\,dS$ *where* dS *is the surface measure on* M_λ *and* $v\in L^2(dS)$. *Conversely, this implies that* $u\in B^*$, $(P(D)-\lambda)\,u=0$ *and*

$$C_K^{-1}\,\|u\|_{B^*}^2\leqq\int_{M_\lambda}|v|^2\,dS\leqq C_K\,\|u\|_{B^*}^2. \tag{14.3.5}$$

We have

$$(2\pi)^{-n}\int_{M_\lambda}|\hat{f}|^2\,dS\leqq C_K\,\|f\|_B^2,\quad f\in B. \tag{14.3.6}$$

Proof. Since $(P_0(D)-\lambda)\,u=0$ implies $(P_0(\xi)-\lambda)\,\hat{u}=0$, hence $\operatorname{supp}\hat{u}\subset M_\lambda$, it follows from Theorem 7.1.27 when $u\in B^*$ that $\hat{u}=v\,dS$, $v\in L^2(M_\lambda,dS)$, and that the second part of (14.3.5) is valid. The proof of Theorem 14.1.1 shows that the first part of (14.3.5) is equivalent to (14.3.6). To prove (14.3.6) we note that by (14.3.4) and (14.3.3) we have

$$\|F^{-1}(Q(.)\,\delta_\lambda(P_0)\,\hat{f})\|_{B^*}\leqq C\sup\tilde{Q}/\tilde{P}_0\,\|f\|_B,\quad f\in B. \tag{14.3.7}$$

Hence the part of (14.3.5) which is already proved gives when $Q=P_0^{(\alpha)}$

$$\int_{M_\lambda} |P_0^{(\alpha)}(\xi)|^2 |\hat{f}(\xi)|^2 |P_0'(\xi)|^{-2}\, dS \leqq C\|f\|_B^2.$$

If we sum over all α with $|\alpha|=1$, the estimate (14.3.6) is proved.

Remark. We could also have proved the first part of (14.3.5) by means of Theorem 7.1.26 and a localization based on Theorem 14.1.7.

It is now easy to describe the asymptotic behavior of the solutions in B^* of the equation $(P(D)-\lambda)u=f\in B$.

Theorem 14.3.4. *Let Q_1, Q_2 be differential operators weaker than P_0, let $f_1, f_2 \in B$, $\lambda \in \mathbb{R} \setminus Z(P_0)$ and $u \in B^*$, $(P_0(D)-\lambda)u=0$, thus $\hat{u}=v\,dS$ where dS is the surface measure on M_λ. Then we have for $\phi \in C_0(\mathbb{R}^n)$*

(14.3.8) $$\lim_{R\to\infty} \int Q_1(D)R(\lambda \pm i0)f_1 \overline{Q_2(D)R(\lambda \pm i0)f_2}\, \phi(./R)\,dx/R$$
$$=(2\pi)^{1-n} \int_{M_\lambda} Q_1 \hat{f}_1 \overline{Q_2 \hat{f}_2}/|P_0'|^2 \Big(\int_{\mathbb{R}^\pm} \phi(tP_0'/|P_0'|)\,dt\Big)\,dS,$$

(14.3.9) $$\lim_{R\to\infty} \int Q_1(D)R(\lambda \pm i0)f_1 \overline{Q_2(D)R(\lambda \mp i0)f_2}\, \phi(./R)\,dx/R=0$$

(14.3.10) $$\lim_{R\to\infty} \int Q_j(D)R(\lambda \pm i0)f_j\, \bar{u}\phi(./R)\,dx/R$$
$$=\pm i(2\pi)^{-n} \int_{M_\lambda} Q_j \hat{f}_j/|P_0'|\; \bar{v}\Big(\int_{\mathbb{R}^\pm} \phi(tP_0'/|P_0'|)\,dt\Big)\,dS,$$

(14.3.11) $$\lim_{R\to\infty} \int |u|^2 \phi(./R)\,dx/R=(2\pi)^{-1-n} \int_{M_\lambda} |v|^2 \Big(\int_{\mathbb{R}} \phi(tP_0'/|P_0'|)\,dt\Big)\,dS.$$

Proof. For f_j in the dense subset $F^{-1}C_0^\infty$ of B the limits (14.3.8), (14.3.9) were given in Theorem 14.2.2. If $v \in C_0^\infty$ we can choose g with $\hat{g} \in C_0^\infty$ and $v=2\pi i \hat{g}/|P_0'|$ on M_λ, thus $u=(R_0(\lambda+i0)-R_0(\lambda-i0))g$. Hence (14.3.10) and (14.3.11) are then consequences of (14.3.8) and (14.3.9). (See also Theorem 7.1.28 and the remark following Corollary 14.2.3.) By hypothesis we have $\tilde{P}_0 \leqq C|P_0'|$ on M_λ, so $Q_j/|P_0'|$ are bounded functions on M_λ. Hence it follows from (14.3.6) that the right-hand sides of (14.3.8)–(14.3.11) can be estimated by $\|f_1\|_B \|f_2\|_B$, $\|f_j\|_B \|v\|_{L^2}$ and $\|v\|_{L^2}^2$. Since $\|Q_j(D)R_0(\lambda \pm i0)f_j\|_{B^*}$ and $\|u\|_{B^*}$ can be estimated by $\|f_j\|_B$ and $\|v\|_{L^2(M_\lambda)}$, the limits (14.3.8)–(14.3.11) are valid for general f_j and v.

Intuitively the limits (14.3.8)–(14.3.11) mean that the contributions to $R_0(\lambda \pm i0)f$ which emanate from the point $\xi \in M_\lambda$ are mainly concentrated in the direction $\pm P_0'(\xi)$ whereas the contributions to u spread equally in the directions $\pm P_0'(\xi)$. This motivates the following definition.

Definition 14.3.5. A function $u \in B^*$ is called λ-outgoing (λ-incoming) if $\lambda \in \mathbb{R} \setminus Z(P_0)$, $(P_0(D)-\lambda)u=f \in B$ and $u=R(\lambda+i0)f$ (resp. $u=R(\lambda-i0)f$).

(14.3.4) shows that the outgoing and incoming solutions of the equation $(P_0(D)-\lambda)u=f \in B$ coincide precisely when $\hat{f}=0$ on M_λ. It is particularly easy to recognize such solutions u by their asymptotic behavior (cf. Corollary 14.2.3):

Theorem 14.3.6. *Let $u \in B^*$, $\lambda \in \mathbb{R} \setminus Z(P_0)$ and assume that $(P_0(D)-\lambda)u = f \in B$. Then the following conditions are equivalent:*

(a) *u is both outgoing and incoming, thus $\hat{f}=0$ on M_λ.*

(b) *$u \in \mathring{B}^*$.*

(c) *$Q(D)u \in \mathring{B}^*$ for every Q weaker than P_0.*

Proof. (c)$\Rightarrow$(b) is trivial, and (a)$\Rightarrow$(c) by (14.3.9) where we choose $\phi \geqq 0$ with $\phi(0)>0$. Assume now that (c) is valid. Choose $\psi \in C_0^\infty(\mathbb{R}^n)$ equal to 1 in the unit ball and set

$$u_R(x)=\psi(x/R)\,u(x).$$

Then

$$(P_0(D)-\lambda)\,u_R(x)=\psi(x/R)\,f(x)+\sum_{\alpha \neq 0} R^{-|\alpha|}(D^\alpha \psi)(x/R)\,P_0^{(\alpha)}(D)\,u(x),$$

and the sum tends to 0 in B when $R \to \infty$ by virtue of (c). Hence the right-hand side converges to f in B. Since the Fourier transform of the left-hand side vanishes on M_λ we conclude using Theorem 14.1.1 that $\hat{f}=0$ on M_λ. If only (b) is assumed then $\chi(D)u$ satisfies (c) for every $\chi \in \mathscr{S}$, which gives $\hat{f}=0$ on M_λ again. Thus $u=u_0+R_0(\lambda \pm i0)f$ where $u_0 \in \mathring{B}^*$ and $(P_0(D)-\lambda)u_0=0$. By (14.3.11) it follows that $u_0=0$, so (c) $\Rightarrow$(b)$\Rightarrow$(a).

The conditions in Theorem 14.3.6 imply that u is almost as small as $|x|\,f(x)$ at infinity:

Theorem 14.3.7. *Let $\lambda \in K \Subset \mathbb{R} \setminus Z(P_0)$ and let $\mu \in C^1(\mathbb{R})$ be any increasing function such that*

$$(1+t)\,\mu'(t) \leqq N\mu(t), \qquad t \geqq 0. \tag{14.3.12}$$

Then there is a constant C independent of λ, μ and f such that if $\tilde{\mu} f \in B$, where $\tilde{\mu}(x)=\mu(|x|)$, and $\hat{f}=0$ on M_λ we have

$$\|\tilde{\mu}Q(D)\,R_0(\lambda \pm i0)\,f\|_{B^*} \leqq C \sup \tilde{Q}/\tilde{P}_0\, \|\tilde{\mu} f\|_B. \tag{14.3.13}$$

Proof. We just have to repeat the proof of Theorem 14.3.2 with the reference to Theorem 14.2.2 replaced by a reference to Theorem 14.2.4. The details are left as an exercise for the reader.

The following result is the main tool in proving that the scattering matrix is unitary (see Section 14.6).

Theorem 14.3.8. *Let $u \in B^*$, $\lambda \in \mathbb{R} \setminus Z(P_0)$ and assume that $(P_0(D)-\lambda)u = f \in B$. Then we can write*

$$u = R_0(\lambda \mp i0)f + u_\pm, \qquad \hat{u}_\pm = v_\pm \, dS$$

where dS is the surface element on M_λ, $v_\pm \in L^2(M_\lambda, dS)$ and

$$\int_{M_\lambda} (|v_+|^2 - |v_-|^2)\,|P_0'|\,dS = 2(2\pi)^{n+1}\,\mathrm{Im}\,(u,f). \tag{14.3.14}$$

We have $P_0^{(\alpha)}(D)u \in B^*$ for all α if and only if $|P_0'|\,v_\pm \in L^2(M_\lambda, dS)$.

Proof. By Theorems 14.3.2 and 14.3.3 we can write

$$u = (R_0(\lambda + i0) + R_0(\lambda - i0))f/2 + u_0, \qquad \hat{u}_0 = v_0 \, dS,$$

where $v_0 \in L^2(M_\lambda)$. Then we have

$$\mathrm{Im}\,(u,f) = \mathrm{Im}\,(u_0, f) = \mathrm{Im}\,(2\pi)^{-n} \int v_0 \bar{\hat{f}}\, dS,$$

for this is true when $f \in \mathscr{S}$, and both sides are continuous functions of $f \in B$. Since

$$(R_0(\lambda + i0) + R_0(\lambda - i0))f/2 = R_0(\lambda \mp i0)f \pm \pi i F^{-1}(\hat{f}\, dS/|P_0'|)$$

we have $v_\pm = v_0 \pm \pi i \hat{f}/|P_0'|$ and $|v_+|^2 - |v_-|^2 = 4\pi\, \mathrm{Im}\, v_0 \bar{\hat{f}}/|P_0'|$, which proves (14.3.14) and the convergence of the integral there. Since $P_0^{(\alpha)}(D)u \in B^*$ for all α if and only if this is true for $u_\pm$, the last statement follows from Theorem 14.3.3.

Corollary 14.3.9. *If (u,f) is real then u is λ-outgoing if and only if u is λ-incoming.*

Proof. When the right-hand side of (14.3.14) is zero it is obvious that $v_+ = 0$ if and only if $v_- = 0$.

Corollary 14.3.10. *If $f \in B$, $\lambda \in \mathbb{R} \setminus Z(P_0)$ and $u = R_0(\lambda \pm i0)f$ then*

$$(2\pi)^{1-n} \int_{M_\lambda} |\hat{f}|^2\, dS/|P_0'| = \pm 2\,\mathrm{Im}\,(u,f). \tag{14.3.15}$$

Proof. In Theorem 14.3.8 we have $v_\mp = 0$ and $v_+ - v_- = 2\pi i \hat{f}/|P_0'|$, hence

$$|v_+|^2 - |v_-|^2 = \pm (2\pi)^2 |\hat{f}|^2/|P_0'|^2.$$

14.4. Short Range Perturbations

In Section 14.3 we have made a rather detailed study of the resolvent $R_0(z)$ of a simply characteristic operator $P_0(D)$ with constant coefficients and $\Lambda(P_0)=\{0\}$. We shall now examine when the results can be carried over to a perturbation

$$P=P_0(D)+V(x,D)$$

by a differential operator $V(x,D)$. Since

$$P-z=P_0-z+V$$

we obtain the resolvent equation

$$R_0(z)=R(z)+R_0(z)\,VR(z)=R(z)+R(z)\,VR_0(z), \tag{14.4.1}$$

where $R(z)=(P-z)^{-1}$, if it is legitimate to multiply left and right by $R(z)$ and $R_0(z)$. Thus we should have

$$R(z)=R_0(z)(I+VR_0(z))^{-1}. \tag{14.4.1$'$}$$

To be able to justify (14.4.1) and (14.4.1)$'$ we need to know that $VR_0(z)$ is a compact operator in an appropriate space. Now Theorem 14.3.2 states that $R_0(z)$ for every $z\in\mathbb{C}^+\cup\mathbb{C}^-\setminus Z(P_0)$ maps B into

$$\{u;\,Q(D)u\in B^* \text{ for all } Q\prec P_0\}. \tag{14.4.2}$$

It is therefore natural to require that V maps this space compactly back into B. We need only take for Q the derivatives of P_0 so we set

$$B^*_{P_0}=\{u;\,P_0^{(\alpha)}(D)u\in B^* \text{ for every } \alpha\},\qquad \|u\|_{B^*_{P_0}}=\sum_\alpha \|P_0^{(\alpha)}(D)u\|_{B^*}.$$

Definition 14.4.1. A differential operator $V(x,D)$ with coefficients in $L^2_{\mathrm{loc}}(\mathbb{R}^n)$ is said to be a short range perturbation of $P_0(D)$ if $V(x,D)$ maps the intersection of $C^\infty(\mathbb{R}^n)$ and the unit ball in $B^*_{P_0}$ into a precompact subset of B.

The restriction to C^∞ only serves to make $V(x,D)$ well defined. It will be removed in the following theorem. There Ω is the unit ball in $\mathbb{R}^n$.

Theorem 14.4.2. *V is a short range perturbation of $P_0(D)$ if and only if*

i) *$V(x+y,D)\{u\in C_0^\infty(\Omega);\ \|P_0(D)u\|_{L^2}\leqq 1\}$ is a precompact subset of L^2 for every fixed $y\in\mathbb{R}^n$.*
ii) *With the notation in* (14.1.2) *we have*

$$\|V(\,.\,+y,D)u\|_{L^2}\leqq M_j\|P_0(D)u\|_{L^2},\qquad u\in C_0^\infty(\Omega),\ y\in X_j, \tag{14.4.3}$$

$$\sum_{1}^{\infty} R_j M_j < \infty. \tag{14.4.4}$$

$C^\infty \cap B^*_{P_0}$ is dense in $B^*_{P_0}$ so the closure of $V(x, D)$ is then a compact operator V from $B^*_{P_0}$ to B.

Proof. The necessity of i) is obvious since

$$\|P_0^{(\alpha)}(D)u\|_{L^2} \leqq C\|P_0(D)u\|_{L^2}, \qquad u \in C_0^\infty(\Omega),$$

by Theorem 10.3.7, for example. Let M_j be the smallest constant such that (14.4.3) is valid and choose $u_j \in C_0^\infty(\Omega)$ with $\|P_0(D)u_j\|_{L^2} = 1$ and $\|V(.+y_j, D)u_j\|_{L^2} > M_j/2$ for some $y_j \in X_j$. Then the terms in the series

$$u = \sum u_{2j}(.-y_{2j}) R_{2j}^{1/2}, \qquad v = \sum u_{2j+1}(.-y_{2j+1}) R_{2j+1}^{1/2}$$

have disjoint supports. We have $u, v \in B^*_{P_0}$ since the support of $u_j(.-y_j)$ is contained in $X_{j-1} \cup X_j \cup X_{j+1}$. If V has short range it follows that Vu, $Vv \in B$, thus

$$\sum R_j^{\frac12} M_j R_j^{\frac12} < \infty.$$

This proves (14.4.4).

To prove the converse we use Theorem 1.4.6 and a change of scales to choose a partition of unity

$$1 = \sum \phi(x - y_k)$$

with $\phi \in C_0^\infty(\Omega)$ and $\{\sqrt{n}\, y_k\}$ equal to the set of lattice points in $\mathbb{R}^n$. Let $u \in C^\infty$ and $\|u\|_{B^*_{P_0}} \leqq 1$. With $u_k = \phi(.-y_k)u$ we have by (14.4.3) when $y_k \in X_j$

$$\|Vu_k\|_{L^2}^2 \leqq M_j^2 \|\sum D^\alpha \phi(.-y_k) P_0^{(\alpha)}(D) u/\alpha!\|_{L^2}^2.$$

Since $|Vu|^2 \leqq 2^n \sum |Vu_k|^2$ it follows that

$$\begin{aligned}\|Vu\|_{L^2(X_j)}^2 &\leqq C(M_{j-1}^2 + M_j^2 + M_{j+1}^2) \sum_\alpha \|P_0^{(\alpha)}(D)u\|_{L^2(X_{j-1}\cup X_j \cup X_{j+1})}^2 \\ &\leqq C' R_j (M_{j-1} + M_j + M_{j+1})^2.\end{aligned}$$

Hence

$$\|Vu\|_B \leqq C'^{\frac12} \sum R_j (M_{j-1} + M_j + M_{j+1}) \leqq C'' \sum R_j M_j.$$

If U_J is the sum of all u_k with $|y_k| > R_J + 1$ we also obtain

$$\|VU_J\|_B \leqq C'' \sum_J^\infty R_j M_j \to 0 \quad \text{as } J \to \infty. \tag{14.4.5}$$

Now assume that we have a sequence $u^\nu \in C^\infty$ with $\|u^\nu\|_{B^*_{P_0}} \leqq 1$. It follows from i) in Theorem 14.4.2 that we can replace it by a (diagonal) subsequence such that $V(x,D)u_k^\nu$ is L^2 convergent for every k if $u_k^\nu=\phi(.-y_k)u^\nu$. In view of (14.4.5) it follows that Vu^ν is a Cauchy sequence in B. This completes the proof of the compactness.

If $u\in B_{B^*_{P_0}}$ and $\varepsilon>0$ we can for every k choose a regularization $v_k\in C_0^\infty(\Omega+\{y_k\})$ such that $\|u_k-v_k\|_{B^*_{P_0}}<\varepsilon/2^k$. Then $v=\sum v_k\in C^\infty$ and $\|u-v\|_{B^*_{P_0}}<\varepsilon$, which proves that $C^\infty\cap B^*_{P_0}$ is dense in $B^*_{P_0}$. These proof is complete.

From now on we always assume that V is a short range perturbation of $P_0(D)$ and use the same notation for the closure mapping $B^*_{P_0}$ to B. For later reference we observe that we have in fact proved that

$$\|Vu\|_B\leqq C''\sum_J^K R_j M_j\|u\|_{B^*_{P_0}} \quad \text{if } u\in B^*_{P_0} \text{ and } \operatorname{supp} u\subset\bigcup_J^K \bar{X}_j. \tag{14.4.5)'}$$

The following lemma also follows easily:

Lemma 14.4.3. *If $u^\nu\in B^*_{P_0}$ is bounded and $u^\nu\to u$ in $\mathscr{D}'$, then $u\in B^*_{P_0}$ and*

$$\|Vu^\nu-Vu\|_B\to 0.$$

Proof. The hypothesis means that $P^{(\alpha)}(D)u^\nu\to P^{(\alpha)}(D)u$ weakly in $L^2(X_j)$ for every j. Thus

$$\|P^{(\alpha)}(D)u\|_{L^2(X_j)}\leqq \varliminf_{\nu\to\infty}\|P^{(\alpha)}(D)u^\nu\|_{L^2(X_j)}\leqq CR_j^{\frac{1}{2}}$$

which proves that $u\in B^*_{P_0}$. When proving the remaining statement we may therefore assume that $u=0$. With the notations in the proof of Theorem 14.4.2 we have for every k

$$\|Vu_k^\nu\|_{L^2}\to 0 \quad \text{as } \nu\to\infty$$

for a compact operator in a Hilbert space maps weakly convergent sequences to strongly convergent sequences. In view of (14.4.5) it follows that $\|Vu^\nu\|_B\to 0$, which completes the proof.

Before leaving the discussion of the meaning of short range perturbations we shall give some sufficient conditions in addition to Theorem 14.4.2. It follows from Theorems 10.3.2 and 10.1.10 that

$$\{Q(D)u;\ \|P_0(D)u\|\leqq 1,\ u\in C_0^\infty(\Omega)\},$$

is precompact in L^2 if and only if

$$\tilde{Q}(\xi)/\tilde{P}_0(\xi)\to 0 \quad \text{as } \xi\to\infty \text{ in } \mathbb{R}^n.$$

If $Q_1, \dots, Q_r$ is a basis in the vector space of such polynomials Q, then

$$V(x,D)=\sum_1^r a_j(x)\,Q_j(D)$$

is a short range perturbation if there is a decreasing function $M(t)$ with $M(0)<\infty$ and

$$|a_j(x)|\leqq M(|x|), \qquad \int_0^\infty M(t)\,dt<\infty.$$

In fact, we have (14.4.3), (14.4.4) then with $M_j=CM(R_{j-2})$. Note that our hypothesis that $\Lambda(P_0)=\{0\}$ implies that $\tilde{Q}(\xi)/\tilde{P}_0(\xi)\to 0$ as $\xi\to\infty$ at least if $Q(\xi)=1$ (Proposition 10.2.9).

One can also allow sufficiently mild local singularities. If P_0 is elliptic and of order m, or of real principal type and order $m+1$, we may for example take

$$V(x,D)=\sum_{|\alpha|<m} V_\alpha(x)\,D^\alpha$$

if for every α we have

$$\sum R_j \sup_{y\in X_j}\Big(\int_{|x|<1} |V_\alpha(x+y)|^p\,dx\Big)^{1/p}<\infty \tag{14.4.6}$$

with $p=n/(m-|\alpha|)$ if $n>2(m-|\alpha|)$, $p>2$ if $n=2(m-|\alpha|)$ and $p=2$ if $n<2(m-|\alpha|)$. We simplify the notation in the proof by assuming that $\alpha=0$; the general statement follows at once from this special case. Then

$$\int_\Omega |V_0(x+y)\,u(x)|^2\,dx\leqq\Big(\int_\Omega |V_0(x+y)|^p\,dx\Big)^{2/p}\Big(\int_\Omega |u(x)|^q\,dx\Big)^{2/q}$$

by Hölder's inequality if $p>2$ and $2/q=1-2/p$. If $n>2m$ and $p=n/m$ then $1/q=1/2-m/n$, and it follows from Theorem 4.5.8 (cf. Theorem 4.5.13) that

$$\|u\|_{L^q}\leqq C\sum_{|\alpha|\leqq m}\|D^\alpha u\|_{L^2}, \qquad u\in C_0^\infty(\Omega).$$

When $n=2m$ and $p>2$ we have

$$\|u\|_{L^q}\leqq C\sum_{|\alpha|\leqq m}\|D^\alpha u\|_{L^r}, \qquad u\in C_0^\infty(\Omega)$$

where $1/r=1/q+m/n>m/n$, thus $r<n/m=2$; we may therefore replace r by 2 in the right-hand side. Finally if $n<2m$ it follows from Theorems 4.5.9 and 4.5.11 that

$$\sup|u|\leqq C\sum_{|\alpha|\leqq m}\|D^\alpha u\|_{L^2}, \qquad u\in C_0^\infty(\Omega).$$

In all three cases we conclude that (14.4.3) and (14.4.4) follow from (14.4.6). We also obtain condition i) in Theorem 14.4.2 by writing V_0

$=W_s'+W_s''$ where $W_s'=V_0$ when $|V_0|<s$ and $W_s'=0$ otherwise. The compactness is clear if V_0 is replaced by W_s', and if V_0 is replaced by W_s'' we obtain a norm which tends to 0 when $s\to\infty$.

From now on we also assume that V is *symmetric*, that is,

$$(Vf,g)=(f,Vg); \qquad f,g\in C_0^\infty. \tag{14.4.7}$$

Then it follows that more generally

$$(Vf,g)=(f,Vg); \qquad f,g\in B_{P_0}^*. \tag{14.4.7$'$}$$

In the proof we may assume that $f,g\in C^\infty$. Choose $\chi\in C_0^\infty$ with $\chi(0)=1$ and set when $\varepsilon>0$

$$f_\varepsilon(x)=\chi(\varepsilon x)f(x), \qquad g_\varepsilon(x)=\chi(\varepsilon x)g(x).$$

Then $f_\varepsilon,g_\varepsilon\in C_0^\infty$; f_ε and g_ε are bounded in $B_{P_0}^*$ and converge to f and g in $\mathscr{D}'$ when $\varepsilon\to 0$. Hence Lemma 14.4.3 shows that $Vf_\varepsilon\to Vf$ in B, $Vg_\varepsilon\to Vg$ in B. Since

$$(Vf_\varepsilon,g_\varepsilon)=(f_\varepsilon,Vg_\varepsilon)$$

we obtain (14.4.7)′ when $\varepsilon\to 0$ if we note that say $(Vf,g_\varepsilon)\to(Vf,g)$ by dominated convergence, and that $|(Vf_\varepsilon-Vf,g_\varepsilon)|\leqq C\|Vf_\varepsilon-Vf\|_B\to 0$.

We shall now prove that P is essentially self-adjoint.

Theorem 14.4.4. *Suppose that P_0 is real and simply characteristic, $\Lambda(P_0)=\{0\}$, and that V is a symmetric short range perturbation. Then the operator $P(x,D)=P_0(D)+V(x,D)$ with domain $\mathscr{S}$ is essentially self-adjoint, that is, the closure is self-adjoint in $L^2(\mathbb{R}^n)$.*

Proof. Since P is symmetric in $\mathscr{S}$, it suffices to prove that the range of $P-z$ is dense in L^2 for some z in each half plane, for this means that the defect indices of the closure will be zero. Now

$$(P-z)u=(I+VR_0(z))(P_0-z)u, \qquad u\in\mathscr{S},$$

and P_0-z is a bijection on $\mathscr{S}$ if $\operatorname{Im} z\neq 0$. Hence it suffices to show that $(I+VR_0(z))\mathscr{S}$ is dense in L^2. Since $\mathscr{S}$ is dense in B, this follows if we show that $I+VR_0(z)$ is bijective on B for suitable z, which is a consequence of the following

Lemma 14.4.5. *If V is a short range perturbation of P_0 then*

$$\|VR_0(it)\|_{L(B,B)}\to 0 \quad \text{as } t\to\infty \text{ in } \mathbb{R}.$$

Proof. Let $t_j\to\infty$ be a sequence in $\mathbb{R}$ and choose $f_j\in B$ with $\|f_j\|_B=1$ so that

$$\|VR_0(it_j)\|_{L(B,B)}<2\|VR_0(it_j)f_j\|_B.$$

Put $u_j = R_0(it_j) f_j \in B^*_{P_0}$. The proof of Theorem 14.3.2 shows that $\|u_j\|_{B^*_{P_0}}$ is uniformly bounded. Since

$$\|u_j\|_{L^2} \leqq |t_j|^{-1} \|f_j\|_{L^2} \leqq 1/|t_j| \to 0$$

it follows from Lemma 14.4.3 that $\|Vu_j\|_B \to 0$. This proves the lemma.

The arguments which led to the resolvent equation (14.4.1) can now be justified. If $f \in \mathscr{S}$ we have

$$(P-z)f = (P_0 - z)f + Vf.$$

When $\operatorname{Im} z \neq 0$ we can set $f = R_0(z)u$ with $u \in \mathscr{S}$ and obtain after multiplication by $R(z)$

$$R_0(z)u = R(z)u + R(z)VR_0(z)u, \qquad u \in \mathscr{S}.$$

Both sides are continuous from B to L^2 so the identity remains valid for $u \in B$. It follows that $I + VR_0(z)$ has kernel $\{0\}$ on B, so it is invertible in $L(B, B)$ by the Fredholm theory. Hence

$$R(z)u = R_0(z)(I + VR_0(z))^{-1} u \in B^*_{P_0}, \qquad u \in B.$$

In the relation

$$(P-z)f = (P_0 - z)f + Vf, \quad f \in B^*_{P_0},$$

we can now replace f by $R(z)u$, $u \in B$, and obtain

$$u = (P_0 - z)R(z)u + VR(z)u.$$

Since $\operatorname{Im} z \neq 0$ we have $R_0(z)(P_0 - z) = \text{identity on } \mathscr{S}'$, hence

$$R_0(z)u = R(z)u + R_0(z)VR(z)u, \qquad u \in B.$$

We shall now prove the existence of the *wave operators* introduced in (14.3).

Theorem 14.4.6. *If V is a symmetric short range perturbation of P_0, then*

$$W_\pm u = \lim_{t \to \pm\infty} e^{itH} e^{-itH_0} u, \qquad u \in L^2(\mathbb{R}^n), \tag{14.4.8}$$

exist and are isometric operators intertwining the closures H and H_0 of P and P_0.

Proof Since $e^{itH} e^{-itH_0}$ is unitary, the existence and isometry of the limits $W_\pm$ follows if we prove the existence for u in a dense subset of L^2. If $u \in \mathscr{S}$, then $e^{-itH_0}u$ is a C^∞ function of t with values in $\mathscr{S}$, which proves the existence of the derivative

$$\frac{d}{dt}(e^{itH} e^{-itH_0} u) = e^{itH}(iH - iH_0)e^{-itH_0} u = i e^{itH} V e^{-itH_0} u$$

in L^2. The limits (14.4.8) exist if the norm is integrable, that is,

$$\int_{|t|>1} \|Ve^{-itH_0}u\|\, dt < \infty. \tag{14.4.9}$$

This will be proved when

$$\hat{u} \in C_0^\infty(\Omega), \qquad \Omega = \{\xi;\, dP_0(\xi) \neq 0\}.$$

This condition guarantees that there are positive numbers r and R such that

$$r < |P_0'(\xi)| < R \quad \text{ if } \xi \in \operatorname{supp} \hat{u}.$$

We shall now estimate

$$e^{-itH_0}u(x) = (2\pi)^{-n} \int e^{i(\langle x,\xi\rangle - tP_0(\xi))}\, \hat{u}(\xi)\, d\xi$$

by means of Theorem 7.7.1. To do so we observe that we have uniform bounds for the ξ derivatives of $(\langle x, \xi\rangle - tP_0(\xi))/(|x|+|t|)$ when ξ is near $\operatorname{supp} \hat{u}$. If $|x| < rt$ or $|x| > Rt$ we also have a fixed lower bound for the differential in a neighborhood of $\operatorname{supp} \hat{u}$. Hence

$$|e^{-itH_0}u(x)| \leqq C_N(|x|+|t|)^{-N}, \tag{14.4.10}$$
$$N = 1, 2, \ldots;\ |x| < rt \quad \text{or} \quad |x| > Rt.$$

Now choose $\chi \in C_0^\infty(\mathbb{R}^n \setminus 0)$ equal to 1 when $r < |x| < R$ and equal to 0 when $|x| < r/2$ or $|x| > 2R$. Set

$$u_t(x) = \chi(x/t)\, e^{-itH_0}u(x), \qquad v_t(x) = (1-\chi(x/t))\, e^{-itH_0}u(x).$$

$v_t(x)$ and its derivatives can then be estimated by any negative power of $(|x|+|t|)$ so

$$\|Vv_t\|_{L^2} \leqq \|Vv_t\|_B \leqq C\|v_t\|_{B^*_{P_0}} \leqq C_N' |t|^{-N}$$

for any N. For every $Q(D)$ we have

$$Q(D)\, u_t = \sum t^{-|\alpha|} D^\alpha \chi(x/t)\, e^{-itH_0}\, Q^{(\alpha)}(D)\, u/\alpha!,$$

and since $rt/2 \leqq |x| \leqq 2Rt$ in the support it follows that

$$\|Q(D)\, u_t\|_{B^*} \leqq C t^{-\frac{1}{2}} \|Q(D)\, u_t\|_{L^2} \leqq C' t^{-\frac{1}{2}}.$$

(Our constants depend on u of course.) From (14.4.5)′ we now obtain

$$t^{\frac{1}{2}} \|Vu_t\|_{L^2} \leqq C_1 \|Vu_t\|_B \leqq C_2 t^{-\frac{1}{2}} \sum_{r|t|/4 < R_j < 4R|t|} R_j M_j.$$

Hence

$$\int_{|t|>1} \|Vu_t\|_{L^2}\, dt \leqq \sum C R_j M_j \int_{R_j/4R}^{4R_j/r} dt/t = C' \sum R_j M_j < \infty.$$

This completes the proof of (14.4.9) and the existence of wave operators.

The intertwining property

$$e^{isH} W_{\pm} = W_{\pm} e^{isH_0}, \qquad s \in \mathbb{R}, \tag{14.4.11}$$

follows immediately if we replace t by $s+t$ in the definition of $W_{\pm}$. The proof of the theorem is therefore complete.

(14.4.11) means that H is the direct sum of an operator unitarily equivalent with H_0 in the range of $W_{\pm}$ and a self-adjoint operator in the orthogonal complement. In Section 14.6 we shall prove that W_+ and W_- have the same range and that on the orthogonal complement the spectrum of H is discrete except at $Z(P_0)$. This will show that the *scattering operator*

$$S = W_+^* W_- = W_+^{-1} W_-$$

is a unitary operator commuting with H_0. The equation $Su_- = u_+$ means that the solution of the perturbed time dependent "Schrödinger equation"

$$i\,\partial v/\partial t = Hv$$

with $v(0) = W_- u_- = W_+ u_+$ is asymptotic to the solution $e^{-itH_0} u_-$ of the unperturbed equation at $-\infty$ and the solution $e^{-itH_0} u_+$ at $+\infty$. Thus S describes the asymptotic effects of the perturbation.

We shall end this section by a very elementary example which shows why the short range condition has been so important. Let $P_0(D) = D$ on $\mathbb{R}$ and let V be a smooth real valued function. Then $D+V$ with domain C_0^∞ is essentially self-adjoint; the closure H is

$$H = e^{-iF} H_0 e^{iF},$$

if F is a primitive function of V. Hence

$$e^{itH} = e^{-iF} e^{itH_0} e^{iF}, \qquad e^{itH_0} u(x) = u(x+t).$$

It follows that

$$e^{itH} e^{-itH_0} u(x) = e^{-iF(x)} e^{iF(x+t)} u(x).$$

If V is integrable we obtain

$$W_{\pm} u(x) = e^{i(F(\pm\infty) - F(x))} u(x)$$

where $F(x) - F(\pm\infty)$ is the primitive function of V which tends to 0 at $\pm\infty$. We have $S = \exp(-i\int_{\mathbb{R}} V dx)$. It is clear that these operators will not exist under much weaker conditions than integrability of V. However,

$$\lim_{t \to \pm\infty} e^{itH} e^{-itH_0 - iF(t)} = e^{-iF(x)}$$

if $V \to 0$ at ∞, and this will be the starting point for the definition of modified wave operators in Chapter XXX.

14.5. The Boundary Values of the Resolvent and the Point Spectrum

Throughout the section we assume that P_0 is real and simply characteristic, $\Lambda(P_0)=\{0\}$, and that $V(x, D)$ is a symmetric short range perturbation of P_0. By $R_0(z)$ we denote the resolvent of P_0 which is continuous from B to $B_{P_0}^*$ if $z\in\mathbb{C}^\pm\setminus Z(P_0)$. It is weak* continuous as a function of z. When $\operatorname{Im} z\neq 0$ the resolvent $R(z)$ of P is expressed in terms of (14.4.1)′ as an operator from B to $B_{P_0}^*$. This was verified after the proof of Lemma 14.4.5. We shall now examine the limit of (14.4.1)′ when z approaches a real value.

Lemma 14.5.1. *If $f\in B$ then*

$$z\to VR_0(z)f\in B$$

is a continuous function of $z\in\mathbb{C}^+\setminus Z(P_0)$ or $z\in\mathbb{C}^-\setminus Z(P_0)$. If K is a compact subset of $\mathbb{C}^\pm\setminus Z(P_0)$ then

$$\{VR_0(z)f;\ \|f\|_B\leqq 1, z\in K\}$$

is precompact in B.

Proof. By Theorem 14.3.2 we know that

$$\{R_0(z)f;\ \|f\|_B\leqq 1, z\in K\}$$

is bounded in $B_{P_0}^*$ so the last statement is obvious. If $\mathbb{C}^+\ni z_j\to z\notin Z(P_0)$ and if $f\in B$, we know that $R_0(z_j)f\to R_0(z)f$ in $\mathscr{S}'$, and the sequence is bounded in $B_{P_0}^*$. Hence Lemma 14.4.3 gives

$$\|VR_0(z_j)f-VR_0(z)f\|_B\to 0$$

as claimed.

In particular Lemma 14.5.1 shows that $VR_0(\lambda\pm i0)$ is a compact operator in B if $\lambda\in\mathbb{R}\setminus Z(P_0)$. As a first step towards the extension of (14.4.1)′ we shall now study the null space of $I+VR_0(\lambda\pm i0)$. Assume that $f\in B$ and that for example

$$f+VR_0(\lambda+i0)f=0.$$

If we set $u=R_0(\lambda+i0)f$ then $u\in B_{P_0}^*$, $(P_0(D)-\lambda)u=f=-Vu$, thus

$$(P_0(D)-\lambda+V)u=0. \tag{14.5.1}$$

By Definition 14.3.5 we know that u is λ-outgoing. However, (u, Vu) is real by (14.4.7)′ so it follows from Corollary 14.3.9 that u is also λ-incoming, thus $f+VR_0(\lambda-i0)f=0$.

Theorem 14.5.2. *If $\lambda \in \mathbb{R} \setminus Z(P_0)$ and $u \in B^*_{P_0}$ is a solution of* (14.5.1) *which is both λ-incoming and λ-outgoing, then u is rapidly decreasing,*

$$\int (1+|x|^2)^N |P_0^{(\alpha)}(D)u|^2 dx < \infty \quad \textit{for all } N \textit{ and } \alpha, \tag{14.5.2}$$

and $(H-\lambda)u=0$. The set Λ of all $\lambda \in \mathbb{R} \setminus Z(P_0)$ for which (14.5.1) *has a solution $\neq 0$ satisfying* (14.5.2) *is discrete in $\mathbb{R} \setminus Z(P_0)$, and for $\lambda \in \Lambda$ the dimension of the space of solutions of* (14.5.1) *satisfying* (14.5.2) *is finite.*

Proof. With a fixed positive integer N and some $\varepsilon \in (0, 1)$ which will later tend to 0 we set

$$\mu_\varepsilon(t) = (1+t)^N (1+\varepsilon t)^{-N}, \qquad t \geqq 0.$$

This is a bounded function, and since

$$\mu'_\varepsilon(t)/\mu_\varepsilon(t) = N(1/(1+t) - \varepsilon/(1+\varepsilon t)) = N(1-\varepsilon)/((1+t)(1+\varepsilon t))$$

we have

$$0 < (1+t)\mu'_\varepsilon(t) < N\mu_\varepsilon(t).$$

Set $\tilde{\mu}_\varepsilon(x) = \mu_\varepsilon(|x|)$ and

$$U_\varepsilon = \sum_\alpha \|\tilde{\mu}_\varepsilon P^{(\alpha)}(D)u\|_{B^*}$$

which is finite since $\tilde{\mu}_\varepsilon$ is bounded. By Theorem 14.3.7

$$U_\varepsilon \leqq C\|\tilde{\mu}_\varepsilon Vu\|_B. \tag{14.5.3}$$

With the notation used in Theorem 14.4.2 we have as in the proof of (14.4.5)

$$\|Vu\|_{L^2(X_j)} \leqq C' U_\varepsilon R_j^{\frac{1}{2}} \mu_\varepsilon(R_j)^{-1}(M_{j-1}+M_j+M_{j+1}).$$

For any $\delta > 0$ we have

$$\sum_{J+1}^{\infty} R_j(M_{j-1}+M_j+M_{j+1}) \leqq \sum_J^{\infty} 4R_j M_j < \delta \qquad \text{if } J > J(\delta).$$

Hence

$$\|\tilde{\mu}_\varepsilon Vu\|_B \leqq \sum \mu_\varepsilon(R_j) R_j^{\frac{1}{2}} \|Vu\|_{L^2(X_j)} \leqq \delta C' U_\varepsilon + \sum_1^{J(\delta)} \mu_0(R_j) R_j^{\frac{1}{2}} \|Vu\|_{L^2(X_j)}.$$

If we combine this estimate with (14.5.3) and choose δ so small that $\delta C C' < 1/2$, we obtain

$$U_\varepsilon/2 \leqq C \sum_1^J \mu_0(R_j) R_j^{\frac{1}{2}} \|Vu\|_{L^2(X_j)}.$$

Letting $\varepsilon \to 0$ now we have proved that $U_0 < \infty$. This proves (14.5.2) with N replaced by $N-1$.

If $P_0^{(\alpha)}(D)v \in L^2$ for every α then v is in the domain of H and $Hv=(P_0(D)+V(x,D))v$, for $P_0(D)+V(x,D)$ is symmetric with this domain. Hence u is in the domain of H and $(H-\lambda)u=0$.

For fixed $\lambda \in \Lambda$ the space of solutions $u \in B_{P_0}^*$ of the equation

$$u+R_0(\lambda+i0)\,Vu=0$$

is finite dimensional since $R_0(\lambda+i0)V$ is compact in $B_{P_0}^*$. To prove the discreteness of Λ we assume that for some $\lambda \notin Z(P_0)$ there is a sequence $\lambda_j \in \Lambda$, $\lambda_j \neq \lambda$, such that $\lambda_j \to \lambda$. Choose $u_j \in B_{P_0}^*$ with norm 1 so that $u_j + R_0(\lambda_j \pm i0)\,Vu_j=0$. The proof of (14.5.2) gives a uniform bound

$$\sum_\alpha \int (1+|x|^2)^N |P_0^{(\alpha)}(D)u_j|^2\,dx \leqq C_N. \tag{14.5.2$'$}$$

If $\chi \in C_0^\infty$ then $\|P_0(D)(\chi u_j)\| \leqq C_\chi \|u_j\|_{B_{P_0}^*} = C_\chi$, so it follows from Theorem 10.3.7 and Proposition 10.2.9 that χu_j has a subsequence converging in L^2. Hence u_j has a subsequence which has a limit u in L^2. Since $Hu_j=\lambda_j u_j$ we have $Hu=\lambda u$, so u is orthogonal to u_j, hence

$$(u,u)=\lim\,(u_j,u)=0.$$

By Lemma 14.4.3 it follows that $\|Vu_j\|_B \to 0$, so

$$\|R_0(\lambda_j \pm i0)\,Vu_j\|_{B_{P_0}^*} \to 0,$$

which is a contradiction since $\|u_j\|_{B_{P_0}^*}=1$. The proof is complete.

We can now discuss the boundary values of $R(z)$ by applying the following well known result from Fredholm theory:

Lemma 14.5.3. *Let K be any compact metrizable space and let*

$$K \ni z \to T(z) \in L(B,B),$$

where B is any Banach space, be a strongly continuous family of operators which is uniformly compact, that is, assume that

$$\{T(z)u;\ \|u\|_B \leqq 1,\ z \in K\}$$

is precompact in B. Then $K_0=\{z \in K;\ I+T(z)$ is not invertible$\}$ is compact, and $z \to (I+T(z))^{-1}$ is strongly continuous when $z \in K \setminus K_0$.

Proof. Assume that $K \ni z_j \to z$ and that for some $u_j \in B$ with $\|u_j\|=1$ we have

$$(I+T(z_j))u_j=f_j \to 0.$$

The sequence $T(z_j)u_j$ has a convergent subsequence so u_j has one. If u is a limit we obtain $(I+T(z))u=0$ and $\|u\|=1$, so $z \in K_0$. Hence K_0 is closed and $(I+T(z))^{-1}$ has locally bounded norm in $K\setminus K_0$. If $z_j\to z\notin K_0$ and $(I+T(z_j))u_j=f$ then the sequence u_j is bounded, hence $T(z_j)u_j$ is precompact so u_j is precompact. If u is a limit then $(I+T(z))u=f$, thus $u=(I+T(z))^{-1}f$. Hence $(I+T(z_j))^{-1}f\to(I+T(z))^{-1}f$ which proves the lemma.

Theorem 14.5.4. *The map $z\to(I+VR_0(z))^{-1}f\in B$ is a continuous function of $z\in\mathbb{C}^{\pm}\setminus(Z(P_0)\cup\Lambda)$ if $f\in B$. (Λ was defined in Theorem 14.5.2.)*

Proof. This is an immediate consequence of Lemmas 14.5.1 and 14.5.3 together with the interpretation of Λ given in Theorem 14.5.2.

We shall now identify the solutions of (14.5.1) with the eigenfunctions of H in L^2.

Theorem 14.5.5. *Let $\lambda\in\mathbb{R}\setminus Z(P_0)$. If $f_1,\dots,f_r$ is a basis for the solutions in B of the equation $(I+VR_0(\lambda+i0))f=0$, then $u_j=R_0(\lambda+i0)f_j$ satisfy* (14.5.2) *and are a basis for the solutions in L^2 of the equation $(H-\lambda)u=0$. Moreover, the equation*

$$(I+VR_0(\lambda+i0))f=g\in B \tag{14.5.4}$$

has a solution $f\in B$ if and only if

$$(g,u_j)=0,\quad j=1,\dots,r. \tag{14.5.5}$$

Proof. That u_j satisfies (14.5.2) follows from Theorem 14.5.2. Next we prove that if $f\in B$ and (14.5.4) is valid, then

$$(g,u)=0 \quad \text{if } u\in L^2 \text{ and } (H-\lambda)u=0. \tag{14.5.5$'$}$$

To do so we set $v=R_0(\lambda+i0)f\in B^*_{P_0}$ and obtain

$$(P_0(D)-\lambda)v+Vv=g.$$

Let $\chi\in C_0^\infty(\mathbb{R}^n)$ be equal to 1 in $\{x;|x|<1/2\}$ and have support in the unit ball. Set $\chi_\nu(x)=\chi(x/R_\nu)$ with R_ν defined by (14.1.2). Then

$$v_\nu(x)=\chi_\nu(x)v(x)$$

has compact support, and $P_0^{(\alpha)}(D)v_\nu\in L^2$ for every α. As observed in the proof of Theorem 14.5.2 this implies that v_ν is in the domain of H, and

$$(H-\lambda)v_\nu=\chi_\nu g+[P_0(D),\chi_\nu]v+(V\chi_\nu-\chi_\nu V)v$$

where

$$[P_0(D),\chi_\nu]v=\sum_{\alpha\neq 0}(D^\alpha\chi)(x/R_\nu)R_\nu^{-|\alpha|}P_0^{(\alpha)}(D)v.$$

Since $v \in B_{P_0}^*$ and the support is in X_ν, the factors $R_\nu^{-|\alpha|}$ guarantee that this commutator tends to 0 in L^2 when $\nu \to \infty$. In addition $V\chi_\nu v \to Vv$ in L^2 by Lemma 14.4.3. Hence

$$0 = (u, (H-\lambda)v_\nu) \to (u, g)$$

which proves (14.5.5)'. By Theorem 14.5.2 u_k is in the domain of H and $(H-\lambda)u_k = 0$, so g is in particular orthogonal to $u_1, \dots, u_r$. By the Fredholm alternative the equation (14.5.4) can be solved for all g in a space of codimension r. Hence it follows that u is a linear combination of $u_1, \dots, u_r$, and that (14.5.4) can be solved if and only if (14.5.5) is fulfilled. The proof is complete.

Combining Theorems 14.5.2 and 14.5.5 we have proved

Corollary 14.5.6. *In $\mathbb{R} \setminus Z(P_0)$ the point spectrum is discrete and of finite multiplicity. The eigenfunctions satisfy* (14.5.2).

14.6. The Distorted Fourier Transforms and the Continuous Spectrum

Theorem 14.5.4 enables us to determine the spectrum of the closure H of $P_0(D) + V(x, D)$ outside the closed countable set $\tilde{\Lambda} = Z(P_0) \cup \Lambda$. We shall see that it is absolutely continuous and equivalent to the unperturbed operator H_0. The argument starts from the following general fact.

Lemma 14.6.1. *If $\chi \in C_0(\mathbb{R})$ and dE_λ is the spectral measure of H, then*

$$\int \chi(\lambda)(dE_\lambda f, f) = \lim_{\varepsilon \to +0} \pm 1/\pi \int \chi(\lambda) \operatorname{Im}(R(\lambda \pm i\varepsilon) f, f)\, d\lambda, \quad f \in L^2. \tag{14.6.1}$$

Proof. By the spectral theorem the right-hand side is the limit of

$$\begin{aligned} &\pm 1/\pi \int\int \chi(\lambda) \operatorname{Im}(t - \lambda \mp i\varepsilon)^{-1} (dE_t f, f)\, d\lambda \\ &= \int (dE_t f, f) \int \chi(t + \varepsilon s)\, ds / (\pi(1+s^2)). \end{aligned}$$

The last integral converges uniformly to $\chi(t)$, which proves the lemma.

When $\tilde{\Lambda} \cap \operatorname{supp} \chi = \emptyset$ and $f \in B$, it follows from Theorem 14.5.4 and (14.6.1) that

$$\int \chi(\lambda)(dE_\lambda f, f) = \pm 1/\pi \int \chi(\lambda) \operatorname{Im}(R(\lambda \pm i0) f, f)\, d\lambda. \tag{14.6.1$'$}$$

By (14.4.1)′, which was justified after Lemma 14.4.5, we have

$$R(\lambda \pm i0) f = R_0(\lambda \pm i0) f_{\lambda \pm i0}$$

where

(14.6.2) $$f_z = (I + V R_0(z))^{-1} f$$

is a continuous function of $z \in \mathbb{C}^{\pm} \setminus \tilde{\Lambda}$ with values in B. Since

$$f = f_{\lambda \pm i0} + V R_0(\lambda \pm i0) f_{\lambda \pm i0}$$

it follows from (14.4.7)′ and Corollary 14.3.10 that

$$2 \operatorname{Im}(R(\lambda \pm i0) f, f) = 2 \operatorname{Im}(R_0(\lambda \pm i0) f_{\lambda \pm i0}, f_{\lambda \pm i0})$$
$$= \pm (2\pi)^{1-n} \int_{M_\lambda} |\hat{f}_{\lambda \pm i0}(\xi)|^2 \, dS / |P_0'(\xi)|, \qquad f \in B.$$

Here $M_\lambda = \{\xi;\, P_0(\xi) = \lambda\}$. Hence

(14.6.1)″ $$\int \chi(\lambda)(dE_\lambda f, f) = (2\pi)^{-n} \int \chi(\lambda) d\lambda \int_{M_\lambda} |\hat{f}_{\lambda \pm i0}(\xi)|^2 \, dS / |P_0'(\xi)|. \qquad f \in B.$$

Letting χ increase to the characteristic function of $\mathbb{R} \setminus \tilde{\Lambda}$ we find that the equality remains valid for this choice of χ. Note that $d\xi = d\lambda \, dS / |P_0'(\xi)|$.

The proof of (14.6.1)″ shows that the inner integral is a continuous function of λ. However, since $\hat{f}_{\lambda \pm i0}$ is only defined almost everywhere in M_λ we have to be somewhat careful about nullsets when considering $\hat{f}_{P_0(\xi) \pm i0}(\xi)$.

Lemma 14.6.2. *If $f \in B$ there exist measurable functions $F_\pm f$ in $\mathbb{R}^n$ such that for every $\lambda \in \mathbb{R} \setminus \tilde{\Lambda}$ we have $F_\pm f(\xi) = \hat{f}_{\lambda \pm i0}(\xi)$ for almost every $\xi \in M_\lambda$ with respect to the surface measure. $F_\pm f$ are uniquely defined almost everywhere with respect to Lebesgue measure in $\mathbb{R}^n$.*

Proof. The last statement follows from Fubini's theorem (recall that $P_0^{-1} \tilde{\Lambda}$ is of measure 0). For every $\lambda \notin \tilde{\Lambda}$ we can choose g_λ^j with $\hat{g}_\lambda^j \in C_0^\infty$ and

(14.6.3) $$\| f_{\lambda + i0} - g_\lambda^j \|_B < 2^{-j},$$

for $F^{-1} C_0^\infty$ is dense in B. Since $f_{\lambda + i0}$ is a continuous function of λ we can take g_λ^j locally independent of λ. Piecing together such choices by a sufficiently fine partition of unity in λ we can make $\hat{g}_\lambda^j$ continuous with respect to λ also. By (14.3.6) it follows from (14.6.3) that

(14.6.3)′ $$\int_{M_\lambda} |\hat{f}_{\lambda + i0}(\xi) - \hat{g}_\lambda^j(\xi)|^2 \, dS(\xi) \leqq C(\lambda) 2^{-j}$$

where $C(\lambda)$ is locally bounded in $\mathbb{R}\setminus\tilde{\Lambda}$. Taking $P_0(\xi)$ as a local coordinate we conclude that if $K\subset\mathbb{R}^n\setminus P_0^{-1}\tilde{\Lambda}$ is compact then

$$\int_K |\hat{g}^{j+1}_{P_0(\xi)}(\xi)-\hat{g}^{j}_{P_0(\xi)}(\xi)|^2\,d\xi \leqq C_K 2^{-j}.$$

Hence $F_+ f(\xi)=\lim \hat{g}^{j}_{P_0(\xi)}(\xi)$ exists almost everywhere in $\mathbb{R}^n$. We define $F_+ f(\xi)=0$ when $P_0(\xi)\in\tilde{\Lambda}$ or the limit does not exist. Then $F_+ f$ is measurable and (14.6.3)′ shows that $\hat{f}_{\lambda+i0}(\xi)=F_+ f(\xi)$ almost everywhere in M_λ.

Definition 14.6.3. If $f\in B$ then the L^2 functions defined by

$$F_\pm f(\xi)=F(I+VR_0(\lambda\pm i0))^{-1}f(\xi)$$

almost everywhere in M_λ are called the distorted Fourier transforms of f.

From now on we shall use the notation

$$E^d=\int_{\tilde{\Lambda}} dE_\lambda,\qquad E^c=\int_{\mathbb{R}\setminus\tilde{\Lambda}} dE_\lambda. \tag{14.6.4}$$

E^dL^2 is spanned by L^2 eigenfunctions of H since $\tilde{\Lambda}$ is countable. From Corollary 14.5.6 we know that the restriction of H to E^cL^2 has a continuous spectrum. We shall now prove that it is absolutely continuous.

Theorem 14.6.4. *For all $f\in B$ we have*

$$\|E^cf\|^2_{L^2}=(2\pi)^{-n}\int |F_\pm f|^2\,d\xi. \tag{14.6.5}$$

Thus the maps $f\to F_\pm f$ can be extended to isometric maps from E^cL^2 to $L^2(d\xi/(2\pi)^n)$ which vanish on E^dL^2. They intertwine H and the operator $\dot{P}_0=FH_0F^{-1}$ defined by multiplication with P_0,

$$F_\pm e^{itH}=e^{it\dot{P}_0}F_\pm,\qquad t\in\mathbb{R}. \tag{14.6.6}$$

Proof. (14.6.5) is the limit of (14.6.1)″ when $\chi\uparrow 1$ in $\complement\tilde{\Lambda}$. Next we prove

$$F_\pm Hf=\dot{P}_0F_\pm f,\qquad f\in\mathscr{S}. \tag{14.6.7}$$

To do so we note that $(H-\lambda)f=(H_0-\lambda)f+Vf\in B$, thus

$$(H-\lambda)f=(I+VR_0(\lambda\pm i0))(H_0-\lambda)f,\qquad \lambda\notin Z(P_0),$$
$$((H-\lambda)f)_{\lambda\pm i0}=(H_0-\lambda)f,\qquad \lambda\notin\tilde{\Lambda}.$$

Taking Fourier transforms we obtain

$$F_\pm(H-\lambda)f(\xi)=(P_0(\xi)-\lambda)\hat{f}(\xi)=0\qquad\text{when } P_0(\xi)=\lambda$$

which proves (14.6.7). For arbitrary f in the domain of H we can choose a sequence $f_k \in \mathscr{S}$ such that $f_k \to f$, $Hf_k \to Hf$ in L^2. Then

$$\dot{P}_0 F_\pm f_k = F_\pm H f_k \to F_\pm H f,$$

so $F_\pm f$ is in the domain of $\dot{P}_0$ and $\dot{P}_0 F_\pm f = F_\pm H f$.

Now consider

$$g(t) = e^{-it\dot{P}_0} F_\pm e^{itH} f$$

where f is in the domain of H. Then $e^{itH} f$ is in the domain of H so $F_\pm e^{itH} f$ is in the domain of $\dot{P}_0$. Hence differentiation gives

$$g'(t) = e^{-it\dot{P}_0} i(F_\pm H - \dot{P}_0 F_\pm) e^{itH} f = 0$$

so $g(t) = g(0) = F_\pm f$. This proves (14.6.6).

Since $\dot{P}_0$ has absolutely continuous spectrum it follows from Theorem 14.6.4 that the restriction of H to $E^c L^2$ is absolutely continuous. We shall now prove that it is unitarily equivalent to H_0 and at the same time show that the range of the wave operators defined in Theorem 14.4.6 is equal to $E^c L^2$.

Theorem 14.6.5. *$W_\pm$ and $F_\pm$ are unitary operators $L^2(dx) \to E^c L^2(dx)$ and $E^c L^2(dx) \to L^2(d\xi/(2\pi)^n)$ with composition equal to the Fourier transformation F. Thus the scattering operator $S = W_+^* W_-$ is a unitary operator in $L^2(dx)$.*

Proof. Let us sum up in a diagram the isometric operators we have introduced and the self-adjoint operators intertwined by them:

(14.6.8)

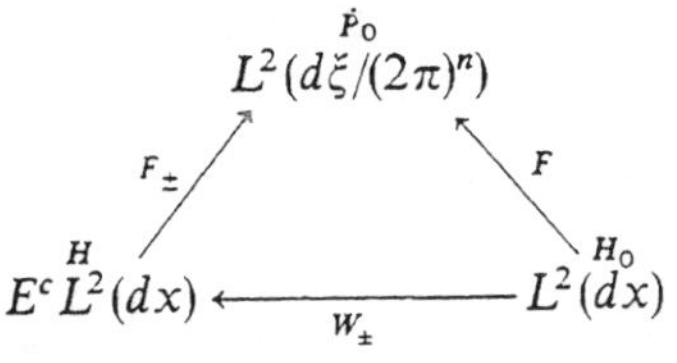

(That the range of $W_\pm$ is contained in $E^c L^2$ follows from the fact that H_0 has no square integrable eigenfunction.) Since F is surjective the other isometric maps must also be surjective if the diagram commutes, as we shall now prove. Take $f \in \mathscr{S}$ so that $\hat{f} \in C_0^\infty$ and $P_0'(\xi) \neq 0$ in supp $\hat{f}$. Then we know from the proof of Theorem 14.4.6 that for example

$$W_- f = f - \int_{-\infty}^{0} e^{itH} i V e^{-itH_0} f \, dt$$

where the integral is absolutely convergent in L^2. Since F_- is continuous it follows if we use the intertwining property (14.6.6) that

$$F_- W_- f = F_- f - \int_{-\infty}^{0} e^{it\dot{P}_0} F_-(iVe^{-itH_0}f)\,dt$$
$$= F_- f - \lim_{\varepsilon \to +0} \int_{-\infty}^{0} e^{\varepsilon t + it\dot{P}_0} F_-(iVe^{-itH_0}f)\,dt.$$

When $P_0(\xi)=\lambda$ the integral here is equal to

$$F_- \int_{-\infty}^{0} e^{\varepsilon t + it\lambda} iVe^{-itH_0} f\,dt = -F_- V(H_0 + i\varepsilon - \lambda)^{-1} f$$

for the integral is absolutely convergent in B. Hence

$$F_- W_- f(\xi) = F_-(I + VR_0(\lambda - i0)) f(\xi) = \hat{f}(\xi) \quad \text{if } P_0(\xi)=\lambda,$$

in view of Lemma 14.5.1, which proves the commutativity of one of the diagrams. For the other sign the proof is of course the same.

We can sum up our conclusions as follows: $L^2 = E^c L^2 \oplus E^d L^2$ where $E^d L^2$ is spanned by the eigenvectors of H and the restriction of H to $E^c L^2$ is unitarily equivalent to the unperturbed operator H_0. In fact, both W_+ and W_- give such a unitary equivalence so the scattering operator $S = W_+^* W_-$ commutes with H_0. We shall study its properties further which is motivated by the physical interpretation indicated in Section 14.4. To do so it is preferable to consider the Fourier transform (momentum space representation)

$$FSF^{-1} = FW_+^{-1} W_- F^{-1} = F_+ F_-^{-1}$$

which commutes with $\dot{P}_0$. Hence it commutes with multiplication by the characteristic function of the sets $\{\xi; a < P_0(\xi) < b\}$ and induces a unitary operator in the corresponding L^2 spaces. Now $d\xi = dS\,dt/|P_0'(\xi)|$, if $t = P_0(\xi)$ and dS is the area element on M_t, so it is natural to expect that FSF^{-1} even induces a unitary operator in $L^2(M_\lambda, dS/|P_0'|)$. We shall prove this by comparing the λ-incoming and λ-outgoing representations of solutions of the equation $(P-\lambda)u=0$.

Lemma 14.6.6. *If* $u \in B^*_{P_0}$, $\lambda \notin Z(P_0)$ *and* $(P_0(D)+V-\lambda)u=0$, *then*

(14.6.9) $$u = u_\pm - R_0(\lambda \mp i0) Vu,$$
$$\hat{u}_\pm = v_\pm \delta(P_0 - \lambda) = v_\pm\, dS/|P_0'|, \qquad v_\pm \in L^2(M_\lambda, dS),$$

(14.6.10) $$\int_{M_\lambda} (|v_+|^2 - |v_-|^2)\, dS/|P_0'| = 0.$$

If $\lambda \notin \tilde{\Lambda}$ *then*

(14.6.11) $$(F_+ f, \hat{u}_+) = (F_- f, \hat{u}_-) = (2\pi)^n (f, u) \quad \textit{if } f \in B.$$

Proof. Since $(P_0(D)-\lambda)u=-Vu\in B$, the decomposition (14.6.9) follows from Theorem 14.3.8, and (14.6.10) is a consequence of (14.3.14) since

$$\operatorname{Im}(u, Vu)=0$$

by (14.4.7)'. With the notation (14.6.2) we have

$$(F_\pm f, \hat{u}_\pm)=(2\pi)^n(f_{\lambda\pm io}, u_\pm)$$

because $F_\pm f=\hat{f}_{\lambda\pm io}$ on M_λ. Since

$$(f_{\lambda\pm io}, (I+R_0(\lambda\mp i0)V)u)=((I+VR_0(\lambda\pm i0))f_{\lambda\pm io}, u)=(f, u)$$

this proves (14.6.11). (This computation is the reason for the choice of signs in Theorem 14.3.8.)

In view of (14.6.11) it is natural to expect that $FSF^{-1}=F_+F_-^{-1}$ induces the unitary map $v_-\to v_+$ in $L^2(M_\lambda, dS/|P_0'|)$. However, so far we have only proved that this is a partial isometry. To study the domain and the range we note that (14.6.9) can be considered as an equation for u

$$(I+R_0(\lambda\mp i0)V)u=u_\pm$$

where $R_0(\lambda\mp i0)V$ is a compact operator in $B^*_{P_0}$; when $(P_0-\lambda)u_\pm=0$ it implies $(P_0+V-\lambda)u=0$.

Lemma 14.6.7. *For any $\lambda\in\mathbb{R}\setminus Z(P_0)$ the kernel N_λ of $(I+R_0(\lambda\mp i0)V)$ as an operator in $B^*_{P_0}$ consists of all $u\in B^*_{P_0}$ satisfying the equation $(P_0(D)-V-\lambda)u=0$ which are both λ-outgoing and λ-incoming. The range is the orthogonal space of $(P_0(D)-\lambda)N_\lambda=VN_\lambda\subset B$.*

Proof. If $u\in N_\lambda$ we obtain $(P_0(D)+V-\lambda)u=0$ if we apply $P_0(D)-\lambda$ to the equation

$$(I+R_0(\lambda\mp i0)V)u=0.$$

With the notation in Lemma 14.6.6 we have $u_\pm=0$, hence also $u_\mp=0$ by (14.6.10) so u is both λ-outgoing and λ-incoming. Conversely, if $(P_0(D)+V-\lambda)u=0$ and u is both λ-outgoing and λ-incoming, then $u=R_0(\lambda\pm i0)(P_0(D)-\lambda)u=-R_0(\lambda\pm i0)Vu$ for both choices of sign. Note that V is injective on N_λ by this formula. In view of the Fredholm alternative it just remains to show that

$$((I+R_0(\lambda\pm i0)V)u, v)=0 \quad \text{if } u\in B^*_{P_0} \text{ and } v\in VN_\lambda.$$

Since $v\in B$ the scalar product is equal to $(u, (I+VR_0(\lambda\mp i0))v)$, and if $v=Vw$, $w\in N_\lambda$, then

$$(I+VR_0(\lambda\mp i0))v=(I+VR_0(\lambda\mp i0))Vw=Vw-Vw=0.$$

The lemma is proved.

We are now ready for the main result on the structure of the scattering operator.

Theorem 14.6.8. *For every* $\lambda \in \mathbb{R} \setminus Z(P_0)$ *we set*

$$u = u_\pm - R_0(\lambda \mp i0)Vu$$

when $u \in B_{P_0}^*$ *satisfies the equation* $(P_0 + V - \lambda)u = 0$. *Then*

$$\hat{u}_\pm = v_\pm \delta(P_0 - \lambda) \quad \text{where } v_\pm \in L^2(M_\lambda, dS).$$

The map

$$v_- \to v_+$$

is a continuous bijection of $L^2(M_\lambda, dS)$ *which extends by continuity to a unitary map* Σ_λ *in* $L^2(M_\lambda, dS/|P_0'|)$ *and can be extended to a continuous bijection in* $L^2(M_\lambda, dS/|P_0'|^\kappa)$ *for every* $\kappa \in [0, 2]$. *In all these spaces* $I - \Sigma_\lambda$ *is compact. The scattering operator* S *is defined by* Σ_λ *through*

$$(14.6.12) \qquad (FSF^{-1})f|_{M_\lambda} = \Sigma_\lambda f|_{M_\lambda}, \quad f \in L^2(\mathbb{R}^n)$$

for almost all λ. *One calls* Σ_λ *the scattering matrix for the energy* λ.

Proof. Given $\hat{u}_+ = v_+ \delta(P_0 - \lambda)$ where $v_+ \in L^2(M_\lambda, dS)$ we first prove that the equation

$$(I + R_0(\lambda - i0)V)u = u_+$$

has a solution $u \in B_{P_0}^*$. Since $u_+ \in B_{P_0}^*$ it follows from Lemma 14.6.7 that we only have to verify that

$$(14.6.13) \qquad (u_+, VU) = 0 \quad \text{if } U \in N_\lambda.$$

But since $(P_0(D) - \lambda)U = -VU$ and U is both incoming and outgoing we know that $\widehat{VU} = 0$ on M_λ (cf. Theorem 14.3.6). This proves (14.6.13). Since u_- can be discussed in the same way we have now found that Σ_λ: $v_- \to v_+$ is a bijection on $L^2(M_\lambda, dS)$. Next we prove that $\Sigma_\lambda - I$ is compact there. To do so we note that

$$(\Sigma_\lambda - I)v_- = v_+ - v_-, \qquad \hat{u}_+ - \hat{u}_- = (v_+ - v_-)\delta(P_0 - \lambda),$$

where

$$u_+ - u_- = (R_0(\lambda - i0)V - R_0(\lambda + i0)V)u, \qquad (I + R_0(\lambda + i0)V)u = u_-.$$

Since u_- lies in the range of $I + R_0(\lambda + i0)V$, which is a Fredholm operator, we can choose u as a continuous linear function of u_- in the norm of $B_{P_0}^*$, so $u_- \to u_+ - u_- \in B_{P_0}^*$ is compact. But the $B_{P_0}^*$ norm on u_- (or $u_+ - u_-$) is equivalent to the L^2 norm on M_λ of v_- (or $v_+ - v_-$) (Theorem 14.3.3). Hence $\Sigma_\lambda - I$ is compact in $L^2(M_\lambda)$. The other continuity statements on Σ_λ will follow from Lemma 14.6.9 below.

To prove (14.6.12) we write (14.6.11) in the form

$$(F_+ f|_{M_\lambda}, \Sigma_\lambda v_-)_\lambda = (F_- f|_{M_\lambda}, v_-)_\lambda, \qquad v_- \in L^2(M_\lambda),\ f \in B,\ \lambda \notin \tilde{\Lambda},$$

where the scalar product is taken with respect to the measure $dS/|P_0'|$ on M_λ. Since Σ_λ is unitary for this scalar product and $F_\pm f \in L^2(M_\lambda)$, it follows that

$$(F_+ f)|_{M_\lambda} = \Sigma_\lambda (F_- f)|_{M_\lambda}.$$

Thus we have proved (14.6.12) for the dense subset $F_- B$ of L^2, and (14.6.12) follows in general by continuity.

What remains is just to prove the following lemma.

Lemma 14.6.9. *Let a be a positive bounded continuous function on a locally compact space M with a positive measure $d\mu$. Assume that T is compact in $L^2(d\mu)$ and that*

$$\int |f + Tf|^2 a\, d\mu = \int |f|^2 a\, d\mu, \qquad f \in L^2(d\mu). \tag{14.6.14}$$

Then it follows that T can be extended to a compact operator in $L^2(a^\kappa d\mu)$ for $0 \leqq \kappa \leqq 2$ and that $I + T$ is bijective in all these spaces.

Proof. It follows from (14.6.14) that $f + Tf \neq 0$ if $0 \neq f \in L^2(d\mu)$, so $I + T$ has an inverse $I + S$ in $L^2(d\mu)$ with S compact in $L^2(d\mu)$. Polarization of (14.6.14) gives the identity

$$\int (f + Tf)\overline{(g + Tg)}\, a\, d\mu = \int f \bar{g} a\, d\mu; \qquad f, g \in L^2(d\mu).$$

If g is replaced by $(I + S)g$ we can rewrite it in the form

$$\int Tf \bar{g} a\, d\mu = \int f \overline{Sg}\, a\, d\mu; \qquad f, g \in L^2(d\mu).$$

Now the spaces $L^2(d\mu)$ and $L^2(a^2 d\mu)$ are dual with respect to the scalar product $\int f \bar{g} a\, d\mu$, and the adjoint of S in $L^2(a^2 d\mu)$ is then an extension of T. Thus T extends to a compact operator in $L^2(a^2 d\mu)$. If the norm of T in $L^2(d\mu)$ and $L^2(a^2 d\mu)$ is denoted by N_0 and N_2, then the norm in $L^2(a^{2\kappa} d\mu)$ is at most $N_0^{1-\kappa} N_2^\kappa$. This is another case of the Riesz-Thorin interpolation theorem (cf. Theorem 7.1.12). The hypothesis means that

$$|N_2^{-z} N_0^{z-1} \int T(f a^{-z}) g a^{z}\, d\mu| \leqq 1 \quad \text{if } \operatorname{Re} z = 0 \text{ or } \operatorname{Re} z = 1 \tag{14.6.15}$$

provided that f vanishes when a is small and that

$$\int |f|^2 d\mu = \int |g|^2 d\mu = 1.$$

In the left-hand side of (14.6.15) we then have a bounded analytic function of z when $0 \leqq \operatorname{Re} z \leqq 1$, so the Phragmén-Lindelöf theorem shows that the inequality (14.6.15) remains valid when $0 \leqq \operatorname{Re} z \leqq 1$.

When $z=\kappa$ we obtain

$$|\int TF\,G\,d\mu|\leqq N_2^\kappa N_0^{1-\kappa} \quad \text{if } \int |F|^2 a^{2\kappa} d\mu \leqq 1,\ \int |G|^2 a^{-2\kappa} d\mu \leqq 1$$

and F, G vanish for small a. Hence the norm of T in $L^2(a^{2\kappa}d\mu)$ is at most equal to $N_2^\kappa N_0^{1-\kappa}$.

Let χ_j be the multiplication operator by the characteristic function of the set where $a>1/j$. Then $\chi_j \to 1$ strongly as $j\to\infty$. Since T is compact it follows that $\|\chi_j T - T\|_{L^2(a^\kappa d\mu)} \to 0$ as $j\to\infty$ if $\kappa=0$ or 2. Taking adjoints we obtain the same conclusion for $T\chi_j - T$, hence for $\chi_j T\chi_j - T = \chi_j(T\chi_j - T) + \chi_j T - T$. From the logarithmic convexity of norms proved above it follows that the norm of $\chi_j T \chi_j - T$ in $L^2(a^\kappa d\mu)$ tends to 0 when $j\to\infty$ for every $\kappa\in[0,2]$. But $\chi_j T\chi_j$ is obviously compact in $L^2(a^\kappa d\mu)$ when $0\leqq\kappa\leqq 2$ for all the $L^2(a^\kappa d\mu)$ norms are equivalent in the range of χ_j. This completes the proof of the lemma.

Example 14.6.10. It is instructive to make Theorem 14.6.8 explicit in the very elementary case of the second order differential operator

$$Pu = -u'' + Vu = D^2u + Vu$$

where $V\in L^2\cap \mathscr{E}'$. When $\operatorname{Im} z\neq 0$ the operator $R_0(z)$ is then convolution with the fundamental solution $E_z(x) = (i/2\sqrt{z})\exp(i|x|\sqrt{z})$ of $D^2 - z$, where $\sqrt{z}$ is chosen in the upper half plane. If now $-u'' + Vu - \lambda u = 0$ and $\lambda>0$ we have

$$u(x) = C_+^+ e^{ix\sqrt{\lambda}} + C_+^- e^{-ix\sqrt{\lambda}}, \qquad x \gg 0,$$
$$u(x) = C_-^+ e^{ix\sqrt{\lambda}} + C_-^- e^{-ix\sqrt{\lambda}}, \qquad x \ll 0,$$

where $\sqrt{\lambda}$ is the positive square root. With $z=\lambda+i\varepsilon$ we have

$$R_0(z)Vu = R_0(z)(u''+\lambda u) = -u - i\varepsilon R_0(z)u,$$
$$-i\varepsilon R_0(z)u(x) = \varepsilon/2\sqrt{z} \int_{-\infty}^{\infty} e^{i|x-y|\sqrt{z}} u(y)\,dy.$$

For large $y>0$ the integrand is

$$e^{-ix\sqrt{z}}(C_+^+ e^{iy(\sqrt{\lambda}+\sqrt{z})} + C_+^- e^{iy(\sqrt{z}-\sqrt{\lambda})}).$$

The integral of the first term is bounded as $\varepsilon\to+0$, and since

$$\varepsilon i/(2\sqrt{z}(\sqrt{z}-\sqrt{\lambda})) \to 1, \qquad \varepsilon\to+0,$$

we obtain after discussing large negative y in the same way

$$R_0(\lambda+i0)Vu = -u + C_+^- e^{-ix\sqrt{\lambda}} + C_-^+ e^{ix\sqrt{\lambda}}.$$

Thus

$$u_- = C_+^- e^{-ix\sqrt{\lambda}} + C_-^+ e^{ix\sqrt{\lambda}}.$$

A similar calculation gives

$$u_+ = C_-^- e^{-ix\sqrt{\lambda}} + C_+^+ e^{ix\sqrt{\lambda}},$$

so the scattering matrix maps (C_+^-, C_-^+) to (C_-^-, C_+^+). The unitarity is equivalent to

$$|C_+^+|^2 - |C_+^-|^2 = |C_-^+|^2 - |C_-^-|^2$$

which also follows directly from the fact that the Wronski determinant of u and $\bar{u}$ is a constant. If (C_+^-, C_-^+) is an eigenvector of the scattering matrix with eigenvalue $e^{i\theta}$, thus $C_+^+ = e^{i\theta} C_-^+$, $C_-^- = e^{i\theta} C_+^-$ then u has the same form at $+\infty$ as at $-\infty$ but with x replaced by $x + \theta/\sqrt{\lambda}$. The effect of the perturbation V is thus at infinity only a phase shift by θ.

14.7. Absence of Embedded Eigenvalues

Theorem 14.6.5 shows that the perturbed operator H is unitarily equivalent to the direct sum of the unperturbed operator H_0 and an operator with pure point spectrum accumulating only at the critical values $Z(P_0)$ of P_0. In this section we shall supplement this information by proving that for second order operators the point spectrum must usually lie outside the continuous spectrum except at $Z(P_0)$. We begin with the elliptic case.

Proposition 14.7.1. *If $\Delta = \sum \partial_j^2$ is the Laplace operator in $\mathbb{R}^n$ and $\lambda > 0$, $\tau > 0$, then*

$$(14.7.1) \quad 2\lambda\tau \int |u|^2 |x|^\tau dx \leq \int |(\Delta + \lambda)u|^2 |x|^{2+\tau} dx, \qquad u \in C_0^\infty(\mathbb{R}^n \setminus \{0\}).$$

Proof. We shall introduce polar coordinates $r = |x|$ and $\omega = x/|x| \in S^{n-1}$. Note that $\partial/\partial x_j = \omega_j \partial/\partial r + r^{-1}\Omega_j$ where Ω_j is a vector field on S^{n-1}, and

$$\sum \omega_j \Omega_j = 0, \quad \sum \Omega_j \omega_j = r \sum \partial\omega_j/\partial x_j = n-1.$$

Hence

$$\Delta = \sum (\omega_j \partial/\partial r + r^{-1}\Omega_j)^2 = \partial^2/\partial r^2 + (n-1) r^{-1} \partial/\partial r + r^{-2} \Delta_S$$

where $\Delta_S = \sum \Omega_j^2$ is the Laplace-Beltrami operator on the unit sphere. With $r = e^t$ we have $\partial/\partial r = e^{-t}\partial/\partial t$, $\partial^2/\partial r^2 = e^{-2t}(\partial^2/\partial t^2 - \partial/\partial t)$, hence

$$\Delta + \lambda = e^{-2t}(\partial^2/\partial t^2 + (n-2)\partial/\partial t + \Delta_S + \lambda e^{2t}).$$

The integral on the right-hand side of (14.7.1) is therefore equal to

$$M=\iint_{|\omega|=1} |(\partial^2/\partial t^2+(n-2)\partial/\partial t+\Delta_S+\lambda e^{2t})u|^2 e^{t(\tau+n-2)}\,dt\,d\omega.$$

To remove the exponential we set

$$v(t,\omega)=e^{at}u(e^t\omega), \qquad 2a=\tau+n-2,$$

and note that

$$(\partial/\partial t-a)^2+(n-2)(\partial/\partial t-a)=\partial^2/\partial t^2-\tau\partial/\partial t+(\tau^2-(n-2)^2)/4.$$

With the notation

$$L_1=\partial^2/\partial t^2+(\tau^2-(n-2)^2)/4+\Delta_S+\lambda e^{2t}, \qquad L_2=-\tau\partial/\partial t$$

we now obtain with L^2 norms

$$M=\|L_1 v+L_2 v\|^2=\|L_1 v\|^2+\|L_2 v\|^2+([L_1,L_2]v,v)$$

since L_1 is symmetric and L_2 is skew symmetric. Now (14.7.1) follows since

$$[L_1,L_2]=2\lambda\tau e^{2t}$$

and $\|e^t v\|^2=\int |u|^2|x|^\tau dx$. The proof is complete.

Before proceeding we note that (14.7.1) is valid for every $u\in\mathscr{E}'(\mathbb{R}^n\setminus\{0\})$ such that $(\Delta+\lambda)u\in L^2$. This follows at once if we apply (14.7.1) to regularizations of u.

Proposition 14.7.1 leads immediately to a uniqueness theorem:

Theorem 14.7.2. *Assume that u is a solution of the equation*

$$(\Delta+\lambda+V)u=0$$

where $\lambda>0$ and V is multiplication by a function $V(x)$ satisfying

$$|V(x)|\leqq C/|x|. \tag{14.7.2}$$

If $(1+|x|)^\tau D^\alpha u\in L^2$ for all τ when $|\alpha|\leqq 1$, it follows that $u=0$.

Proof. Choose $\chi\in C_0^\infty$ so that $\chi(x)=1$ when $|x|<1$ and $\chi(x)=0$ when $|x|>2$, and set with r small and R large

$$u_{r,R}(x)=\chi(x/R)u_r(x), \qquad u_r(x)=(1-\chi(x/r))u(x).$$

If $2r<R$ we have

$$|(\Delta+\lambda)u_{r,R}-\chi(.\,/R)(\Delta+\lambda)u_r|\leqq C/R\sum_{|\alpha|\leqq 1}|D^\alpha u|$$

and the left hand side vanishes when $|x|<R$. If we apply (14.7.1) to $u_{r,R}$ and let $R \to \infty$ it follows that

$$2\lambda\tau \int |u_r|^2 |x|^\tau dx \leqq \int |(\Delta+\lambda)u_r|^2 |x|^{2+\tau} dx$$
$$\leqq 2\int |Vu_r|^2 |x|^{2+\tau} dx + C_1(2r)^{\tau-2}$$

if $2r<1$. Combining this estimate with (14.7.2) we obtain

$$2(\lambda\tau - C^2)\int |u_r|^2 |x|^\tau dx \leqq C_1(2r)^{\tau-2}.$$

Letting $\tau \to \infty$ we conclude that $u=0$ when $|x|>2r$, and since r is any small positive number this completes the proof.

Recall that by Theorems 14.5.2 and 14.5.5 the hypotheses on u in Theorem 14.7.2 are fulfilled if V is a short range perturbation of Δ and $u \in L^2$, $(H+\lambda)u=0$ where H is the self adjoint closure of $\Delta+V$. Thus no eigenvalues are embedded in the continuous spectrum of H if V is a short range perturbation satisfying (14.7.2). Similar conclusions can be drawn from the following results in the non-elliptic case.

As in Section 6.2 we now denote by B and A two dual quadratic forms and consider the differential operator $B(\partial)=-B(D)$ in the cone defined by $A>0$. We assume now that A is indefinite so this is not an empty set. We can introduce polar coordinates corresponding to A,

$$r=A(x)^{\frac{1}{2}}, \qquad \omega=x/r, \qquad \text{thus } A(\omega)=1.$$

For the sake of simplicity let $A(x)=\sum a_j x_j^2$. Then

$$2r\,\partial r/\partial x_j = \partial A(x)/\partial x_j = 2a_j x_j$$

which proves that

$$\partial/\partial x_j = a_j\omega_j \partial/\partial r + r^{-1}\Omega_j$$

where Ω_j is a vector field in the hyperboloid $A(x)=1$. We have

$$\sum \omega_j \Omega_j = 0, \qquad \sum \Omega_j \omega_j = r\sum \partial\omega_j/\partial x_j = n-1$$

since $\sum a_j\omega_j^2=1$ and $\sum x_j\partial/\partial x_j = r\partial/\partial r$. With $b_j=1/a_j$ we obtain

$$B(\partial)=\sum b_j(a_j\omega_j\partial/\partial r + r^{-1}\Omega_j)^2$$
$$=\sum a_j\omega_j^2 \partial^2/\partial r^2 + r^{-2}\sum b_j\Omega_j^2 + (n-1)r^{-1}\partial/\partial r.$$

Thus we have as before

$$B(\partial)=\partial^2/\partial r^2 + (n-1)r^{-1}\partial/\partial r + r^{-2}\Delta_S$$

where $\Delta_S=\sum b_j\Omega_j^2$ is the Laplace-Beltrami operator in the hyperboloid $A(\omega)=1$. It is symmetric with respect to the natural density $d\omega$

such that $dx = r^{n-1}\,dr\,d\omega$. Now the proof of (14.7.1) gives with no change

Proposition 14.7.3. *If* $u \in C_0^\infty(\{x; A(x) > 0\})$ *then*

$$2\lambda\tau \int |u|^2 A^{\tau/2}\,dx \leqq \int |(B(\partial)+\lambda)u|^2 A^{1+\tau/2}\,dx, \quad \tau > 0,\ \lambda > 0. \tag{14.7.3}$$

Again the estimate leads to a uniqueness theorem.

Theorem 14.7.4. *Let B be an indefinite non-singular real quadratic form with dual form A, and let u be a solution of the equation*

$$(B(\partial)+\lambda+V)u = 0$$

where $\lambda \in \mathbb{R} \setminus \{0\}$ *and V is just multiplication by a function* $V(x)$ *such that*

$$|V(x)| \leqq C(|A(x)|+1+|x|)^{-\frac{1}{2}}. \tag{14.7.4}$$

If $(1+|x|)^\tau D^\alpha u \in L^2$ *for all* τ *when* $|\alpha| \leqq 1$ *it follows that* $u = 0$.

Proof. Changing the signs of B and V if necessary we can assume that $\lambda > 0$. Let $\psi \in C^\infty(\mathbb{R})$ be equal to 1 in $(2, \infty)$ and 0 in $(-\infty, 1)$ and set

$$u_{r,R}(x) = \chi(x/R)u_r(x), \qquad u_r(x) = \psi(A(x)/r)u(x)$$

where r is small, R is large and $\chi \in C_0^\infty(\{x; |x| < 2\})$, $\chi(x) = 1$ when $|x| < 1$. As in the proof of Theorem 14.7.2

$$|(B(\partial)+\lambda)u_{r,R} - \chi(\,.\,/R)(B(\partial)+\lambda)u_r| \leqq C/R \sum_{|\alpha| \leqq 1} |D^\alpha u_r|,$$

and the left-hand side vanishes when $|x| < R$. Furthermore

$$|(B(\partial)+\lambda)u_r + \psi(A/r)Vu| \leqq C \sum_{|\alpha| \leqq 1} |D^\alpha u(x)|(1+|x|)^{2-|\alpha|}/r^2$$

and the left-hand side vanishes except when $r < A < 2r$. We can apply (14.7.3) to $u_{r,R}$ and conclude when $R \to \infty$ that (14.7.3) is also valid for u_r. Thus we have when $2r < 1$

$$2\lambda\tau \int |u_r|^2 A^{\tau/2}\,dx \leqq 2\int |Vu_r|^2 A^{1+\tau/2}\,dx + C_1(2r)^{(\tau-6)/2}.$$

The hypothesis (14.7.4) implies $|A(x)|\,|V(x)|^2 \leqq C^2$, hence

$$2(\lambda\tau - C^2)\int |u_r|^2 A^{\tau/2}\,dx \leqq C_1(2r)^{(\tau-6)/2}.$$

As before we conclude that $u = 0$ when $A(x) > 2r$, hence $u = 0$ when $A(x) > 0$.

To complete the proof it suffices to observe that (14.7.4) implies that $V(x)^2 A(x-y)$ is bounded for any fixed y. Placing the origin at y

we have therefore proved that $u(x)=0$ if $A(x-y)>0$, so $u=0$ everywhere. The proof is complete.

Notes

The subject discussed in this chapter originates from a classical paper by Weyl [3]. He proved that for the Sturm-Liouville operator $-d^2/dx^2+q$ on $(0, \infty)$ with a self adjoint boundary condition at 0 there is an eigenfunction expansion with eigenfunctions asymptotic to those of the unperturbed problem at ∞ apart from a phase shift, if q is an integrable function. A natural extension to several variables was made possible by the introduction of the Møller wave operators (see Møller [1], Cook [1], Hack [1], Hörmander [34], Jauch-Zinnes [1], Jörgens-Weidmann [1], Kuroda [1], Veselič-Weidmann [1, 2]). The asymptotic completeness of these operators, that is, the identity of the range with the (absolutely) continuous subspace was first established by Ikebe [1] for the Schrödinger equation in $\mathbb{R}^3$ with potential $O(|x|^{-2-\varepsilon})$ at infinity. The basic structure of his proof goes back to Povsner [1] and is still used here although the technicalities are quite different. (See also Friedrichs [4].) Singular eigenfunction expansions are avoided altogether since they require a rate of decrease of the perturbation at infinity which increases with the dimension. It was Agmon [2, 3] who showed how scattering theory could be carried out for general elliptic operators with perturbations $O(|x|^{-1-\varepsilon})$ at infinity. (The presence of an obstacle gives rise to a situation similar to that when the perturbing potential has compact support. However, it will not be considered here since we have not yet developed the basic theory of elliptic boundary problems.) Agmon-Hörmander [1] developed the technicalities required to study perturbations of the general operators called simply characteristic here. This chapter is a full development of those ideas combined with the methods of Agmon [3]. Fur further references to the literature and related results the reader should consult Reed-Simon [1]. Closely related results are due to Deič, Korotjaev and Jafaev [1].

The uniqueness theorem in Scction 14.7 is essentially due to Kato [1]. A difference is that we just require the potential to be $O(1/|x|)$ but that the solution is rapidly decreasing. This is natural in the application to scattering theory. However, there is a famous example due to von Neumann-Wigner [1] of a potential which is $O(1/|x|)$ but for which there is an embedded eigenvalue. The proof used here follows that of Aronszajn [2] and Cordes [1] for the unique continuation theorem for solutions of second order elliptic operators. We shall return to it in Chapter XVII. Actually it is quite close to the proof of uniqueness for certain long range perturbations by Agmon, Kato and Simon as presented in Reed-Simon [1].

We have followed here the so called stationary method in scattering theory. There is also a time dependent method where one works with the operator $i\partial/\partial t+H$ instead of the operator H and its resolvent; we did so here only in connection with the wave operators. Recently Enss [1, 2] has made this an attractive alternative to the stationary approach. However, the time dependent method does not seem to be applicable when $P_0(\xi)$ does not tend to ∞ at ∞, which has decided in favor of the stationary approach here.

Chapter XV. Analytic Function Theory and Differential Equations

Summary

If K is a compact convex set in $\mathbb{R}^n$ then the Paley-Wiener-Schwartz theorem describes the Fourier transform of $C_0^\infty(K)$ completely, including the topology. One can therefore transfer the study of the action of a constant coefficient partial differential operator $P(D)$ on $C_0^\infty(K)$ to the study of multiplication by $P(\zeta)$ on a space of entire analytic functions. However, to profit from this we need some technical tools. In particular, we want to be able to cut off an analytic function having the required bounds in a subset of $\mathbb{C}^n$ and then modify it to an analytic function with the appropriate bounds in all of $\mathbb{C}^n$. This requires that one can solve the Cauchy-Riemann system

$$\partial u/\partial \bar{z}_j = f_j, \quad j=1,\dots,n, \tag{15.1}$$

when f_j satisfies the necessary compatibility conditions

$$\partial f_j/\partial \bar{z}_k - \partial f_k/\partial \bar{z}_j = 0. \tag{15.2}$$

The solution u should have bounds of the same type as f. Such techniques will be developed in Section 15.1 with conditions on u of the form

$$\int |u|^2 e^{-\phi} \, d\lambda < \infty \tag{15.3}$$

where $d\lambda$ is the Lebesgue measure in $\mathbb{C}^n$ and ϕ is plurisubharmonic. Roughly speaking this means that $2\log|u|$ is bounded by ϕ, so assuming ϕ plurisubharmonic is quite natural since $\log|u|$ is plurisubharmonic if u is analytic.

In Section 15.2 we discuss the inductive limit topology in $B_{2,k}^c(X) = B_{2,k} \cap \mathscr{E}'(X)$ which has the dual space $B_{2,1/\check{k}}^{\mathrm{loc}}(X)$. When X is convex we prove that the topology can be described by semi-norms such as (15.3) on the Fourier-Laplace transforms, with ϕ plurisubharmonic. The analogue for $C_0^\infty(X)$ is proved in Section 15.4. We discuss this topic at some length since the literature on it is scant and it has even been claimed that it is impossible to take ϕ plurisubharmonic in general. In Section 15.3 we then prove a theorem on the representation of solutions of the equation $P(D)u=0$ by integrals of exponential solutions. As corollaries we obtain some regularity theorems already proved in Chapter XI with other methods. If the existence theory for the Cauchy-Riemann system is extended to forms of higher degree and is combined with some local algebraic geometry, one can prove general existence theorems for overdetermined system of differential equations with constant coefficients and results on the decomposition of solutions in exponential solutions. That is undoubtedly the main merit of analytic function theory in this context. However, several presentations of this topic are available in monograph form, so we content ourselves with references to the literature in the notes.

15.1. The Inhomogeneous Cauchy-Riemann Equations

The main result in this section is the following existence theorem. We use the notation $z_j=x_j+iy_j$ for the coordinates in $\mathbb{C}^n$, write $\partial/\partial z_j=(\partial/\partial x_j-i\partial/\partial y_j)/2$ and $d\lambda=dx\,dy$ for the Lebesgue measure in $\mathbb{C}^n$.

Theorem 15.1.1. *Let $\phi\in C^2(\mathbb{C}^n)$ be strictly plurisubharmonic, that is,*

$$\kappa(z)=\inf_t \sum \partial^2\phi(z)/\partial z_j\partial\bar{z}_k\, t_j\bar{t}_k/\sum|t_j|^2>0. \tag{15.1.1}$$

For all $f_j\in L^2(\mathbb{C}^n,e^{-\phi}\kappa^{-1}d\lambda)$ satisfying

$$\partial f_j/\partial\bar{z}_k=\partial f_k/\partial\bar{z}_j,\quad j,k=1,\dots,n, \tag{15.1.2}$$

one can then find $u\in L^2(\mathbb{C}^n,e^{-\phi}d\lambda)$ such that

$$\partial u/\partial\bar{z}_j=f_j,\quad j=1,\dots,n, \tag{15.1.3}$$

$$\int|u|^2e^{-\phi}d\lambda\leqq\int|f|^2e^{-\phi}\kappa^{-1}d\lambda. \tag{15.1.4}$$

Proof. With the notation

$$(u,v)_\phi=\int u\bar{v}e^{-\phi}d\lambda;\quad u,v\in L^2_\phi=L^2(\mathbb{C}^n,e^{-\phi}d\lambda);$$

the equations (15.1.3) are equivalent to

$$(f,g)_\phi=\sum_1^n(f_j,g_j)_\phi=-\Big(u,\sum_1^n\delta_jg_j\Big)_\phi,\quad g_j\in C_0^\infty,$$

where we have written

$$\delta_jw=e^\phi\partial(e^{-\phi}w)/\partial z_j=\partial w/\partial z_j-\partial\phi/\partial z_jw,\quad j=1,\dots,n. \tag{15.1.5}$$

If the theorem were proved it would follow from (15.1.4) that

$$|(f,g)_\phi| \leqq \Big\| \sum_1^n \delta_j g_j \Big\|_\phi \, \|f\|_{\phi+\log\kappa}, \qquad g_j \in C_0^\infty. \tag{15.1.6}$$

Conversely, if we prove (15.1.6) then the map

$$\sum \delta_j g_j \to -(g,f)_\phi, \qquad g_j \in C_0^\infty,$$

can be extended to a linear form on C_0^0 with norm $\leqq \|f\|_{\phi+\log\kappa}$ with respect to the norm $\|\ \|_\phi$. Hence it must be of the form $v \to (v,u)_\phi$ where $\|u\|_\phi \leqq \|f\|_{\phi+\log\kappa}$. This will prove the theorem.

The key to the proof of (15.1.6) is the identity

$$\begin{aligned} \|\sum \delta_j g_j\|_\phi^2 &+ \tfrac{1}{2}\sum \|\partial g_j/\partial \bar z_k - \partial g_k/\partial \bar z_j\|_\phi^2 \\ &= \sum \|\partial g_j/\partial \bar z_k\|_\phi^2 + \int \sum g_j \bar g_k \, \partial^2\phi/\partial z_j \partial \bar z_k \, e^{-\phi}\, d\lambda. \end{aligned} \tag{15.1.7}$$

To prove it we expand the squares of norms in the left-hand side. Since

$$(\delta_j g_j, \delta_k g_k)_\phi - (\partial g_j/\partial \bar z_k, \partial g_k/\partial \bar z_j)_\phi = ([\delta_j, \partial/\partial \bar z_k] g_j, g_k)_\phi$$

we obtain (15.1.7) by computing the commutator

$$[\delta_j, \partial/\partial \bar z_k] = \delta_j \partial/\partial \bar z_k - \partial/\partial \bar z_k \delta_j = \partial^2 \phi/\partial z_j \partial \bar z_k. \tag{15.1.8}$$

If we drop the first sum in the right-hand side of (15.1.7) and estimate the other from below by means of (15.1.1) we have proved that

$$\int |g|^2 \kappa e^{-\phi} d\lambda \leqq \|\sum \delta_j g_j\|_\phi^2 + \sum \|\partial g_j/\partial \bar z_k - \partial g_k/\partial \bar z_j\|_\phi^2/2 \tag{15.1.9}$$

if $g_j \in C_0^\infty$ for all j. The estimate remains valid for all $g_j \in L_\phi^2$ with $\sum \delta_j g_j \in L_\phi^2$ and $\partial g_j/\partial \bar z_k - \partial g_k/\partial \bar z_j \in L_\phi^2$ for all j and k. If $\operatorname{supp} g$ is compact this follows if we apply (15.1.9) to $g_j^\varepsilon = g_j * \psi_\varepsilon$, where $\psi \in C_0^\infty$, $\int \psi \, dx = 1$ and $\psi_\varepsilon(x) = \varepsilon^{-n}\psi(x/\varepsilon)$. When $\varepsilon \to 0$ we have with convergence in L_ϕ^2

$$g_j^\varepsilon \to g_j, \qquad \partial g_j^\varepsilon/\partial \bar z_k - \partial g_k^\varepsilon/\partial \bar z_j \to \partial g_j/\partial \bar z_k - \partial g_k/\partial \bar z_j,$$
$$\sum \partial g_j^\varepsilon/\partial z_j \to \sum \partial g_j/\partial z_j,$$

which proves that (15.1.9) is valid for g. If the support of g is not compact we apply (15.1.9) to $\chi(\varepsilon x) g(x)$ where $\chi \in C_0^\infty$ and $\chi(0)=1$. Terms where χ is differentiated contain a factor ε so their L_ϕ^2 norm is $O(\varepsilon)$. When $\varepsilon \to 0$ we obtain (15.1.9) in general. By Cauchy-Schwarz' inequality it follows that

$$\begin{aligned} &|(f,g)_\phi|^2 \\ &\leqq \int |f|^2 e^{-\phi} \kappa^{-1} d\lambda \Big(\|\sum \delta_j g_j\|_\phi^2 + \sum_{j<k} \|\partial g_j/\partial \bar z_k - \partial g_k/\partial \bar z_j\|_\phi^2\Big) \end{aligned} \tag{15.1.10}$$

if $g_j \in L_\phi^2$, $\sum \delta_j g_j \in L_\phi^2$ and $\partial g_j/\partial \bar z_k - \partial g_k/\partial \bar z_j \in L_\phi^2$.

Assuming that $f_j \in L^2_\phi$ for all j we shall now prove that (15.1.10) implies (15.1.6). To do so we denote by N the set of all $g=(g_1,\dots,g_n)$ with $g_j \in L^2_\phi$ such that $\partial g_j/\partial \bar{z}_k - \partial g_k/\partial \bar{z}_j = 0$. Thus $f \in N$ by (15.1.2), and $(\partial\psi/\partial\bar{z}_1,\dots,\partial\psi/\partial\bar{z}_n) \in N$ if $\psi \in C_0^\infty$. If H is in the orthogonal complement $N^\perp$ of N in $L^2_\phi \oplus \dots \oplus L^2_\phi$ we therefore have

$$0=\sum (H_j, \partial\psi/\partial\bar{z}_j)_\phi = (\sum \delta_j H_j, \psi)_\phi,$$

hence $\sum \delta_j H_j = 0$. If $g_j \in C_0^\infty$ we can now write

$$g=G+H, \qquad G\in N \text{ and } H \in N^\perp.$$

Then it follows from (15.1.10) that

$$\begin{aligned}|(f,g)_\phi|^2 = |(f,G)_\phi|^2 &\leqq \int |f|^2 e^{-\phi}\kappa^{-1}\, d\lambda \,\|\sum \delta_j G_j\|^2_\phi \\ &= \int |f|^2 e^{-\phi}\kappa^{-1}\, d\lambda \,\|\sum \delta_j g_j\|^2_\phi.\end{aligned}$$

This completes the proof of the theorem when $f_j \in L^2_\phi$.

To eliminate the hypothesis that $f_j \in L^2_\phi$ we choose a positive convex C^2 function Φ such that $\Phi(z) \geqq |z| \log\kappa(z)$. Then $\log\kappa(z) - \varepsilon\Phi(z) \leqq 0$ if $|\varepsilon z| > 1$, hence

$$\int |f|^2 e^{-\phi-\varepsilon\Phi}\, d\lambda < \infty$$

for every $\varepsilon > 0$. If we apply the part of the theorem which is already proved with ϕ replaced by $\phi+\varepsilon\Phi$, then κ is replaced by a larger function so we obtain a solution u_ε of (15.1.3) with

$$\int |u_\varepsilon|^2 e^{-\phi-\varepsilon\Phi}\, d\lambda \leqq \int |f|^2 e^{-\phi}\kappa^{-1}\, d\lambda.$$

When $\varepsilon \to 0$ we can take a weak limit u of u_ε in $L^2_{\text{loc}}(\mathbb{C}^n)$ and conclude that u satisfies (15.1.3) and (15.1.4). The proof is complete.

The following variant of Theorem 15.1.1 is often more convenient to use because it does not assume that ϕ is smooth or strictly plurisubharmonic.

Theorem 15.1.2. *Let ϕ be a plurisubharmonic function in $\mathbb{C}^n$ and let $f_j \in L^2(\mathbb{C}^n, e^{-\phi}\, d\lambda)$. If* (15.1.2) *is valid one can find a solution u of* (15.1.3) *such that*

$$2\int |u|^2 e^{-\phi}(1+|z|^2)^{-2}\, d\lambda \leqq \int |f|^2 e^{-\phi}\, d\lambda. \tag{15.1.11}$$

Proof. First assume that $\phi \in C^2$. We can then apply Theorem 15.1.1 with ϕ replaced by $\phi + 2\log(1+|z|^2)$, for

$$\begin{aligned}\sum t_j \bar{t}_k \partial^2(\log(1+|z|^2))/\partial z_j \partial\bar{z}_k &= (1+|z|^2)^{-2}(|t|^2(1+|z|^2) - |(t,z)|^2) \\ &\geqq (1+|z|^2)^{-2}|t|^2,\end{aligned}$$

and this implies $\kappa(z)\geqq 2(1+|z|^2)^{-2}$. Thus (15.1.11) follows from (15.1.4).

If we regularize ϕ as in the proof of Theorem 4.1.8 we obtain plurisubharmonic functions $\phi_\varepsilon \in C^\infty$ decreasing to ϕ as $\varepsilon\to 0$. From the part of the theorem which is already proved it follows that (15.1.3) has a solution u_ε with

$$2\int |u_\varepsilon|^2 e^{-\phi_\varepsilon}(1+|z|^2)^{-2}\, d\lambda \leqq \int |f|^2 e^{-\phi_\varepsilon}\, d\lambda \leqq \int |f|^2 e^{-\phi}\, d\lambda.$$

Hence we can find a sequence $\varepsilon_j\to 0$ such that u_{ε_j} converges weakly in L^2 on every compact set. It follows that the limit u satisfies (15.1.3) and that

$$2\int |u|^2 e^{-\phi_\varepsilon}(1+|z|^2)^{-2}\, d\lambda \leqq \int |f|^2 e^{-\phi}\, d\lambda$$

for every $\varepsilon>0$. This completes the proof.

Remark. It is not surprising that $2^{-1}\log(1+|z|^2)$ is plurisubharmonic, for it is the logarithm of the norm of the vector valued analytic function $\mathbb{C}^n \ni z \to (1,z)\in \mathbb{C}^{n+1}$.

In what follows we shall use the standard notation

$$\bar\partial u = \sum \partial u/\partial \bar z_j\, d\bar z_j.$$

The equations (15.1.3) can then be written $\bar\partial u = f$ where $f=\sum f_j\, d\bar z_j$. We set

$$\bar\partial f = \sum \bar\partial f_j \wedge d\bar z_j = \sum \partial f_j/\partial \bar z_k\, d\bar z_k \wedge d\bar z_j$$

with the notations of exterior differential calculus. The compatibility condition (15.1.2) can then be written $\bar\partial f=0$, and it just reflects that $\bar\partial\bar\partial=0$.

The following extension theorem gives a simple but typical example of how Theorem 15.1.2 is used

Theorem 15.1.3. *Let ϕ be a plurisubharmonic function in $\mathbb{C}^n$ such that*

$$|\phi(z)-\phi(z')|<C \quad \text{if } |z-z'|<1,$$

for some constant C. Let W be a complex linear subspace of $\mathbb{C}^n$ of codimension k. For every analytic function u in W such that

$$\int_W |u|^2 e^{-\phi}\, dS < \infty,$$

where dS denotes the surface area in W, there exists an anylytic function U in $\mathbb{C}^n$ such that $U=u$ in W and

$$\int |U|^2 e^{-\phi}(1+|z|^2)^{-3k}\, d\lambda \leqq (6\pi e^C)^k \int_W |u|^2 e^{-\phi}\, dS. \tag{15.1.12}$$

Proof. Since $\log(1+|z|^2)$ is plurisubharmonic by the remark above, it is enough to prove the theorem when W is a hyperplane and then iterate this special case k times. We may assume that W is the hyperplane $z_n=0$. Then u is an analytic function of $z'=(z_1,\dots,z_{n-1})$ and may be regarded as an analytic function in $\mathbb{C}^n$ which is independent of z_n. We have

$$\int_{|z_n|<1} |u|^2 e^{-\phi}\, d\lambda \leqq \pi e^C \int_W |u|^2 e^{-\phi}\, dS \tag{15.1.13}$$

but when $|z_n|$ is large we can no longer ignore that ϕ depends on z_n. Thus we choose a cut off function $\psi(z_n)$ with support in the unit disc which is 1 when $|z_n|<1/2$ and $2(1-|z_n|)$ when $1/2\leqq|z_n|\leqq 1$. Then $|\partial\psi/\partial\bar{z}_n|\leqq 1$. Writing

$$U(z)=\psi(z_n)\,u(z')-z_n\,v(z)$$

we have $U(z)=u(z')$ when $z_n=0$, and U is analytic, that is, $\bar\partial U=0$, if and only if v satisfies the equation

$$\bar\partial v=z_n^{-1}\,u(z')\,\bar\partial\psi(z_n)=z_n^{-1}\,u(z')\,\partial\psi/\partial\bar{z}_n\,d\bar{z}_n=f. \tag{15.1.14}$$

It is clear that $\bar\partial f=0$, and from (15.1.13) we obtain

$$\int |f|^2 e^{-\phi}\, d\lambda \leqq 4\pi e^C \int_W |u|^2 e^{-\phi}\, dS.$$

Application of Theorem 15.1.2 now gives a solution of (15.1.14) with

$$\int |v|^2 e^{-\phi}(1+|z|^2)^{-2}\, d\lambda \leqq 2\pi e^C \int_W |u|^2 e^{-\phi}\, dS.$$

Combining this estimate with (15.1.13) we obtain (15.1.12).

Corollary 15.1.4. *Let ϕ and W satisfy the hypothesis in Theorem 15.1.3 and let u be an analytic function in W such that*

$$|u(z)|\leqq C_1\, e^{\phi(z)}, \qquad z\in W.$$

Then one can find U analytic in $\mathbb{C}^n$ with $U=u$ in W and

$$|U(z)|\leqq C_2(1+|z|)^{n+2k+1}\, e^{\phi(z)}, \qquad z\in\mathbb{C}^n.$$

Proof. The hypothesis implies that

$$\int_W |u(z)|^2 e^{-\psi(z)}\, dS(z)<\infty,$$

where $\psi(z)=2\phi(z)+(n-k+1)\log(1+|z|^2)$. Hence we can find U so that

$$\int |U(z)|^2(1+|z|^2)^{-(n+2k+1)} e^{-2\phi(z)}\, d\lambda(z)<\infty.$$

U is a harmonic function in $\mathbb{R}^{2n}$ so $U(z)$ is equal to the mean value of U over the unit ball with center at z. By Cauchy-Schwarz' inequality this proves the corollary.

Corollary 15.1.4 allows us to prove the analogue for hyperfunctions of the Paley-Wiener-Schwartz theorem, mentioned at the end of Section 9.1.

Theorem 15.1.5. *Let K be a convex compact set in $\mathbb{R}^n$ with supporting function H and let $u \in A'(K)$ be an analytic functional with support in K. Then the Fourier-Laplace transform*

$$\hat{u}(\zeta) = u(\exp(-i\langle .,\zeta\rangle)), \quad \zeta \in \mathbb{C}^n,$$

is an entire analytic function such that for every $\varepsilon > 0$

$$|\hat{u}(\zeta)| \leqq C_\varepsilon \exp(H(\operatorname{Im}\zeta) + \varepsilon|\zeta|), \quad \zeta \in \mathbb{C}^n. \tag{15.1.15}$$

Conversely, every entire function satisfying these bounds is the Fourier-Laplace transform of a unique $u \in A'(K)$.

Proof. Only the converse remains to be proved. Let K_ε be the set of points in $\mathbb{C}^n$ at distance $\leqq \varepsilon$ from K. The theorem will be proved if we show that there exists a distribution u_ε with support in $K_\varepsilon \subset \mathbb{R}^{2n}$ which defines u. This is true if and only if for every $\zeta \in \mathbb{C}^n$

$$u(e^{-i\langle z,\zeta\rangle}) = u_\varepsilon(e^{-i(\langle x,\zeta\rangle + i\langle y,\zeta\rangle)})$$

where we have written $z = x + iy$. The condition can be written in the form

$$\hat{u}(\zeta) = \hat{u}_\varepsilon(\zeta, i\zeta), \quad \zeta \in \mathbb{C}^n. \tag{15.1.16}$$

By Corollary 15.1.4 we can find an entire function U in $\mathbb{C}^{2n}$ such that

$$U(\zeta, i\zeta) = \hat{u}(\zeta), \quad \zeta \in \mathbb{C}^n, \tag{15.1.17}$$

$$|U(\zeta_1,\zeta_2)| \leqq C'_\varepsilon(1+|\zeta_1|+|\zeta_2|)^{4n+1}\exp(H(\operatorname{Im}\zeta_1) + \varepsilon|(\operatorname{Im}\zeta_1, \operatorname{Im}\zeta_2)|),$$
$$\zeta_1 \in \mathbb{C}^n, \ \zeta_2 \in \mathbb{C}^n.$$

By the Paley-Wiener-Schwartz theorem (Theorem 7.3.1) there exists a distribution u_ε with support in K_ε and Fourier-Laplace transform U. (15.1.17) means that (15.1.16) is fulfilled so the theorem is proved.

In Section 15.3 we shall prove additional extension theorems with applications to differential operators with constant coefficients. However, we shall close this section by proving a characterization of plurisubharmonic functions already mentioned in Section 4.1. (It will not be used later.)

Theorem 15.1.6. *Let P_A be the set of all functions of the form $N^{-1}\log|f(z)|$ where N is a positive integer and f an entire function $\not\equiv 0$. Then the closure of P_A in $L^1_{\mathrm{loc}}(\mathbb{C}^n)$ consists of all plurisubharmonic functions.*

The following lemma is useful in the proof.

Lemma 15.1.7. *If ϕ is a continuous subharmonic function in $X\subset\mathbb{R}^n$ and ϕ_j is a sequence of subharmonic functions $\leqq\phi$ such that $\phi_j(x)\to\phi(x)$ as $j\to\infty$ for every x in a dense set $E\subset X$, then $\phi_j\to\phi$ in $L^1_{\mathrm{loc}}(X)$.*

Proof. By Theorem 4.1.9 the sequence ϕ_j is precompact in L^1_{loc} and every limit is defined by a subharmonic function ψ. It is clear that $\psi\leqq\phi$, and (4.1.8) shows that $\psi(x)\geqq\phi(x)$ when $x\in E$. Hence

$$\phi(x)\leqq \int_{|y-x|<r}\psi(y)\,dy\Big/\int_{|y|<r}dy, \qquad r>0,$$

if $x\in E$. Since both sides are continuous functions of x this must remain true for all x. When $r\to 0$ we conclude that $\phi\leqq\psi$. Hence $\phi=\psi$ which completes the proof.

To pass from estimates in L^2 norm to estimates in maximum norm we shall also need the following simple lemma.

Lemma 15.1.8. *If $u\in L^2(B_r)$, $B_r=\{z\in\mathbb{C}^n, |z|<r\}$ and $\bar\partial u\in L^\infty(B_r)$ then u is continuous in B_r and*

$$|u(0)|\leqq C(\sup_{B_r} r|\bar\partial u|+r^{-n}\|u\|_{L^2(B_r)}). \tag{15.1.18}$$

Proof. By introducing z/r as a new variable we reduce the proof to the case where $r=1$. Let $\chi\in C_0^\infty(B_1)$, $\chi(z)=1$ when $|z|<1/2$. If E is the fundamental solution of the Laplacean in $\mathbb{R}^{2n}$ (cf. Theorem 3.3.2) then

$$\chi u=E*\Delta(\chi u)=4\sum\partial E/\partial z_j*\partial(\chi u)/\partial\bar z_j,$$

$$\partial(\chi u)/\partial\bar z_j=\chi\,\partial u/\partial\bar z_j+u\,\partial\chi/\partial\bar z_j.$$

Since $\partial\chi/\partial\bar z_j=0$ in $B_{\frac12}$ it follows at once that u is continuous there and that (15.1.18) is valid. The conclusion can be applied to a ball $B\subset B_r$ with center at any point in B_r, which completes the proof.

Proof of Theorem 15.1.6. If ϕ is plurisubharmonic and

$$0\leqq\chi\in C_0^\infty(\mathbb{C}^n), \quad \int\chi\,d\lambda=1, \quad \chi_\varepsilon(z)=\varepsilon^{-2n}\chi(z/\varepsilon),$$

then $\phi * \chi_\varepsilon(z)+\varepsilon|z|^2\to\phi$ in $L^1_{\text{loc}}(\mathbb{C}^n)$ when $\varepsilon\to 0$. It is therefore sufficient to prove that the closure of P_A contains all plurisubharmonic functions $\phi\in C^\infty(\mathbb{C}^n)$ such that the function κ defined by (15.1.1) has a positive lower bound. Let $z_1, z_2, \ldots$ be a dense sequence in $\mathbb{C}^n$. It is then sufficient to construct a sequence of analytic functions f_j and integers $N_j\to\infty$ such that

$$|f_j(z)| \leqq \exp(N_j\phi(z)) \quad \text{when } |z|<j; \tag{15.1.19}$$
$$|f_j(z_\nu)|>2^{-1}\exp(N_j\phi(z_\nu)) \quad \text{if } \nu\leqq j,$$

for Lemma 15.1.7 will then show that $N_j^{-1}\log|f_j|\to\phi$ in L^1_{loc}.

Set $\phi_k=\partial\phi/\partial z_k$, $\phi_{jk}=\partial^2\phi/\partial z_j\partial z_k$. Then Taylor's formula and the strict plurisubharmonicity of ϕ gives for fixed ν

$$\phi(z)-\operatorname{Re}P_\nu(z)\geqq\kappa(z_\nu)|z-z_\nu|^2-C_\nu|z-z_\nu|^3\geqq c_\nu|z-z_\nu|^2,$$

in a neighborhood U_ν of z_ν. Here

$$P_\nu(z)=\phi(z_\nu)+2\sum\phi_k(z_\nu)(z_k-z_{\nu k})+\sum\phi_{kl}(z_\nu)(z_k-z_{\nu k})(z_l-z_{\nu l}).$$

Choose $\chi_\nu\in C_0^\infty(U_\nu)$, $\nu=1,\ldots,j$, so that $\chi_\nu=1$ in a neighborhood of z_ν and the supports are disjoint. We set

$$f(z)=\sum_1^j\chi_\nu(z)\exp(NP_\nu(z))-v$$

where v shall be chosen so that f is analytic, that is,

$$\bar\partial v=\sum_1^j\bar\partial\chi_\nu\exp(NP_\nu(z))=g. \tag{15.1.20}$$

With positive constants c and C_r which may depend on j but not on N we have

$$|ge^{-N\phi}|\leqq C_1e^{-cN}, \quad \text{hence} \quad \|g\|_{2N\phi}\leqq C_2e^{-cN}.$$

By Theorem 15.1.1 we can therefore choose v so that (15.1.20) is valid and

$$\|v\|_{2N\phi}\leqq C_3e^{-cN}.$$

When $|z|<j$ we conclude using Lemma 15.1.8 that

$$|v(z)|\leqq C_4e^{-cN/2}e^{N\phi}.$$

We just have to take the radius r so small that the oscillation of ϕ in $B_r+\{z\}$ is less than $c/2$ if $|z|<j$. Choose N_j so that $C_4e^{-cN_j/2}<1/3$ and set $f_j=3f/4$ where f is defined as above with $N=N_j$. Then the condition (15.1.19) is fulfilled so the proof is complete.

15.2. The Fourier-Laplace Transform of $B^c_{2,k}(X)$ when X is Convex

Let X be an open set in $\mathbb{R}^n$ and let $k\in\mathscr{K}$ (Definition 10.1.1). We can then introduce in $B^c_{2,k}(X)=B_{2,k}\cap\mathscr{E}'(X)$ the *inductive limit topology* which makes the dual equal to $B^{\text{loc}}_{2,1/\check{k}}(X)$. This topology is defined by means of all semi-norms q in $B^c_{2,k}(X)$ such that for every compact set $K\subset X$ the restriction of q to $B_{2,k}\cap\mathscr{E}'(K)$ is bounded by a constant times $\|\ \|_{2,k}$. If $1=\sum\chi_j$ is a partition of unity in X, thus $\chi_j\in C_0^\infty(X)$ and no compact subset of X meets more than a finite number of the supports, we then obtain

$$q(u)\leqq\sum q(\chi_j u)\leqq\sum C_j\,\|\chi_j u\|_{2,k},\qquad u\in B^c_{2,k}(X)$$

for some constants C_j. On the other hand, when $\operatorname{supp} u\subset K$ we have

$$\sum C_j\,\|\chi_j u\|_{2,k}\leqq C_K\,\|u\|_{2,k}$$

by Theorem 10.1.15 for $K\cap\operatorname{supp}\chi_j=\emptyset$ except for finitely many j. Hence

$$u\to\sum C_j\,\|\chi_j u\|_{2,k}$$

is a continuous semi-norm in $B^c_{2,k}(X)$ for any sequence of positive C_j, and these semi-norms define the inductive limit topology in $B^c_{2,k}(X)$.

If L is a continuous linear form on $B^c_{2,k}(X)$ it follows that

$$|L(u)|\leqq\sum C_j\,\|\chi_j u\|_{2,k},\qquad u\in B^c_{2,k}(X),$$

for some positive C_j. By the Hahn-Banach theorem it follows that the map

$$\{\chi_j u\}\to L(u)$$

can be extended to a continuous linear form of norm at most 1 on the space of all sequences $f_j\in B_{2,k}(\mathbb{R}^n)$ with the norm $\sum C_j\,\|f_j\|_{2,k}<\infty$. By Theorem 10.1.14 this means that

$$L(u)=\sum\langle v_j,\chi_j u\rangle,\qquad u\in B^c_{2,k}(X),$$

where $v_j\in B_{2,1/\check{k}}$ and $\|v_j\|_{2,1/\check{k}}\leqq C_j$. Thus $L(u)=\langle v,u\rangle$ where

$$v=\sum\chi_j v_j\in B^{\text{loc}}_{2,1/\check{k}}(X).$$

Conversely, if $v\in B^{\text{loc}}_{2,1/\check{k}}(X)$ then $u\to|\langle v,u\rangle|$ is a continuous semi-norm on $B^c_{2,k}(X)$ by the definition of the topology, so $B^{\text{loc}}_{2,1/\check{k}}(X)$ is indeed the dual space of $B^c_{2,k}(X)$. (Note that the preceding arguments are essentially just a repetition of the discussion in Section 2.1.)

If X is an open convex set in $\mathbb{R}^n$ we can recognize the Fourier-Laplace transforms of the elements in $B_{2,k}\cap\mathscr{E}'(X)$ by means of the

Paley-Wiener-Schwartz theorem. The following main theorem of this section describes the topology in a way which is suitable for the application of the results in Section 15.1. (This is the reason why we only consider the spaces $B_{p,k}$ with $p=2$.)

Theorem 15.2.1. *Let X be an open convex set in $\mathbb{R}^n$ and let $k\in\mathscr{K}$. Every continuous semi-norm in $B^c_{2,k}(X)$ is then bounded by a semi-norm of the form*

$$u\to\left(\int|\hat{u}(\zeta)|^2 e^{-2\phi(\zeta)}\,d\lambda(\zeta)\right)^{\frac{1}{2}} \tag{15.2.1}$$

where ϕ has the following properties:

(i) *if K is a convex compact subset of X with supporting function H_K then*

$$e^{-\phi(\zeta)}\leqq C_K e^{-H_K(\operatorname{Im}\zeta)}k(\operatorname{Re}\zeta)$$

(ii) *for every $A>0$ there is a constant C_A such that*

$$k(\operatorname{Re}\zeta)\leqq C_A e^{-\phi(\zeta)} \quad \text{if } |\operatorname{Im}\zeta|<A.$$

(iii) *ϕ is locally Lipschitz continuous,*

$$|d\phi(\zeta)|\leqq C+\log(1+|\operatorname{Im}\zeta|)$$

(iv) *ϕ is plurisubharmonic; more precisely we have for every $w\in\mathbb{C}^n$*

$$c(1+|\operatorname{Im}\zeta|^2)^{-\frac{1}{2}}|w|^2\leqq\sum w_j\bar{w}_k\,\partial^2\phi(\zeta)/\partial\zeta_j\partial\bar{\zeta}_k$$

in the distribution sense. Here $c>0$.

The proof starts from the following expressions of the $B_{2,k}$ norms in terms of weighted L^2 norms of the Fourier-Laplace transform. We assume that $k\in\mathscr{K}$, and let N be the constant in (10.1.1)'. By K we denote a convex compact set in $\mathbb{R}^n$ with supporting function H_K.

Lemma 15.2.2. *For every $u\in\mathscr{E}'(K)\cap B_{2,k}$ we have*

$$\int|\hat{u}(\zeta)|^2 e^{-2H_K(\operatorname{Im}\zeta)}k(\operatorname{Re}\zeta)^2(1+|\operatorname{Im}\zeta|^2)^{-N-2n}\,d\lambda(\zeta)\leqq C_{K,k}\|u\|^2_{2,k}. \tag{15.2.2}$$

If $\psi\in C_0^\infty(K)$ we have for every $u\in\mathscr{E}'\cap B_{2,k}$ and $\eta\in\mathbb{R}^n$

$$\|\psi u\|^2_{2,k}\leqq C_{k,\psi}e^{2H_K(-\eta)}\int|\hat{u}(\xi+i\eta)|^2 k(\xi)^2\,d\xi. \tag{15.2.3}$$

Finally

$$\|u\|^2_{2,k}\leqq C_k\int_{|\operatorname{Im}\zeta|<1}|\hat{u}(\zeta)|^2 k(\operatorname{Re}\zeta)^2\,d\lambda(\zeta), \quad u\in\mathscr{E}'(\mathbb{R}^n)\cap B_{2,k}. \tag{15.2.4}$$

Proof. To prove (15.2.2) we choose as in the proof of Theorem 7.3.1 a function $\chi_\delta\in C_0^\infty$ equal to 1 in a neighborhood of K and 0 outside a δ

neighborhood of K, so that

$$|\chi_\delta^{(\alpha)}(x)| \leqq C_\alpha \delta^{-|\alpha|} \quad \text{for all } \alpha.$$

Since $\hat{u}(\xi+i\eta)$ for fixed η is the Fourier transform of $u e^{\langle \cdot, \eta \rangle} \chi_\delta$, we obtain using Theorem 10.1.15

$$(2\pi)^n \int |\hat{u}(\xi+i\eta)|^2 k(\xi)^2 \, d\xi \leqq \|u\|^2_{2,k} (\int |\hat{\chi}_\delta(\xi+i\eta)| (1+C|\xi|)^N d\xi)^2.$$

From the estimate

$$|\hat{\chi}_\delta(\xi+i\eta)| (1+\delta|\xi+i\eta|)^{N+n+1} \leqq C_K e^{H_K(\eta)+\delta|\eta|}$$

it follows that

$$\int |\hat{\chi}_\delta(\xi+i\eta)| (1+C|\xi|)^N d\xi \leqq C'_K e^{H_K(\eta)+\delta|\eta|} \delta^{-N-n}, \qquad \delta<1,$$

if we introduce $\delta\xi$ as a new integration variable. When $\delta = 1/(|\eta|+1)$ we obtain

$$\int |\hat{u}(\xi+i\eta)|^2 k(\xi)^2 \, d\xi \leqq C''_K \|u\|^2_{2,k} e^{2H_K(\eta)} (1+|\eta|)^{2(N+n)}$$

which gives (15.2.2).

If we write $\psi u = \psi e^{-\langle \cdot, \eta \rangle} e^{\langle \cdot, \eta \rangle} u$ we obtain by Theorem 10.1.15

$$(2\pi)^{3n} \|\psi u\|^2_{2,k} \leqq \int |\hat{u}(\xi+i\eta)|^2 k(\xi)^2 \, d\xi (\int |\hat{\psi}(\xi-i\eta)| (1+C|\xi|)^N d\xi)^2.$$

This gives (15.2.3) if $\hat{\psi}$ is estimated by means of Theorem 7.3.1.

Since $|\hat{u}(\zeta)|^2$ is subharmonic we have

$$\begin{aligned} \int |\hat{u}(\theta)|^2 k(\theta)^2 \, d\theta &\leqq \iint_{|\zeta|<1} |\hat{u}(\theta+\zeta)|^2 k(\theta)^2 \, d\theta \, d\lambda(\zeta) \Big/ \int_{|\zeta|<1} d\lambda(\zeta) \\ &\leqq C'_k \iint_{|\zeta|<1} |\hat{u}(\theta+\zeta)|^2 k(\theta+\operatorname{Re}\zeta)^2 \, d\theta \, d\lambda(\zeta) \\ &\leqq C_k \int_{|\operatorname{Im}\zeta|<1} |\hat{u}(\zeta)|^2 k(\operatorname{Re}\zeta)^2 \, d\lambda(\zeta). \end{aligned}$$

This proves (15.2.4) and completes the proof.

It follows from (15.2.2) and condition (i) in Theorem 15.2.1, applied to a neighborhood of K, that the semi-norm (15.2.1) can be estimated by $\|u\|_{2,k}$ when $u \in \mathscr{E}'(K) \cap B_{2,k}$. Hence (15.2.1) is a continuous semi-norm in $B^c_{2,k}(X)$ as soon as condition (i) is fulfilled. From (ii) and (15.2.4) it follows that (15.2.1) restricted to $\mathscr{E}'(K) \cap B_{2,k}$ is in fact equivalent to $\|u\|_{2,k}$. However, to prove the theorem we must also show how to replace integrals of the form (15.2.2) by integrals of the form (15.2.1) with a plurisubharmonic ϕ.

Let $0 \leqq \chi \in C_0^\infty$ have support in the unit ball, $\int \chi(\xi) \, d\xi = 1$. If $k \in \mathscr{K}$ we set with a large constant t and $M(\eta, t) = (t^2+|\eta|^2)^{\frac{1}{2}}$

(15.2.5) $$\log k_t(\xi, \eta) = \int \chi(\theta) \log k(\xi + M(\eta, t)\theta) \, d\theta.$$

From (10.1.1)′ it follows that

$$(15.2.6)\qquad (1+C|\xi'|)^{-N}\leqq k_t(\xi+\xi',\eta)/k_t(\xi,\eta)\leqq(1+C|\xi'|)^N,$$

and that

$$|\log k_t(\xi,\eta)-\log k(\xi)|\leqq N\int\chi(\theta)\log(1+CM(\eta,t)|\theta|)\,d\theta\leqq N'\log M(\eta,t).$$

Hence

$$(15.2.7)\qquad M(\eta,t)^{-N'}\leqq k_t(\xi,\eta)/k(\xi)\leqq M(\eta,t)^{N'}.$$

The advantage of the regularization is of course that $k_t\in C^\infty$ since

$$\log k_t(\xi,\eta)=\int\chi((\theta-\xi)/M(\eta,t))\log k(\theta)\,d\theta/M(\eta,t)^n.$$

By induction with respect to $|\beta|$ it follows that $D_\eta^\beta D_\xi^\alpha\log k_t(\xi,\eta)$ is a linear combination of terms of the form

$$M(\eta,t)^\mu\prod_1^J D^{\gamma_j}M(\eta,t)\int((\theta-\xi)/M(\eta,t))^\gamma\chi^{(\alpha+\gamma)}((\theta-\xi)/M(\eta,t))\log k(\theta)\,d\theta$$

where $\mu=\sum|\gamma_j|-n-|\alpha|-|\beta|-J$. Since $\int\theta^\gamma\chi^{(\alpha+\gamma)}(\theta)\,d\theta=0$ if $\alpha\neq 0$ we can then replace $\log k(\theta)$ by $\log k(\theta)-\log k(\xi)$. When $\alpha=0$ we do so before the differentiation, noting that $D_\eta^\beta\log k(\xi)=0$. We have

$$|D^\gamma M(\eta,t)|\leqq C_\gamma M(\eta,t)^{1-|\gamma|}$$

in view of the homogeneity. Since

$$|\log k(\theta)-\log k(\xi)|\leqq N\log(1+C|\xi-\theta|)$$

we obtain for every α by returning to the original integration variables

$$(15.2.8)\qquad |D_{\xi,\eta}^\alpha\log k_t(\xi,\eta)|\leqq C_\alpha M(\eta,t)^{-|\alpha|}\log M(\eta,t).$$

Using this estimate for $|\alpha|=2$ we shall now show that $-\log k_t$ can be made plurisubharmonic by adding a rather unimportant function of η.

If $f\in C^2(\mathbb{R})$ and $\eta=\operatorname{Im}\zeta$ then

$$\sum w_j\bar w_k\,\partial^2 f(|\eta|^2+t^2)/\partial\zeta_j\overline{\partial\zeta_k}=f'(|\eta|^2+t^2)|w|^2/2+f''(|\eta|^2+t^2)|\langle\eta,w\rangle|^2.$$

Here $0\leqq|\langle\eta,w\rangle|^2\leqq|\eta|^2|w|^2$. It follows that

$$-\log k_t(\operatorname{Re}\zeta,\operatorname{Im}\zeta)+f(|\operatorname{Im}\zeta|^2+t^2)$$

is plurisubharmonic if for a sufficiently large C and $s\geqq t^2$

$$f'(s)\geqq Cs^{-1}\log s,\qquad f'(s)+2f''(s)(s-t^2)\geqq Cs^{-1}\log s.$$

In fact, if $s=M(\eta,t)^2$ then $0\leqq|\langle\eta,w\rangle|^2\leqq(s-t^2)|w|^2$. The second condition follows from the first when $f''(s)>0$, so it may be replaced by

$$f'(s)+2sf''(s)\geqq Cs^{-1}\log s.$$

Set

$$f(s)=t^{-\frac{1}{2}}s^{\frac{1}{2}}-s^{\frac{1}{4}}.$$

Then

$$f'(s)+2sf''(s)=s^{-\frac{3}{4}}/8\geqq Cs^{-1}\log s+s^{-\frac{3}{4}}/9,$$

$$f'(s)=2^{-1}t^{-\frac{1}{2}}s^{-\frac{1}{2}}-4^{-1}s^{-\frac{3}{4}}\geqq 4^{-1}s^{-\frac{3}{4}}\geqq Cs^{-1}\log s+s^{-\frac{3}{4}}/9$$

if $s\geqq t^2$ and t is large. Hence we have proved

Lemma 15.2.3. *If $k\in\mathscr{K}$ and k_t is defined by* (15.2.5), *then* (15.2.6)–(15.2.8) *are valid and*

(15.2.9) $\quad \psi(\zeta)=-\log k_t(\operatorname{Re}\zeta, \operatorname{Im}\zeta)+t^{-\frac{1}{2}}M(\operatorname{Im}\zeta, t)-M(\operatorname{Im}\zeta, t)^{\frac{1}{4}}$

is strictly plurisubharmonic for large t; more precisely,

(15.2.10) $\quad \sum w_j\bar{w}_k\partial^2\psi(\zeta)/\partial\zeta_j\overline{\partial\zeta_k}\geqq|w|^2M(\operatorname{Im}\zeta, t)^{-\frac{3}{4}}/18.$

Proof of Theorem 15.2.1. Let q be a continuous semi-norm in $B^c_{2,k}(X)$. We must construct a function ϕ satisfying the conditions in the theorem so that

(15.2.11) $\quad q(u)\leqq(\int|\hat{u}(\zeta)|^2e^{-2\phi(\zeta)}d\lambda(\zeta))^{\frac{1}{2}},\qquad u\in B^c_{2,k}(X).$

To do so we first choose a sequence of convex compact subsets K_j of X with union X so that K_j is in the interior of K_{j+1}. Set

$$\psi_j(\zeta)=H_{K_j}(\operatorname{Im}\zeta)-\log k_{t_j}(\operatorname{Re}\zeta, \operatorname{Im}\zeta)+t_j^{-\frac{1}{2}}M(\operatorname{Im}\zeta, t_j)-M(\operatorname{Im}\zeta, t_j)^{\frac{1}{4}}$$

with t_j so large that Lemma 15.2.3 gives

(15.2.12) $\quad \sum w_j\bar{w}_k\partial^2\psi_j(\zeta)/\partial\zeta_j\partial\bar{\zeta}_k\geqq|w|^2M(\operatorname{Im}\zeta, t_j)^{-\frac{3}{4}}/18$

and $2t_j^{-\frac{1}{2}}$ is less than the distance from K_j to the complement of K_{j+1}. We shall prove that sequences G_j and A_j tending to $+\infty$ with j can be chosen so that if

$$\phi_j(\zeta)=\max_{1\leqq k\leqq j}(\psi_k(\zeta)-G_k)$$

we have $\phi_{j+1}(\zeta)=\phi_j(\zeta)$ when $|\operatorname{Im}\zeta|<A_j$ and

(15.2.13) $\quad q(u)\leqq j/(j+1)(\int_{|\operatorname{Im}\zeta|<A_j}|\hat{u}(\zeta)|^2e^{-2\phi_j(\zeta)}d\lambda(\zeta))^{\frac{1}{2}},$

$$u\in B_{2,k}\cap\mathscr{E}'(K_{j+3}).$$

The functions ϕ_j will satisfy (iii) and (iv) with constants independent of j. If $\phi=\lim_{j\to\infty}\phi_j$ we have $\phi(\zeta)=\phi_j(\zeta)$ when $|\operatorname{Im}\zeta|<A_j$. Hence (15.2.11) follows from (15.2.13). Condition (i) follows from the fact that $\phi(\zeta)\geqq\psi_j(\zeta)-G_j$ for every j, and (ii)–(iv) will obviously be fulfilled. The

theorem will therefore be proved if G_j and A_j can be chosen with the required properties.

If f and g are Lipschitz continuous functions then $\max(f,g)$ is Lipschitz continuous with the same Lipschitz constant. If f and g are functions in $\mathbb{C}$ with $\Delta f \geqq m$, $\Delta g \geqq m$ where m is a continuous function, then $\Delta \max(f,g) \geqq m$. In fact, if v is a solution of the equation $\Delta v = m$ then $f-v$ and $g-v$ are subharmonic, so $\max(f-v, g-v) = \max(f,g) - v$ is subharmonic (Corollary 16.1.5), which proves the statement.

Now choose C so that

$$|d\psi_1(\zeta)| \leqq C.$$

Since ψ_j is Lipschitz continuous by (15.2.8) we can then choose B_j so that

$$|d\psi_j(\zeta)| \leqq \log(1+|\operatorname{Im}\zeta|) \quad \text{if } |\operatorname{Im}\zeta| > B_j.$$

Similarly we have by (15.2.10) for suitable B_j and $c>0$

$$\sum w_l \bar{w}_k \partial^2 \psi_j(\zeta)/\partial\zeta_l \partial\bar{\zeta}_k \geqq c|w|^2 (1+|\operatorname{Im}\zeta|^2)^{-\frac{3}{2}} \quad \text{if } |\operatorname{Im}\zeta| > B_j.$$

We shall choose G_1 in a moment. If $G_j - G_1$ is sufficiently large we have

$$\psi_j(\zeta) - G_j < \phi_1(\zeta) \quad \text{if } |\operatorname{Im}\zeta| < B_j + 1. \tag{15.2.14}$$

Since $\phi_j(\zeta) = \phi_{j-1}(\zeta)$ when $|\operatorname{Im}\zeta| < B_j + 1$, and (iii), (iv) are satisfied by ψ_j when $|\operatorname{Im}\zeta| > B_j$, these conditions follow successively for ϕ_j in $\mathbb{C}^n$.

When $j=1$ the estimate (15.2.13) is valid if G_1 is large and $A_1 = 1$, for $q(u) \leqq C_1 \|u\|_{2,k}$ and the right-hand side of (15.2.13) has a lower bound $C_0 e^{G_1} \|u\|_{2,k}$ by (15.2.4). Assume now that (15.2.13) has already been proved for a certain j. Choose G_{j+1} satisfying (15.2.14) and

$$\psi_{j+1}(\zeta) - G_{j+1} < \phi_j(\zeta) \quad \text{when } |\operatorname{Im}\zeta| < A_j. \tag{15.2.14$'$}$$

We must then prove (15.2.13) with j replaced by $j+1$ if A_{j+1} is large enough. To do so we choose a partition of unity $1 = \sum_0^\mu \chi_\nu$ near K_{j+4} so that $\chi_0 \in C_0^\infty(K_{j+3})$ and $K_{j+2} \cap ch \operatorname{supp} \chi_\nu = \emptyset$ when $\nu \neq 0$. Then we have

$$\begin{aligned} q(u) &\leqq \sum q(\chi_\nu u) \leqq j/(j+1) \Big(\int_{|\operatorname{Im}\zeta| < A_j} |\widehat{\chi_0 u}(\zeta)|^2 e^{-2\phi_j(\zeta)} \, d\lambda(\zeta) \Big)^{\frac{1}{2}} \\ &\quad + C_j \sum_{\nu \neq 0} \|\chi_\nu u\|_{2,k} \\ &\leqq j/(j+1) \Big(\int_{|\operatorname{Im}\zeta| < A_j} |\hat{u}(\zeta)|^2 e^{-2\phi_j(\zeta)} \, d\lambda(\zeta) \Big)^{\frac{1}{2}} + C_j' \sum_{\nu \neq 0} \|\chi_\nu u\|_{2,k}. \end{aligned}$$

Here we have first applied (15.2.13) to $\chi_0 u$ and then used the triangle inequality and (15.2.2). When $|\operatorname{Im}\zeta| < A_j$ we have $\phi_j = \phi_{j+1}$. What remains to prove is therefore that

$$C'_j \sum_{\nu \neq 0} \|\chi_\nu u\|_{2,k} \leqq 1/((j+1)(j+2))\Big(\int_{|\mathrm{Im}\,\zeta|<A_{j+1}} |\hat{u}(\zeta)|^2 e^{-2\phi_{j+1}(\zeta)}\, d\lambda\Big)^{\frac{1}{2}} \tag{15.2.15}$$

if A_{j+1} is sufficiently large.

Recall that we have chosen t_{j+1} so large that

$$2t_{j+1}^{-\frac{1}{2}}|\mathrm{Im}\,\zeta| + H_{K_{j+1}}(\mathrm{Im}\,\zeta) \leqq H_{K_{j+2}}(\mathrm{Im}\,\zeta).$$

In view of (15.2.7) it follows that

$$\phi_{j+1}(\zeta) \leqq H_{K_{j+2}}(\mathrm{Im}\,\zeta) - \log k(\mathrm{Re}\,\zeta) + C''_j. \tag{15.2.16}$$

Now (15.2.3) gives, if h_ν is the supporting function of $ch\,\mathrm{supp}\,\chi_\nu$,

$$\begin{aligned}\|\chi_\nu u\|^2_{2,k} &\leqq C_k e^{2h_\nu(-\eta)} \int |\hat{u}(\xi+i\eta)|^2 k(\xi)^2\, d\xi \\ &\leqq C_k \exp 2(h_\nu(-\eta) + H_{K_{j+2}}(\eta) + C''_j) \int |\hat{u}(\xi+i\eta)|^2 e^{-2\phi_{j+1}(\xi+i\eta)}\, d\xi.\end{aligned}$$

Since $\mathrm{supp}\,\chi_\nu$ and K_{j+2} can be separated by a hyperplane we have for some θ

$$\inf_{x\in \mathrm{supp}\,\chi_\nu} \langle x, \theta\rangle > \sup_{x\in K_{j+2}} \langle x, \theta\rangle.$$

This means that $h_\nu(-\theta) + H_{K_{j+2}}(\theta) < 0$. Let $|\theta| = 1$. If we integrate the preceding estimate over all η with $|\eta - (A_{j+1}-1)\theta| < 1$, we conclude that for some $c>0$

$$\|\chi_\nu u\|^2_{2,k} \leqq C_k e^{-cA_{j+1}} \int_{|\mathrm{Im}\,\zeta|<A_{j+1}} |\hat{u}(\zeta)|^2 e^{-2\phi_{j+1}(\zeta)}\, d\lambda(\zeta).$$

When A_{j+1} is large enough the estimate (15.2.15) follows and the proof of the theorem is complete.

Theorem 15.2.1 can be restated as a representation theorem for $B^{\mathrm{loc}}_{2,1/\check{k}}(X)$. In fact, if $v \in B^{\mathrm{loc}}_{2,1/\check{k}}(X)$ then

$$B^c_{2,k}(X) \ni u \to \langle v, u\rangle$$

is a continuous linear form on $B^c_{2,k}(X)$. Thus

$$|\langle v, u\rangle| \leqq C\Big(\int |\hat{u}(\zeta)|^2 e^{-2\phi(\zeta)}\, d\lambda(\zeta)\Big)^{\frac{1}{2}}$$

for some ϕ satisfying the conditions in Theorem 15.2.1. We can therefore find a measurable function V with

$$\int |V(\zeta)|^2 e^{2\phi(-\zeta)}\, d\lambda(\zeta) \leqq C \tag{15.2.17}$$

such that

$$\langle v, u\rangle = \int V(-\zeta)\, \hat{u}(\zeta)\, d\lambda(\zeta), \qquad u \in B^c_{2,k}(X). \tag{15.2.18}$$

If we introduce the definition of the Fourier-Laplace transform of u we obtain formally

$$v(x)=\int V(\zeta)\, e^{i\langle x,\zeta\rangle}\, d\lambda(\zeta). \tag{15.2.18$'$}$$

When $k\in L^2$ it follows from condition (i) in Theorem 15.2.1 and (15.2.17) that the integral in (15.2.18)$'$ is absolutely convergent for every $x\in X$, and (15.2.18)$'$ follows then from (15.2.18) for every $x\in X$. Note that $k\in L^2$ is precisely the condition which guarantees that $B^{\mathrm{loc}}_{2,1/\check{k}}(X)\subset C(X)$. (See Theorem 10.1.25.) As is customary in distribution theory we shall sometimes use the suggestive notation (15.2.18)$'$ even when it has no pointwise sense so that the interpretation must be given by (15.2.18).

For later reference we shall now prove the converse statement in a somewhat more general form. (Compare with the discussion of (7.3.14).)

Theorem 15.2.4. *Let ϕ be a function satisfying the conditions* (i) *and* (iii) *in Theorem 15.2.1, and let μ be a measure in $\mathbb{C}^n$ such that*

$$\int_{|\theta-\zeta|<1} |d\mu(\theta)|< C, \qquad \zeta\in\mathbb{C}^n. \tag{15.2.19}$$

If V is μ measurable and

$$\int |V(\zeta)|^2\, e^{2\phi(-\zeta)}\,|d\mu(\zeta)|<\infty, \tag{15.2.20}$$

it follows that there is precisely one $v\in B^{\mathrm{loc}}_{2,1/\check{k}}(X)$ such that

$$\langle v, u\rangle=\int V(\zeta)\,\hat{u}(-\zeta)\, d\mu(\zeta), \qquad u\in B^c_{2,k}(X). \tag{15.2.21}$$

We write formally

$$v(x)=\int V(\zeta)\, e^{i\langle x,\zeta\rangle}\, d\mu(\zeta). \tag{15.2.21$'$}$$

Proof. Let $u\in B_{2,k}\cap\mathscr{E}'(K)$ where K is a fixed compact subset of X. The statement means that the right-hand side of (15.2.21) is a convergent integral which can be estimated by a constant times $\|u\|_{2,k}$, so it is sufficient to prove that

$$\int |\hat{u}(\zeta)|^2\, e^{-2\phi(\zeta)}\,|d\mu(-\zeta)|\leqq C\,\|u\|^2_{2,k}, \qquad u\in B_{2,k}\cap\mathscr{E}'(K).$$

Since $|\hat{u}|^2$ is subharmonic we have

$$|\hat{u}(\zeta)|^2\leqq C_1 \int_{|\theta|<1} |\hat{u}(\zeta+\theta)|^2\, d\lambda(\theta)$$

where $1/C_1$ is the volume of the unit ball in $\mathbb{C}^n$. Hence

$$\begin{aligned}&\int |\hat{u}(\zeta)|^2 e^{-2\phi(\zeta)} |d\mu(-\zeta)|\\ &\leqq C_1 \iint_{|\theta|<1} |\hat{u}(\zeta+\theta)|^2 e^{-2\phi(\zeta)} d\lambda(\theta)|d\mu(-\zeta)|\\ &\leqq C_2 \iint_{|\theta|<1} |\hat{u}(\zeta)|^2 e^{-2\phi(\zeta)}(1+|\operatorname{Im}\zeta|)^2 d\lambda(\theta)\,|d\mu(\theta-\zeta)|.\end{aligned}$$

Here we have used condition (iii) in Theorem 15.2.1. The measure in the integral is equal to $d\lambda(\zeta)\,|d\mu(\theta-\zeta)|$ which is perfectly clear when $d\mu$ has a smooth density and therefore by continuity in general. We can now estimate the integral with respect to θ by means of (15.2.19) and obtain

$$\begin{aligned}\int |\hat{u}(\zeta)|^2 e^{-2\phi(\zeta)} |d\mu(-\zeta)| &\leqq C_3 \int |\hat{u}(\zeta)|^2 e^{-2\phi(\zeta)}(1+|\operatorname{Im}\zeta|)^2 d\lambda(\zeta)\\ &\leqq C_4 \|u\|_{2,k}^2, \quad u\in B_{2,k}\cap\mathscr{E}'(K).\end{aligned}$$

Here the last estimate follows from (15.2.2) and condition (i) in Theorem 15.2.1, with K replaced by a compact neighborhood of K. The proof is complete.

15.3 Fourier-Laplace Representation of Solutions of Differential Equations

Solutions of ordinary differential equations with constant coefficients can be written as sums of exponential solutions. In this section we shall prove a similar fact for partial differential equations. However, we shall begin with a weaker result where all exponentials close to exponential solutions are allowed, for this permits a stronger statement from the point of view of regularity. By $P(D)$ we denote a partial differential operator with constant coefficients and by N the set of zeros of $P(\zeta)$ in $\mathbb{C}^n$,

$$N(r)=\{\zeta\in\mathbb{C}^n;\ |\zeta-\theta|<r \text{ for some } \theta\in N\}.$$

Theorem 15.3.1. *Let X be an open convex set in $\mathbb{R}^n$, $k\in\mathscr{K}$, and let $u\in B^{\mathrm{loc}}_{2,1/\tilde{k}}(X)$ be a solution of the differential equation $P(D)u=0$ in X. Then one can find a function ϕ satisfying all conditions in Theorem 15.2.1 and for any $r>0$ a measurable function U_r such that*

$$\int |U_r(\zeta)|^2 e^{2\phi(-\zeta)} d\lambda(\zeta)<\infty$$

and in the precise sense given in (15.2.18)

(15.3.1) $$u(x)=\int_{N(r)} U_r(\zeta)e^{i\langle x,\zeta\rangle}\,d\lambda(\zeta),\qquad x\in X.$$

Proof. A brief outline of the proof is as follows. Since $u\in B^{\mathrm{loc}}_{2,1/\tilde{k}}(X)$ we have

(15.3.2) $$|\langle u,v\rangle|^2\leqq C\int|\hat{v}(\zeta)|^2e^{-2\phi(\zeta)}\,d\lambda(\zeta),\qquad v\in B^c_{2,k}(X),$$

for some ϕ satisfying the conditions in Theorem 15.2.1. In addition

$$\langle u,P(-D)w\rangle=0,\qquad w\in B^c_{2,k_1}(X)$$

where $k_1(\xi)=k(\xi)\tilde{P}(-\xi)$. Thus $\langle u,v\rangle=0$ if $v\in B^c_{2,k}(X)$ and $\hat{v}(\zeta)/P(-\zeta)$ is entire analytic (Theorems 7.3.2 and 10.3.2). We shall write a general $v\in B^c_{2,k}(X)$ in the form

$$v=v_1+v_2$$

where $\hat{v}_2(\zeta)/P(-\zeta)$ is entire and v_1 is constructed from the restriction of $\hat{v}$ to $N(r)$. This will prove the theorem.

If we regularize the characteristic function of $N(3r/4)$ as in the proof of Theorem 1.4.1 we obtain a function $\chi\in C^\infty(\mathbb{C}^n)$ with $\chi=1$ in $N(r/2)$, $\operatorname{supp}\chi\subset N(r)$, and $D^\alpha\chi$ bounded for every α. Set $P(-\zeta)=\check{P}(\zeta)$ and

$$V_1=\check{\chi}\hat{v}-\check{P}w,\qquad V_2=(1-\check{\chi})\hat{v}+\check{P}w.$$

V_1 and $V_2/\check{P}$ will be entire analytic functions with $V_1+V_2=\hat{v}$ if

(15.3.3) $$\bar{\partial}w=f;\qquad f=\hat{v}\check{P}^{-1}\bar{\partial}\check{\chi}.$$

To be able to estimate $\langle u,v_1\rangle$, where v_1 is the inverse Fourier-Laplace transform of V_1, we can use (15.3.2) if we can find an estimate for

$$\int|V_1(\zeta)|^2e^{-2\phi(\zeta)}\,d\lambda(\zeta).$$

Thus we need an estimate of

$$\int|w(\zeta)|^2\tilde{P}(-\zeta)^2e^{-2\phi(\zeta)}\,d\lambda(\zeta).$$

This we shall obtain from Theorem 15.1.1 with a plurisubharmonic minorant of $2(\phi(\zeta)-\log\tilde{P}(-\zeta))$ instead of ϕ in the exponent.

In order to preserve the plurisubharmonicity of ϕ we must use a regularization of $\tilde{P}$. It could be constructed by means of (15.2.5) but we prefer to use the function $\tilde{P}(\xi,M)$ defined by (10.4.2). Thus we set

(15.3.4) $$\psi(\zeta)=\phi(\zeta)-\log\tilde{P}(-\zeta,M(\operatorname{Im}\zeta,t)),$$
$$M(\eta,t)=(t^2+|\eta|^2)^{\frac{1}{2}}$$

where t is a large positive number. Then

$$\psi(\zeta)+\log\tilde{P}(-\zeta)\leqq\phi(\zeta)\leqq\psi(\zeta)+\log\tilde{P}(-\zeta)+m\log M(\operatorname{Im}\zeta,t). \tag{15.3.5}$$

Now it follows from Taylor's formula that the oscillation of

$$2\log\tilde{P}(-\xi-i\eta,M(\eta,t))=\log\left(\sum P^{(\alpha)}(-\xi-i\eta)\bar{P}^{(\alpha)}(-\xi+i\eta)M(\eta,t)^{2|\alpha|}\right)$$

is uniformly bounded in a complex ball in $\mathbb{C}^{2n}$ with real center (ξ,η) and radius $cM(\eta,t)$ where c is a small positive number depending only on n and m. Hence the derivatives are $O(M(\eta,t)^{-|\alpha|})$. If t is large enough it follows from condition (iv) in Theorem 15.2.1 that

$$(c/2)(1+|\operatorname{Im}\zeta|^2)^{-\frac{3}{4}}|w|^2\leqq\sum w_j\bar{w}_k\partial^2\psi(\zeta)/\partial\zeta_j\partial\bar{\zeta}_k.$$

Here it is important that $3/2<2$.

If ψ were in C^2 it would follow at once from Theorem 15.1.1 that the equation (15.3.3) has a solution w with

$$c/2\int|w(\zeta)|^2e^{-2\psi(\zeta)}d\lambda(\zeta)\leqq\int|f(\zeta)|^2e^{-2\psi(\zeta)}(1+|\operatorname{Im}\zeta|^2)^{\frac{3}{4}}d\lambda(\zeta),$$

provided that the right-hand side is finite. This is also true if ψ is not in C^2; we just have to replace ψ by a regularization $\psi_\varepsilon\geqq\psi$ and let $\varepsilon\to 0$ as in the proof of Theorem 15.1.2, noting that the regularization of $(1+|\operatorname{Im}\zeta|^2)^{-\frac{3}{4}}$ is $\geqq(1+|\operatorname{Im}\zeta|^2)^{-\frac{3}{4}}(1-O(\varepsilon))$.

By Lemma 11.1.4 we have $\tilde{P}(\zeta)\leqq C_r|P(\zeta)|$ if $\zeta\notin N(r/2)$. In view of (15.3.5) it follows that

$$\int|V_1(\zeta)|^2e^{-2\phi(\zeta)}d\lambda(\zeta)\leqq C\int_{-N(r)}|\hat{v}(\zeta)|^2e^{-2\phi(\zeta)}M(\operatorname{Im}\zeta,t)^{2m+2}d\lambda(\zeta). \tag{15.3.6}$$

With a change of the constants it is clear that

$$\tilde{\phi}(\zeta)=\phi(\zeta)-(m+1)\log M(\operatorname{Im}\zeta,t)$$

satisfies the conditions in Theorem 15.2.1 if t is large enough. If V_1 were the Fourier-Laplace transform of a distribution $v_1\in B^c_{2,k}(X)$ we would now obtain from the arguments at the beginning of the proof

$$|\langle u,v\rangle|^2=|\langle u,v_1\rangle|^2\leqq C\int_{-N(r)}|\hat{v}(\zeta)|^2e^{-2\tilde{\phi}(\zeta)}d\lambda(\zeta), \tag{15.3.7}$$

and this would prove the statement with ϕ replaced by $\tilde{\phi}$.

To remove the unjustified assumption on V_1 we recall from the proof of Theorem 15.2.1 that we may assume that ϕ is the increasing limit of a sequence of functions ϕ_j such that

a) conditions (iii) and (iv) in Theorem 15.2.1 are satisfied uniformly,

b) condition (i) is fulfilled by ϕ_j for every convex compact set K in X when j exceeds some integer depending on K,

c) for fixed j there is a compact subset K of X and some $c>0$ such that

$$\exp(-\phi_j(\zeta)) \geqq c k(\operatorname{Re}\zeta) \exp(-H_K(\operatorname{Im}\zeta)).$$

For given $v \in B^c_{2,k}(X)$ we can apply the constructions above with ϕ replaced by ϕ_j when j is sufficiently large. This gives (15.3.6) with ϕ replaced by ϕ_j; the choice of V_1 may of course depend on j. By the Paley-Wiener-Schwartz theorem and c) we conclude that V_1 is the Fourier-Laplace transform of a distribution in $B^c_{2,k}(X)$. Hence (15.3.7) is valid with ϕ replaced by ϕ_j and a constant C independent of j. Letting $j \to \infty$ we obtain

$$|\langle u, v\rangle|^2 \leqq C \int_{-N(r)} |\hat{v}(\zeta)|^2 e^{-2\tilde{\Phi}(\zeta)} d\lambda(\zeta)$$

which completes the proof of the theorem.

The following corollary gives a typical example of the regularity theorems which can be derived from Theorem 15.3.1.

Corollary 15.3.2. *Let $k, k_1 \in \mathscr{K}$ and assume that for some C and N*

$$k(\operatorname{Re}\zeta) \leqq C k_1(\operatorname{Re}\zeta)(1+|\operatorname{Im}\zeta|)^N, \quad \textit{if } P(\zeta)=0. \tag{15.3.8}$$

If X is an open set in $\mathbb{R}^n$ and $u \in B^{\mathrm{loc}}_{2,k_1}(X)$ satisfies the equation $P(D)u=0$ it follows then that $u \in B^{\mathrm{loc}}_{2,k}(X)$.

Proof. (15.3.8) remains valid with another constant C when ζ is in the set $N(1)$ of points at distance at most 1 from the zeros of P. Assuming as we may that X is convex we use Theorem 15.3.1 to write u in the form

$$u(x) = \int_{N(1)} U(\zeta) e^{i\langle x,\zeta\rangle} d\lambda(\zeta)$$

where

$$\int_{N(1)} |U(\zeta)|^2 e^{2\phi(-\zeta)} d\lambda(\zeta) < \infty.$$

Here ϕ satisfies condition (i) in Theorem 15.2.1 with k replaced by $1/\check{k}_1$. By (15.3.8) it follows that

$$\int_{N(1)} |U(\zeta)|^2 e^{2\tilde{\Phi}(-\zeta)} d\lambda(\zeta) < \infty$$

where

$$e^{\tilde{\Phi}(-\zeta)} = e^{\phi(-\zeta)}(1+|\operatorname{Im}\zeta|)^{-N} k(\operatorname{Re}\zeta)/k_1(\operatorname{Re}\zeta).$$

By condition (i) of Theorem 15.2.1 with k replaced by $1/\check{k}_1$

$$\begin{aligned} 1 &\leqq C_K e^{\phi(\zeta) - H_K(\operatorname{Im}\zeta)}/k_1(-\operatorname{Re}\zeta) \\ &= C_K e^{\tilde{\Phi}(\zeta) - H_K(\operatorname{Im}\zeta)}(1+|\operatorname{Im}\zeta|)^N/k(-\operatorname{Re}\zeta), \end{aligned}$$

for an arbitrary compact set $K \subset X$. Hence condition (i) in Theorem 15.2.1 is satisfied by $\tilde{\phi}$ with k replaced by $1/\check{k}$. Condition (iii) is obvious for $\tilde{\phi}$ since $|k'/k|$ is bounded for every $k \in \mathscr{K}$. By Theorem 15.2.4 it follows that $u \in B_{2,k}^{\mathrm{loc}}(X)$, which completes the proof.

We leave as an exercise for the reader to verify that (15.3.8) is equivalent to (11.1.7). Thus Corollary 15.3.2 is essentially equivalent to Theorem 11.1.7 with $p=2$. It is also an easy exercise to prove Theorem 11.4.1 using Theorem 15.3.1 and Theorem 11.4.8.

We shall now prove that the decomposition (15.3.1) of solutions of the equation $P(D)u=0$ can be improved so that the integral runs only over the zeros of $P(\zeta)$ and the integrand consists of solutions of the equation. However, we must then admit exponential polynomial solutions and will in general have some loss of regularity in the decomposition.

We write P as a product of relatively prime polynomials which are powers of irreducible polynomials

$$P(\zeta)=P_1(\zeta)^{m_1} \dots P_r(\zeta)^{m_r}$$

and denote by N_j the set of zeros of P_j in $\mathbb{C}^n$. From Theorems 4.1.15 and 4.1.12 we obtain the estimate

$$\int_{|\theta-\zeta|<r} dS_j(\theta) \leqq r^{2n-2} C_{2n-2} \deg P_j \tag{15.3.9}$$

for the surface measure dS_j on N_j. In particular it satisfies (15.2.19). As in the proof that one can approximate by exponential solutions (Theorem 7.3.6 and Lemma 7.3.7) we shall only use polynomials in one variable $\langle x, \tau \rangle$ where $\tau \in \mathbb{C}^n$ is non-characteristic with respect to P.

Theorem 15.3.3. *Let X be an open convex set in $\mathbb{R}^n$, $k \in \mathscr{K}$, and let $u \in B_{2,1/\check{k}}^{\mathrm{loc}}(X)$ be a solution of the differential equation $P(D)u=0$ in X. Then one can find a function ϕ satisfying all conditions in Theorem 15.2.1, and dS_j measurable functions U_k^j on N_j, $0 \leqq k < m_j$, such that*

$$\int |U_k^j(\zeta)|^2 e^{2\phi(-\zeta)}(1+|\zeta|^2)^{-\kappa} dS_j(\zeta) < \infty \tag{15.3.10}$$

and in the sense of (15.2.21)′

$$u(x)=\sum_1^r \sum_{k=0}^{m_j-1} \langle x, \tau \rangle^k \int_{N_j} U_k^j(\zeta) e^{i\langle x, \zeta \rangle} dS_j(\zeta). \tag{15.3.11}$$

Here κ depends only on the dimension n and the degree of P.

Roughly speaking the result means that the decomposition in exponential solutions succeeds with a loss of κ derivatives. This result is not very precise but some loss cannot always be avoided since at infinity the polynomials P_j may be close to having multiple or common factors.

The proof of Theorem 15.5.3 is to a large extent parallel to that of Theorem 15.3.1. Thus we consider the form

$$B^c_{2,k}(X)\ni v \to \langle u, v\rangle$$

and wish to show that $\langle u, v\rangle$ can be estimated using only the restriction to $-N_j$ of $\langle D, \tau\rangle^k \hat{v}$ when $k<m_j$, $j=1, \dots, r$. To do so we must first modify $\hat{v}$ locally by means of an interpolation formula; we can take $\tau=(1, 0, \dots, 0)$.

Lemma 15.3.4. *Let $t_1, \dots, t_\nu$ be complex numbers in the unit disc with $\delta = \prod_{i\neq j} |t_i - t_j| \neq 0$, and let $\mu_1, \dots, \mu_\nu$ be positive integers with sum μ and maximum $\bar{\mu}$. If q is a polynomial of degree $\mu-1$ then*

$$\delta^{2\bar{\mu}-1} \sup_{|t|<1} |q(t)| \leqq C_\mu \sum_{j=1}^{\nu} \sum_{k<\mu_j} |q^{(k)}(t_j)|. \tag{15.3.12}$$

Proof. Using an expansion in partial fractions we can write

$$q(t)=\sum q_j(t)p_j(t), \quad p_j(t)=\prod_{k\neq j} (t-t_k)^{\mu_k},$$

where $q_j(t)$ is a polynomial of degree $<\mu_j$. Taylor expansion of q/p_j gives

$$q_j(t)=\sum_{k<\mu_j} \partial^k q(t_j)/k! \sum_{l<\mu_j-k} \partial^l(1/p_j)(t_j)(t-t_j)^{k+l}/l!.$$

Here $\partial^l(1/p_j)(t_j)$ is a sum of terms of the form $\prod_{i\neq j}(t_i-t_j)^{-d_i}$ where $d_i \leqq \mu_i + l \leqq 2\bar{\mu}-1$. Hence there is a bound for the product by $\delta^{2\bar{\mu}-1}$, which proves the lemma.

Lemma 15.3.5. *Let $Q_1, \dots, Q_r$ be polynomials of fixed degree in $\mathbb{C}^n$ and principal part equal to 1 at $(1, 0, \dots, 0)$. Let $m_1, \dots, m_r$ be fixed positive integers and denote by m the degree of $Q=Q_1^{m_1} \dots Q_r^{m_r}$ while $\bar{m} = \max m_j$. Assume that with a fixed constant M*

$$\tilde{Q}(0) \leqq M \sup_{|\zeta_1|<1} |Q(\zeta_1, 0)|,$$

and that the discriminant $R(\zeta')$ of $\prod Q_j(\zeta)$ as a polynomial in ζ_1 is not identically 0. Then there are constants c, C, κ depending only on n, m, M such that every analytic function f in $\{\zeta; |\zeta|<1\}$ can be written

$$f(\zeta)=Q(\zeta)g(\zeta)+h(\zeta), \quad |\zeta|<c,$$

where g and h are analytic when $|\zeta|<c$, h is a polynomial in ζ_1 of degree $<m$, and

$$\tilde{R}(0)^{\bar{m}-\frac{1}{2}} \sup_{|\zeta|<c} |h(\zeta)| \leqq C\tilde{Q}(0)^{\kappa} \sum_{j=1}^{r} \sum_{k<m_j} \int_{|\zeta|<1} |\partial^k f/\partial\zeta_1^k| dS_j. \tag{15.3.13}$$

Here dS_j is the surface measure in $Q_j^{-1}(0)$.

Proof. We shall essentially just have to elaborate the proofs of Theorems 7.5.1 and 7.5.2. Since $Q(\zeta_1, 0)/\tilde{Q}(0)$ belongs to a compact set of non-zero polynomials of fixed degree m we can choose ρ_1, depending on Q, with $1/3<\rho_1<2/3$, so that for some constant C independent of Q

$$\tilde{Q}(0) \leqq C|Q(\zeta_1, 0)|, \quad |\zeta_1|=\rho_1.$$

We can then choose $\rho'<1/3$ so that

$$\tilde{Q}(0) \leqq 2C|Q(\zeta_1, \zeta')| \quad \text{if } ||\zeta_1|-\rho_1|<\rho' \text{ and } |\zeta'|<\rho'. \tag{15.3.14}$$

This implies that the equation $Q_j(\zeta_1, \zeta')=0$ has a fixed number of roots $\zeta_1=t_\nu(\zeta')$ with $|\zeta_1|<\rho_1$ if $|\zeta'|<\rho'$. We set

$$q_j(\zeta_1, \zeta')=\prod(\zeta_1-t_\nu(\zeta')), \quad |\zeta'|<\rho',$$

with the product taken over these roots, and we set $q=\prod q_k^{m_k}$. Since $|\partial Q_j/\partial\zeta_1|/|Q_j| \leqq (\deg Q_j)/\rho'$ when $|\zeta_1|=\rho_1$, it follows from the proof of Theorem 7.5.1 that there are uniform bounds for the coefficients of q_j when $|\zeta'|<\rho'$, and that they are analytic functions of ζ'.

As in the proof of Theorem 7.5.2 we now obtain the required decomposition of f with

$$h(\zeta)=(2\pi i)^{-1} \int_{|z|=\rho_1} f(z, \zeta')(q(z, \zeta')-q(\zeta_1, \zeta'))/((z-\zeta_1)q(z, \zeta'))dz.$$

This is a polynomial in ζ_1 of degree less than the degree μ of q. The discriminant $R(\zeta')$ is the product of the discriminant of $\prod q_j(\zeta)$ and certain squares of differences of zeros of $Q(\zeta_1, \zeta')$. These can all be estimated by $C\tilde{Q}(0)$ since the coefficient of ζ_1^m in Q is 1 and the other coefficients can be estimated by $C\tilde{Q}(0)$. Hence the hypothesis of Lemma 15.3.4 is fulfilled if $R(\zeta') \neq 0$, and

$$|R(\zeta')| \leqq C\delta^2 \tilde{Q}(0)^{2\kappa/(2\bar{m}-1)}$$

for some κ. It follows that

$$\sup_{|\zeta_1|<\rho_1} |h(\zeta)| |R(\zeta')|^{\bar{m}-\frac{1}{2}} \leqq C\tilde{Q}(0)^{\kappa} \sum_{j=1}^{r} \sum_{k<m_j} \sum_{q_j(\zeta)=0} |\partial^k f(\zeta)/\partial\zeta_1^k|.$$

We integrate this estimate for $|\zeta'|<\rho'$ recalling that $d\lambda(\zeta') \leqq dS_j$ (see the proof of Theorem 4.1.12). In view of Lemma 7.3.12 it follows that (15.3.13) holds.

Proof of Theorem 15.3.3. Since $u \in B^{\text{loc}}_{2,1/\tilde{k}}(X)$ we know that for some ϕ satisfying the conditions in Theorem 15.2.1 we have

(15.3.15) $$|\langle u, v\rangle|^2 \leqq C \int |\hat{v}(\zeta)|^2 e^{-2\phi(\zeta)} d\lambda(\zeta), \qquad v \in B^c_{2,k}(X).$$

Our problem is to write v as a sum $v_1 + v_2$ where $\hat{v}_2$ is divisible by $\check{P}(\zeta) = P(-\zeta)$ and v_1 can be estimated in terms of $\partial^k \hat{v}/\partial \zeta_1^k$ restricted to $\check{N}_j$ when $k < m_j$.

For any $\theta \in \mathbb{C}^n$ we can apply Lemma 15.3.5 to

$$Q(\zeta) = \check{P}(\theta_1 + \zeta_1, \theta' + (1+|\theta|)^{1-m} \zeta')$$

where m is the degree of P. In fact, the coefficient of ζ_1^m in $Q(\zeta)$ is equal to that in $\check{P}(\zeta)$ which may be assumed equal to 1. Hence

$$\sup_{|\zeta_1|<1} |Q(\zeta_1, 0)| \geqq \max |\partial_1^j Q(0)|/j! \geqq 1$$

by Cauchy's inequalities. On the other hand, $Q^{(\alpha)}(0)$ has a uniform bound if $\alpha' \neq 0$ for differentiation gives some factor $(1+|\theta|)^{1-m}$ then. We define Q_j in the same way by means of P_j. The discriminant R of $\prod \check{P}_j$ is not identically 0 since P_j are irreducible and relatively prime, so some coefficient of the discriminant $R(\theta' + (1+|\theta|)^{1-m}\zeta')$ of $\prod Q_j$ can be bounded from below by a power of $(1+|\theta|)^{-1}$. To return to the original variables we set

$$B_r(\theta) = \{\zeta;\ |\zeta_1 - \theta_1|^2 + |\zeta' - \theta'|^2 (1+|\theta|)^{2m-2} < r^2\}.$$

Lemma 15.3.5 then gives for every θ a decomposition

(15.3.16) $$\hat{v}(\zeta) = P(-\zeta) g_\theta(\zeta) + h_\theta(\zeta), \qquad \zeta \in B_c(\theta),$$

and for some new constants κ and C the estimate

(15.3.17) $$\sup_{B_c(\theta)} |h_\theta| \leqq C(1+|\theta|)^\kappa \sum_{j=1}^{r} \sum_{k<m_j} \int_{\check{N}_j \cap B_1(\theta)} |\partial^k \hat{v}/\partial \zeta_1^k| \, dS_j.$$

Note that $h_\theta = 0$ if $\check{N}_j \cap B_1(\theta) = \emptyset$ for every j.

Using Lemma 1.4.9 and Theorem 1.4.10 we can choose a partition of unity $1 = \sum \chi_\nu(\zeta)$ where χ_ν has support in $B_{c/2}(\theta_\nu)$, at most a fixed number of balls $B_1(\theta_\nu)$ can overlap and

$$|\partial \chi_\nu / \partial \zeta_1| + |\partial \chi_\nu / \partial \zeta'| (1+|\zeta|)^{1-m} \leqq C.$$

Set

$$V_1 = \sum \chi_\nu h_{\theta_\nu} - \check{P} w$$

where w will be chosen so that V_1 is analytic, that is,

(15.3.18) $$\bar{\partial} w = f; \quad f = \sum (h_{\theta_\nu}/\check{P}) \bar{\partial} \chi_\nu = \sum \chi_\mu ((h_{\theta_\nu} - h_{\theta_\mu})/\check{P}) \bar{\partial} \chi_\nu.$$

Here we have used that $\sum \bar{\partial}\chi_\nu = 0$ and that $\sum \chi_\mu = 1$. By (15.3.16) we know that

$$(h_{\theta_\nu} - h_{\theta_\mu})/\check{P} = g_{\theta_\mu} - g_{\theta_\nu}$$

is analytic in $B_c(\theta_\nu) \cap B_c(\theta_\mu)$. A bound for $h_{\theta_\nu} - h_{\theta_\mu}$ in this set is provided by (15.3.17). By Lemma 7.3.12 division by $P(-\zeta)$ does not increase the bound by more than a constant factor in $B_{c/2}(\theta_\nu) \cap B_{c/2}(\theta_\mu)$. Hence

$$\int |\chi_\mu(h_{\theta_\nu} - h_{\theta_\mu})/\check{P}|^2 |\bar{\partial}\chi_\nu|^2 d\lambda(\zeta) \leqq M_\nu + M_\mu, \tag{15.3.19}$$

$$M_\nu = C(1+|\theta_\nu|^2)^\kappa \sum_{j=1}^{r} \sum_{k<m_j} \int_{\check{N}_j \cap B_1(\theta_\nu)} |\partial^k \hat{v}/\partial \zeta_1^k|^2 dS_j$$

by the Cauchy-Schwarz inequality for we have a fixed bound for the mass of dS_j in $\check{N}_j \cap B_1(\theta)$. (The constants may be changed of course.) Similarly

$$\int |\chi_\nu h_{\theta_\nu}|^2 d\lambda(\zeta) \leqq M_\nu. \tag{15.3.20}$$

From the proof of Theorem 15.3.1 we know that (15.3.18) has a solution w with

$$\begin{aligned}\int |w(\zeta)|^2 \tilde{P}(-\zeta)^2 e^{-2\phi(\zeta)} d\lambda(\zeta) \\ \leqq C \int |f(\zeta)|^2 e^{-2\phi(\zeta)} \tilde{P}(-\zeta)^2 (1+|\operatorname{Im}\zeta|^2)^{m+1} d\lambda(\zeta),\end{aligned} \tag{15.3.21}$$

provided that the right-hand side is finite. To estimate it we use that

$$|f|^2 \leqq N^2 \sum |\chi_\mu((h_{\theta_\nu} - h_{\theta_\mu})/\check{P})\bar{\partial}\chi_\nu|^2$$

if N is an upper bound for the number of $B_1(\theta_j)$ which can have a point in common. It follows from (iii) in Theorem 15.2.1 that we can insert a factor $\tilde{P}(-\zeta)^2 e^{-2\phi(\zeta)}$ in the integrals in (15.3.19) if κ is increased by 2. For every μ (resp. ν) the left-hand side of (15.3.19) vanishes except for at most N values of ν (resp. μ). Hence the right-hand side of (15.3.21) can be estimated by

$$C \sum_{j=1}^{r} \sum_{k<m_j} \int_{\check{N}_j} |\partial^k \hat{v}/\partial \zeta_1^k|^2 e^{-2\phi(\zeta)} \tilde{P}(-\zeta)^2 (1+|\zeta|^2)^\kappa dS_j(\zeta)$$

for some new constants C and κ. By (15.3.20) we have a similar estimate for $\sum \chi_\nu h_{\theta_\nu}$. Hence

$$\int |V_1(\zeta)|^2 e^{-2\phi(\zeta)} d\lambda(\zeta) \leqq CG, \tag{15.3.22}$$

$$G = \sum_{j=1}^{r} \sum_{k<m_j} \int_{\check{N}_j} |\partial^k \hat{v}/\partial \zeta_1^k|^2 e^{-2\phi(\zeta)} (1+|\zeta|^2)^\kappa dS_j(\zeta).$$

We have now obtained a decomposition $\hat{v} = V_1 + V_2$ where

$$V_2/\check{P} = (\hat{v} - V_1)/\check{P} = \sum \chi_\nu g_{\theta_\nu} + w$$

is a locally square integrable function, hence analytic. The end of the proof of Theorem 15.3.1 can therefore be applied to pass from (15.3.15) and (15.3.22) to the estimate

$$|\langle u, v\rangle|^2 \leqq C'G. \tag{15.3.23}$$

Hence

$$\langle u, v\rangle = \sum_{j=1}^{r} \sum_{k<m_j} \int_{\tilde{N}_j} U_k^j(\zeta)\partial^k \hat{v}(\zeta)/\partial\zeta_1^k \, dS_j(\zeta), \qquad v \in C_0^\infty(X),$$

where

$$\int_{\tilde{N}_j} |U_k^j(\zeta)|^2 e^{2\phi(\zeta)}(1+|\zeta|^2)^{-\kappa} dS_j(\zeta) < \infty.$$

Here $\partial^k \hat{v}(\zeta)/\partial\zeta_1^k$ is the Fourier-Laplace transform of $(-ix_1)^k v$, so we have in the sense of (15.2.21)′

$$u = \sum_{j=1}^{r} \sum_{k<m_j} (-ix_1)^k \int_{N_j} e^{i\langle x,\zeta\rangle} U_k^j(-\zeta) dS_j(-\zeta)$$

which apart from differences in notation is Theorem 15.3.3.

15.4. The Fourier-Laplace Transform of $C_0^\infty(X)$ when X is Convex

Let X be an open set in $\mathbb{R}^n$. As in Section 15.2 the inductive limit topology in $C_0^\infty(X)$ is defined by the semi-norms q in $C_0^\infty(X)$ which have continuous restriction to the Fréchet space $C_0^\infty(K)$ for every compact set $K \subset X$. Equivalently, a *convex* set $V \subset C_0^\infty(X)$ is a neighborhood of 0 if and only if $V \cap C_0^\infty(K)$ is a neighborhood of 0 in $C_0^\infty(K)$ for every compact set $K \subset X$. The following theorem is just a slight addition to Theorem 2.1.5.

Theorem 15.4.1. *Let $K_1 \subset K_2 \subset \dots$ be compact subsets of X such that each K_j is in the interior of K_{j+1} and $\bigcup K_j = X$. Then the following conditions on a convex subset V of $C_0^\infty(X)$ are all equivalent:*

(i) *V is a neighborhood of* 0.

(ii) *There exist integers N_j and numbers $\varepsilon_j > 0$ such that V contains any finite sum $\sum u_j$ with $u_j \in C_0^\infty(K_j)$ and $\sup|D^\alpha u_j| < \varepsilon_j$, $|\alpha| \leqq N_j$.*

(iii) *There exist functions $\rho_\alpha \in C^0(X)$ such that each compact set in X meets only finitely many* $\operatorname{supp}\rho_\alpha$, *and V contains every $u \in C_0^\infty(X)$ satisfying the estimate*

$$\sum_\alpha \sup|\rho_\alpha D^\alpha u| < 1. \tag{15.4.1}$$

(iv) *There exist integers* N_j *and numbers* $\varepsilon_j>0$ *such that* V *contains all* $u\in C_0^\infty(X)$ *satisfying*

$$|D^\alpha u|\leqq \varepsilon_j \quad \text{in } \complement K_j \text{ for } |\alpha|\leqq N_j,\ j=0,1,\ldots. \tag{15.4.2}$$

Here K_0 *denotes the empty set.*

Proof. (i) $\Rightarrow$ (ii). By hypothesis N_j and ε_j can be chosen so that V contains every $u\in C_0^\infty(K_j)$ with $\sup|D^\alpha u|<2^j\varepsilon_j$ when $|\alpha|\leqq N_j$. Hence V contains $2^j u_j$ if u_j has the properties stated in (ii), and this implies

$$u_j=\sum 2^{-j}2^j u_j\in V,$$

since $\sum 2^{-j}<1$ and $0\in V$.

(ii) $\Rightarrow$ (iii). Choose $\chi_j\in C_0^\infty(K_j)$ so that $\chi_j=1$ in K_{j-1}, and set $\psi_1=\chi_1$, $\psi_j=\chi_j-\chi_{j-1}$ when $j>1$. Then $u=\sum u_j$ if $u_j=\psi_j u$, and the sum is finite. We have $|D^\alpha u_j|<\varepsilon_j$ when $|\alpha|\leqq N_j$ if

$$2^{N_j}|\partial^\beta\psi_j||\partial^\alpha u|<\varepsilon_j, \qquad |\alpha+\beta|\leqq N_j.$$

Set

$$\rho_\alpha(x)=\sum_{|\alpha+\beta|\leqq N_j} 2^{N_j}|\partial^\beta\psi_j(x)|/\varepsilon_j.$$

On any compact set the sum is finite and vanishes for large $|\alpha|$. It follows from (15.4.1) that u_j satisfies the conditions in (ii).

(iii) $\Rightarrow$ (iv). Define N_j so that $|\alpha|\leqq N_j$ when $\operatorname{supp}\rho_\alpha\cap K_{j+1}\neq\emptyset$. Then it follows from (15.4.2) that $|\rho_\alpha D^\alpha u|<2^{-n-|\alpha|}$ in $K_{j+1}\setminus K_j$ if ε_j is chosen small enough. Hence (15.4.1) is valid in $K_{j+1}\setminus K_j$ for every j and therefore in X.

(iv) $\Rightarrow$ (i). This is clear since the set of all $u\in C_0^\infty(K)$ satisfying (15.4.2) is only restricted by bounds on a finite number of derivatives.

In the convex case we shall now give a description of the neighborhoods in terms of Fourier transforms. This supplements topologically the characterization of the Fourier transform given in Theorem 7.3.1.

Theorem 15.4.2. *Let* X *be convex and choose* K_j *as in Theorem 15.4.1 and in addition convex. Then the conditions* (i)–(iv) *are equivalent to*

(v) *There exist positive numbers* δ_j *and* M_j *such that* V *contains all* $u\in C_0^\infty(X)$ *such that*

$$|\hat{u}(\zeta)|\leqq\sum \delta_j(1+|\zeta|)^{-M_j}\exp H_{K_j}(\operatorname{Im}\zeta). \tag{15.4.3}$$

Proof. (v) $\Rightarrow$ (i). The derivation of (7.3.3) from (7.3.1) shows that (15.4.3) is fulfilled with just one term in the right-hand side if $u\in C_0^\infty(K_j)$ and $\|D^\alpha u\|_{L^1}$ is small enough when $|\alpha|\leqq M_j$. Thus the intersection of $C_0^\infty(K_j)$ and the set defined by (15.4.3) is a neighborhood of 0 in $C_0^\infty(K_j)$.

(iv) ⇒ (v). We must show that (15.4.2) follows from (15.4.3) for suitable δ_j and M_j. To do so we shall use the proof of Theorem 7.3.8 with the definition of Γ_η modified to

$$\xi \to \xi + i\eta \log(2+|\xi|^2)$$

to make the imaginary part $\neq 0$ everywhere. If $u \in C_0^\infty$ we have for every η

$$u(x) = (2\pi)^{-n} \int_{\Gamma_\eta} \hat{u}(\zeta) e^{i\langle x,\zeta\rangle} d\zeta_1 \wedge \ldots \wedge d\zeta_n.$$

We shall prove that if $\delta_1, \ldots, \delta_{j-1}, M_1, \ldots, M_{j-1}$ are already fixed then (15.4.3) implies (15.4.2) for this particular value of j provided that δ_k and M_k are respectively small and large enough for $k \geqq j$. This will show that δ_k, M_k can be chosen successively so that (15.4.3) implies (15.4.2) and thus complete the proof.

First we observe that for some $c>0$ one can for every $x \notin K_j$ find $\eta \in \mathbb{R}^n$ with $|\eta|=1$ and

$$\langle x, \eta\rangle > H_{K_{j-1}}(\eta) + c.$$

We can take for c the distance from K_{j-1} to the complement of K_j. This implies that

$$\exp(H_{K_{j-1}}(\operatorname{Im}\zeta) - \langle x, \operatorname{Im}\zeta\rangle) < (2+|\xi|^2)^{-cR}, \qquad \zeta \in \Gamma_{R\eta}.$$

Hence it follows from (15.4.3) that if $|\alpha| \leqq N_j$ and $x \notin K_j$ then

$$|D^\alpha u(x)| \leqq C \sum_{k=1}^{\infty} \delta_k \int (1+|\xi + iR\eta \log(2+|\xi|^2)|)^{N_j - M_k}$$
$$\cdot (2+|\xi|^2)^{Rc(k,\eta)} (1+R)^n d\xi,$$
$$c(k,\eta) = -c + H_{K_k}(\eta) - H_{K_{j-1}}(\eta).$$

The terms with $k<j$ decrease exponentially as $R \to \infty$ since $c(k,\eta)<0$. Fix R now so that their sum is $<\varepsilon_j/2$. If M_k and $1/\delta_k$ are large enough for $k \geqq j$ then the sum of terms with $k \geqq j$ is also $<\varepsilon_j/2$. Thus (15.4.2) is valid and the proof is complete.

Condition (v) does not yet have the appropriate form for applications of the estimates in Section 15.1 so we must make some further modifications similar to those in the proof of Theorem 15.2.1.

Theorem 15.4.3. *Let X be an open convex subset of $\mathbb{R}^n$ and V a convex subset of $C_0^\infty(X)$. Then V is a neighborhood of 0 in $C_0^\infty(X)$ if and only* *if V contains all $u\in C_0^\infty(X)$ with*

$$\int |\hat{u}(\zeta)|^2 e^{-2\phi(\zeta)}d\lambda(\zeta)<1 \tag{15.4.4}$$

where ϕ is a function in $\mathbb{C}^n$ with the following properties:

(a) *For every convex compact set $K\subset X$ there are constants C_K, N_K such that*

$$e^{-\phi(\zeta)}\leqq C_K(1+|\zeta|)^{N_K}e^{-H_K(\operatorname{Im}\zeta)},$$

where H_K is the supporting function of K.

(b) *ϕ is locally Lipschitz continuous,*

$$|d\phi(\zeta)|\leqq C_b+\log(1+|\operatorname{Im}\zeta|).$$

(c) *ϕ is plurisubharmonic; more precisely we have for every $w\in\mathbb{C}^n$*

$$c(1+|\operatorname{Im}\zeta|^2)^{-\frac{1}{2}}|w|^2\leqq\sum w_j\bar{w}_k\partial^2\phi(\zeta)/\partial\zeta_j\partial\bar{\zeta}_k$$

in the distribution sense. Here c is a positive constant.

Proof. If K is a compact subset of X and $u\in C_0^\infty(K)$ then

$$|\zeta^\alpha\hat{u}(\zeta)|\leqq e^{H_K(\operatorname{Im}\zeta)}\|D^\alpha u\|_{L^1}.$$

Hence it follows from (a) that (15.4.4) is fulfilled if $\|D^\alpha u\|_{L^1}$ is sufficiently small when $|\alpha|\leqq N_K+n+1$. Thus the set of all $u\in C_0^\infty(X)$ satisfying (15.4.4) is a neighborhood of 0.

Assume now that V is a neighborhood of 0 in $C_0^\infty(X)$. By Theorem 15.4.2 we can choose K_j, δ_j and M_j so that (15.4.3) implies $u\in V$. We shall construct ϕ so that (15.4.4) implies (15.4.3). Since $|\hat{u}(\zeta)|^2$ is subharmonic it follows from (15.4.4) and (b) that

$$\begin{aligned}|\hat{u}(\zeta)|^2\pi^n/n!&\leqq\int_{|z|<1}|\hat{u}(\zeta+z)|^2d\lambda(z)\\&\leqq(2+|\zeta|)^2e^{2\phi(\zeta)+2C_b}\int_{|z|<1}|\hat{u}(\zeta+z)|^2e^{-2\phi(\zeta+z)}d\lambda(z).\end{aligned}$$

It is therefore sufficient to choose ϕ so that, in addition to (a), (b), (c), we have

(d) $\quad (n!/\pi^n)^{\frac{1}{2}}e^{\phi(\zeta)}(2+|\zeta|)e^{C_b}<\sum\delta_j(1+|\zeta|)^{-M_j}\exp H_{K_j}(\operatorname{Im}\zeta).$

Set

$$\begin{aligned}\psi_j(\zeta)=&H_{K_j}(\operatorname{Im}\zeta)-2^{-1}(M_{j+1}+1)\log(t_j^2+|\zeta|^2)\\&+t_j^{-\frac{1}{2}}(t_j^2+|\operatorname{Im}\zeta|^2)^{\frac{1}{2}}-(t_j^2+|\operatorname{Im}\zeta|^2)^{\frac{1}{4}}.\end{aligned}$$

Since H_{K_j} is plurisubharmonic and

$$\sum w_j\bar{w}_k\partial^2\log(t^2+|\zeta|^2)/\partial\zeta_j\partial\bar{\zeta}_k\leqq|w|^2/(t^2+|\zeta|^2)$$

we obtain from the computations which led to (15.2.10)

$$\begin{aligned}\sum w_j\bar{w}_k\partial^2\psi_j(\zeta)/\partial\zeta_j\partial\bar{\zeta}_k \\ &\geqq -(M_{j+1}+1)|w|^2/2(t_j^2+|\zeta|^2)+|w|^2(t_j^2+|\operatorname{Im}\zeta|^2)^{-\frac{3}{4}}/16 \\ &\geqq |w|^2(t_j^2+|\operatorname{Im}\zeta|^2)^{-\frac{3}{4}}/32\end{aligned}$$

if $t_j^{1/2}\geqq 16(M_{j+1}+1)$. This implies that ψ_j satisfies (c) uniformly when $|\operatorname{Im}\zeta|>t_j$. If C_j is a Lipschitz constant for H_{K_j} we have

$$|d\psi_j(\zeta)|\leqq(M_j+1)/t_j+C_j+2t_j^{-\frac{1}{4}}<\log t_j$$

if t_j is large enough. We fix t_j now so that the preceding conditions are fulfilled and

$$H_{K_j}(\eta)+t_j^{-\frac{1}{4}}|\eta|\leqq H_{K_{j-1}}(\eta)$$

and let C_b be a bound for $|d\psi_1(\zeta)|$. Then

$$\begin{aligned}\psi_j(\zeta)\leqq &-(M_{j+1}+1)\log(2+|\zeta|)\\ &+H_{K_{j-1}}(\operatorname{Im}\zeta)-C_b+\tfrac{1}{2}\log(\delta_{j+1}^2\pi^n/n!)+G_j\end{aligned}$$

for some G_j. We choose G_j so large that in addition

$$\psi_j(\zeta)-G_j<\psi_1(\zeta)-G_1; \quad |\operatorname{Im}\zeta|<t_j+1;$$

and set

$$\phi_j(\zeta)=\max_{k\leqq j}(\psi_k(\zeta)-G_k).$$

Then we have $\phi_j(\zeta)=\phi_{j-1}(\zeta)$ when $|\operatorname{Im}\zeta|<t_j+1$. Since ψ_j-G_j satisfies (b), (c) uniformly for $|\operatorname{Im}\zeta|>t_j$ it follows inductively that ϕ_j does so too. We have chosen G_j so that (d) is satisfied by ϕ_j for every j. Hence $\phi(\zeta)=\lim\phi_j(\zeta)$ satisfies (b), (c), (d). Since $\phi(\zeta)\geqq\psi_j(\zeta)-G_j$ the condition (a) follows for $K=K_j$, and this completes the proof.

Representation formulas for distribution solutions of an equation $P(D)u=0$ can now be proved by repeating the proof of Theorem 15.3.3, and we can also prove an analogue of the representation (15.2.18) for arbitrary distributions. We leave the details as an exercise for the reader.

Notes

L^2 methods for the study of the overdetermined Cauchy-Riemann equations were first developed by Morrey [1] and by Kohn [1]. These methods work not only in $\mathbb{C}^n$ but in arbitrary pseudo-convex domains or even in Stein manifolds. The exposition in Section 15.1 is mainly taken from Hörmander [19] where the reader can find a systematic development of complex analysis in several variables with similar techniques and also additional references.

The applications to differential operators taken up in Section 15.3 require that the relevant norm on the Fourier transforms of the test functions can be expressed as a L^2 norm in $\mathbb{C}^n$ with respect to a weight $e^{-2\phi}$ where ϕ is plurisubharmonic. Ehrenpreis [3, 5, 6, 7] has studied systematically for which spaces of distributions of compact support that the Fourier-Laplace transform can be described including the topology by conditions on the absolute value alone. However, he did not examine the possibility of using L^2 norms with respect to weights $e^{-2\phi}$ where ϕ is plurisubharmonic. In his monograph on differential operators with constant coefficients Palamodov [1, p. 423] dismissed the possibility of using results such as Theorem 15.1.1 altogether on the ground that the function

$$\phi(\zeta) = -k \log(1+|\zeta|) + H_K(\operatorname{Im}\zeta)$$

which turns up in the description of the Fourier transform of $C_0^k(K)$ is not plurisubharmonic. For this reason we have devoted Sections 15.2 and 15.4 to a perhaps excessively long discussion of how one can construct equivalent strictly plurisubharmonic functions. The main idea is just that the plurisubharmonicity of $H_K(\operatorname{Im}\zeta)$ is quite strong when $\operatorname{Im}\zeta = 0$ and can be spread out so that it overrules the lack of plurisubharmonicity in the first term. If the results are combined with Theorem 7.6.11 in Hörmander [19] as in the proof of Theorem 7.6.13 there and also with Theorem 1.5 in Björk [1, Chapter 8] one obtains the complete fundamental principle as proved in Ehrenpreis [6], Malgrange [5], and Palamodov [1]. (See also the third edition of Hörmander [19], to be published in 1989). However, in order to avoid a long excursion into local algebraic geometry and cohomology with bounds for forms of higher degree we have just included a study of a single differential operator here.

Chapter XVI. Convolution Equations

Summary

A partial differential operator has constant coefficients if and only if it is translation invariant. We have seen in Chapter IV that every translation invariant operator is a convolution operator. It is therefore natural to examine to what extent the results obtained for partial differential operators with constant coefficients can be extended to convolution operators. This is the purpose of the present chapter. It is natural that we must then drop the finer points and concentrate on questions concerning existence and regularity in the spaces $\mathscr{D}'$, $\mathscr{D}'_F$ and C^∞.

The Fourier-Laplace transformation replaces convolution of distributions with compact support by multiplication of analytic functions of exponential type. To take advantage of this we must continue the study of (pluri-)subharmonic functions begun in Section 4.1. In Section 16.1 we discuss representation of subharmonic functions by Newtonian potentials or more precisely by means of Green's potentials and the Poisson kernel. The main result is that if a subharmonic function is bounded above in the half space $x_n>0$ in $\mathbb{R}^n$, then

$$u(tx)/t$$

converges in L^1_{loc} to a linear function of x_n as $t\to+\infty$. The applications to plurisubharmonic functions are given in Section 16.2: If u is a plurisubharmonic function in $\mathbb{C}^n$ such that

$$u(z)\leqq C+A|\operatorname{Im} z|, \qquad z\in\mathbb{C}^n,$$

then there is a smallest supporting function H such that

$$u(z)\leqq C+H(\operatorname{Im} z), \qquad z\in\mathbb{C}^n,$$

and $u(tz)/t$ approaches $H(\operatorname{Im} z)$ as $t\to\infty$ if $z\in\mathbb{C}\mathbb{R}^n=\{wx;\ w\in\mathbb{C}, x\in\mathbb{R}^n\}$. This allows us to give in Section 16.3 another proof of the theorem of supports for convolutions. Similar but much more complex results are also proved concerning the singular supports of convolutions. In particular we give a number of characterizations of distributions $\mu \in \mathscr{E}'$ such that

$$u \in \mathscr{E}', \quad \mu * u \in C^\infty \Rightarrow u \in C^\infty.$$

These are called invertible. When μ is invertible we always have

$$ch \operatorname{sing\,supp} u \subset ch \operatorname{sing\,supp} (\mu * u) - ch \operatorname{sing\,supp} \mu, \qquad u \in \mathscr{E}'.$$

For invertible distributions an existence theory for the inhomogeneous convolution equation $\mu * u = f$ analogous to that in Sections 10.6 and 10.7 is given in Section 16.5. However, solutions also exist for every $\mu \neq 0$ if f is real analytic. This is proved in Section 16.4 where we also prove a general theorem on approximation of solutions of the homogeneous equation $\mu * u = 0$ by exponential solutions.

In Sections 16.6 and 16.7 finally we give an analogue for convolution equations of the study of hypoelliptic and hyperbolic partial differential equations in Chapters XI and XII.

16.1. Subharmonic Functions

As already stated in Theorem 4.1.8 a function u_0 in an open set $X \subset \mathbb{R}^n$ is called subharmonic if it is upper semicontinuous with values in $[-\infty, \infty)$ and the mean value

$$M(x, r) = \int_{|\omega|=1} u_0(x + r\omega)\, d\omega \Big/ \int_{|\omega|=1} d\omega$$

is an increasing function of r for $x \in X$ and $0 \leq r < d(x, \complement X)$. When u_0 is not identically $-\infty$ in a component of X we proved that $u_0 \in L^1_{\text{loc}}$ and that $\Delta u_0 \geqq 0$. Conversely, every distribution with non-negative Laplacean is defined by precisely one subharmonic function.

Let $E(x)$ be the fundamental solution of the Laplacean defined in Theorem 3.3.2. If μ is a positive measure with compact support then

$$u(x) = \int E(x - y)\, d\mu(y)$$

is a subharmonic function defining the convolution $E * d\mu$ in the sense of distribution theory. In fact, if ψ_ε is defined as in the proof of Theorem 4.1.8 then $\psi_\varepsilon * E = E_\varepsilon \downarrow E$ when $\varepsilon \downarrow 0$, hence

$$\psi_\varepsilon * E * d\mu(x) \downarrow \int E(x - y)\, d\mu(y) \quad \text{when } \varepsilon \downarrow 0.$$

Thus u is upper semi-continuous. Since $E_\varepsilon * d\mu \to E * d\mu$ in L^1_{loc} it also follows that u defines the distribution $E * d\mu$, with $\Delta(E * d\mu) = d\mu \geqq 0$.

Since $E\in L^p_{\text{loc}}$ if $p<n/(n-2)$ (which we read as $p<\infty$ when $n=2$), we have $E*d\mu\in L^p_{\text{loc}}$ for such values of p. If $f\in\mathscr{E}'$ and $f*E$ is a continuous function it also follows that $f*(E*d\mu)\in C$. These observations can be carried over to arbitrary subharmonic functions:

Proposition 16.1.1. *If v is subharmonic in the open set $X\subset\mathbb{R}^n$ and not identically $-\infty$ in any component, then $v\in L^p_{\text{loc}}$ if $p<n/(n-2)$. For every $f\in\mathscr{E}'$ such that $f*E$ is continuous, the convolution $f*v$ is also continuous in the open set $\{x;\ \{x\}-\operatorname{supp} f\subset X\}$ where it is defined.*

Proof. Set $d\mu=\Delta v$ which is a positive measure. For every open set $Y \Subset X$ we can choose a non-negative $\chi\in C_0^\infty(X)$ equal to 1 in Y and obtain

$$v=E*(\chi d\mu)+w$$

where $\Delta w=(1-\chi)d\mu=0$ in Y. Hence $w\in C^\infty(Y)$. Since $\chi d\mu$ is a positive measure the statement follows from the observations made before the proposition.

Proposition 16.1.2. *If v_j and v are subharmonic functions in the open set $X\subset\mathbb{R}^n$, not identically $-\infty$ in any component, and $v_j\to v$ in $\mathscr{D}'(X)$, then $v_j\to v$ in $L^p_{\text{loc}}(X)$ if $p<n/(n-2)$. For every $x\in X$ we have*

$$\overline{\lim_{j\to\infty}}\, v_j(x)\leqq v(x). \tag{16.1.1}$$

If dv is a positive measure with compact support in X such that $dv*E$ is continuous, we have for v almost every $x\in X$

$$\overline{\lim_{j\to\infty}}\, v_j(x)=v(x)>-\infty. \tag{16.1.2}$$

Proof. Set $d\mu_j=\Delta v_j$ and $d\mu=\Delta v$. Then $d\mu_j\to d\mu$ in $\mathscr{D}'$, hence in the weak topology of measures (Theorem 2.1.9). Writing

$$v_j=E*(\chi d\mu_j)+w_j, \qquad v=E*(\chi d\mu)+w,$$

with χ chosen as in the proof of Proposition 16.1.1, we have $\Delta w_j=0$ in Y and $w_j\to w$ in $\mathscr{D}'(Y)$. By Theorem 4.4.2 it follows that $w_j\to w$ in $C^\infty(Y)$. If E_ε is defined as in the argument which led to Proposition 16.1.1, we have $E_\varepsilon(x)=E(x)$ when $|x|>c\varepsilon$, hence $E_\varepsilon-E\to 0$ in L^p when $\varepsilon\to 0$. Since for fixed ε

$$E_\varepsilon*(\chi(d\mu_j-d\mu))\to 0$$

locally uniformly as $j\to\infty$ and

$$\|(E-E_\varepsilon)*(\chi(d\mu_j-d\mu))\|_{L^p}\leqq C\,\|E-E_\varepsilon\|_{L^p}\to 0 \qquad \text{as } \varepsilon\to 0,$$

it follows that $v_j\to v$ in $L^p_{\text{loc}}(Y)$ as $j\to\infty$.

For $x \in Y$ we have

$$v_j(x) = E*(\chi d\mu_j)(x) + w_j(x) \leqq E_\varepsilon*(\chi d\mu_j)(x) + w_j(x) \\ \to E_\varepsilon*(\chi d\mu)(x) + w(x), \quad j \to \infty,$$

which gives (16.1.1) if we let $\varepsilon \to 0$. (This was proved in Theorem 4.1.9 by another method.) From the mean value property it follows that the sequence v_j has a uniform upper bound on any compact subset of X. Hence (16.1.1) and Fatou's lemma give

$$\int v(x)\,dv(x) \geqq \int (\overline{\lim}\, v_j(x))\,dv(x) \geqq \overline{\lim} \int v_j(x)\,dv(x).$$

The last statement in the proposition will therefore be proved if we show that

$$\int v_j(x)\,dv(x) \to \int v(x)\,dv(x), \quad j \to \infty.$$

If we choose Y so that $\operatorname{supp} dv \subset Y$, this follows from the fact that

$$\int (\int E(x-y)\chi(y)\,d\mu_j(y))\,dv(x) \\ = \int (dv*E)(y)\chi(y)\,d\mu_j(y) \to \int (dv*E)(y)\chi(y)\,d\mu(y).$$

The proof is complete.

The representation of subharmonic functions by Newtonian potentials of positive measures can be exploited for a further study of the exceptional set where $\overline{\lim}\, v_j(x) < v(x)$. However, we shall not make any use of such refinements here but only of the simple consequence of Proposition 16.1.2 that it is of measure 0 in any C^1 hypersurface. In fact, if dv is a continuous density of compact support on a C^1 hypersurface, the continuity of $E*dv$ is obvious if the surface is a hyperplane and follows by a change of variables in general.

We shall now elaborate the maximum principle further and show that it characterizes subharmonic functions. To do so we shall solve the Dirichlet problem in the unit ball using the Poisson kernel

$$P_1(x, y) = (1-|x|^2)|x-y|^{-n}/c_n \quad \text{if } x, y \in \mathbb{R}^n,\ |x| < 1 = |y|.$$

Here c_n is the area of S^{n-1}. This is a harmonic function of x, for

$$P_1(x, y) = -(2\langle y, z\rangle |z|^{-n} + |z|^{2-n})/c_n, \quad \text{if } x = y + z,$$

and both terms are harmonic functions of z. Writing $x = r\omega$, $|\omega| = 1$, $r < 1$, we have since P_1 is harmonic

$$P_1(0, y) = \int P_1(r\omega, y)\,d\omega/c_n = \int P_1(ry, \omega)\,d\omega/c_n.$$

Here we have used that $|r\omega - y|^2 = 1 + r^2 - 2r\langle \omega, y\rangle = |\omega - ry|^2$. Hence, with $x = ry$,

$$\int P_1(x, \omega)\,d\omega = 1 \quad \text{if } |x| < 1. \tag{16.1.3}$$

We are now ready to prove

Lemma 16.1.3. *If h is a continuous function on the sphere S^{n-1}, then*

$$H(x) = \int_{|\omega|=1} P_1(x,\omega)h(\omega)\,d\omega, \quad |x|<1; \qquad H(x)=h(x), \quad |x|=1,$$

is a continuous function which is harmonic in the interior of the unit ball.

Proof. That H is harmonic follows from the fact that P_1 is harmonic by differentiation under the integral sign. If $r<1$ we have by (16.1.3) if $|y|=1$

$$|H(ry)-h(y)| \leqq \int P_1(ry,\omega)\,|h(\omega)-h(y)|\,d\omega$$
$$\leqq \sup_{|y-\omega|<\varepsilon} |h(\omega)-h(y)| + \int_{|y-\omega|>\varepsilon} P_1(ry,\omega)\,|h(\omega)-h(y)|\,d\omega.$$

The last integral converges uniformly to 0 when $r\to 1$, and the other term on the right hand side is as small as we please when ε is small enough. This proves that $H(ry)\to h(y)$ uniformly in y when $r\to 1$, which completes the proof.

The maximum principle shows that H is the only harmonic function with the boundary values h; it is said to be the solution of the Dirichlet problem in the unit ball with Dirichlet data h.

We can now give a number of equivalent definitions of subharmonicity:

Proposition 16.1.4. *Let u be an upper semi-continuous function in the open set $X\subset\mathbb{R}^n$ with values in $[-\infty,\infty)$. Then each of the following conditions is necessary and sufficient for u to be subharmonic:*

(i) *If $x\in X$ has distance $>r$ to ∂X then*

$$u(x)\int_{|\omega|=1} d\omega \leqq \int_{|\omega|=1} u(x+r\omega)\,d\omega.$$

(ii) *If $x\in X$ has distance $>r$ to $\complement X$ there is some positive measure $d\mu$ on $[0,r]$ with strictly positive mass on $(0,r]$ such that*

$$u(x)\iint_{|\omega|=1} d\omega\,d\mu(r) \leqq \iint_{|\omega|=1} u(x+r\omega)\,d\omega\,d\mu(r).$$

(iii) *If K is a compact subset of X and h a continuous function on K with $u\leqq h$ on ∂K, which is harmonic in the interior of K, then $u\leqq h$ in K.*

(iv) *If B is a closed ball $\subset X$ and h a harmonic function in $\mathbb{R}^n$ with $u\leqq h$ on ∂B then $u\leqq h$ in B.*

Proof. Theorem 4.1.8 shows that subharmonicity is equivalent to (i). Integration with respect to $d\mu(r)$ shows that (i) $\Rightarrow$ (ii) for every μ. To prove that (ii) $\Rightarrow$ (iii) we shall elaborate the proof of the maximum principle in Theorem 4.1.8. Assume that $M=\sup_K(u-h)>0$. The supremum M is attained in a closed subset F of K since u is upper semi-continuous. By hypothesis F has positive distance δ to ∂K. Let x be a point in F with distance δ to ∂K. Every sphere $\{x;|x-y|=r\}$ with $0<r\leqq\delta$ must then contain a point y with distance $\delta-r<\delta$ to ∂K, and then we have $v=u-h<M$ in a neighborhood of y. Hence

$$\iint v(x+r\omega)\,d\omega\,d\mu(r)<M\iint d\omega\,d\mu(r)=v(x)\iint d\omega\,d\mu(r).$$

Since

$$\int h(x+r\omega)\,d\omega=h(x)\int d\omega$$

if $r<\delta$, hence also when $r=\delta$, this contradicts (ii). That (iii) $\Rightarrow$ (iv) is trivial. To prove that (iv) $\Rightarrow$ (i) we let U be a continuous function strictly larger than u defined on the boundary of a closed ball B with center x contained in X. Lemma 16.1.3 shows that there is a continuous function H in B with boundary values U which is harmonic in the interior. H is the uniform limit of $H(x+t(.-x))$ in B as $t\nearrow 1$, and this is a harmonic function in a neighborhood of B. Hence it follows from Theorem 4.4.5 that H is the uniform limit in B of a sequence of functions H_j which are harmonic in $\mathbb{R}^n$. For large j we have $H_j>u$ on ∂B, so it follows from (iv) that $H_j\geqq u$ in B. Hence

$$u(x)\leqq H_j(x)=\int_{|\omega|=1}H_j(x+r\omega)\,d\omega\Big/\int_{|\omega|=1}d\omega$$
$$\to\int_{|\omega|=1}U(x+r\omega)\,d\omega\Big/\int_{|\omega|=1}d\omega.$$

Taking the infimum of the right-hand side for all U considered we have proved (i) since u is upper semi-continuous. The proof is complete.

Corollary 16.1.5. *If u_ι, $\iota\in I$, is a family of subharmonic functions in X and $u=\sup u_\iota$ is upper semi-continuous and $<+\infty$, then u is subharmonic.*

Proof. Let K be a compact subset of X and h a continuous function in K with $u\leqq h$ on ∂K, which is harmonic in the interior of K. Then $u_\iota\leqq h$ on ∂K so $u_\iota\leqq h$ in K since u_ι is subharmonic. Hence $u\leqq h$ in K, and u is subharmonic by condition (iii) in Proposition 16.1.4.

If I is finite then u is of course automatically semi-continuous. Corollary 16.1.5, which was already used in Section 15.2, is the only reason why we had to prove Proposition 16.1.4.

The main topic of this section is the study of subharmonic functions v in

$$\mathbb{R}^n_+=\{x\in\mathbb{R}^n, x_n>0\}$$

such that for some constants C_0 and C_1

(16.1.4) $$v(x) \leqq C_0 + C_1 x_n, \quad x_n > 0.$$

Note that it follows from the Paley-Wiener-Schwartz theorem that $v(z) = \log|\hat{u}(z)|$ satisfies this condition if u is a measure with compact support on $\mathbb{R}$ and $\mathbb{C}$ is identified with $\mathbb{R}^2$. That is the reason for our interest in functions satisfying (16.1.4). We shall only need them when $n=2$ but no additional difficulties occur for larger values of n.

Lemma 16.1.6. *If v is a subharmonic function in $\mathbb{R}^n_+$ satisfying (16.1.4) it follows that*

$$M(x_n) = \sup_{x'} v(x', x_n)$$

is a convex function of x_n.

Proof. Let $0 < a < b$ and let $L(x_n)$ be a linear function with $M(a) \leqq L(a)$ and $M(b) \leqq L(b)$. Then the subharmonic function

$$v_\varepsilon(x) = v(x) - L(x_n) - \varepsilon(x_1^2 + \ldots + x_{n-1}^2 - (n-1)(x_n^2 - b^2))$$

is $\leqq 0$ on the boundary of the slab $\{x; a \leqq x_n \leqq b\}$ and $\to -\infty$ at ∞. If we apply the maximum principle to a large compact subset we conclude that $v_\varepsilon(x) \leqq 0$ in the slab. Letting $\varepsilon \to 0$ we conclude that $v(x) \leqq L(x_n)$ when $a < x_n < b$, hence $M(x_n) \leqq L(x_n)$ when $a < x_n < b$. This proves the lemma.

Since $M(x_n) \leqq C_0 + C_1 x_n$ and $(M(x_n) - M(1))/(x_n - 1)$ is increasing, the limit

(16.1.5) $$\gamma = \lim_{x_n \to +\infty} M(x_n)/x_n$$

exists and $-\infty < \gamma \leqq C_1$. From the convexity of M it follows that

$$M(x_n + y_n) \leqq M(x_n) + \gamma y_n, \quad x_n > 0, \; y_n > 0.$$

The number γ will occur in the Riesz representation formula for v which we shall now derive.

With E still denoting the fundamental solution of Δ we set

$$G(x,y) = E(x-y) - E(x-y^*) = E(x-y) - E(x^*-y) \quad \text{for } x, y \in \mathbb{R}^n_+.$$

Here $y^* = (y', -y_n)$ is the reflection of $y = (y', y_n)$ in the boundary plane. G is called the Green's function of $\mathbb{R}^n_+$. If x is fixed in $\mathbb{R}^n_+$ we have $\Delta_y G = \delta_x$, $G \leqq 0$, and G vanishes when y tends to the boundary. Set

$$P(x, y') = -\partial G(x,y)/\partial y_n|_{y_n=0} = 2x_n |x-y|^{-n}/c_n|_{y_n=0},$$

where c_n is the area of S^{n-1}. One calls P the Poisson kernel of $\mathbb{R}^n_+$. Green's formula gives if $u \in C_0^\infty(\mathbb{R}^n)$

$$(16.1.6) \quad u(x) = \int_{y_n > 0} G(x, y) \Delta u(y) dy + \int P(x, y') u(y', 0) dy', \qquad x_n > 0.$$

If $x_n \to 0$ the first integral converges to 0 (by dominated convergence after a change of variables for each term in G). If $x' = 0$ the second integral is

$$2/c_n \int x_n (x_n^2 + |y'|^2)^{-n/2} u(y', 0) dy'$$
$$= 2/c_n \int (1 + |y'|^2)^{-n/2} u(x_n y', 0) dy' \to u(0) 2/c_n \int (1 + |y'|^2)^{-n/2} dy'.$$

It follows that the last integral must be one, that is,

$$(16.1.3)' \qquad \int P(x, y') dy' = 1, \qquad x_n > 0.$$

If $\psi \in C_0^\infty(\mathbb{R}^n_+)$ and $\phi \in C_0^\infty(\mathbb{R}^{n-1})$, then

$$v(x) = \int G(x, y) \psi(y) dy + \int P(x, y') \phi(y') dy'$$

satisfies the equation $\Delta v = \psi$ in $\mathbb{R}^n_+$ and $v(., x_n) - \phi \to 0$ uniformly as $x_n \to 0$. To prove the first statement it suffices to note that $E(x - y^*)$ is a harmonic function of x in $\mathbb{R}^n_+$ if $y \in \mathbb{R}^n_+$ and that $P(x, y')$ is also harmonic, for this gives in $\mathbb{R}^n_+$

$$\Delta v(x) = \Delta \int E(x - y) \psi(y) dy = \psi(x).$$

Since G vanishes when $x_n = 0$ the second statement follows by repeating the proof of Lemma 16.1.3:

$$|\int P(x, y') \phi(y') dy' - \phi(x')|$$
$$= |\int P(0, 1, y')(\phi(x' + x_n y') - \phi(x')) dy'| \leqq C x_n^{\frac{1}{2}}$$

since $|\phi(x' + x_n y') - \phi(x')|$ can be estimated both by $2 \sup |\phi|$ and by $x_n |y'| \sup |\text{grad}\, \phi|$, hence also by the geometric mean of these bounds.

Finally we note that if $x \to \infty$ while y remains bounded then

$$(16.1.7) \quad G(x, y) = -|x|^{1-n} P(x/|x|, 0) y_n + O(y_n^2 x_n |x|^{-n-1}) = O(x_n |x|^{-n}).$$

This follows from the homogeneity

$$G(x, y) = t^{2-n} G(x/t, y/t), \qquad t > 0,$$

if we take $t = |x|$ and use Taylor's formula.

We are now ready to prove a representation theorem for subharmonic functions satisfying (16.1.4).

Theorem 16.1.7. *Let v be a subharmonic function in $\mathbb{R}^n_+$ satisfying* (16.1.4) *which is not identically $-\infty$. Then the measure $v(x', x_n)dx'$ converges weakly to a measure $d\sigma$ in $\mathbb{R}^{n-1}$ as $x_n \to 0$. If $d\mu = \Delta v$ we have*

$$(16.1.8) \qquad \int (1+|y'|)^{-n}|d\sigma(y')| < \infty, \qquad \int y_n(1+|y|)^{-n} d\mu(y) < \infty,$$

$$(16.1.9) \qquad v(x) = \int P(x, y')d\sigma(y') + \int_{y_n>0} G(x, y)d\mu(y) + \gamma x_n, \qquad x \in \mathbb{R}^n_+,$$

where γ is defined by (16.1.5).

Proof. We may assume that $C_0 = \gamma = 0$ since we can otherwise just subtract $C_0 + \gamma x_n$ from v. Thus $v \leqq 0$. If $0 \leqq \chi \leqq 1$ and $\chi \in C_0^\infty(\mathbb{R}^n_+)$ then

$$v(x) = \int G(x, y)\chi(y)d\mu(y) + v_\chi(x), \qquad x \in \mathbb{R}^n_+,$$

where v_χ is subharmonic and $\leqq 0$. In fact, the integral is the Newtonian potential of $\chi(y)d\mu(y) - \chi(y^*)d\mu(y^*)$ so $\Delta v_\chi = (1-\chi)d\mu \geqq 0$. For any $\varepsilon > 0$ we can find a compact set $K \subset \mathbb{R}^n_+$ such that the integral is $\geqq -\varepsilon$ in $\mathbb{R}^n_+ \setminus K$. Hence $v_\chi \leqq \varepsilon$ in $\complement K$ so $v_\chi \leqq \varepsilon$ in K by the maximum principle. This proves that $v_\chi \leqq 0$. Choose now an increasing non-negative sequence $\chi_j \in C_0^\infty(\mathbb{R}^n_+)$ such that $\chi_j = 1$ on any compact set in $\mathbb{R}^n_+$ when j is large. Then v_{χ_j} increases and is harmonic on any compact set for large j, so v_{χ_j} converges to a harmonic function $v_1 \leqq 0$ such that

$$v(x) = v_1(x) + v_2(x), \qquad v_2(x) = \int G(x, y)d\mu(y).$$

If we choose x so that $v(x) > -\infty$ we conclude that the second condition in (16.1.8) is fulfilled.

We shall now prove that $v_2(\cdot, x_n) \to 0$ weakly as a measure when $x_n \to 0$. Since $v_2 \leqq 0$ it is sufficient by Theorem 2.1.9 to prove convergence in $\mathscr{D}'$. Take $\phi \in C_0^\infty(\mathbb{R}^{n-1})$ and form

$$\int G(x', x_n, y', y_n)\phi(x')dx'.$$

This integral converges to 0 when $x_n \to 0$ for every fixed $y \in \mathbb{R}^n_+$, and by (16.1.7) we have a majorant $Cy_n|y|^{-n}$ outside a compact set. Now the integral

$$F(y) = \int E(y' - x', y_n)\phi(x')dx'$$

is a continuous function of y and so are the derivatives of all orders with respect to y' since we can let them act on ϕ. Since $\Delta F = 0$ when $y_n \neq 0$ we also have a local bound for the second order derivatives with respect to y_n there so F is locally Lipschitz continuous. (See also Theorem 4.4.8.) Thus

$$\int G(x', x_n, y', y_n)\phi(x')dx' = F(y', x_n - y_n) - F(y', x_n + y_n) = O(y_n),$$

so the integral is uniformly bounded by $Cy_n/(1+|y|)^n$ and $\to 0$ as $x_n \to 0$. Hence

$$\int v_2(x', x_n)\phi(x')dx' = \int d\mu(y) \int G(x', x_n, y', y_n)\phi(x')dx' \to 0, \qquad x_n \to 0.$$

If $\varepsilon > 0$ and $0 \leqq \psi \leqq 1$, $\psi \in C_0^\infty(\mathbb{R}^{n-1})$, then

$$h_\psi(x) = v_1(x', x_n + \varepsilon) - \int P(x, y')\psi(y')v_1(y', \varepsilon)dy'$$

is a harmonic function when $x_n > 0$ and in C^∞ when $x_n \geqq 0$. The boundary values $v_1(x', \varepsilon)(1 - \psi(x'))$ when $x_n = 0$ are $\leqq 0$, and $\overline{\lim_{x\to\infty}}\, h_\psi(x) \leqq 0$. Hence the maximum principle gives as above that $h_\psi \leqq 0$. If we extend the definition of h_ψ to $\mathbb{R}^n$ by writing $h_\psi(x', x_n) = -h_\psi(x', -x_n)$ when $x_n < 0$, then

$$\Delta h_\psi = 2v_1(x', \varepsilon)(1 - \psi(x'))\delta'(x_n),$$

so h_ψ is harmonic at $(x', 0)$ if $\psi = 1$ in a neighborhood of x' (Schwarz' reflection principle). If we choose an increasing non-negative sequence $\psi_j \in C_0^\infty(\mathbb{R}^{n-1})$ which is equal to 1 on any compact set in $\mathbb{R}^{n-1}$ for large j, it follows that h_{ψ_j} increases (decreases) to a harmonic function h in $\mathbb{R}^n$ which is $\leqq 0$ ($\geqq 0$) in the upper (lower) half space. We have

$$v_1(x', x_n + \varepsilon) = h(x) + \int P(x, y')v_1(y', \varepsilon)dy', \qquad x_n > 0.$$

h must be a linear function of x_n. To prove this we apply Lemma 16.1.3 to $x \to h(Rx)$ with R large. This gives

$$h(x) = \int_{|y|=1, y_n>0} (P_1(x/R, y) - P_1(x^*/R, y))h(Ry)dS(y), \qquad |x| < R,$$

where dS is the surface element on the unit sphere. Since

$$|x^*/R - y|^2 - |x/R - y|^2 = 4x_n y_n/R$$

and the derivative of $t^{-n/2}$ is $-n/2$ when $t = 1$, we obtain for fixed x

$$P_1(x/R, y) - P_1(x^*/R, y) = 2nx_n y_n(1 + O(1/R))/Rc_n.$$

Hence

$$h(x) = 2nx_n(1 + O(1/R)) \int_{|y|=1, y_n>0} h(Ry)y_n R^{-1} dy.$$

When $R \to \infty$ it follows that $h(x) = ax_n$ with a constant a.

Since $v(x', x_n + \varepsilon) \leqq ax_n$ and $a \leqq 0$ it follows that $a = 0$, for we have assumed that $\gamma = 0$. Thus

(16.1.10) $$v_1(x', x_n) = \int P(x', x_n - \varepsilon, y')v_1(y', \varepsilon)dy', \qquad x_n > \varepsilon.$$

Since $\gamma = 0$ we can for any $\delta > 0$ choose x with x_n large so that

$$v(x', x_n) > -\delta x_n.$$

Thus

$$2/c_n \int |v_1(y', \varepsilon)|/(|x'-y'|^2+x_n^2)^{-n/2} dy' < \delta x_n/(x_n-\varepsilon).$$

If M_δ is sufficiently large this means that for all small ε

$$\int_{|y'|>M_\delta} |v_1(y', \varepsilon)| \, |y'|^{-n} dy' \leqq c_n \delta.$$

Hence we can choose a weak limit $d\sigma$ of $v_1(y', \varepsilon)dy'$ when $\varepsilon \to 0$ and pass to the limit in (16.1.10) in spite of the non-compact range of integration. The measure $d\sigma$ satisfies (16.1.8) and we obtain

$$v_1(x', x_n) = \int P(x, y') d\sigma(y'), \quad x_n > 0. \tag{16.1.10$'$}$$

It follows from (16.1.10)′ that $v_1(x', x_n)dx'$ converges weakly to $d\sigma$ when $x_n \to 0$. In fact, if $\phi \in C_0^\infty(\mathbb{R}^{n-1})$ we have

$$\int v_1(x', x_n) \phi(x') dx' = \int d\sigma(y') \int P(x, y') \phi(x') dx'.$$

The inner integral converges uniformly to $\phi(y')$ when $x_n \to 0$ and has a uniform bound of the form $C/(1+|y'|)^n$. Hence the limit when $x_n \to 0$ of the double integral is $\int \phi(y') d\sigma(y')$ which completes the proof.

We can now show that $v(x)$ is close to γx_n in the mean as $x \to \infty$. This is the main result of the section.

Theorem 16.1.8. *Let v be a subharmonic function in $\mathbb{R}^n_+$ satisfying* (16.1.4) *and define γ by* (16.1.5). *Then we have*

$$\int_K |v(tx)/t - \gamma x_n| dx \to 0 \quad \textit{as } t \to \infty, \tag{16.1.11}$$

if K is a compact subset of the closure of $\mathbb{R}^n_+$.

Proof. The statement is valid if v is a constant so we may assume that $C_0 = 0$, hence that $d\sigma \leqq 0$. Then we have $v(tx)/t - \gamma x_n \leqq 0$ so it is sufficient to prove (16.1.11) with the absolute values removed. By (16.1.9) we have

$$\int_K (v(tx)/t - \gamma x_n) dx = \int K_1(t, y') d\sigma(y') - \int K_2(t, y) d\mu(y)$$

where

$$0 \leqq K_1(t, y') = t^{-1} \int_K P(tx, y') dx = t^{-n} \int_K P(x, y'/t) dx \leqq C t^{-n}(1+|y'/t|)^{-n},$$

$$\begin{aligned} 0 \leqq K_2(t, y) &= t^{-1} \int_K |G(tx, y)| dx = t^{1-n} \int_K |G(x, y/t)| dx \\ &\leqq C t^{1-n}(y_n/t)(1+|y/t|)^{-n}. \end{aligned}$$

The last estimate follows from (16.1.7) when y/t is large and is quite obvious when y/t is in a bounded set. In view of (16.1.8) we now obtain (16.1.11) by dominated convergence.

We close the section by giving two examples of applications of Theorem 16.1.8 which will not be needed later on but give a good idea of the significance of Theorem 16.1.8.

Theorem 16.1.9. *Let u be a measure on $\mathbb{R}$ with $ch\operatorname{supp} u=[a,b]$. Then*

$$\Delta(\log|\hat{u}(t\zeta)|)/t \to (b-a)\delta(\operatorname{Im}\zeta), \qquad t\to\infty.$$

If $N(R)$ is the number of zeros of $\hat{u}$ with $|z|<R$, counted with multiplicity, then $N(R)/R \to (b-a)/\pi$ when $R\to\infty$.

Proof. By Theorem 7.3.1 we have

$$\log|\hat{u}(\zeta)| \leqq C + b\operatorname{Im}\zeta, \qquad \operatorname{Im}\zeta>0,$$

and b cannot be replaced by any smaller constant here. Similarly

$$\log|\hat{u}(\zeta)| \leqq C + a\operatorname{Im}\zeta, \qquad \operatorname{Im}\zeta<0,$$

and a is the best possible constant here. Hence Theorem 16.1.8 proves that

$$t^{-1}\log|\hat{u}(t\zeta)| \to h(\operatorname{Im}\zeta)$$

in L^1_{loc}, where $h(t)=bt$ for $t>0$ and $h(t)=at$ for $t<0$. We apply the Laplacean to both sides and obtain in view of Example 4.1.10 if z_j are the zeros of $\hat{u}$ with multiple zeros repeated

$$2\pi/t\sum\delta_{z_j/t} \to (b-a)\delta(\operatorname{Im} z)$$

in the distribution sense, hence also with weak convergence of measures.

If μ_j are positive measures converging weakly to μ, then

$$\mu_j(\phi)\to\mu(\phi) \tag{16.1.12}$$

by definition for all $\phi\in C_0$. However, (16.1.12) remains true for every Borel measurable ϕ which is Riemann integrable with respect to μ. In fact, if ϕ is real valued and $\varepsilon>0$ we can find ϕ_1, $\phi_2\in C_0$ with $\phi_1\leqq\phi\leqq\phi_2$ and $\int(\phi_2-\phi_1)\,d\mu<\varepsilon$. Hence

$$\overline{\lim}\int\phi\,d\mu_j \leqq \lim\int\phi_2\,d\mu_j=\int\phi_2\,d\mu\leqq\int\phi\,d\mu+\varepsilon,$$
$$\underline{\lim}\int\phi\,d\mu_j \geqq \lim\int\phi_1\,d\mu_j=\int\phi_1\,d\mu\geqq\int\phi\,d\mu-\varepsilon$$

which proves the assertion. If we take for ϕ the characteristic function of the unit disc, the theorem follows.

Next we give a somewhat weaker form of Theorem 1.3.8.

Theorem 16.1.10. *Let* $u \in C_0^\infty(\mathbb{R})$ *and let*

$$|u^{(k)}(x)| \leqq M_1 \dots M_k, \qquad k=1,2,\dots$$

where M_k *is an increasing sequence. If* u *is not identically* 0 *then*

$$\sum 1/M_k < \infty.$$

Proof. Application of Theorem 16.1.7 to $\log|\hat{u}|$ in the upper half plane gives if u is not identically 0 that

$$\int \log|\hat{u}(\xi)|\, d\xi/(1+\xi^2) > -\infty.$$

Now the hypothesis implies

$$|\hat{u}(\xi)| \leqq C M_1/|\xi| \dots M_k/|\xi|, \qquad k=1,2,\dots.$$

This estimate is favorable in the interval $M_k < |\xi| < M_{k+1}$. Hence

$$\int_{M_1}^{\infty} \log|\hat{u}(\xi)|\, d\xi/\xi^2 \leqq (\log C)/M_1 + \sum_1^\infty \int_{M_k}^{\infty} \log(M_k/\xi)\, d\xi/\xi^2$$
$$= (\log C)/M_1 - \sum_1^\infty M_k^{-1} \int_1^\infty \log\xi\, d\xi/\xi^2.$$

This proves the theorem.

16.2. Plurisubharmonic Functions

The notion of plurisubharmonic function was introduced in Theorem 4.1.11. Since the cone of plurisubharmonic functions is contained in the cone of subharmonic functions and is closed in the distribution topology, all that we have proved concerning limits of sequences of subharmonic functions remains valid for sequences of plurisubharmonic functions. We shall now prove an analogue of Lemma 16.1.6.

Lemma 16.2.1. *Let* v *be a plurisubharmonic function in* $\mathbb{C}^n$ *such that for some constants* C_0 *and* C_1

(16.2.1) $$v(x+iy) \leqq C_0 + C_1|y|; \quad x, y \in \mathbb{R}^n.$$

If $v \not\equiv -\infty$ *the function of* $y \in \mathbb{R}^n$ *defined by*

(16.2.2) $$M(y) = \sup_x v(x+iy)$$

is convex. The limit

(16.2.3) $$H(y) = \lim_{t\to+\infty} M(ty)/t$$

exists and is a supporting function. We have $M(x+y)\leqq M(x)+H(y)$ *for all* $x, y\in\mathbb{R}^n$.

Proof. Since

$$M(y+ty_1)=\sup_x v(x+i(y+ty_1))=\sup_x \sup_{\operatorname{Im} w=t} v(x+iy+wy_1)$$

and the supremum of a family of convex functions is convex, it follows from Lemma 16.1.6 that $M(y+ty_1)$ is a convex function of t, that is, M is convex. Here we have used that $M(y)\leqq C_0+C_1|y|<\infty$ and that M is not identically $-\infty$. From the convexity it follows that the limit

$$H(y)=\lim_{t\to+\infty} M(ty)/t=\lim_{t\to+\infty}(M(ty)-M(0))/t$$

exists, and it is obviously convex and positively homogeneous, $H(y)\leqq C_1|y|$. The monotonicity of the difference quotient shows that $M(y)\leqq M(0)+H(y)$. Hence

$$\begin{aligned}M(x+y)-M(x)&\leqq \lim_{t\to+\infty}(M(x+ty)-M(x))/t\\&\leqq \lim_{t\to+\infty}(M(0)+H(x+ty)-M(x))/t\\&\leqq \lim_{t\to+\infty}(H(x)+tH(y))/t=H(y)\end{aligned}$$

which proves the lemma.

We shall call H the supporting function of v. (When $v\equiv 0$ we define $H=-\infty$, which is the supporting function of the empty set.) In Section 16.3 we shall use the Paley-Wiener-Schwartz theorem to identify the supporting function of $ch\operatorname{supp} u$ with the supporting function of $\log|\hat{u}|$ when u is a measure of compact support (see also the proof of Theorem 16.1.9). This will justify the terminology. However, for the moment we just continue the study of plurisubharmonic functions satisfying (16.2.1) in order to determine how well they are approximated by $H(\operatorname{Im}\zeta)$ at ∞.

Lemma 16.2.2. *Let* v *be plurisubharmonic and bounded from above in* $\mathbb{C}^n$. *Then it follows that* v *is a constant.*

Proof. It suffices to prove the statement when $n=1$. Let $v(z)\leqq C$ for every z, and let z_0 be a point where $v(z_0)<a$ say. Then we can choose $\varepsilon>0$ so that $v(z)<a$ when $|z-z_0|\leqq\varepsilon$. It follows that

$$v(z)\leqq a+(C-a)(\log(|z-z_0|/\varepsilon))/\log(R/\varepsilon)$$

in the annulus $\varepsilon \leqq |z-z_0| \leqq R$ for this is true on the boundary. When $R \to \infty$ it follows that $v(z) \leqq a$, hence that $v(z) \leqq v(z_0)$. Interchanging the roles of z and z_0 we obtain $v(z)=v(z_0)$ and the lemma is proved.

Lemma 16.2.3. *Let v_k be a sequence of plurisubharmonic functions in $\mathbb{C}^n$ which are uniformly bounded from above on every compact set. If $v = \overline{\lim}\, v_k$ is bounded from above in the whole of $\mathbb{C}^n$, it follows that $v(z) = \sup v$ for almost every z.*

Proof. If $\sup v = -\infty$ there is nothing to prove. On the other hand, if $\sup v > A > -\infty$ it follows from Theorem 4.1.9 that there is a subsequence v_{j_k} such that $\overline{\lim}\, v_{j_k}(z) > A$ for some z and v_{j_k} converges in L^1_{loc}. The limit is defined by a plurisubharmonic function V with $V(z) > A$ for some z. Since V is a constant by Lemma 16.2.2 this is true for every z and Theorem 4.1.9 gives that

$$v(z) \geqq \overline{\lim}\, v_{j_k}(z) = V(z) > A \qquad \text{for almost every } z.$$

This proves the lemma.

Note that the same proof is applicable if k is a real parameter. We shall use this in extending Theorem 16.1.8.

Theorem 16.2.4. *Let v satisfy the hypotheses of Lemma 16.2.1 and define H by* (16.2.2), (16.2.3). *If $y \in \mathbb{R}^n$ we have for almost all $\zeta \in \mathbb{C}^n$*

$$\lim_{t \to +\infty} \int_K |v(\zeta + twy)/t - H(\operatorname{Im} wy)|\, d\lambda(w) = 0 \tag{16.2.4}$$

if K is a compact subset of $\mathbb{C}$.

Proof. We may assume that $\operatorname{Im} w \geqq 0$ in K. If $v \in C^\infty$ it is clear that

$$v_t(\zeta) = \int_K v(\zeta + twy)/t\, d\lambda(w)$$

is plurisubharmonic, for the Laplacean along any complex line can be applied under the integral sign so it is non-negative. The same conclusion is valid for arbitrary plurisubharmonic v since v is the decreasing limit of C^∞ plurisubharmonic functions. By hypothesis we have $v(\zeta) \leqq H(\operatorname{Im} \zeta) + C_0$, hence

$$v_t(\zeta) \leqq m(K) C_0/t + \int_K H(t^{-1} \operatorname{Im} \zeta + \operatorname{Im} wy)\, d\lambda(w) \to \int_K H(\operatorname{Im} wy)\, d\lambda(w)$$

as $t \to \infty$. In view of Lemma 16.2.3 it follows that there is a constant A such that

$$A \leqq H(y) \int_K \operatorname{Im} w\, d\lambda(w), \qquad \overline{\lim}\, v_t(\zeta) \leqq A,$$

with equality for almost every ζ. When the function $w \to v(\zeta + wy)$, $\operatorname{Im} w > 0$, is not identically $-\infty$, it follows from Theorem 16.1.8 that

$$\lim v_t(\zeta) = \gamma(\zeta) \int_K \operatorname{Im} w \, d\lambda(w)$$

where $\gamma(\zeta)$ is defined according to (16.1.5). If we write

$$A = \gamma \int_K \operatorname{Im} w \, d\lambda(w)$$

we have $\gamma \leqq H(y)$ and $\gamma(\zeta) \leqq \gamma$ for every ζ, with equality for almost every ζ. Hence

$$v(\zeta + wy) \leqq M(\operatorname{Im} \zeta) + \gamma(\zeta) \operatorname{Im} w \leqq M(\operatorname{Im} \zeta) + \gamma \operatorname{Im} w, \qquad \operatorname{Im} w \geqq 0,$$

so $M(ty) \leqq M(0) + \gamma t$. This proves that $H(y) \leqq \gamma$, thus $H(y) = \gamma$. Now (16.2.4) follows from (16.1.11) when $\gamma(\zeta) = H(y)$, which completes the proof.

Corollary 16.2.5. *If the hypotheses of Theorem 16.2.4 are fulfilled and B is a ball in $\mathbb{C}^n$, then*

$$\lim_{t \to +\infty} \iint_{K \times B} |v(\zeta + twy)/t - H(\operatorname{Im} wy)| \, d\lambda(w) \, d\lambda(\zeta) = 0. \tag{16.2.5}$$

Proof. We can replace B by a larger ball with center at a point ζ_0 for which (16.2.4) is valid. Since

$$\begin{aligned} v(\zeta + twy)/t - H(\operatorname{Im} wy) &\leqq M(\operatorname{Im} \zeta + t \operatorname{Im} wy)/t - H(\operatorname{Im} wy) \\ &\leqq M(\operatorname{Im} \zeta)/t \to 0 \end{aligned}$$

it is sufficient to prove that

$$\varliminf_{t \to +\infty} \iint_{K \times B} (v(\zeta + twy)/t - H(\operatorname{Im} wy)) \, d\lambda(w) \, d\lambda(\zeta) \geqq 0. \tag{16.2.6}$$

But v is subharmonic so the quotient of the double integral by the volume of B is at least equal to

$$\int_K (v(\zeta_0 + twy)/t - H(\operatorname{Im} wy)) \, d\lambda(w).$$

Thus (16.2.6) follows from (16.2.4).

Corollary 16.2.6. *If the hypotheses of Theorem 16.2.4 are fulfilled and B^t are balls in $\mathbb{C}^n$ with fixed center and radius bounded from below by a positive constant, then*

$$\varliminf_{t \to +\infty} m(B^t)^{-1} \iint_{K \times B^t} (v(\zeta + twy)/t - H(\operatorname{Im} wy)) \, d\lambda(w) \, d\lambda(\zeta) \geqq 0, \tag{16.2.7}$$

(16.2.8) $$\varlimsup_{t\to+\infty} m(B^t)^{-1} \iint_{K\times B^t} |v(\zeta+twy)/t-H(\operatorname{Im} wy)|\,d\lambda(w)\,d\lambda(\zeta)$$
$$\leqq \varlimsup_{t\to+\infty} 2m(B^t)^{-1} \iint_{K\times B^t} |H(\operatorname{Im}\zeta)|\,t^{-1}\,d\lambda(w)\,d\lambda(\zeta).$$

Proof. Since the left-hand side of (16.2.7) increases with the radius of B^t the assertion follows from Corollary 16.2.5. To prove (16.2.8) we just have to note again that

$$v(\zeta+twy)/t-H(\operatorname{Im} wy)\leqq M(\operatorname{Im}\zeta)/t\leqq C/t+H(\operatorname{Im}\zeta)/t,$$

hence that

$$|v(\zeta+twy)/t-H(\operatorname{Im} wy)|$$
$$\leqq 2(|C|/t+|H(\operatorname{Im}\zeta)|/t)-(v(\zeta+twy)/t-H(\operatorname{Im} wy))$$

(16.2.8) is therefore a consequence of (16.2.7).

Remark. If the radius of B^t is $o(t)$ then the right-hand side of (16.2.8) is 0. Thus $v(\zeta)$ is close to $H(\operatorname{Im}\zeta)$ in the mean when ζ is in a small conic neighborhood of

$$\mathbb{C}\mathbb{R}^n=\{wy;\ w\in\mathbb{C},\ y\in\mathbb{R}^n\}.$$

However, this is not always true elsewhere as is shown by the example

$$v(\zeta)=|\operatorname{Im}(\langle\zeta,\zeta\rangle^{\frac{1}{2}})|$$

or if one prefers $v(\zeta)=\log|\cos(\langle\zeta,\zeta\rangle^{\frac{1}{2}})|$.

The following consequence of Corollary 16.2.5 will be very important in Section 16.3.

Theorem 16.2.7. *Let v_j, $j=1, 2, 3$, be plurisubharmonic functions in $\mathbb{C}^n$ such that $v_3=v_1+v_2$ and*

$$v_j(z)\leqq C_j+A_j|\operatorname{Im} z|,\qquad z\in\mathbb{C}^n,$$

for some constants C_j and A_j. If H_j is the supporting function of v_j defined according to Lemma 16.2.1, then

$$H_3=H_1+H_2.$$

Proof. Since for $y\in\mathbb{R}^n$, $\zeta\in\mathbb{C}^n$, $w\in\mathbb{C}$, $\operatorname{Im} w\geqq 0$ we have

$$\begin{aligned}\operatorname{Im} w(H_3(y)-H_1(y)-H_2(y))=H_3(\operatorname{Im} wy)-v_3(\zeta+twy)/t\\ +v_1(\zeta+twy)/t-H_1(\operatorname{Im} wy)\\ +v_2(\zeta+twy)/t-H_2(\operatorname{Im} wy),\end{aligned}$$

the statement follows immediately from Corollary 16.2.5.

16.3. The Support and Singular Support of a Convolution

If u_1 and u_2 are arbitrary distributions in $\mathbb{R}^n$, one of which has compact support, we have the fairly obvious inclusion

$$\operatorname{supp}(u_1 * u_2) \subset \operatorname{supp} u_1 + \operatorname{supp} u_2 \tag{16.3.1}$$

(cf. (4.2.2)). In Section 4.3 we proved the far less obvious theorem of supports which states that when u_1 and u_2 both have compact support, then

$$ch \operatorname{supp}(u_1 * u_2) = ch \operatorname{supp} u_1 + ch \operatorname{supp} u_2 \tag{16.3.2}$$

where ch denotes convex hull. We shall begin this section by giving another proof with the methods developed in Section 16.2.

Lemma 16.3.1. *If u is a measure of compact support then*

$$\log|\hat{u}(\zeta)| \leqq C_0 + C_1 |\operatorname{Im} \zeta|$$

for some C_0 and C_1. The supporting function of the set $\operatorname{supp} u$ is equal to the supporting function of the plurisubharmonic function $\log|\hat{u}|$ defined in Lemma 16.2.1.

Proof. By Theorem 7.3.1 we have

$$\log|\hat{u}(\zeta)| \leqq C_0 + H(\operatorname{Im} \zeta), \qquad \zeta \in \mathbb{C}^n$$

if e^{C_0} is the total mass of u and H is the supporting function of $\operatorname{supp} u$. Conversely, if this estimate is valid for the supporting function of a convex set K then $\operatorname{supp} u \subset K$. This proves the assertion.

We can now prove (16.3.2) as follows. Assume first that u_j are measures and set $u_3 = u_1 * u_2$. If H_j is the supporting function of $\operatorname{supp} u_j$ then H_j is the supporting function of $\log|\hat{u}_j|$ by Lemma 16.3.1. Since

$$\log|\hat{u}_3| = \log|\hat{u}_1| + \log|\hat{u}_2|$$

it follows from Theorem 16.2.7 that $H_3 = H_1 + H_2$, and this proves (16.3.2).

To pass to the general case we take $\phi \in C_0^\infty(\mathbb{R}^n)$ with support in a small convex neighborhood K of 0. Then $u_j * \phi \in C_0^\infty$ so we have proved that

$$\begin{aligned} ch \operatorname{supp}(u_1 * \phi) + ch \operatorname{supp}(u_2 * \phi) &= ch \operatorname{supp}(u_1 * \phi * u_2 * \phi) \\ &= ch \operatorname{supp}(u_3 * \phi * \phi) \\ &\subset ch \operatorname{supp} u_3 + 2K \end{aligned}$$

where the inclusion follows from (16.3.1). When K shrinks to the origin and $\phi \to \delta$ it follows that

$$ch\operatorname{supp} u_1 + ch\operatorname{supp} u_2 \subset ch\operatorname{supp} u_3,$$

and the opposite inclusion follows from (16.3.1).

For singular supports the analogue

$$\operatorname{sing\,supp}(u_1 * u_2) \subset \operatorname{sing\,supp} u_1 + \operatorname{sing\,supp} u_2 \tag{16.3.3}$$

of (16.3.1) is valid (cf. (4.2.3)). The rest of this section will be devoted to the study of analogues of (16.3.2). These turn out to be much more complicated. The reason for this is that the singularities of a distribution $u \in \mathscr{E}'$ are reflected by the behavior of the Fourier transform $\hat{u}$ at infinity in $\mathbb{R}^n$, and this depends very much on how one approaches infinity.

When $u \in \mathscr{E}'$ we introduce for real ξ the plurisubharmonic function of z defined by

$$L_u(z, \xi) = (\log |\hat{u}(\xi + z \log |\xi|)|)/\log |\xi|, \quad |\xi| > 2. \tag{16.3.4}$$

If N is the order of u we have by Theorem 7.3.1 for some constants C, N, A

$$|\hat{u}(\zeta)| \leqq C(1 + |\zeta|)^N e^{A|\operatorname{Im}\zeta|}.$$

This gives the estimate

$$L_u(z, \xi) \leqq A|\operatorname{Im} z| + (\log C + N \log(1 + |\xi + z \log |\xi||))/\log |\xi|.$$

Hence $L_u(z, \xi)$ is bounded from above when $\xi \to \infty$ for z in any compact set, and we have

$$\varlimsup_{\xi \to \infty} L_u(z, \xi) \leqq A|\operatorname{Im} z| + N. \tag{16.3.5}$$

By Theorem 4.1.9 we can therefore from every sequence $\xi_j \to \infty$ in $\mathbb{R}^n$ extract a subsequence ξ_{j_k} such that $L_u(z, \xi_{j_k})$ converges when $k \to \infty$ to a plurisubharmonic function v with $v(z) \leqq N + A|\operatorname{Im} z|$. The convergence is in L^1_{loc} if v is finite, and if $v = -\infty$ identically the convergence to $-\infty$ is locally uniform. The limit depends only on the class of u in $\mathscr{E}'(\mathbb{R}^n)/C_0^\infty(\mathbb{R}^n)$, that is, on the singularities of u. In fact, if $a > v(z_0)$ then

$$|\hat{u}(\xi_{j_k} + z \log |\xi_{j_k}|)| \leqq |\xi_{j_k}|^a$$

for all z in a neighborhood of z_0 and all large k. If $u_1 \in C_0^\infty$ then

$$|\hat{u}_1(\xi_{j_k} + z \log |\xi_{j_k}|)| < |\xi_{j_k}|^a$$

for such z and large enough k, which shows that

$$\varlimsup L_{u+u_1}(z, \xi_{j_k}) \leqq a$$

in a neighborhood of z_0. This proves the assertion.

According to Lemma 16.2.1 there is associated with every limit v an element of the set $\mathscr{H}$ consisting of supporting functions of compact convex sets, including the empty set with supporting function $-\infty$. This construction is formalized in the following definition where we consider several distributions simultaneously since we intend to study distributions satisfying $u_1 * u_2 = u_3$.

Definition 16.3.2. If $u_1, \ldots, u_k \in \mathscr{E}'$ we denote by $\mathscr{H}(u_1, \ldots, u_k)$ the set of elements $(h_1, \ldots, h_k) \in \mathscr{H}^k$ such that there is a sequence $\xi_\nu \to \infty$ in $\mathbb{R}^n$ with $L_{u_j}(z, \xi_\nu)$ converging to a plurisubharmonic function with supporting function h_j for $j=1, \ldots, k$.

We shall see that $\mathscr{H}(u)$ gives rather precise information on $\operatorname{sing\,supp} u$. A set $\mathscr{H}_\xi(u)$ giving information on $WF(u)$ could also have been defined by adding the condition $\xi_\nu/|\xi_\nu| \to \xi$ in Definition 16.3.2. However, in order not to prolong the section we shall leave such simple modifications aside.

If $j<k$ it is clear that $\mathscr{H}(u_1, \ldots, u_k)$ projects into $\mathscr{H}(u_1, \ldots, u_j)$ if we drop the last $k-j$ components. The projection is surjective:

Lemma 16.3.3. *If $u_1, \ldots, u_k \in \mathscr{E}'$ and $(h_1, \ldots, h_j) \in \mathscr{H}(u_1, \ldots, u_j)$ for some $j<k$, then one can choose $h_{j+1}, \ldots, h_k$ so that $(h_1, \ldots, h_k)$ belongs to $\mathscr{H}(u_1, \ldots, u_k)$.*

Proof. Let ξ_ν be a sequence $\to \infty$ such that $L_{u_i}(z, \xi_\nu)$ converges when $\nu \to \infty$ to a plurisubharmonic function with supporting function h_i for every $i \leqq j$. Passing if necessary to a subsequence we may assume that the sequences $L_{u_i}(z, \xi_\nu)$ also converge when $i=j+1, \ldots, k$. If we define h_i for these indices as the supporting functions of the corresponding limits, the lemma follows.

Remark. Working with ultrafilters instead we would not need to take k finite. Thus we can define $\mathscr{H}(\mathscr{E}') \subset \mathscr{H}^{\mathscr{E}'}$ so that the projection on the $u_1, \ldots, u_k$ coordinates is always $\mathscr{H}(u_1, \ldots, u_k)$. One can regard $\mathscr{H}(\mathscr{E}')$ as a compactification of $\mathbb{R}^n$ at ∞, but it would take us too far to develop this aspect here.

Theorem 16.3.4. *If $u \in \mathscr{E}'$ and H denotes the supporting function of $\operatorname{ch} \operatorname{sing\,supp} u$, then*

$$H(\xi) = \sup\{h(\xi);\, h \in \mathscr{H}(u)\}. \tag{16.3.6}$$

Proof. If $\operatorname{sing\,supp} u$ is empty, that is, if $u \in C_0^\infty$, it follows from (7.3.3) that $L_u(z, \xi) \to -\infty$ uniformly on every compact subset of $\mathbb{C}^n$ when

$\xi \to \infty$, so $\mathscr{H}(u)$ consists of the function $-\infty$ only. The theorem is therefore true in this case. Now assume that sing supp u is not empty. By (7.3.9) and the proof of (16.3.5) we have

$$\varlimsup_{\xi \to \infty} L_u(z, \xi) \leqq N + H(\operatorname{Im} z). \tag{16.3.5$'$}$$

Hence $h \leqq H$ for every $h \in \mathscr{H}(u)$. On the other hand, let $-\infty \neq H' \in \mathscr{H}$ and $H' \geqq h$ for every $h \in \mathscr{H}(u)$. Then we claim that (7.3.9) is valid with H replaced by H', N replaced by $N+1$ (N is the order of u) and suitable constants C_m. If this is false we can for some fixed m find a sequence $\zeta_\nu \to \infty$ such that $|\operatorname{Im} \zeta_\nu| \leqq m \log(|\zeta_\nu|+1)$ and

$$|\hat{u}(\zeta_\nu)| \geqq (1+|\zeta_\nu|)^{N+1} e^{H'(\operatorname{Im} \zeta_\nu)}.$$

By passing to a subsequence we may also assume that $L_u(z, \operatorname{Re} \zeta_\nu)$ converges to a plurisubharmonic limit V. Since $V \leqq N$ in $\mathbb{R}^n$ and the supporting function of V is $\leqq H'$, we obtain (see Lemma 16.2.1)

$$V(z) \leqq N + H'(\operatorname{Im} z)$$

for all z. For large values of ν we have $|\operatorname{Im} \zeta_\nu| \leqq (m+1) \log |\operatorname{Re} \zeta_\nu|$, and it follows from Theorem 4.1.9 that

$$L_u(z, \operatorname{Re} \zeta_\nu) < N + 1 + H'(\operatorname{Im} z) \quad \text{when } |z| \leqq m+1.$$

If we take $z = i \operatorname{Im} \zeta_\nu / \log |\operatorname{Re} \zeta_\nu|$ this gives

$$|\hat{u}(\zeta_\nu)| < |\zeta_\nu|^{N+1} e^{H'(\operatorname{Im} \zeta_\nu)},$$

which contradicts the assumption. Hence (7.3.9) is in fact valid with H replaced by H' which proves that $H \leqq H'$ and that (16.3.6) is true.

The second part of the proof can be modified so that one gets additional information on sing supp u from $\mathscr{H}(u)$, The result is not required later on in this chapter except for some refinements but will be given for the sake of completeness.

Theorem 16.3.5. *If* $u \in \mathscr{E}'$ *then* sing supp u *is contained in the closed union of the convex compact sets with supporting function in* $\mathscr{H}(u)$.

Proof. First we shall prove that for every fixed m we can find $h_\xi \in \mathscr{H}(u)$ with

$$|\hat{u}(\xi + \zeta)| \leqq |\xi|^{N+1} \exp h_\xi(\operatorname{Im} \zeta), \quad |\zeta| < m \log |\xi|, \tag{16.3.7}$$

provided that ξ is large enough. The proof is again by contradiction. If this is false we can find $\xi_j \to \infty$ so that for every $h \in \mathscr{H}(u)$

$$\sup_{|\zeta| < m \log |\xi_j|} |\hat{u}(\xi_j + \zeta)| \, |\xi_j|^{-N-1} \exp -h(\operatorname{Im} \zeta) > 1. \tag{16.3.8}$$

This means that for every $h\in\mathscr{H}(u)$

$$\sup_{|z|<m} L_u(z,\xi_j)-h(\operatorname{Im} z)>N+1.$$

By passing to a subsequence we can assume that $L_u(z,\xi_j)$ has a plurisubharmonic limit V. If h is the supporting function of V we have as above

$$V(z)\leqq N+h(\operatorname{Im} z).$$

By Theorem 4.1.9 it follows that for large j

$$L_u(z,\xi_j)<N+1+h(\operatorname{Im} z), \qquad |z|<m,$$

and since $h\in\mathscr{H}(u)$ this is a contradiction proving (16.3.7).

To simplify notation we assume now that 0 has distance $\geqq r>0$ to all convex sets with supporting function in $\mathscr{H}(u)$ and shall then prove that $u\in C^\infty$ in a neighborhood of 0. This hypothesis means that for every $h\in\mathscr{H}(u)$ one can find $\theta\in\mathbb{R}^n$ with $|\theta|=1$ so that $h(\theta)\leqq -r$, for disjoint convex sets can be separated by a hyperplane. To complete the proof we must localize the second part of the proof of Theorem 7.3.8 so that near ξ we can go out into the complex domain in the direction θ corresponding to the supporting function h_ξ which occurs in (16.3.7).

Fix a small $\varepsilon>0$. The metric

$$|\theta|_\xi=\varepsilon|\theta|/\log(2+|\xi|), \qquad \theta\in\mathbb{R}^n,$$

is slowly varying in the sense of Definition 1.4.7. It follows from Theorem 1.4.10 that we can find a corresponding partition of unity $1=\sum\phi_\nu$ in $\mathbb{R}^n$ such that

(i) there is a fixed bound for the number of overlapping supports

(ii) $|\xi-\xi_\nu|_{\xi_\nu}<1$ in $\operatorname{supp}\phi_\nu$ for suitable ξ_ν

(iii) $|D^\alpha\phi_\nu|\leqq(C\varepsilon)^{|\alpha|}$, $|\alpha|\leqq\log(3+|\xi_\nu|)$.

Here we have used the remark after the proof of Theorem 1.4.10 and chosen $d_j=1/2\log(3+|\xi|)$ when $j<\log(2+|\xi|)$. For large ν we choose $h_\nu\in\mathscr{H}(u)$ so that (16.3.7) is fulfilled when $\xi=\xi_\nu$ and $m=2/\varepsilon$. Then we choose $\theta_\nu\in\mathbb{R}^n$ with $|\theta_\nu|=1$ so that $h_\nu(\theta_\nu)\leqq -r$. Now we have $u=\sum u_\nu$ where

$$u_\nu(x)=(2\pi)^{-n}\int e^{i\langle x,\xi\rangle}\phi_\nu(\xi)\hat{u}(\xi)\,d\xi.$$

We shall estimate u_ν by moving the integration into the complex domain in the direction θ_ν. This requires that we extend the definition of ϕ_ν to $\mathbb{C}^n$.

As in the proof of Theorem 8.4.8 we set

$$\phi_\nu(\xi+i\eta)=\sum_{|\alpha|\leqq k_\nu}(i\eta)^\alpha\partial^\alpha\phi_\nu(\xi)/\alpha!; \qquad \xi,\eta\in\mathbb{R}^n;$$

where k_ν is the largest integer $<\log(3+|\xi_\nu|)-1$. Recall that this implies

$$2\partial\phi_\nu(\zeta)/\partial\bar{\zeta}_j = \sum_{|\alpha|=k_\nu} (i\eta)^\alpha \partial_j \partial^\alpha \phi_\nu(\xi)/\alpha!, \qquad \zeta=\xi+i\eta,$$

or if we use (iii)

$$|\partial\phi_\nu(\zeta)/\partial\bar{\zeta}_j| \leqq (C_1\varepsilon|\operatorname{Im}\zeta|)^{k_\nu}/k_\nu!. \tag{16.3.9}$$

By Stokes' formula (or as in the proof of (8.1.15))

$$u_\nu(x)=(2\pi)^{-n}\int e^{i\langle x,\xi+ik_\nu\theta_\nu/\varepsilon\rangle}(\phi_\nu\hat{u})(\xi+ik_\nu\theta_\nu/\varepsilon)\,d\xi$$
$$+2i(2\pi)^{-n}\int\int_0^{k_\nu/\varepsilon} e^{i\langle x,\xi+it\theta_\nu\rangle}\langle\bar{\partial}\phi_\nu(\xi+it\theta_\nu),\theta_\nu\rangle\,\hat{u}(\xi+it\theta_\nu)\,d\xi\,dt.$$

In the second integral we estimate the exponential by $e^{|x|t}$ and estimate $\hat{u}(\xi+it\theta_\nu)$ by $|\xi_\nu|^{N+1}\exp(-tr)$ which is possible for large ν by (16.3.7) with $h=h_\nu$ since $h_\nu(\theta_\nu)\leqq -r$. In view of (16.3.9) this gives the bound $|\xi_\nu|^{N+1}$ times

$$2n(2\pi)^{-n}\int_0^\infty (C_1\varepsilon t)^{k_\nu} e^{-rt/2}\,dt/k_\nu! = C_2(2C_1\varepsilon/r)^{k_\nu}$$

for the integral with respect to t, provided that $|x|<r/2$ which we assume from now on. In the first integral we estimate $|\phi_\nu|$ by

$$\sum C^{|\alpha|}k_\nu^{|\alpha|}/\alpha! = e^{nCk_\nu}$$

which gives the bound

$$e^{nCk_\nu}e^{-rk_\nu/2\varepsilon}|\xi_\nu|^{N+1} < e^{-rk_\nu/3\varepsilon}|\xi_\nu|^{N+1}$$

for the integrand if ε is small enough. Summing up, we have for $|x|<r/2$

$$|u_\nu(x)| < C_3|\xi_\nu|^{N+1}k_\nu^n(e^{-rk_\nu/3\varepsilon}+(2C_1\varepsilon/r)^{k_\nu}).$$

When ε is so small that $r/3\varepsilon>M$ and $2C_1\varepsilon/r<e^{-M}$ this means that

$$|u_\nu(x)| \leqq C_4|\xi_\nu|^{N+1-M}.$$

Hence $\sum u_\nu(x)$ is uniformly convergent when $|x|<r/2$. Differentiation of u_ν will only cause a factor ξ to appear, so we also have

$$|D^\alpha u_\nu(x)| \leqq C_{\alpha,M}|\xi_\nu|^{N+1-M}$$

for any M. Hence $u\in C^\infty$ when $|x|<r/2$, which completes the proof.

It is clear that the sets of supporting functions introduced in Definition 16.3.2 give much more information than the singular support alone since they take into account which frequencies give rise to the singularities. (The idea is thus quite close to that of the wave front set, and the connection could be developed further.) In terms of the quantities in Definition 16.3.2 we can therefore prove a rather precise theorem on the singular supports of convolutions from which more explicit special cases will be derived later on.

Theorem 16.3.6. *Let* $u_j', u_j'' \in \mathscr{E}'$, $j=1, \dots, k$. *Then*

$$\mathscr{H}(u_1' * u_1'', \dots, u_k' * u_k'') \\ = \{(h_1' + h_1'', \dots, h_k' + h_k''); (h_1', h_1'', \dots, h_k', h_k'') \in \mathscr{H}(u_1', u_1'', \dots, u_k', u_k'')\}.$$

Proof. Let ξ_ν be a sequence $\to \infty$ in $\mathbb{R}^n$ such that $L_{u_j}(z, \xi)$ converges for $j=1, \dots, k$ if $u_j = u_j' * u_j''$. Denote the limits by V_j. By passing to a subsequence we may assume that $L_{u_j'}(z, \xi_\nu)$ and $L_{u_j''}(z, \xi_\nu)$ also converge. Denoting the limits by V_j' and V_j'' we have

$$V_j = V_j' + V_j''.$$

From Theorem 16.2.7 it follows now that the set in the left is contained in the one on the right. The opposite inclusion is proved in precisely the same way and is left for the reader.

Before starting to give weaker but more useful versions of Theorem 16.3.6 we shall study the convergence of L_u more closely. First note that if $\xi_j \to \infty$ and $L_u(z, \xi_j) \to V$, then every limit of $L_u(z, \eta_j)$ when $\eta_j - \xi_j = O(\log|\xi_j|)$ and $j \to \infty$ is a translation of V. In fact, the limit is $V(z+\theta)$ if $(\eta_j - \xi_j)/\log|\xi_j| \to \theta$. A slightly less obvious result is given in the following

Lemma 16.3.7. *If* $u \in \mathscr{E}'$, $\xi_j \to \infty$ *and* $L_u(., \xi_j) \to V$ *where* V *is a plurisubharmonic function with supporting function* h, *then there exist numbers* $R_j \to \infty$ *and* C *such that if* $E = \{\xi \in \mathbb{R}^n; |\xi - \xi_j| < R_j \log|\xi_j|$ *for some* $j\}$ *then*

$$\varlimsup_{E \ni \xi \to \infty} L_u(z, \xi) \leqq C + h(\operatorname{Im} z).$$

Proof. We assume in the proof that $h \not\equiv -\infty$; the case $h = -\infty$ can be handled in the same way. If N is the order of u we have $V(z) \leqq N + h(\operatorname{Im} z)$, and it follows from Theorem 4.1.9 that for every k one can find j_k so that

$$L_u(z, \xi_j) < N + 1 + h(\operatorname{Im} z) \quad \text{if } |z| < 2k \text{ and } j > j_k.$$

We can take j_k increasing. If we introduce the inverse of the function $k \to j_k$ this means that we can find $R_j \to \infty$ when $j \to \infty$ so that

$$L_u(z, \xi_j) < N + 1 + h(\operatorname{Im} z) \quad \text{if } |z| < 2R_j.$$

We may choose R_j so that $R_j \log|\xi_j| = o(|\xi_j|)$ when $j \to \infty$. When $|\xi - \xi_j| < R_j \log|\xi_j|$ we write $\xi + z \log|\xi| = \xi_j + w_j \log|\xi_j|$ and find for fixed z that

$$|w_j| = |\zeta - \xi_j + z \log|\xi||/|\log|\xi_j|| < 2R_j$$

if j is large. Hence the lemma follows with $C = N+1$.

Lemma 16.3.8. *Let ξ_j and R_j be sequences $\to \infty$ in $\mathbb{R}^n$ and in $\mathbb{R}_+$ respectively, and set $E = \{\xi \in \mathbb{R}^n;\ |\xi - \xi_j| < R_j \log|\xi_j| \text{ for some } j\}$. Then there exists a function $u \in \mathscr{E}' \cap C^0$ with $\operatorname{sing\,supp} u = \{0\}$ and $u \notin C^1$ such that*

(16.3.10) $$L_u(z, \xi) \to -\infty \quad \text{when } E \not\ni \xi \to \infty,$$

with uniform convergence on compact subsets of $\mathbb{C}^n$, whereas a subsequence of $L_u(z, \xi_j)$ converges to 0.

Proof. Let $\mathscr{F}$ be the Fréchet space of all $u \in C_0^0$ with support in the unit ball such that $u \in C^\infty(\complement\{0\})$ and

$$p_{N,m}(u) = \sup\{|\hat{u}(\zeta)||\xi|^N;\ \xi \notin E, |\zeta - \xi| < m \log|\xi|\} < \infty$$

for all positive integers N and m. It is clear that the latter condition implies (16.3.10). Let t_j be a positive sequence such that $\log t_j / \log|\xi_j| \to 0$ but $t_j/(\log|\xi_j|)^n \to \infty$ as $j \to \infty$. Set

$$\mathscr{F}_0 = \{u \in \mathscr{F};\ \lim t_j |\hat{u}(\xi_j)| = 0\}.$$

We shall prove that $\mathscr{F}_0 \neq \mathscr{F}$. Every $u \in \mathscr{F} \setminus \mathscr{F}_0$ has the required properties. That $u \notin C^1$ is clear since $u \in C^1$ implies that $|\xi_j||\hat{u}(\xi_j)|$ is bounded, hence that $t_j |\hat{u}(\xi_j)| \to 0$. Moreover, $u \notin \mathscr{F} \setminus \mathscr{F}_0$ implies $\overline{\lim}\, L_u(0, \xi_j) \geqq 0$, so the sequence $L_u(z, \xi_j)$ has a subsequence with a plurisubharmonic limit V such that $V(0) \geqq 0$. Since $u \in C^0$ and $\operatorname{sing\,supp} u = \{0\}$ we have $V \leqq 0$ everywhere, hence $V = 0$ identically by Lemma 16.2.2.

If $\mathscr{F}_0$ were equal to $\mathscr{F}$ we would have a closed map

$$\mathscr{F} \ni u \to \{t_j \hat{u}(\xi_j)\} \in l^\infty.$$

By the closed graph theorem it must be continuous, so

(16.3.11) $$\sup t_j |\hat{u}(\xi_j)| \leqq C(\sup|u| + \sum_{|\alpha| < N_1} \sup_K |D^\alpha u| + p_{N,m}(u)), \qquad u \in \mathscr{F},$$

for some constants N, m, N_1, C and some compact set $K \subset \complement\{0\}$. Let X be a convex neighborhood of 0 contained in the unit ball such that $X \cap K = \emptyset$, and choose $\psi \in C_0^\infty(X)$ with $\psi \geqq 0$ and $\int \psi\, dx = 1$. We define $u_j \in C_0^\infty(X)$ by

$$\hat{u}_j(\zeta) = (\hat{\psi}((\zeta - \xi_j)/k_j))^{k_j}$$

where k_j is the largest integer $< \log|\xi_j|$ and j is large. (This is essentially the same construction as in the proof of Theorem 1.3.5.) Then we obtain

$$\sup|u_j| \leqq \int |\hat{\psi}(\xi/k_j)| d\xi = C k_j^n = o(t_j).$$

The left-hand side of (16.3.11) is at least equal to t_j when $u=u_j$ so the lemma will be proved if we show that $p_{N,m}(u_j)$ is bounded.

Let $\xi \notin E$, $\zeta \in \mathbb{C}^n$ and $|\zeta - \xi| < m \log|\xi|$. Then $|\xi - \xi_j| \geqq R_j \log|\xi_j|$ by the definition of E. If $z = (\zeta - \xi_j)/k_j$ it follows that

$$\begin{aligned}|\xi| &\leqq |\zeta - \xi_j| + |\zeta - \xi| + |\xi_j| \leqq k_j|z| + m\log|\xi| + e^{k_j+1}\\ &\leqq k_j|z| + |\xi|/2 + C + e^{k_j+1},\end{aligned}$$

where C and the following constants depend on m. Hence

$$\begin{aligned}|\xi| &< C_1 e^{k_j}(1 + k_j|z|) < (C_2(1+|z|))^{k_j},\\ k_j|z| &= |\zeta - \xi_j| \geqq |\xi - \xi_j| - |\zeta - \xi| \geqq R_j \log|\xi_j| - m\log|\xi|\\ &\geqq k_j(R_j - m\log(C_2(1+|z|)),\end{aligned}$$

which implies $|z| > R_j/2$ if j is large. Since

$$|\operatorname{Im} z| = |\operatorname{Im}\zeta|/k_j \leqq m(\log|\xi|)/k_j$$

it also follows that

$$e^{|\operatorname{Im} z|} \leqq (C_2(1+|z|))^m. \tag{16.3.12}$$

Altogether we have found that

$$|\hat{u}_j(\zeta)|\,|\xi|^N \leqq (|\hat{\psi}(z)|\, C_2^N(1+|z|)^N)^{k_j}$$

where z satisfies (16.3.12) and $|z| > R_j/2$. Since $\hat{\psi}$ is rapidly decreasing in the set defined by (16.3.12) we conclude that $p_{N,m}(u_j) \to 0$. The lemma is proved.

We shall now examine when a convolution with no C^∞ factor can be in C^∞.

Theorem 16.3.9. *If $u \in \mathscr{E}'$ the following conditions are equivalent:*

(i) $-\infty \in \mathscr{H}(u)$.

(ii) *for every $x \in \mathbb{R}^n$ one can find $w \in \mathscr{E}' \cap (C^0 \setminus C^1)$ with* $\operatorname{sing\,supp} w = \{x\}$ *and $w * u \in C^\infty$.*

(iii) *there exists some $w \in \mathscr{E}'$ such that $w * u \in C^\infty$ but $w \notin C^\infty$.*

Proof. (i) $\Rightarrow$ (ii). We may assume that $x = 0$. Choose $\xi_j \to \infty$ so that $L_u(.,\xi_j) \to -\infty$ and then choose E by means of Lemma 16.3.7 so that $L_u(.,\xi) \to -\infty$ if $E \ni \xi \to \infty$. By Lemma 16.3.8 we can find $w \in C^0 \setminus C^1$ with compact support and $\operatorname{sing\,supp} w = \{0\}$ so that $L_w(.,\xi) \to -\infty$ when $\xi \to \infty$ in $\complement E$. Hence

$$L_{u*w}(.,\xi) = L_u(.,\xi) + L_w(.,\xi) \to -\infty \quad \text{if } \xi \to \infty$$

so $u*w\in C^\infty$. (ii) $\Rightarrow$ (iii) is trivial. (iii) $\Rightarrow$ (i). We can choose $h_w \neq -\infty$ in $\mathscr{H}(w)$ (Theorem 16.3.4) and then h_u so that $(h_u, h_w)\in\mathscr{H}(u,w)$ (Lemma 16.3.3). By Theorem 16.3.6 we have $h_u+h_w\in\mathscr{H}(u*w)$ so $h_u+h_w=-\infty$, hence $h_u=-\infty$ which proves (i).

The negation of the conditions in Theorem 16.3.9 will be very important:

Theorem 16.3.10. *If* $u\in\mathscr{E}'$ *the following conditions are equivalent:*

(i) $-\infty\notin\mathscr{H}(u)$

(ii) *there is a constant* $A>0$ *such that*

$$\sup\{|\hat{u}(\zeta)|;\ \zeta\in\mathbb{C}^n, |\zeta-\xi|<A\log(2+|\xi|)\}>(A+|\xi|)^{-A}, \qquad \xi\in\mathbb{R}^n.$$

(iii) *for every* $a>0$ *there is a constant* $A>0$ *such that*

$$\sup\{|\hat{u}(\xi+\eta)|;\ \eta\in\mathbb{R}^n, |\eta|<a\log(2+|\xi|)\}>(A+|\xi|)^{-A}.$$

(iv) *for every* $a>0$ *there is a constant* $A>0$ *such that*

$$\int_{|\zeta|<a}\log|\hat{u}(\xi+\zeta\log|\xi|)|\,d\lambda(\zeta)>-A\log|\xi| \quad \text{if } \xi\in\mathbb{R}^n,\ |\xi|>2.$$

(v) *if* $w\in\mathscr{E}'$ *and* $\hat{w}/\hat{u}$ *is an analytic function then it is the Fourier transform of a distribution in* $\mathscr{E}'$.

Proof. It is obvious that (iii) $\Rightarrow$ (ii) $\Rightarrow$ (i). The implication (i) $\Rightarrow$ (iv) follows from Theorem 4.1.9 a). To prove that (iv) $\Rightarrow$ (iii) we assume that (iii) is not valid. This means that there is a sequence $\xi_j\to\infty$ such that $L_u(x,\xi_j)\to-\infty$ if $|x|<a$ and x is real. If $y\in\mathbb{R}^n$ it follows that $L_u(x+wy,\xi_j)\to-\infty$ for all complex w, for by Proposition 16.1.2 a subsequence would otherwise have a finite upper limit for almost all real w. Hence $L_u(z,\xi_j)\to-\infty$ when $|z|<a$, so (iv) is not valid.

We shall now prove that (iv) $\Rightarrow$ (v). Let F be the entire function $\hat{w}/\hat{u}$. We shall then prove in order that

$$|F(\zeta)|<Ce^{A|\zeta|}, \qquad \zeta\in\mathbb{C}^n, \tag{16.3.13}$$

$$|F(\xi)|<C(1+|\xi|)^N, \qquad \xi\in\mathbb{R}^n, \tag{16.3.14}$$

$$|F(\zeta)|<C(1+|\zeta|)^N e^{A|\operatorname{Im}\zeta|}, \qquad \zeta\in\mathbb{C}^n. \tag{16.3.15}$$

By Theorem 7.3.1 the last estimate means that F is the Fourier-Laplace transform of a distribution in $\mathscr{E}'$.

To prove (16.3.13) we use that $|\hat{u}(\zeta)|<\exp(A|\zeta|+C)$ for some A and C. Assuming as we may that $\hat{u}(0)\neq 0$, the mean value of $\log|\hat{u}|$ over the ball $|\zeta|<R$ is at least $\log|\hat{u}(0)|$ so the mean value of the non-negative function $A|\zeta|+C-\log|\hat{u}(\zeta)|$ is $\leqq AR+C-\log|\hat{u}(0)|$. It follows

that the mean value of $|\log|\hat{u}(\zeta)||$ is $\leqq 2(AR+C)-\log|\hat{u}(0)|$. Hence

$$\sup_{|\zeta|<R/2} \log|F(\zeta)| = O(R)$$

for $\log|F(\zeta)|$ is bounded by the mean value of $\log|\hat{w}|-\log|\hat{u}|$ over the ball of radius $R/2$ and center at ζ. (So far we have proved a classical theorem of Lindelöf which does not depend on the condition (iv).)

To prove (16.3.14) we use condition (iv). Since $\log|F|$ is subharmonic we can use the mean value property to estimate $\log|F(\xi)|$. If m is the volume of the ball of radius a in $\mathbb{C}^n$ then

$$\begin{aligned} m\log|F(\xi)| &\leqq \int_{|\zeta|<a} \log|F(\xi+\zeta\log|\xi|)|\,d\lambda(\zeta) \\ &= \int_{|\zeta|<a} \log|\hat{w}(\xi+\zeta\log|\xi|)|\,d\lambda(\zeta) \\ &\quad - \int_{|\zeta|<a} \log|\hat{u}(\xi+\zeta\log|\xi|)|\,d\lambda(\zeta) \end{aligned}$$

so we obtain $\log|F(\xi)|<N\log|\xi|$ for some N, which proves (16.3.14). The proof of (16.3.15) requires a classical case of the Phragmén-Lindelöf principle.

Lemma 16.3.11. *Let v be a subharmonic function in $\mathbb{C}$ such that $v\leqq 0$ on $\mathbb{R}$, $v\leqq C$ on the positive imaginary axis and $v(z)\leqq C+A|z|$ when* $\operatorname{Im} z\geqq 0$. *Then $v(z)\leqq 0$ when* $\operatorname{Im} z\geqq 0$.

Proof. If $\varepsilon>0$ the subharmonic function

$$v(z)-\varepsilon\operatorname{Re}(ze^{-\pi i/4})^{\frac{3}{2}}$$

is $\leqq 0$ on the positive real axis, $\leqq C$ on the positive imaginary axis and $\to -\infty$ at ∞ in the first quadrant. Hence the maximum principle shows that it is $\leqq C^+$ in the whole quadrant where $C^+=\max(C,0)$. When $\varepsilon\to 0$ we obtain $v\leqq C^+$ in the first quadrant. This is of course also true in the second quadrant. Thus the subharmonic function

$$v(z)-\varepsilon\log|z+i|$$

tends to $-\infty$ at ∞ in the upper half plane and it is $\leqq 0$ on the boundary so it is $\leqq 0$ in the upper half plane. When $\varepsilon\to 0$ we obtain $v\leqq 0$.

End of proof of Theorem 16.3.10. To prove that (16.3.15) follows from (16.3.13) and (16.3.14) we consider for $\xi,\eta\in\mathbb{R}^n$ the subharmonic function

$$v(z)=\log|F(\xi+z\eta)|-N\log|2+|2\xi|-iz|2\eta||-A|\eta|\operatorname{Im} z-\log C$$

in the upper half plane. For real z we have $v(z)\leqq 0$ by (16.3.14), and on the positive imaginary axis we have $v(z)\leqq A|\xi|$ by (16.3.13). From (16.3.13) we also obtain the general bound required in Lemma 16.3.11 and conclude that $v(i)\leqq 0$. Apart from the size of the constant this is the estimate (16.3.15).

Finally assume that (v) is valid. Let H be the supporting function of $\operatorname{supp} u$ and denote by B the Banach space of all entire functions F such that

$$\|F\|=\sup|F(\zeta)||\hat{u}(\zeta)|\exp(-|\operatorname{Im}\zeta|-H(\operatorname{Im}\zeta))<\infty.$$

Set

$$E_N=\{F\in B;\ |F(\xi)|\leqq N(1+|\xi|)^N,\ \xi\in\mathbb{R}^n\}.$$

By hypothesis $\bigcup_N E_N=B$. Since every E_N is closed, convex and symmetric, it follows from Baire's theorem that E_N is a neighborhood of 0 for some N. Hence

$$\sup|F(\xi)|(1+|\xi|)^{-N}\leqq C\|F\|, \qquad F\in B.$$

If $w\in\mathscr{E}'$ has support in the unit ball and $w*u\in C$ we have $|\hat{w}(\zeta)\hat{u}(\zeta)|\leqq C_1\exp(|\operatorname{Im}\zeta|+H(\operatorname{Im}\zeta))$ by the Paley-Wiener-Schwartz theorem. Hence we can take $F=\hat{w}$ and obtain $\hat{w}(\xi)=O(|\xi|^N)$ as $\xi\to\infty$. If $w*u\in C^\infty$ we can apply this to the derivatives of w and obtain that $\hat{w}$ is rapidly decreasing, hence $w\in C_0^\infty$. Thus condition (ii) in Theorem 16.3.9 is false, so Theorem 16.3.9 shows that $-\infty\notin\mathscr{H}(u)$. We have therefore proved that (v) $\Rightarrow$ (i) which completes the proof.

Definition 16.3.12. The distribution u is called invertible and $\hat{u}$ is called slowly decreasing if the equivalent conditions in Theorem 16.3.10 are fulfilled.

In Section 16.5 we shall see that this property plays an essential role in the study of the inhomogeneous convolution equation $u*w=f$. A first example of this is condition (v) in Theorem 16.3.10. Note that invertibility of u only depends on the class of u in $\mathscr{E}'/C_0^\infty$. More generally, if u is invertible there is a number N such that $u+v$ is invertible for every $v\in C_0^N$. This follows immediately from say condition (ii) in Theorem 16.3.10.

We shall now discuss inclusions opposite to (16.3.3). In view of the applications to convolution equations it is natural to keep one of the distributions u_1 and u_2 fixed.

Theorem 16.3.13. *If $u \in \mathscr{E}'$ one can for every $h \in \mathscr{H}(u)$ find $w \in \mathscr{E}'$ with* $\operatorname{sing\,supp} w = \{0\}$ *so that h is equal to the supporting function of* $ch \operatorname{sing\,supp} u*w$. *If K and K' are compact convex non-empty sets with supporting functions H and H' then*

$$w \in \mathscr{E}', \quad \operatorname{sing\,supp} u*w \subset K \Rightarrow \operatorname{sing\,supp} w \subset K' \tag{16.3.16}$$

if and only if for every $h \in \mathscr{H}(u)$ and $x \in \mathbb{R}^n$

$$h(\xi) + \langle x, \xi \rangle \leqq H(\xi) \quad \text{for all } \xi \in \mathbb{R}^n \Rightarrow x \in K'. \tag{16.3.17}$$

In particular this requires that $-\infty \notin \mathscr{H}(u)$.

Proof. If $h \in \mathscr{H}(u)$ we can choose a sequence $\xi_j \to \infty$ so that $L_u(.,\xi_j)$ has a limit with supporting function h. By Lemma 16.3.7 and Lemma 16.3.8 we can then choose a neighborhood E of the sequence ξ_j such that

$$\varlimsup_{E \ni \xi \to \infty} L_u(z, \xi) \leqq C + h(\operatorname{Im} z)$$

and a function $w \in \mathscr{E}' \cap C^0$ with $\operatorname{sing\,supp} w = \{0\}$ such that

$$\varlimsup_{\complement E \ni \xi \to \infty} L_w(.,\xi) = -\infty$$

but a subsequence of $L_w(.,\xi_j)$ converges to 0. It follows that $\mathscr{H}(u*w)$ contains h and otherwise only supporting functions bounded by h, so $ch \operatorname{sing\,supp} u*w$ has the supporting function h (Theorem 16.3.4). This proves the first statement. If $h(\xi) + \langle x, \xi \rangle \leqq H(\xi)$ we obtain

$$\operatorname{sing\,supp} u*(w*\delta_x) \subset K.$$

Hence (16.3.16) implies (16.3.17). On the other hand, assume that (16.3.17) is fulfilled. If $h_w \in \mathscr{H}(w)$ we can find $h \in \mathscr{H}(u)$ so that $(h_w, h) \in \mathscr{H}(w, u)$. Hence $h_w + h \in \mathscr{H}(w*u)$ so $h_w + h \leqq H$ if $\operatorname{sing\,supp} u*w \subset K$. If h_w is the supporting function of the compact convex set k, and $x \in k$, then

$$h(\xi) + \langle x, \xi \rangle \leqq h(\xi) + h_w(\xi) \leqq H(\xi).$$

Hence (16.3.17) gives $x \in K'$, so $\operatorname{sing\,supp} w \subset K'$ by Theorem 16.3.4.

Remark. Using Theorem 16.3.5 we obtain the stronger result that $\operatorname{sing\,supp} u*w \subset K$ implies that $\operatorname{sing\,supp} w$ is contained in the closure of the set of points x such that

$$h(\xi) + \langle x, \xi \rangle \leqq H(\xi), \qquad \xi \in \mathbb{R}^n,$$

for some $h \in \mathscr{H}(u)$.

Corollary 16.3.14. *Let $u \in \mathscr{E}'$. In order that for every $v \in \mathscr{E}'$*

$$\text{(16.3.18)} \quad ch \operatorname{sing\,supp}(u*v) = ch \operatorname{sing\,supp} u + ch \operatorname{sing\,supp} v$$

it is necessary and sufficient that $\mathscr{H}(u)$ consists of the supporting function of sing supp u *only.*

Proof. We can choose v with $\operatorname{sing\,supp} v = \{0\}$ so that the supporting function of the left-hand side is any $h \in \mathscr{H}(u)$. This proves the necessity, and the sufficiency is also an immediate consequence of Theorem 16.3.6, Theorem 16.3.4 and Lemma 16.3.3

Corollary 16.3.15. *Let $u \in \mathscr{E}'$ be invertible. Then we have*

$$\text{(16.3.19)} \quad ch \operatorname{sing\,supp} v \subset ch \operatorname{sing\,supp}(u*v) - ch \operatorname{sing\,supp} u, \qquad v \in \mathscr{E}'.$$

Proof. If h is a supporting function $\neq -\infty$ and

$$h(\xi) + \langle x, \xi \rangle \leqq H(\xi)$$

we obtain since $h(\xi) + h(-\xi) \geqq h(0) = 0$ that

$$\langle x, \xi \rangle \leqq H(\xi) + h(-\xi).$$

Hence (16.3.17) is fulfilled if $K' = K - ch \operatorname{sing\,supp} u$, which proves (16.3.19)

We shall now discuss some examples.

Example 16.3.16. If $u \neq 0$ is a distribution with support at 0, then the only limits of $L_u(.,\xi)$ as $\xi \to \infty$ are constants $\in [0, N]$ where N is the order of u. Thus the supporting functions in $\mathscr{H}(u)$ are all 0.

Proof. Since $\hat{u}$ is a polynomial of degree N, we have by Taylor's formula

$$\hat{u}(\xi + z \log|\xi|) = a(\xi) p_\xi(z)$$

where p_ξ is a polynomial of degree N with $\sum |D^\alpha p_\xi(0)| = 1$ and

$$C_1 \leqq |a(\xi)| < C_2 |\xi|^N.$$

It is obvious that $(\log|p_\xi|)/\log|\xi| \to 0$ as $\xi \to \infty$, and this proves the assertion.

Note that the "localizations at infinity" used in this section are so much cruder than those in Section 10.2 that all information on the polynomial except the order is wiped out.

The following result allows us to study more general examples.

Proposition 16.3.17. *Let* $u_1, \dots, u_k \in \mathscr{E}'$ *have disjoint singular supports and set* $u = \sum u_j$. *Then*

$$(h, h_1, \dots, h_k) \in \mathscr{H}(u, u_1, \dots, u_k)$$

implies that $h = \sup_j h_j$. *We have*

$$\mathscr{H}(u) = \{\sup_j h_j; (h_1, \dots, h_k) \in \mathscr{H}(u_1, \dots, u_k)\}. \tag{16.3.20}$$

Proof. Choose a sequence $\xi_\nu \to \infty$ such that $L_{u_j}(\cdot, \xi_\nu)$ and $L_u(\cdot, \xi_\nu)$ converge to plurisubharmonic functions with supporting functions h_j and h respectively. By Lemmas 16.3.7 and 16.3.8 we can then find $w \in \mathscr{E}'$ with $\operatorname{sing\,supp} w = \{0\}$ so that $ch \operatorname{sing\,supp} u_j * w$ and $ch \operatorname{sing\,supp} u * w$ have the supporting functions h_j and h. (See the first part of the proof of Theorem 16.3.13.) Since $u * w = \sum u_j * w$ and the sets $\operatorname{sing\,supp} u_j * w \subset \operatorname{sing\,supp} u_j$ are disjoint, it follows that $ch \operatorname{sing\,supp}(u * w)$ is the convex hull of the sets $ch \operatorname{sing\,supp} u_j * w$. Hence $h = \sup_j h_j$. This proves the first part of the statement, and (16.3.20) follows in view of Lemma 16.3.3.

Corollary 16.3.18. *If* H *is the supporting function of the closed convex hull of all isolated points in* $\operatorname{supp} u$, *then* $H \leqq h$ *for every* $h \in \mathscr{H}(u)$.

Proof. If x is an isolated point in $\operatorname{supp} u$, we can write $u = u_1 + u_2$ where $u_1, u_2 \in \mathscr{E}'$ and $\operatorname{supp} u_1 = \{x\}$ while $\operatorname{sing\,supp} u_2$ does not contain the point x. By Proposition 16.3.17 and Example 16.3.16 it follows that $\langle x, \xi \rangle \leqq h(\xi)$, hence $H \leqq h$.

In particular Corollary 16.3.18 shows that $\mathscr{H}(u)$ contains only the supporting function of $ch \operatorname{supp} u$ if u has finite support.

Example 16.3.19. For every convex compact set K one can find a measure u with $\operatorname{supp} u \subset K$ such that the supporting function of K is the only element in $\mathscr{H}(u)$. In fact, let π be the smallest affine hyperplane containing K and choose a sequence x_j in the interior of K relative to π such that the limit points of the sequence are precisely all points in the boundary of K relative to π. Then $u = \sum 2^{-j} \delta_{x_j}$ has the required property by Corollary 16.3.18.

On the other hand, the following theorem shows that (16.3.18) is not valid for some distributions with very simple singularities.

Theorem 16.3.20. *Let* X *be a bounded open convex set with* C^∞ *boundary* ∂X *of strictly positive curvature. When* $\xi \in S^{n-1}$ *we denote by* D_ξ *the line segment between the two points on* ∂X *where* ξ *is a normal of* ∂X.

If $u = a\,dS$ where a is a positive C^∞ function and dS the area element on ∂X, then $\mathscr{H}(u)$ consists of the supporting functions of the intervals D_ξ.

Proof. Let us first consider a part of the surface which can be represented in the form

$$x_n = \psi(x')$$

and assume that v is a C_0^∞ density $a(x')\,dx'$ on this surface. Thus

$$\hat{v}(\xi) = \int e^{-i(\langle x', \xi'\rangle + \psi(x')\xi_n)} a(x')\,dx',$$
$$\hat{v}(\xi + z\log|\xi|) = \int e^{-i(\langle x', \xi'\rangle + \psi(x')\xi_n)} |\xi|^{-i(\langle x', z'\rangle + \psi(x')z_n)} a(x')\,dx'.$$

When z has a fixed bound then

$$|D_{x'}^\alpha a(x')|\xi|^{-i(\langle x', z'\rangle + \psi(x')z_n)}| \leqq C_\alpha |\xi|^{\langle x', \operatorname{Im} z'\rangle + \psi(x')\operatorname{Im} z_n} (\log|\xi|)^{|\alpha|}.$$

Hence it follows from Theorem 7.7.1 that $\hat{v}(\xi + z\log|\xi|)$ can be estimated by any power of $1/|\xi|$ if $|\xi_n|\sup|\psi'| < |\xi'|/2$. Let us therefore assume that

$$|\xi'| < 2\sup|\psi'|\,|\xi_n|.$$

Since ψ is strictly convex the equation for the critical point

$$\xi' + \xi_n \partial\psi(x')/\partial x' = 0$$

has at most one solution $x' \in \operatorname{supp} a$, and it must be a homogeneous function of ξ of degree 0. Let Γ be an open cone such that there is a solution with $|a(x')| > c > 0$ when $\xi \in \Gamma$. Then Theorem 7.7.6 gives

$$\begin{aligned}\hat{v}(\xi + z\log|\xi|) = {} & |\det(\xi_n \psi''(x')/2\pi)|^{-\frac{1}{2}} e^{\pm\pi i(n-1)/4} \\ & \cdot e^{-i(\langle x', \xi'\rangle + \psi(x')\xi_n)} |\xi|^{-i(\langle x', z'\rangle + \psi(x')z_n)} \\ & \cdot a(x')(1 + O((\log|\xi|)^2/|\xi|)), \qquad \xi \in \Gamma,\end{aligned}$$

where $x' = x'(\xi)$. If $\xi \to \infty$ in Γ and $\xi/|\xi| \to \xi_0$ then

$$\log|\hat{v}(\xi + z\log|\xi|)|/\log|\xi| + (n-1)/2 \to \langle x'(\xi_0), \operatorname{Im} z'\rangle + \psi(x')\operatorname{Im} z_n = h(\operatorname{Im} z)$$

where h is the supporting function of the point where ξ_0 is normal to the surface.

To prove the theorem we must compute the limit of $L_u(z, \xi)$ as $\xi \to \infty$ and $\xi/|\xi| \to \xi_0$. In doing so we can split u into a sum $u = u_1 + u_2 + u_3$ where u_1 and u_2 have disjoint supports and can be represented just as v above and ξ_0 is never normal to ∂X in $\operatorname{supp} u_3$. Then $\hat{u}_3(\xi + z\log|\xi|)$ is rapidly decreasing when $\xi \to \infty$ in a conic neighborhood of ξ_0, so $L_u(z, \xi)$ and $L_{u_1+u_2}(z, \xi)$ have the same limits then. By Proposition 16.3.17 every supporting function of a limit of L_u when

$\xi \to \infty$ in the direction ξ_0 is therefore of the form $\max(h_1, h_2)$ where h_j corresponds to a limit of $L_{u_j}(z, \xi)$. Thus h_j is the supporting function of a point where ξ_0 is normal to ∂X, which proves the theorem.

Remark. Using a similar splitting of an arbitrary $u \in \mathscr{E}'$ it is easy to see that for every $h \in \mathscr{H}(u)$ there is some $\xi \in \mathbb{R}^n \setminus \{0\}$ such that $h \leqq H_\xi$ if H_ξ is the supporting function of the convex hull of

$$\{x; (x, \xi) \in WF(u)\}.$$

This would also have been quite obvious if we had introduced the microlocal form of the sets $\mathscr{H}(u)$.

16.4. The Approximation Theorem

Let $0 \neq \mu \in \mathscr{E}'(\mathbb{R}^n)$ and let X be an open *convex* set in $\mathbb{R}^n$. If $u \in C^\infty(X)$ the convolution $\mu * u$ is defined in the open set

$$X_\mu = \{x; x - y \in X \text{ when } y \in \operatorname{supp} \mu\}. \tag{16.4.1}$$

We shall study the set $N_\mu(X) = \{u \in C^\infty(X); \mu * u = 0 \text{ in } X_\mu\}$. This is a closed linear subspace of $C^\infty(X)$ which is equal to $C^\infty(X)$ when X_μ is empty.

By E_μ we denote the set of all linear combinations of exponential solutions of the equation $\mu * u = 0$ in $\mathbb{R}^n$, that is, solutions of the form

$$u(x) = f(x)\, e^{i\langle x, \zeta\rangle}$$

where $\zeta \in \mathbb{C}^n$ and f is a polynomial. Since

$$\mu * e^{i\langle \cdot, \zeta\rangle} = \hat{\mu}(\zeta) e^{i\langle \cdot, \zeta\rangle}$$

the equation $\mu * (f e^{i\langle \cdot, \zeta\rangle}) = 0$ is equivalent to

$$f(D_\zeta)(e^{i\langle \cdot, \zeta\rangle} \hat{\mu}(\zeta)) = 0$$

or by Leibniz' rule, if $f^{(\alpha)} = \partial^\alpha f$

$$f^{(\alpha)}(D)\hat{\mu}(\zeta) = 0 \qquad \text{for all } \alpha.$$

In particular, ζ must be a zero of $\hat{\mu}$.

The main theorem of this section is the following

Theorem 16.4.1. *The restrictions to X of elements in E_μ are dense in $N_\mu(X)$ with the topology induced by $C^\infty(X)$, if X is an open convex set.*

Theorem 7.3.6 is of course a special case of this theorem, and the method of proof is basically the same. By the Hahn-Banach theorem the statement means that every $v \in \mathscr{E}'(X)$ which is orthogonal to E_μ is also orthogonal to $N_\mu(X)$. The first step in the proof is

Lemma 16.4.2. *A distribution $v \in \mathscr{E}'$ is orthogonal to E_μ if and only if $\hat{v}(\zeta) = F(\zeta)\hat{\mu}(-\zeta)$ where F is an entire analytic function.*

Proof. No change is required in the proof of Lemma 7.3.7 except that the vector θ may have to be varied; θ is chosen so that the proof works as before in the neighborhood of a given point. We leave for the reader to reconsider the proof.

Proof of Theorem 16.4.1 when μ is invertible. In this case the proof of Theorem 7.3.6 is still valid with minor modifications. In fact, it follows from condition (v) in Theorem 16.3.10 that $F = \hat{\sigma}$ where $\sigma \in \mathscr{E}'$. By the theorem of supports

$$ch \operatorname{supp} v = ch \operatorname{supp} \sigma - ch \operatorname{supp} \mu.$$

Hence $ch \operatorname{supp} \sigma \subset X_\mu$. Since $\check{v} = \check{\sigma} * \mu$ we have

$$v(u) = \check{v} * u(0) = \check{\sigma} * \mu * u(0) = \sigma(\mu * u)$$

if $u \in C^\infty(\mathbb{R}^n)$, hence also when $u \in C^\infty(X)$, for only the restriction of u to a compact subset of X is relevant. If $\mu * u = 0$ in X_μ it follows that $v(u) = 0$ and the theorem is proved.

To prove the theorem in general we must first examine the estimates for F which follow from the results of Section 16.2 although F grows too fast in $\mathbb{R}^n$ to be a Fourier transform of a distribution.

Lemma 16.4.3. *Let v_j, $j = 1, 2, 3$, be plurisubharmonic functions in $\mathbb{C}^n$ which are not identically $-\infty$, and let $v_3 = v_1 + v_2$. Assume that v_1 and v_3 satisfy* (16.2.1) *and denote their supporting functions by H_1 and H_3. Then it follows that $H_2 = H_3 - H_1$ is a supporting function and that for every $\varepsilon > 0$ there is a constant C_ε such that*

$$v_2(z) < H_2(\operatorname{Im} z) + \varepsilon |z| + C_\varepsilon. \tag{16.4.2}$$

Proof. If $y \in \mathbb{R}^n$ we obtain from (16.2.8) applied to v_1 and v_3

$$\varlimsup_{t \to \infty} m(B^t)^{-1} \iint_{K \times B^t} |v_2(\zeta + twy)/t - H_2(\operatorname{Im} wy)| \, d\lambda(w) \, d\lambda(\zeta)$$
$$\leq \varlimsup_{t \to \infty} 2m(B^t)^{-1} \iint_{K \times B^t} (|H_1(\operatorname{Im} \zeta)| + |H_3(\operatorname{Im} \zeta)|) t^{-1} \, d\lambda(w) \, d\lambda(\zeta).$$

Here K can be taken as any compact set in $\mathbb{C}$ and we choose for B' the ball $|\zeta|<2\delta t$ where $\delta>0$. In the right-hand side the mean value is then independent of t. For any given w_0 with $|w_0|=1$ we can choose K as the disc with center w_0 and small positive radius ε. By the mean value property

$$\begin{aligned}&v_2(\zeta+w_0 t y)/t-H_2(\operatorname{Im} w_0 y)\\ &\quad\leqq\int\int(v_2(\zeta'+wty)/t-H_2(\operatorname{Im} w_0 y))\,d\lambda(\zeta')\,d\lambda(w)/M\end{aligned}$$

where the integral is taken over the set where $|\zeta'-\zeta|<\delta t$, $w\in K$, and M is the measure of this set. If A is the Lipschitz constant of H_2 we have

$$H_2(\operatorname{Im} w_0 y)\geqq H_2(\operatorname{Im} w y)-A\varepsilon|y|,$$

and we can estimate $|H_1(\operatorname{Im}\zeta)|+|H_3(\operatorname{Im}\zeta)|$ by $B|\zeta|$. This gives

(16.4.3) $$\varlimsup_{t\to\infty}\sup_{|\zeta|<\delta t} v_2(\zeta+w_0 t y)/t-H_2(\operatorname{Im} w_0 y)\leqq A\varepsilon|y|+2^{2n+2}B\delta.$$

We can let $\varepsilon\to 0$ in the right-hand side since ε does not occur on the left. If we take $\zeta=tz$ in (16.4.3) it follows that

$$v_2(z)\leqq C_1|z|+C_2$$

for some constants C_1 and C_2. This follows also more easily from the proof of (16.3.13). Thus $v_2(tz)/t\leqq C_1|z|+C_2$ when $t>1$.

Let t_j be any sequence $\to+\infty$ such that the sequence

$$v_2(t_j z)/t_j$$

has a plurisubharmonic limit V. Then it follows from (16.4.3) that

(16.4.4) $$\sup_{|z|<\delta} V(z+wy)\leqq H_2(\operatorname{Im} wy)+C\delta,\qquad |w|=1.$$

Hence

$$V(z)\leqq C_1|z|,\quad z\in\mathbb{C}^n;\qquad V(wy)\leqq H_2(\operatorname{Im} wy),\quad y\in\mathbb{R}^n,\ w\in\mathbb{C}.$$

In particular $V\leqq 0$ in $\mathbb{R}^n$, and (16.4.4) gives $V(x+ity)\leqq H_2(ty)+C|x|$ for real x, y, t. If we now apply Lemma 16.3.11 to the subharmonic function

$$\mathbb{C}\ni w\to V(x+wy)-\operatorname{Im} w H_2(y)$$

it follows that $V(x+wy)-\operatorname{Im} w H_2(y)\leqq 0$. Hence

(16.4.5) $$V(z)\leqq H_2(\operatorname{Im} z),\qquad z\in\mathbb{C}^n.$$

It follows from (16.4.5) that the supporting function h of V is defined and $\leqq H_2$. Let $H\in\mathscr{H}$ be the supremum of the supporting functions of all the limits so obtained. Then we have $H\leqq H_2$ and

(16.4.5)′ $$V(z)\leqq H(\operatorname{Im} z).$$

If $\varepsilon>0$ it follows that for sufficiently large t

$$v_2(tz)/t<H(\operatorname{Im} z)+\varepsilon, \quad |z|\leqq 1, \tag{16.4.6}$$

for otherwise (16.4.6) would be false for a sequence $t_j\to\infty$. Passing to a subsequence we may assume that $v_2(t_jz)/t_j$ has a plurisubharmonic limit V. Since V satisfies (16.4.5)′ this is a contradiction in view of Theorem 4.1.9.

We can write (16.4.6) in the equivalent form

$$v_2(z)<H(\operatorname{Im} z)+\varepsilon|z|, \quad |z|>C_\varepsilon. \tag{16.4.6$'$}$$

Hence $v_3(z)<H_1(\operatorname{Im} z)+C_1+H(\operatorname{Im} z)+\varepsilon|z|$, $|z|>C_\varepsilon$. Since $v_3(x)\leqq C_3$, $x\in\mathbb{R}^n$, an application of Lemma 16.3.11 to the subharmonic function

$$\mathbb{C}\ni w\to v_3(x+wy)-\operatorname{Im} w(H_1(y)+H(y)+\varepsilon|y|)$$

gives that

$$v_3(x+iy)\leqq H_1(y)+H(y)+C_3+\varepsilon|y|$$

for every $\varepsilon>0$. Thus $H_3\leqq H_1+H$, that is, $H\geqq H_2$, so $H_2=H$ is a supporting function. This completes the proof.

General Proof of Theorem 16.4.1. Again we assume that $v\in\mathscr{E}'(X)$ is orthogonal to E_μ and have by Lemma 16.4.2

$$\hat{v}(\zeta)=F(\zeta)\hat{\mu}(-\zeta).$$

Choose $\phi\in C_0^\infty$ and multiply both sides by $\hat{\phi}(\zeta)$. We can then apply Lemma 16.4.3 with

$$v_1(\zeta)=\log|\hat{\phi}(\zeta)\hat{\mu}(-\zeta)|, \quad v_2(\zeta)=\log|F(\zeta)|,$$
$$v_3(\zeta)=\log|\hat{\phi}(\zeta)\hat{v}(\zeta)|.$$

If H_1, H_3 and H are the supporting functions of $ch\operatorname{supp}\check{\mu}$, $ch\operatorname{supp} v$ and $ch\operatorname{supp}\phi$, it follows from the theorem of supports and Lemma 16.3.1 that $H+H_3$ and $H+H_1$ are the supporting functions of v_3 and v_1. Hence

$$H_2=H_3+H-(H_1+H)=H_3-H_1$$

is a supporting function, and for every $\varepsilon>0$ there is a constant C_ε such that

$$|F(\zeta)|\leqq C_\varepsilon\exp(H_2(\operatorname{Im}\zeta)+\varepsilon|\zeta|). \tag{16.4.7}$$

As in the proof of Theorem 16.4.1 in the invertible case we see that H_2 is the supporting function of a compact set $K_2\subset X_\mu$,

$$K_2+ch\operatorname{supp}\check{\mu}=ch\operatorname{supp} v\subset X. \tag{16.4.8}$$

It follows from Theorem 15.1.5 that F is the Fourier-Laplace transform of an analytic functional σ with support in K_2. When u is entire analytic we have

$$\nu(u)=\sigma(\mu*u). \tag{16.4.9}$$

It suffices to verify this equality of analytic functionals when $u=e^{-i\langle x,\zeta\rangle}$. Then the left-hand side is $\hat{\nu}(\zeta)$, we have $\mu*u=\hat{\mu}(-\zeta)u$, so the right-hand side is equal to $\hat{\mu}(-\zeta)F(\zeta)$ which proves (16.4.9). We claim that (16.4.9) remains valid if $u\in C_0^\infty$ and $\mu*u$ is analytic in a neighborhood of K_2. For the proof we set $u_j=u*E_j$ where

$$E_j(x)=(j/\pi)^{n/2}\exp(-j\langle x,x\rangle).$$

Then $u_j\to u$ in C^∞ as $j\to\infty$, u_j is entire analytic, and

$$\mu*u_j=E_j*(\mu*u)$$

converges uniformly to $\mu*u$ in a complex neighborhood of K_2 by the proof of Proposition 9.1.2. Since (16.4.9) is valid for u_j it follows when $j\to\infty$ that it is valid for u.

Now assume that $u\in C^\infty(X)$ and that $\mu*u=0$ in X_μ. Choose $\chi\in C_0^\infty(X)$ equal to 1 in a neighborhood of $K_2-\operatorname{supp}\mu$, and set $v=\chi u$. Then $v\in C_0^\infty$ and $\mu*v=0$ in a neighborhood of K_2. We can therefore apply (16.4.9) to v which gives

$$\nu(u)=\nu(v)=0.$$

This completes the proof of the theorem.

Corollary 16.4.4. *If X is an open convex set in $\mathbb{R}^n$ and μ a distribution with compact support such that $\{x\}-\operatorname{supp}\mu$ is not contained in X for any x, then the exponential solutions of $\mu*u=0$ are dense in $C^\infty(X)$.*

The methods developed in this section also allow us to prove an existence theorem for an arbitrary $\mu\in\mathscr{E}'$.

Theorem 16.4.5. *If X is convex and $0\neq\mu\in\mathscr{E}'$, then the equation $\mu*u=f$ has a solution $u\in C^\infty(X)$ for every f which is real analytic in X_μ.*

Proof. Choose a sequence of open convex sets $X^1\Subset X^2\Subset\ldots$ with union equal to X, and set $X_\mu^j=\{x;\{x\}-\operatorname{supp}\mu\subset X^j\}$. Then we have

$$X_\mu^1\Subset X_\mu^2\Subset\ldots,\qquad \bigcup X_\mu^j=X_\mu.$$

Theorem 16.4.5 will follow if for every j we can find $u_j\in C^\infty(\mathbb{R}^n)$ so that $\mu*u_j=f$ in X_μ^j. In fact, since $\mu*(u_j-u_{j-1})=0$ in X_μ^{j-1}, we can by Theorem 16.4.1 subtract from u_j an element in E_μ to make u_j-u_{j-1} small in X^{j-2}, say

$$|D^\alpha(u_j-u_{j-1})|<2^{-j}\quad\text{in } X^{j-2}\quad\text{if } |\alpha|\leqq j.$$

This guarantees the existence of $\lim u_j = u \in C^\infty(X)$. It is clear that $\mu * u = f$ in X_μ. (Compare with the proof of Theorem 10.6.7.)

The equation $\mu * u = f$ in X_μ^j means that

$$f(\phi) = (\mu * u)(\phi) = u(\check{\mu} * \phi) \quad \text{for } \phi \in C_0^\infty(X_\mu^j),$$

so to construct u by the Hahn-Banach theorem means to prove the continuity of the map

$$\check{\mu} * \phi \to f(\phi)$$

in the $\mathscr{E}'$ topology. Choose $\psi \in C_0^\infty$ not identically 0. We shall prove that

$$|f(\phi)| \leqq C \|\check{\mu} * \phi * \psi\|_{L^1}, \quad \phi \in C_0^\infty(X_\mu^j). \tag{16.4.10}$$

Admitting this estimate for a moment we conclude by the Hahn-Banach theorem that there is a function $v \in L^\infty(\mathbb{R}^n)$ such that

$$f(\phi) = \int v(\check{\mu} * \phi * \psi)\, dx, \quad \phi \in C_0^\infty(X_\mu^j).$$

If $u = v * \check{\psi}$ we conclude that $f = \mu * u$ in X_μ^j, and since $u \in C^\infty$ this proves the existence of u_j. (Note that

$$|D^\alpha u(x)| \leqq \|v\|_{L^\infty} \|D^\alpha \psi\|_{L^1},$$

so it would also be easy to find a solution u in any non-quasianalytic class of C^∞ functions.)

To prove (16.4.10) we set $M = \|\check{\mu} * \phi * \psi\|_{L^1}$ and have with $\Phi = \hat{\phi}$

$$|\hat{\mu}(-\zeta)\Phi(\zeta)\hat{\psi}(\zeta)| \leqq M \exp(H(\operatorname{Im}\zeta) + H_j(\operatorname{Im}\zeta)) \tag{16.4.11}$$

where H and H_j are the supporting functions of $ch \operatorname{supp} \check{\mu} * \psi$ and of $\bar{X}_\mu^j$ respectively. By Lemma 16.4.3 we have for every entire function Φ satisfying (16.4.11) and every $\varepsilon > 0$

$$|\Phi(\zeta)| \leqq C_{\Phi,\varepsilon} \exp(H_j(\operatorname{Im}\zeta) + \varepsilon|\zeta|). \tag{16.4.12}$$

Let K be a convex compact neighborhood of $\bar{X}_\mu^j$ in $\mathbb{C}^n$ contained in the set where f is analytic; by Proposition 9.1.2 we can choose it so that f is the uniform limit of entire functions on K. Now it follows from (16.4.12) and Theorem 15.1.5 that Φ is the Fourier-Laplace transform of an analytic functional v with support in $\bar{X}_\mu^j$. Thus we have for some C depending on Φ

$$|v(F)| \leqq C \sup_K |F(z)|, \tag{16.4.13}$$

for every entire analytic F.

Let B be the Banach space of entire functions Φ satisfying (16.4.11), with the norm of Φ defined as the smallest M which can be used in (16.4.11). The set B_C of all $\Phi \in B$ such that $\Phi(\zeta) = \hat{v}(\zeta)$ for some analytic functional satisfying (16.4.13) is for every C a closed, convex and symmetric subset. Since the union of the sets B_C is equal to B it follows that 0 is an interior point of B_C for some C. Hence (16.4.11) implies if F is an entire function

$$|v(F)| \leqq C' M \sup_K |F(z)|, \tag{16.4.13$'$}$$

where $\hat{v} = \Phi$ and C' is independent of Φ. In particular

$$|\phi(F)| \leqq C' M \sup_K |F|$$

when F is entire, hence also when $F = f$ since f can be approximated uniformly on K by entire functions. This completes the proof of (16.4.10) and of the theorem.

If we choose μ in a non-quasianalytic class of C^∞ functions which is invariant under differentiation we find that the equation $\mu * u = f$ cannot have a solution unless f is also in this class. The intersection of all such classes is the analytic class (Bang [1]; see also Boman [1]) so the hypothesis in Theorem 16.4.5 that f is analytic cannot be dropped when no condition is placed on μ. As pointed out in the course of the proof, it would have been possible to obtain a solution in a non-quasianalytic class of functions but even for differential operators it is not always possible to find an analytic solution. (See the references.)

16.5. The Inhomogeneous Convolution Equation

Let $\mu \in \mathscr{E}'(\mathbb{R}^n)$ and let X_1, X_2 be two non-empty open subsets of $\mathbb{R}^n$ such that

$$X_2 - \operatorname{supp} \mu \subset X_1. \tag{16.5.1}$$

Then the convolution $\mu * u$ is a well defined distribution in $\mathscr{D}'(X_2)$ if $u \in \mathscr{D}'(X_1)$. We shall study the existence of solutions of the inhomogeneous convolution equation

$$\mu * u = f \tag{16.5.2}$$

when $f \in \mathscr{D}'(X_2)$ is given. The results extend those of Sections 10.6 and 10.7 concerning differential equations. When the proofs are essentially identical we shall leave them for the reader. However, the first necessary condition did not arise for differential operators:

Theorem 16.5.1. *If* (16.5.2) *has a solution* $u \in \mathscr{D}'(X_1)$ *for every* $f \in C_0^\infty(X_2)$ *it follows that* μ *is invertible* (*Definition 16.3.12*).

Proof. If $v \in C_0^\infty(X_2)$ it follows from (16.5.2) that

$$\int f v\,dx = u(\check{\mu} * v). \tag{16.5.3}$$

We claim that for every compact subset K_2 of X_2 there are constants C and N such that

$$|\int f v\,dx| \leq C \sum_{|\alpha| \leq N} \sup|D^\alpha f| \sum_{|\beta| \leq N} \sup|D^\beta \check{\mu} * v|; \quad f, v \in C_0^\infty(K_2). \tag{16.5.4}$$

To prove (16.5.4) we denote by F the set $C_0^\infty(K_2)$ with the topology defined by the semi-norms $f \to \sup|D^\alpha f|$ and by V the set $C_0^\infty(K_2)$ with the topology defined by the semi-norms $v \to \sup|D^\beta \check{\mu} * v|$. Both are metrizable spaces and F is complete. Furthermore, the bilinear form $\int f v\,dx$ on $F \times V$ is trivially continuous with respect to f when v is fixed. Since (16.5.2) has a solution by hypothesis for every f, it follows from (16.5.3) that it is also continuous with respect to v when f is fixed. Hence the bilinear form is continuous, which means precisely that (16.5.4) is valid.

Let x_0 be an interior point of K_2. By Theorem 16.3.9 it is sufficient to prove that if $w \in \mathscr{E}'(\mathbb{R}^n)$, $\operatorname{sing\,supp} w = \{x_0\}$ and $\check{\mu} * w \in C^\infty$ then $w \in C^\infty$. In doing so we may assume that $\operatorname{supp} w$ is in the interior of K_2. Choose $\chi \in C_0^\infty(\mathbb{R}^n)$ non-negative with $\int \chi\,dx = 1$ and set $\chi_\delta(x) = \delta^{-n}\chi(x/\delta)$. For small δ we have $v_\delta = w * \chi_\delta \in C_0^\infty(K_2)$ and

$$\sup|D^\beta \check{\mu} * v_\delta| \leq \sup|D^\beta(\check{\mu} * w)| < \infty.$$

If γ is any multi-index, application of (16.5.4) to $D^\gamma v_\delta$ gives for small δ

$$|\int f(D^\gamma w * \chi_\delta)\,dx| \leq C_\gamma \sum_{|\alpha| \leq N} \sup|D^\alpha f| \leq C'_\gamma \|f\|_{(N+n)}, \quad f \in C_0^\infty(K_2),$$

where the last estimate follows from Lemma 7.6.3 (cf. Definition 7.9.1). Hence

$$|\langle f, D^\gamma w\rangle| \leq C'_\gamma \|f\|_{(N+n)}, \quad f \in C_0^\infty(K_2).$$

Replacing f by ψf where $\psi \in C_0^\infty(K_2)$ is 1 in a neighborhood of $\operatorname{supp} w$ we conclude that this is true for all $f \in \mathscr{S}$, so $D^\gamma w \in H_{(-N-n)}$ for every γ, hence $w \in C^\infty$ by Lemma 10.7.7. This completes the proof.

Additional conditions are required if the right-hand side f of the convolution equation does not have compact support but is allowed to grow fast at the boundary.

Theorem 16.5.2. *Assume that* (16.5.2) *has a solution* $u \in \mathscr{D}'(X_1)$ *for every* $f \in C^\infty(X_2)$. *For every compact set* $K_1 \subset X_1$ *one can then find a compact set* $K_2 \subset X_2$ *such that* $\operatorname{supp} v \subset K_2$ *if* $v \in C_0^\infty(X_2)$ *and* $\operatorname{supp} \check{\mu} * v \subset K_1$.

Proof. We can essentially use the same arguments as in the beginning of the proof of Theorem 16.5.1. However, we now choose for F the Fréchet space $C^\infty(X_2)$ with the topology defined by the semi-norms $f \to \sup_{K_2} |D^\alpha f|$, where K_2 is a compact subset of X_2, and for V the set of $v \in C_0^\infty(X_2)$ with $\operatorname{supp} \check{\mu} * v \subset K_1$ and the same semi-norms as before. Since the argument is also quite parallel to the proof of Theorem 10.6.6 we leave the details for the reader.

Proposition 16.5.3. *The necessary condition in Theorem 16.5.2 is equivalent to*

$$d(\operatorname{supp} v, \complement X_2) = d(\operatorname{supp} \check{\mu} * v, \complement X_1), \qquad v \in \mathscr{E}'(X_2). \tag{16.5.5}$$

Here $d(A, B) = \inf\{|x-y|; x \in A, y \in B\}$.

Proof. This is just a repetition of the proof of Theorem 10.6.3 with the reference to Theorem 7.3.2 replaced by a reference to the theorem of supports.

Definition 16.5.4. The pair (X_1, X_2) of open sets satisfying (16.5.1) is called μ-convex for supports if (16.5.5) is fulfilled or equivalently the necessary condition in Theorem 16.5.2 is valid.

By a regularization it follows of course that there is no difference between having (16.5.5) for all $v \in \mathscr{E}'(X_2)$ or for all $v \in C_0^\infty(X_2)$.

Example 16.5.5. If X_1 is convex and $X_2 = \{x; \{x\} - \operatorname{supp} \mu \subset X_1\}$ then X_2 is convex and it follows from the theorem of supports that (X_1, X_2) is μ-convex for supports. No other X_2 would do:

Proposition 16.5.6. *If* (X_1, X_2) *is* μ-*convex for supports then* X_2 *is a union of components of* $Y = \{x; \{x\} - \operatorname{supp} \mu \subset X_1\}$.

Proof. X_2 is closed in Y for if $X_2 \ni x_k \to x \in Y$ then $\operatorname{supp} \check{\mu} * \delta_{x_k}$ is for all k contained in a compact subset of X_1, hence $x_k = \operatorname{supp} \delta_{x_k}$ belongs to a compact subset of X_2 for all k. It must also contain the limit x.

Theorem 10.6.4 and Corollary 10.6.5 have obvious analogues for convolution equations which we leave for the reader to state and prove. The following is a strong converse of Theorems 16.5.1 and 16.5.2.

Theorem 16.5.7. *The following conditions on $\mu \in \mathscr{E}'(\mathbb{R}^n)$ and the open sets X_1, X_2 satisfying* (16.5.1) *are equivalent:*

(i) (16.5.2) *has a solution $u \in C^\infty(X_1)$ for every $f \in C^\infty(X_2)$,*

(ii) (16.5.2) *has a solution $u \in \mathscr{D}'(X_1)$ for every $f \in C^\infty(X_2)$.*

(iii) *μ is invertible and (X_1, X_2) is μ-convex for supports.*

Proof. (i) $\Rightarrow$ (ii) is trivial and (ii) $\Rightarrow$ (iii) by Theorems 16.5.1 and 16.5.2. The proof that (iii) $\Rightarrow$ (i) will not be based on the same method as the proof of Corollary 10.6.8. Instead we shall use the following basic facts on duality in Fréchet spaces.

Lemma 16.5.8. *Let E and F be Fréchet spaces, T a continuous linear map of E into F. Then T is surjective if and only if the adjoint T^* from F' to E' is injective and has a weak* closed range in E'.*

Lemma 16.5.9. *If E is a Fréchet space with dual E', a linear subset M of E' is weak* closed if and only if $M \cap U^\circ$ is weak* closed for every neighborhood U of 0 in E. Here U° is the polar set*

$$U^\circ = \{y;\, y \in E',\, |\langle x, y\rangle| \leqq 1 \text{ for every } x \in U\},$$

where $\langle\ ,\ \rangle$ is the canonical bilinear form in $E \times E'$.

For a proof see e.g. Schaefer [1, Chap. IV, 6.4 and 7.7].

End of Proof of Theorem 16.5.7. We apply Lemma 16.5.8 with $E = C^\infty(X_1)$, $F = C^\infty(X_2)$ and $T = \mu *$. Then $T^*: \mathscr{E}'(X_2) \to \mathscr{E}'(X_1)$ is convolution with $\check{\mu}$ so it is injective. Let U be a neighborhood of 0 in $C^\infty(X_1)$. We may assume that it is defined as

$$\{\chi \in C^\infty(X_1);\, C \sum_{|\alpha| \leqq N} \sup_{K_1} |D^\alpha \chi| \leqq 1\}$$

where C and N are constants and K_1 is a compact subset of X_1. Then U° is the set of all $\psi \in \mathscr{E}'(X_1)$ with

$$|\psi(\chi)| \leqq C \sum_{|\alpha| \leqq N} \sup_{K_1} |D^\alpha \chi|, \quad \chi \in C^\infty(X_1). \tag{16.5.6}$$

Thus $\operatorname{supp} \psi \subset K_1$ if $\psi \in U^\circ$. If $\psi \in U^\circ \cap (\check{\mu} * \mathscr{E}'(X_2))$ then $\psi = \check{\mu} * \phi$ for some $\phi \in \mathscr{E}'(X_2)$ and it follows from the μ-convexity for supports that $\operatorname{supp} \phi \subset K_2$ for a fixed compact set $K_2 \subset X_2$. Moreover, it follows from (16.5.6) that

$$|\hat{\psi}(\zeta)| \leqq C'(1+|\zeta|)^N \exp H_{K_1}(\operatorname{Im} \zeta); \quad \psi \in U^\circ,\ \zeta \in \mathbb{C}^n. \tag{16.5.7}$$

Now the proof of (iv) $\Rightarrow$ (v) in Theorem 16.3.10 (or the beginning of the proof that (v) $\Rightarrow$ (i)) shows that there is a constant M such that for entire analytic F

$$\sup |F(\xi)|(1+|\xi|)^{-M} \leqq M \sup |\hat{\mu}(-\zeta) F(\zeta)|(1+|\zeta|)^{-N} \exp(-H_{K_1}(\operatorname{Im} \zeta)).$$

It follows that

$$\Phi=\{\phi\in\mathscr{E}'(X_2);\ \check{\mu}*\phi\in U^\circ\}$$

is bounded in $\mathscr{D}'^{M+n+1}$, and since $\operatorname{supp}\phi\subset K_2$ when $\phi\in\Phi$ it follows that Φ is relatively weak* compact in $\mathscr{E}'(X_2)$.

Now the map

$$\mathscr{E}'(X_2)\ni\phi\to\check{\mu}*\phi\in\mathscr{E}'(X_1)$$

is weak* continuous, and U° is weak* closed, so it follows that Φ is weak* closed, hence weak* compact. This implies that $\check{\mu}*\Phi=U^\circ\cap(\check{\mu}*\mathscr{E}'(X_2))$ is weak* compact. The proof is now completed by Lemma 16.5.9.

We shall next study (16.5.2) when f is a distribution. Since existence theorems have already been given in the C^∞ case we shall gain in generality and clarity by working mod C^∞ as in Section 10.7. First note that if $\mu\in\mathscr{E}'(\mathbb{R}^n)$ and X_1, X_2 are open sets in $\mathbb{R}^n$ with

$$X_2-\operatorname{sing\,supp}\mu\subset X_1 \tag{16.5.1$'$}$$

then a linear map $\mathscr{D}'(X_1)/C^\infty(X_1)\to\mathscr{D}'(X_2)/C^\infty(X_2)$ is induced by convolution with μ. In fact, for any compact neighborhood K of $\operatorname{sing\,supp}\mu$ we can choose $v\in\mathscr{E}'(K)$ with $\mu-v\in C^\infty$. If $u\in\mathscr{D}'(X_1)$ the convolution $v*u$ is defined in $\{x;\{x\}-K\subset X_1\}$ which contains any compact subset of X_2 when K is close to $\operatorname{sing\,supp}\mu$. If $u\in\mathscr{D}'(X_1)$ the convolution $v*u$ can therefore be defined on any relatively compact open set in X_2, and modulo C^∞ it does not depend on the choice of v. If $u\in C^\infty$ the convolution is in C^∞ so it only depends on $u \bmod C^\infty$. We shall denote it by $\mu*$. If $u\in\mathscr{D}'(X_j)$ we write $s_j u$ for the residue class of u in $\mathscr{D}'(X_j)/C^\infty(X_j)$.

Theorem 16.5.10. *If the equation*

$$\mu*s_1u=s_2f \tag{16.5.2$'$}$$

has a solution $u\in\mathscr{D}'(X_1)$ for every $f\in\mathscr{D}'(X_2)$, it follows that μ is invertible.

Proof. If μ is not invertible one can find $v\in\mathscr{E}'$ with $\operatorname{sing\,supp}v=\{0\}$ and $\mu*v\in C^\infty$. From the associativity of the usual convolution and (16.5.2)′ it follows that

$$v*s_2f=(v*\mu)*s_1u=0.$$

Hence (16.5.2)′ cannot be solved unless $v*s_2f=0$ which is not true if f is the Dirac measure at a point in X_2 for example.

Theorem 16.5.11. *Assume that*

$$\mu*(\mathscr{D}'(X_1)/C^\infty(X_1))=\mathscr{D}'(X_2)/C^\infty(X_2).$$

For every compact set $K_1\subset X_1$ *one can then find a compact set* $K_2\subset X_2$ *such that* $\operatorname{sing\,supp} v\subset K_2$ *if* $v\in\mathscr{E}'(X_2)$ *and* $\operatorname{sing\,supp}\check{\mu}*v\subset K_1$.

The proof is so close to that of Theorem 10.7.6 that we leave it for the reader.

Proposition 16.5.12. *The necessary condition in Theorem 16.5.11 implies that* μ *is invertible. It is equivalent to*

(16.5.5)′ $\quad d(\operatorname{sing\,supp} v, \complement X_2)=d(\operatorname{sing\,supp}\check{\mu}*v, \complement X_1), \qquad v\in\mathscr{E}'(X_2),$

when μ *is invertible.*

Proof. The first statement follows from Theorem 16.3.9. On the other hand, if μ is invertible it follows from Corollary 16.3.15 that $\operatorname{sing\,supp} v$ is in a fixed compact set in $\mathbb{R}^n$ if $\operatorname{sing\,supp}\check{\mu}*v$ is. The proof of Theorem 10.7.3 is therefore applicable with no further change.

Definition 16.5.13. The pair (X_1, X_2) of open sets satisfying (16.5.1)′ is called μ-convex for singular supports if (16.5.5)′ is fulfilled or, equivalently, the necessary condition in Theorem 16.5.11 is satisfied.

Theorem 10.7.4 and Corollary 10.7.5 have obvious analogues for convolution equations which we leave for the reader to state and prove. However, the analogue of Example 16.5.5 is somewhat more complicated:

Proposition 16.5.14. *If* X_1 *and* X_2 *are convex sets satisfying* (16.5.1)′ *and* μ *is invertible, then* (X_1, X_2) *is* μ*-convex for singular supports if and only if* X_2 *contains every* $x\in\mathbb{R}^n$ *such that* $\{x\}-K\subset X_1$ *for some compact convex set* K *with* $H_K\in\mathscr{H}(\mu)$.

Proof. a) Necessity. Let K_1 be a convex compact subset of X_1 so large that $\{x\}-\operatorname{sing\,supp}\mu\subset K_1$ for some $x\in X_2$. By the μ-convexity for singular supports there is a compact set $K_2\subset X_2$ such that

$$\operatorname{sing\,supp} v\subset K_2 \quad \text{if } v\in\mathscr{E}'(X_2) \quad \text{and} \quad \operatorname{sing\,supp}\check{\mu}*v\subset K_1.$$

Then it follows from the first part of Theorem 16.3.13 that K_2 must contain every $x\in X_2$ such that $\{x\}-K\subset K_1$, for one can then find v with $\operatorname{sing\,supp} v=\{x\}$ and $\operatorname{sing\,supp}\check{\mu}*v\subset\{x\}-K$. Hence the convex compact set

$$\{x\in\mathbb{R}^n; \{x\}-K\subset K_1\}$$

is contained in K_2, for it contains some points in X_2. When K_1 increases to X_1 it follows that $\{x\}-K\subset X_1$ implies that $x\in X_2$.

b) Sufficiency. If K_1 is a convex compact subset of X_1, then

$$K_2=ch\{x; \{x\}-K\subset K_1 \text{ for some convex compact } K \text{ with } H_K\in\mathscr{H}(\mu)\}$$

is a bounded convex set with $d(K_2, \complement X_2)\geqq d(K_1, \complement X_1)$. Theorem 16.3.13 gives

$$v\in\mathscr{E}'(X_2), \quad \operatorname{sing\,supp}\check{\mu}*v\subset K_1 \Rightarrow \operatorname{sing\,supp} v\subset K_2.$$

This completes the proof.

Corollary 16.5.15. *If X_1 and X_2 are open convex sets satisfying* (16.5.1)′ *and $\mathscr{H}(\mu)$ consists of the supporting function of $ch\operatorname{sing\,supp}\mu$ alone then (X_1, X_2) is μ-convex for singular supports if and only if $X_2=\{x; \{x\}-\operatorname{sing\,supp}\mu\subset X_1\}$.*

Here and below we assume tacitly that $\mu\notin C^\infty$. In particular one can take $X_1=X_2-ch\operatorname{sing\,supp}\mu$ for any open convex set X_2. This is not possible if $\mathscr{H}(\mu)$ contains several elements:

Proposition 16.5.16. *If X_1 and X_2 are open convex sets satisfying* (16.5.1)′ *such that (X_1, X_2) is μ-convex for singular supports, and if ξ is the outer normal of X_2 at a differentiable point, then $h(-\xi)$ is independent of $h\in\mathscr{H}(\mu)$.*

Proof. Let H be the supporting function of $K=ch\operatorname{sing\,supp}\mu$. The assertion is that $h(-\xi)=H(-\xi)$ for every $h\in\mathscr{H}(\mu)$. Let $x\in\partial X_2$ be a differentiable point with outer normal ξ, thus

$$\langle x'-x, \xi\rangle\leqq 0 \quad \text{for every } x'\in X_2.$$

Then $\{x\}-K\subset\bar{X}_1$ but $\{x\}-K$ is not contained in X_1. Let y be a point in K such that $x-y\in\partial X_1$. If η is the outer normal of a supporting plane of ∂X_1 at $x-y$ we have

$$\langle x'-y', \eta\rangle\leqq\langle x-y, \eta\rangle \quad \text{if } x'\in\bar{X}_2 \text{ and } y'\in K.$$

Taking $y'=y$ we conclude that $\langle x', \eta\rangle\leqq\langle x, \eta\rangle$ if $x'\in\bar{X}_2$, and since ∂X_2 is differentiable at x it follows that η is a positive multiple of ξ. We may therefore assume that $\eta=\xi$. With $x'=x$ instead we now obtain $\langle y', -\xi\rangle\leqq\langle y, -\xi\rangle$ if $y'\in K$, that is, $H(-\xi)=\langle y, -\xi\rangle$.

Let k be any convex compact set with $H_k=h\in\mathscr{H}(\mu)$. Then it follows from Proposition 16.5.14 that $\{x\}-k$ is not contained in X_1

for x would not be a boundary point of X_2 then. But $k \subset K$ so this means by what we have just proved that k must contain a point y with $\langle y, -\xi \rangle = H(-\xi)$. Hence $h(-\xi) = H(-\xi)$ which proves the assertion.

Proposition 16.5.17. *Let X_2 be an open convex set $\neq \mathbb{R}^n$ and $\mu \in \mathscr{E}'$; assume that $h(-\xi)$ is independent of $h \in \mathscr{H}(\mu)$ if ξ is the outer normal of X_2 at a differentiable point. Set $X_1 = X_2 - ch\, \operatorname{sing\,supp} \mu$. Then it follows that (X_1, X_2) is μ-convex for singular supports.*

Proof. Since differentiable points are dense in ∂X_2 the hypothesis implies that X_2 is the interior of the intersection of half spaces $X = \{x; \langle x, \xi \rangle < c\}$ such that $h(-\xi)$ is independent of $h \in \mathscr{H}(\mu)$. By the analogue of Theorem 10.7.4 it is therefore sufficient to prove the proposition when $X_2 = X$. But then it is an immediate consequence of Proposition 16.5.14.

The following is a converse of Theorems 16.5.10 and 16.5.11.

Theorem 16.5.18. *Assume that $\mu \in \mathscr{E}'(\mathbb{R}^n)$ (is invertible) and that X_1, X_2 are open sets in $\mathbb{R}^n$ satisfying (16.5.1)$'$ such that (X_1, X_2) is μ-convex for singular supports. Then it follows that*

$$\mu * (\mathscr{D}'(X_1)/C^\infty(X_1)) = \mathscr{D}'(X_2)/C^\infty(X_2).$$

Proof. The proof of Theorem 10.7.8 is applicable with minor changes. First let K_j be an increasing sequence of compact sets in X_1 with union X_1 and $K_0 = \emptyset$. Next determine for every j a compact set $K_j' \subset X_2$ such that

$$v \in \mathscr{E}'(X_2), \quad \operatorname{sing\,supp} \check{\mu} * v \subset K_j \Rightarrow \operatorname{sing\,supp} v \subset K_j',$$

and K_j' is in the interior of K_{j+1}'. If $X_2 - \operatorname{supp} \mu \subset X_1$ one can then repeat the proof of Theorem 10.7.8 with $P(-D)$ replaced by $\check{\mu} *$ throughout. Otherwise let $1 = \sum \chi_j$ be a partition of unity in X_2. For every j we can choose μ_j with $\mu_j - \mu \in C_0^\infty$ and $\operatorname{supp} \mu_j$ so close to $\operatorname{sing\,supp} \mu$ that $\operatorname{supp} \chi_j - \operatorname{supp} \mu_j \subset X_1$. Then the map

$$T \colon C_0^\infty(X_2) \ni v \to \sum \check{\mu}_j * (\chi_j v) \in C_0^\infty(X_1)$$

is well defined, and the adjoint

$$\mathscr{D}'(X_1) \ni u \to \sum \chi_j (\mu_j * u) \in \mathscr{D}'(X_2)$$

induces the map $\mathscr{D}'(X_1)/C^\infty(X_1) \to \mathscr{D}'(X_2)/C^\infty(X_2)$ which we have denoted by $\mu *$. Now the proof of Theorem 10.7.8 can be repeated with $P(-D)$ replaced by T. We leave for the reader to examine the details of the proof.

Corollary 16.5.19. *Let X_1 and X_2 be open sets in $\mathbb{R}^n$ such that* (16.5.1) *is valid. Then $\mu * \mathscr{D}'(X_1) = \mathscr{D}'(X_2)$ if and only if (μ is invertible and) the pair (X_1, X_2) is μ-convex for supports and singular supports.*

We shall finally discuss the equation (16.5.2) when f is of finite order. As usual we write $\mathscr{D}'_F = \bigcup \mathscr{D}'^m$.

Lemma 16.5.20. *If $f \in \mathscr{D}'^m(X_2)$ one can find a distribution $g \in \mathscr{D}'^{m+1}(\mathbb{R}^n)$ such that $f - g \in C^\infty(X_2)$.*

Proof. By means of a partition of unity we can write $f = \sum f_j$ where $f_j \in \mathscr{E}'^m(X_2)$ and the supports of f_j are locally finite in X_2. Choose $\chi \in C_0^\infty(\mathbb{R}^n)$ with $\int \chi\, dx = 1$ and set $\chi_\delta(x) = \delta^{-n}\chi(x/\delta)$. Then $f_j * \chi_\delta \in C_0^\infty(X_2)$ for small δ, and if $\phi \in C_0^\infty(\mathbb{R}^n)$ we have

$$(f_j - f_j * \chi_\delta)(\phi) = f_j(\phi - \check{\chi}_\delta * \phi),$$

$$\sum_{|\alpha| \leqq m} \sup |D^\alpha(\phi - \check{\chi}_\delta * \phi)| \leqq C\delta \sum_{|\alpha| \leqq m+1} \sup |D^\alpha \phi|.$$

In fact, it suffices to verify this when $\alpha = 0$ and then it is an immediate consequence of the mean value theorem (see the proof of Theorem 1.3.2). It follows that

$$|(f_j - f_j * \chi_\delta)(\phi)| \leqq C_j \delta \sum_{|\alpha| \leqq m+1} \sup |D^\alpha \phi|.$$

Choose a sequence $\delta_j \to 0$ such that $\sum C_j \delta_j < \infty$ and the supports of the convolutions $f_j * \chi_{\delta_j}$ are locally finite in X_2. Then the sums $\sum f_j * \chi_{\delta_j}$ and $\sum (f_j - f_j * \chi_{\delta_j})$ converge respectively in $C^\infty(X_2)$ and in $\mathscr{D}'^{m+1}(\mathbb{R}^n)$ which proves the lemma.

Theorem 16.5.21. *If μ is invertible and (X_1, X_2) is μ-convex for supports then the equation $\mu * u = f$ has a solution $u \in \mathscr{D}'(X_1)$ for every $f \in \mathscr{D}'_F(X_2)$.*

Proof. This is an immediate consequence of Lemma 16.5.20, Theorem 16.5.7 and Theorem 16.5.18 which is applicable in $\mathbb{R}^n$ by Proposition 16.5.14.

The solution obtained in Theorem 16.5.21 is of finite order on any bounded subset of X_1. Hence it is in $\mathscr{D}'_F(X_1)$ if X_1 is bounded. However, in general the question whether u can be chosen of finite order requires further study. We shall now discuss it when $X_1 = X_2 = \mathbb{R}^n$, which is the most restrictive case.

Theorem 16.5.22. *Assume that the equation $\mu * u = \delta$ has a solution $u \in \mathscr{D}'^m(\mathbb{R}^n)$. If $A > m$ and $\varepsilon > 0$ it follows then that*

$$\varliminf_{\xi\to\infty} |\xi|^A \sup\{|\hat{\mu}(\xi+\eta)|;\ |\eta| < \varepsilon \log|\xi|\} > 0 \tag{16.5.8}$$

where ξ and η are in $\mathbb{R}^n$.

Proof. Since $u(\check{\mu} * \phi) = \phi(0)$, $\phi \in C_0^\infty$, the hypothesis implies that for every compact set K

$$\begin{aligned} |\phi(0)| &\leqq C_K \sum_{|\alpha|\leqq m} \sup |D^\alpha \check{\mu} * \phi| \\ &\leqq C'_K \int (1+|\eta|)^m |\hat{\mu}(-\eta)\hat{\phi}(\eta)|\, d\eta, \qquad \phi \in C_0^\infty(K). \end{aligned} \tag{16.5.9}$$

We shall now use the construction in the proof of Lemma 16.3.8. Thus we choose an even non-negative function $\psi \in C_0^\infty$ with $\hat{\psi}(0) = 1$ and set

$$\hat{\phi}(\eta) = \hat{\psi}((\eta - \xi)/t)^N.$$

The support of ϕ lies in a fixed compact set when N/t is bounded, and

$$(2\pi)^n \phi(0) = \int \hat{\psi}(\eta/t)^N d\eta = t^n \int \hat{\psi}(\eta)^N d\eta \geqq c(t/\sqrt{N})^n$$

if N is large, for $\hat{\psi}$ is real and $\hat{\psi}(\eta) \geqq \exp(-c_1|\eta|^2)$ for sufficiently small η. To estimate the right-hand side of (16.5.9) we use that

$$|\hat{\mu}(-\eta)| \leqq C(1+|\eta|)^M$$

where M is the order of μ, and that $(1+|\xi-\eta|) \leqq (1+|\xi|)(1+|\eta|)$. Since

$$\int (1+|\eta|)^m |\hat{\phi}(\xi-\eta)|\, d\eta = t^n \int (1+|t\eta|)^m |\hat{\psi}(\eta)|^N d\eta \leqq C t^{n+m},$$

$$\int_{|\eta|>t} (1+|\eta|)^{m+M} |\hat{\phi}(\xi-\eta)|\, d\eta = t^n \int_{|\eta|>1} (1+|t\eta|)^{m+M} |\hat{\psi}(\eta)|^N d\eta < C t^{n+m+M} e^{-\gamma N}$$

where $0 < \gamma = \inf_{|\eta|>1} -\log|\hat{\psi}(\eta)|$, we obtain

$$\begin{aligned} (t/\sqrt{N})^n < C t^{n+m} \{ &(1+|\xi|)^m \sup_{|\eta|<t} |\hat{\mu}(\eta-\xi)| \\ &+ t^M (1+|\xi|)^{M+m} e^{-\gamma N}\}. \end{aligned} \tag{16.5.10}$$

We choose $t = \varepsilon \log|\xi|$ and $N = [\gamma^{-1}(M+m+1)\log|\xi|]$. Then the quotient N/t is bounded as $\xi \to \infty$ and the left-hand side of (16.5.10) tends to infinity whereas the last term tends to 0. Hence we obtain for large ξ

$$1 \leqq |\xi|^A \sup\{|\hat{\mu}(\eta-\xi)|,\ |\eta| < \varepsilon \log|\xi|\}.$$

The proof is complete.

Before proving the converse we shall give (16.5.8) another form which is similar to condition (iv) in Theorem 16.3.10.

Lemma 16.5.23. *If* (16.5.8) *is fulfilled for every* $\varepsilon>0$ *and* M *is the order of* μ, *then*

$$\overline{\lim_{\xi\to\infty}}\, m(B_\delta)^{-1} \int_{B_\delta} |\log|\hat{\mu}(\xi+\zeta\log|\xi|)||\,d\lambda(\zeta)/\log|\xi| \leqq 2M+A+2C\delta \tag{16.5.11}$$

if $B_\delta=\{\zeta\in\mathbb{C}^n; |\zeta|<\delta\}$ *and* $|x|\leqq C$ *when* $x\in\operatorname{supp}\mu$.

Proof. For large $|\xi|$ we have $\log|\hat{\mu}|\leqq C_0+(M+C\delta)\log|\xi|$ in the ball $\{\xi\}+(\log|\xi|)B_\delta$. Hence

$$\log|\hat{\mu}|+|\log|\hat{\mu}||\leqq 2(C_0+(M+C\delta)\log|\xi|)$$

there. Since

$$\log|\hat{\mu}(\xi)|\leqq m(B_\delta)^{-1}\int_{B_\delta}\log|\hat{\mu}(\xi+\zeta\log|\xi|)|\,d\lambda(\zeta)$$

it follows that

$$m(B_\delta)^{-1}\int_{B_\delta}|\log|\hat{\mu}(\xi+\zeta\log|\xi|)||\,d\lambda(\zeta) \leqq 2(C_0+(M+C\delta)\log|\xi|)-\log|\hat{\mu}(\xi)|.$$

For every large ξ we can find ξ' with

$$|\xi-\xi'|<\varepsilon\log|\xi| \quad\text{and}\quad |\hat{\mu}(\xi')|>c|\xi'|^{-A}.$$

If we take $\varepsilon<\delta$ and apply the preceding result with δ replaced by $\varepsilon+\delta$ and ξ replaced by ξ' it follows that the left-hand side of (16.5.11) is bounded by

$$((\varepsilon+\delta)/\delta)^{2n}(2M+2C\delta+A).$$

Letting $\varepsilon\to 0$ we obtain (16.5.11).

Lemma 16.5.24. *If* (16.5.8) *is valid for every* $\varepsilon>0$ *and* K *is a compact set,* k *an integer* $\geqq 0$, *then*

$$|\phi|_k\leqq C|\check{\mu}*\phi|_{k+m}, \qquad \phi\in C_0^\infty(K), \tag{16.5.12}$$

if $m>n+2M+A$ *where* M *is the order of* μ. *Here*

$$|\phi|_k=\sum_{|\alpha|\leqq k}\sup|D^\alpha\phi|.$$

Proof. If $|\check{\mu}*\phi|_{k+m}=1$ we have for some constants C_0, C_1 depending on K, if $\psi=\check{\mu}*\phi$,

$$|\hat{\psi}(\zeta)|\leqq C_0(1+|\zeta|)^{-k-m}\exp C_1|\operatorname{Im}\zeta|.$$

With the notation in Lemma 16.5.23 it follows that

$$m(B_\delta)^{-1} \int_{B_\delta} \log|\hat{\psi}(\xi+\zeta\log(|\xi|+2))|\,d\lambda(\zeta)$$
$$\leqq (C_1\delta-k-m)\log(|\xi|+2)+C_2.$$

Using (16.5.11) we conclude that

$$\log|\hat{\phi}(\xi)| \leqq m(B_\delta)^{-1}\int \log|\hat{\phi}(\xi+\zeta\log(|\xi|+2))|\,d\lambda(\zeta)$$
$$\leqq (C_1\delta-k-m+2M+A+2C\delta)\log(|\xi|+2)+C_3.$$

Since $-k-m+2M+A<-k-n$ we obtain for some $B<-k-n$ by taking δ small

$$|\hat{\phi}(\xi)|<C(1+|\xi|)^B, \quad \xi\in\mathbb{R}^n.$$

Hence $|\phi|_k \leqq C'$ which proves the lemma.

Definition 16.5.25. $\hat{\mu}$ is said to be very slowly decreasing if (16.5.8) is valid for some A and all $\varepsilon>0$, or equivalently if (16.5.11) is valid for all $\delta>0$.

Remark. It is easy to see that the definition means that all the plurisubharmonic limits of $L_\mu(z,\xi)$ when $\xi\to\infty$ have the same lower bound in $\mathbb{R}^n$.

Theorem 16.5.26. *If $\mu\in\mathscr{E}'$, $\hat{\mu}$ is very slowly decreasing and (X_1, X_2) is μ-convex for supports, then the equation $\mu*u=f$ has a solution $u\in\mathscr{D}'_F(X_1)$ for every $f\in\mathscr{D}'_F(X_2)$.*

Proof. In view of Lemma 16.5.20 and Theorem 16.5.7 it is sufficient to prove the theorem when $X_1=X_2=\mathbb{R}^n$. Let K_j be an increasing sequence of compact sets with union $\mathbb{R}^n$ and choose K'_j so that

$$v\in\mathscr{E}', \quad \operatorname{supp}\check{\mu}*v\subset K_j \Rightarrow \operatorname{supp} v\subset K'_j.$$

We may assume that each K'_j is in the interior of K'_{j+1}. If m is chosen as in Lemma 16.5.24 and $f\in\mathscr{D}'^k$ we shall prove that

$$|f(\phi)|\leqq q(\check{\mu}*\phi), \quad \phi\in C_0^\infty(\mathbb{R}^n), \tag{16.5.13}$$

for a suitable choice of positive Lipschitz continuous functions a_α in

$$q(\phi)=\sum_{|\alpha|\leqq m+k+1} \sup a_\alpha(x)|D^\alpha\phi(x)|. \tag{16.5.14}$$

An application of the Hahn-Banach theorem to the map $\check{\mu}*\phi\to f(\phi)$ will then show that there is a linear form u on $C_0^\infty(\mathbb{R}^n)$ with

$$|u(\psi)|\leqq q(\psi), \quad \psi\in C_0^\infty(\mathbb{R}^n),$$

such that

$$f(\phi)=u(\check{\mu}*\phi), \qquad \phi\in C_0^\infty(\mathbb{R}^n).$$

This means that $u\in\mathscr{D}'^{m+k+1}$ and that $\mu*u=f$.

To prove (16.5.13) it suffices to show that if (16.5.13) is valid for all $\phi\in C_0^\infty(K_j')$ and if $\varepsilon>0$, then (16.5.13) remains valid for $\phi\in C_0^\infty(K_{j+1}')$ if a_α are just multiplied by $(1+\varepsilon)$ and increased enough outside K_{j-1}. Now if this were not true we could find a sequence $\phi_\nu\in C_0^\infty(K_{j+1}')$ such that

(16.5.15) $|f(\phi_\nu)|=1+\varepsilon,\ q(\check{\mu}*\phi_\nu)\leqq 1,\ \check{\mu}*\phi_\nu\to 0 \quad \text{in } C^{m+k+1}(\complement K_{j-1}).$

It follows from Lemma 16.5.24 that ϕ_ν is bounded in $C_0^{k+1}(K_{j+1}')$; hence the sequence is relatively compact in $C_0^k(K_{j+1}')$ by Ascoli's theorem. If ϕ_0 is a limit there we have $\check{\mu}*\phi_0=0$ in $\complement K_{j-1}$ so $\operatorname{supp}\phi_0\subset K_{j-1}'$. Since $f\in\mathscr{D}'^k$ we have $|f(\phi_0)|=\lim|f(\phi_\nu)|=1+\varepsilon$.

Take $\chi\in C_0^\infty(\mathbb{R}^n)$ with support in the unit ball, $\chi\geqq 0$ and $\int\chi\,dx=1$. With $\chi_\delta(x)=\delta^{-n}\chi(x/\delta)$ we have $\phi_0*\chi_\delta\in C_0^\infty(K_j')$ for small δ, and

$$f(\phi_0*\chi_\delta)\to f(\phi_0) \qquad \text{as } \delta\to 0$$

since $\phi_0\in C_0^k$ and $f\in\mathscr{D}'^k$. If x is in a compact set then $a_\alpha(x+y)/a_\alpha(x)\leqq 1+C\delta$, $|y|<\delta$. Hence $q(\check{\mu}*\phi_\nu*\chi_\delta)\leqq 1+C\delta$, and

$$q(\check{\mu}*\phi_0*\chi_\delta)\leqq\overline{\lim}\, q(\check{\mu}*\phi_\nu*\chi_\delta)\leqq 1+C\delta$$

since $\check{\mu}*\phi_0*\chi_\delta$ is a limit of $\check{\mu}*\phi_\nu*\chi_\delta$ in C_0^∞. For sufficiently small δ it follows that (16.5.13) is not valid for $\phi=\phi_0*\chi_\delta\in C_0^\infty(K_j')$. This contradicts the inductive hypothesis and completes the proof of the theorem.

16.6. Hypoelliptic Convolution Equations

In this section we shall prove results on the smoothness of solutions of convolution equations which are similar to those proved for differential equations in Chapter XI.

Theorem 16.6.1. *If every $u\in\mathscr{D}'(\mathbb{R}^n)$ satisfying the convolution equation $\mu*u=0$ is in $C^\infty(\mathbb{R}^n)$ then*

(16.6.1) $$|\operatorname{Im}\zeta|/\log|\zeta|\to\infty \quad \textit{if } \zeta\to\infty$$

on the surface $\hat{\mu}(\zeta)=0$.

Proof. Assume that there is a sequence $\zeta_j \to \infty$ such that $\hat{\mu}(\zeta_j)=0$ and $|\operatorname{Im} \zeta_j| \leqq M \log |\zeta_j|$. Then

$$u(x)=\sum a_j e^{i\langle x, \zeta_j\rangle}$$

converges in $\mathscr{D}'(\mathbb{R}^n)$ and $\mu * u=0$ if $\sum |a_j|<\infty$. The convergence in $\mathscr{D}'$ follows since $\hat{\phi}(-\zeta_j)$ is bounded if $\phi \in C_0^\infty$. (See Theorem 7.3.1 and also the discussion of (7.3.14).) The map $a \to u$ is continuous from l^1 to $\mathscr{D}'$ with the weak topology. If all such solutions are in C^∞ it follows that we have a closed and therefore continuous map $l^1 \to C^\infty$. Hence

$$\sum_{|\alpha|=1} |D^\alpha u(0)| \leqq C \sum |a_j|$$

which means that $|\zeta_j|$ is a bounded sequence contrary to the hypothesis. The contradiction proves the theorem.

Corollary 16.6.2. *Let X_1, X_2, X be non-empty open sets in $\mathbb{R}^n$ with*

$$X_2 - \operatorname{sing\,supp} \mu \subset X_1, \qquad X \subset X_1,$$

and assume that $u \in \mathscr{D}'(X_1)$ and $\mu * s_1 u=0$ in $\mathscr{D}'(X_2)/C^\infty(X_2)$ implies $u \in C^\infty(X)$. Then it follows that μ is invertible and that (16.6.1) *holds.*

Proof. This follows from Theorems 16.3.9 and 16.6.1.

When (16.6.1) is fulfilled the plurisubharmonic function $L_\mu(z, \xi)$ defined by (16.3.4) is pluriharmonic on any compact subset of $\mathbb{C}^n$ when ξ is large enough. Since $L_\mu(., \xi)$ is bounded in $L^1_{\text{loc}}(\mathbb{C}^n)$ when μ is invertible, we then obtain a uniform bound for $|L_\mu(0, \xi)|$. If H is the supporting function of $ch\, \operatorname{sing\,supp} \mu$ and N is the order of μ, then the limits of $L_\mu(., \xi)$ are pluriharmonic functions V with

$$V(z) \leqq H(\operatorname{Im} z)+N, \qquad V(0) \geqq -A,$$

for some constant A. If we apply Liouville's theorem to the harmonic function $w \to V(\xi + w\eta)$ it follows that V is linear, that is,

$$V(z)=\operatorname{Im}\langle a, z\rangle + b.$$

Here $a \in \operatorname{sing\,supp} \mu$ by Theorem 16.3.13, and $\operatorname{sing\,supp} \mu$ is the closure of the set of points a which occur for some limit V (Theorem 16.3.5). Since $-A \leqq b \leqq N$ we have

$$-V(z)=\operatorname{Im}\langle a, -z\rangle - b \leqq H(-\operatorname{Im} z)+A,$$

and since $-L_\mu(., \xi)$ is (sub)harmonic it follows from Theorem 4.1.9 that on any compact set in $\mathbb{C}^n$ the estimate

$$-L_\mu(z, \xi) \leqq H(-\operatorname{Im} z)+A+1$$

is valid if $|\xi|$ is large enough. We have therefore proved the following

Lemma 16.6.3. *Suppose that μ is invertible and satisfies* (16.6.1). *Let H be the supporting function of* $ch\,\mathrm{sing\,supp}\,\mu$. *Then there is a constant B and a sequence of constants C_m, $m=1, 2, \ldots$ such that for every m*

$$|1/\hat{\mu}(\zeta)| \leqq |\zeta|^B e^{H(-\operatorname{Im}\zeta)} \quad \text{if } |\operatorname{Im}\zeta| < m\log|\zeta|,\ |\zeta| > C_m. \tag{16.6.2}$$

Conversely, it is obvious that (16.6.2) implies that μ is invertible and satisfies (16.6.1).

Definition 16.6.4. We shall say that μ is hypoelliptic if μ is invertible and satisfies (16.6.1), or equivalently if (16.6.2) is valid.

Theorem 16.6.5. *The following conditions on $\mu\in\mathscr{E}'(\mathbb{R}^n)$ are equivalent:*

(i) $u\in\mathscr{D}'(\mathbb{R}^n)$, $\mu*u\in C^\infty(\mathbb{R}^n)$ *implies* $u\in C^\infty(\mathbb{R}^n)$.

(ii) μ *is hypoelliptic.*

(iii) *there exists a distribution F (parametrix) with $\mu*F-\delta\in C^\infty$ and*

$$ch\,\mathrm{sing\,supp}\,F \subset -ch\,\mathrm{sing\,supp}\,\mu.$$

(iv) *there exists a parametrix $F\in\mathscr{E}'$.*

(v) *there exists a parametrix $F\in\mathscr{E}'$ with* $\mathrm{sing\,supp}\,F = -\mathrm{sing\,supp}\,\mu$.

(vi) *there exists a fundamental solution E, that is, $E\in\mathscr{D}'(\mathbb{R}^n)$ and $\mu*E=\delta$, such that* $\mathrm{sing\,supp}\,E = -\mathrm{sing\,supp}\,\mu$.

(vii) *if X is an open set in $\mathbb{R}^n$, $X+\mathrm{sing\,supp}\,\mu\subset X_2$ and $X_2-\mathrm{sing\,supp}\,\mu\subset X_1$, where X_1, X_2 are also open sets in $\mathbb{R}^n$, and if $u\in\mathscr{D}'(X_1)$ has image s_1u in $\mathscr{D}'(X_1)/C^\infty(X_1)$, then $u\in C^\infty(X)$ if $\mu*s_1u=0$ in $\mathscr{D}'(X_2)/C^\infty(X_2)$.*

Proof. (i) $\Rightarrow$ (ii) by Corollary 16.6.2. To prove that (ii) $\Rightarrow$ (iii) we apply Lemma 16.6.3 and set

$$F(\phi)=(2\pi)^{-n}\int_{|\xi|>C_1}\hat{\phi}(-\xi)/\hat{\mu}(\xi)\,d\xi, \qquad \phi\in C_0^\infty.$$

Then $F\in\mathscr{S}'$ and the Fourier transform is $1/\hat{\mu}$ when $|\xi|>C_1$ and 0 elsewhere. Hence

$$\mu*F-\delta=-(2\pi)^{-n}\int_{|\xi|<C_1} e^{i\langle\cdot,\xi\rangle}\,d\xi\in C^\infty.$$

If $\phi\in C_0^\infty$ we have

$$F*\phi(x)=(2\pi)^{-n}\int_{|\xi|>C_1}\hat{\phi}(\xi)/\hat{\mu}(\xi)\,e^{i\langle x,\xi\rangle}\,d\xi.$$

As in the proof of Theorem 7.3.8 (with (7.3.9) replaced by (16.6.2)) the integration can be shifted to the cycle

$$\Gamma_\eta\colon\ \xi\to\zeta(\xi)=\xi+i\eta\log(1+|\xi|^2)$$

outside a compact set. Since an integral over a compact set only contributes a C^∞ term we conclude as in the proof Theorem 7.3.8 that $\operatorname{sing\,supp} F$ is contained in the set with supporting function $\check{H}$, that is, $-ch \operatorname{sing\,supp} \mu$. That (iii) $\Rightarrow$ (iv) follows if we just replace the parametrix F in (iii) by χF where $\chi \in C_0^\infty$ is equal to 1 in a neighborhood of $\operatorname{sing\,supp} F$. It is obvious that (iv) $\Rightarrow$ (i) for we have $u = F * \mu * u + (\delta - F * \mu) * u \in C^\infty$. Thus the conditions (i)–(iv) are equivalent.

(iv) $\Rightarrow$ (v). Since $F * \mu = \delta + \psi$, $\psi \in C_0^\infty$, it follows from Theorem 16.3.6 and Lemma 16.3.3 that for every $h \in \mathscr{H}(F)$ one can find $h' \in \mathscr{H}(\mu)$ so that $h + h' = 0$. As seen above we have $h'(\eta) = \langle a, \eta \rangle$ for some $a \in \operatorname{sing\,supp} \mu$. Hence $h(\eta) = \langle -a, \eta \rangle$, and it follows from Theorem 16.3.5 that $\operatorname{sing\,supp} F \subset -\operatorname{sing\,supp} \mu$. The opposite inclusion is proved in the same way. Alternatively we may observe that μ can be regarded as a parametrix of F so F is also hypoelliptic.

(v) $\Rightarrow$ (vi). With the notation just used it follows from Theorem 16.5.7 that $\mu * \phi = \psi$ for some $\phi \in C^\infty(\mathbb{R}^n)$. Thus $E = F - \phi$ is a fundamental solution with the required property.

(vi) $\Rightarrow$ (iii) is obvious and so is (vii) $\Rightarrow$ (i). To prove that (v) $\Rightarrow$ (vii) we just have to observe that if $s_X u$ is the image of u in $\mathscr{D}'(X)/C^\infty(X)$ and F is the parametrix in (v), then

$$s_X u = F * (\mu * s_1 u) = 0.$$

The proof is complete.

Remarks. 1) Parametrices differ only by C^∞ functions. 2) It follows from (iv) that μ remains hypoelliptic if a C_0^∞ function is added to μ. If $\mu \in \mathscr{D}'$ it follows that either all or no $\mu_0 \in \mathscr{E}'$ with $\mu - \mu_0 \in C^\infty$ is hypoelliptic; in the first case we say that μ is hypoelliptic. 3) An example of a hypoelliptic distribution is $\mu = (1 - |x|^2 - i0)^{-1}$. In the one dimensional case this follows at once from the fact that the Fourier transform is $\pi i e^{-i|\xi|}$. For any n a similar result can be obtained from Theorem 7.7.14.

16.7. Hyperbolic Convolution Equations

In this section we shall extend to convolution operators the discussion of hyperbolic differential operators given in Chapter XII. Thus assume that $\mu \in \mathscr{E}'$, $E \in \mathscr{D}'$ and that

$$\mu * E = \delta, \qquad \operatorname{supp} E \subset K, \tag{16.7.1}$$

where K is a closed convex set. K cannot be compact unless μ is a δ-function, for since $\hat{\mu}\hat{E}=1$ it follows then that $\hat{\mu}$ has no zeros, hence that $\log\hat{\mu}$ is a linear function.

The supporting function H_K of K,

$$H_K(\xi)=\sup_{x\in K}\langle x,\xi\rangle$$

is a convex, positively homogeneous and lower semi-continuous function with values in $(-\infty,+\infty]$ (see Section 7.4). As in Section 12.5 we want K to be contained in a proper convex cone, and this condition can be expressed in terms of the supporting function:

Lemma 16.7.1. *The following conditions on the closed convex set K and the vector $\theta\in\mathbb{R}^n\setminus\{0\}$ are equivalent:*

(i) *$H_K(\xi)<\infty$ for all ξ in a neighborhood of θ.*

(ii) *The set $K_c=\{x\in K;\langle x,\theta\rangle\geqq c\}$ is compact for every c.*

(iii) *For some $x_0\in\mathbb{R}^n$ and $C>0$ we have*

$$|x-x_0|\leqq C\langle x_0-x,\theta\rangle,\qquad x\in K.$$

Proof. (i) ⇒ (ii) If $x\in K_c$ we have $\langle x,\xi\rangle\leqq H_K(\xi)<\infty$ for all ξ in a neighborhood of θ and also $\langle x,-\theta\rangle\leqq -c$. Since the convex hull of $-\theta$ and a neighborhood of θ is a neighborhood of 0, this proves that K_c is bounded, hence compact. (ii) ⇒ (iii) To simplify notation we may assume that $0\in K$. If $x\in K\setminus K_{-1}$ we can find $t\in(0,1)$ so that $tx\in k$ $=\{y\in K;\ \langle y,\theta\rangle=-1\}$. If $C=\sup_{x\in k}|x|$ it follows that $|x|\leqq -C\langle x,\theta\rangle$, $x\in K\setminus K_{-1}$. Since K_{-1} is compact we obtain (iii) for suitable x_0. (iii) ⇒ (i) is obvious.

Definition 16.7.2. μ is called hyperbolic with respect to $\theta\in\mathbb{R}^n\setminus\{0\}$ if there is a convex set K with $H_K(\xi)<\infty$ for all ξ in a neighborhood of $-\theta$ and a fundamental solution E of $\mu*$ with support in K, that is, E is a distribution satisfying (16.7.1).

Theorem 16.7.3. *Let $\mu\in\mathscr{E}'(\mathbb{R}^n)$, $E\in\mathscr{D}'(\mathbb{R}^n)$ and assume that $\mu*E=\delta$. Denote by H_μ and by H_E the supporting functions of $\operatorname{ch}\operatorname{supp}\mu$ and of the closure K of $\operatorname{ch}\operatorname{supp}E$ respectively. Assume that the interior Γ of the set where $H_E<\infty$ is not empty. Then Γ is an open convex cone and there is a vector $a\in\mathbb{R}^n$ such that*

$$H_\mu(\xi)=-H_E(\xi)=\langle a,\xi\rangle,\qquad \xi\in\Gamma.$$

If $\check{\Gamma}^\circ$ is the negative dual cone defined by

$$\check{\Gamma}^\circ=\{x\in\mathbb{R}^n;\ \langle x,\xi\rangle\leqq 0 \text{ when } \xi\in\Gamma\}$$

then $K=\check{\Gamma}^\circ-\{a\}$ *and* $\operatorname{supp}\mu\subset\check{\Gamma}^\circ+\{a\}$. *For every closed cone* Γ_1 *contained in* $\Gamma\cup\{0\}$ *we have for some constants* C *and* M

$$|1/\hat{\mu}(\zeta)|\leqq C(1+|\zeta|)^M e^{-\langle a,\operatorname{Im}\zeta\rangle} \tag{16.7.2}$$

if $\operatorname{Im}\zeta\in\Gamma_1$, $|\operatorname{Im}\zeta|>C\log(|\zeta|+2)$.

Proof. It is obvious that Γ is a convex cone. If $\xi\in\Gamma$ then

$$\{x\in K;\langle x,\xi\rangle\geqq -H_\mu(\xi)-1\}$$

is compact by Lemma 16.7.1 with θ replaced by ξ. Hence we can choose $\chi\in C_0^\infty$ equal to 1 in this set. With $E=E_1+E_2$, $E_1=\chi E$, we obtain

$$\langle x,\xi\rangle\leqq H_\mu(\xi)-H_\mu(\xi)-1\leqq -1 \qquad \text{if } x\in\operatorname{supp}\mu*E_2.$$

Since $\mu*E_1=\delta-\mu*E_2$ it follows that

$$\sup\{\langle x,\xi\rangle;x\in\operatorname{supp}\mu*E_1\}=0,\qquad \xi\in\Gamma.$$

By the theorem of supports this means that

$$\sup\{\langle x,\xi\rangle;x\in\operatorname{supp}E_1\}+H_\mu(\xi)=0,\qquad \xi\in\Gamma.$$

Hence $H_E(\xi)+H_\mu(\xi)=0$, that is, $H_E(\xi)=-H_\mu(\xi)$, $\xi\in\Gamma$. Since a function which is both convex and concave must be linear, this proves that

$$H_\mu(\xi)=-H_E(\xi)=\langle a,\xi\rangle,\qquad \xi\in\Gamma.$$

We have $x\in K$ if and only if

$$\langle x,\xi\rangle\leqq H_E(\xi),\qquad \xi\in\mathbb{R}^n.$$

This is trivially satisfied if $\xi\notin\bar{\Gamma}$ since $H_E(\xi)=+\infty$ then. If the inequality is valid when $\xi\in\Gamma$ it follows for $\xi\in\bar{\Gamma}$ since H_E is convex. Thus $x\in K$ if and only if $\langle x+a,\xi\rangle\leqq 0$, $\xi\in\Gamma$, which means that $x+a\in\check{\Gamma}^\circ$. That $\operatorname{supp}\mu\subset\check{\Gamma}^\circ+\{a\}$ is obvious.

Choose $\psi\in C_0^\infty(\mathbb{R}^n)$ equal to 1 in a neighborhood of $-a$ and set $F=\psi E\in\mathscr{E}'$. Then

$$\mu*F=\delta+R$$

where $R=\mu*((\psi-1)E)$ has support in $\check{\Gamma}^\circ$, $0\notin\operatorname{supp}R$. Hence $\langle x,\eta\rangle\leqq -c|\eta|$ for some $c>0$ if $\eta\in\Gamma_1$ and $x\in\operatorname{supp}R$. By Theorem 7.3.1 it follows that

$$|\hat{R}(\zeta)|\leqq C(1+|\zeta|)^M e^{-c|\operatorname{Im}\zeta|} \qquad \text{if } \operatorname{Im}\zeta\in\Gamma_1,$$
$$|\hat{F}(\zeta)|\leqq C(1+|\zeta|)^M e^{-\langle a,\operatorname{Im}\zeta\rangle} \qquad \text{if } \operatorname{Im}\zeta\in\Gamma_1.$$

But $1+\hat{R}(\zeta)=\hat{\mu}(\zeta)\hat{F}(\zeta)$ so $|1/\hat{\mu}(\zeta)|\leqq 2|\hat{F}(\zeta)|$ if $|\hat{R}(\zeta)|<1/2$. This proves the statement.

Theorem 16.7.4. *Assume that* $\mu \in \mathscr{E}'$ *and that* Γ *is an open convex cone such that for every closed cone* $\Gamma_1 \subset \Gamma \cup \{0\}$ *there are constants* C, M, A *such that*

$$|1/\hat{\mu}(\zeta)| < C(1+|\zeta|)^M e^{A|\operatorname{Im}\zeta|} \tag{16.7.3}$$

if $\operatorname{Im}\zeta \in \Gamma_1$, $|\operatorname{Im}\zeta| > C\log(|\zeta|+2)$. *Then there is a fundamental solution* $E \in \mathscr{D}'$, *thus* $\mu * E = \delta$, *such that the supporting function of the closed convex hull of* $\operatorname{supp} E$ *is finite in* Γ.

Proof. Define γ_η as the cycle

$$\mathbb{R}^n \ni \xi \to \xi + i\eta \log(2+|\xi|^2).$$

When $\eta \in \Gamma_1$ and $|\eta|$ is sufficiently large it is clear that γ_η lies in the set where (16.7.3) is applicable. In view of the discussion of (7.3.14) in Section 7.3 we can therefore define a distribution E_η by

$$E_\eta(\phi) = (2\pi)^{-n} \int_{\gamma_\eta} \hat{\phi}(-\zeta)/\hat{\mu}(\zeta)\, d\zeta_1 \wedge \ldots \wedge d\zeta_n, \quad \phi \in C_0^\infty(\mathbb{R}^n). \tag{16.7.4}$$

Since the integrand is very small at infinity we see by an application of Stokes' formula that $E_\eta = E_{t\eta}$, $t > 1$, and that E_η is independent of $\eta \in \Gamma_1$ when $|\eta|$ is sufficiently large. But $E_{t\eta}(\phi) \to 0$ as $t \to +\infty$ if

$$H(-\eta) + A|\eta| < 0$$

where H is the supporting function of $ch \operatorname{supp} \phi$. Thus E vanishes in a neighborhood of x if $\langle x, -\eta\rangle + A|\eta| < 0$, that is, $\langle x, \eta\rangle \leqq A|\eta|$ if $x \in \operatorname{supp} E$ and $\eta \in \Gamma_1$. This proves the theorem.

Remark. It is clear that (16.7.3) implies that μ is invertible and that $\hat{\mu}(\zeta) \neq 0$ when $\operatorname{Im}\zeta \in \Gamma_1$ and $|\operatorname{Im}\zeta| > C\log(|\zeta|+2)$. Conversely, if this condition is fulfilled for every closed cone $\Gamma_1 \subset \Gamma \cup \{0\}$ it is easily proved by means of Theorem 4.1.9 that (16.7.3) is valid.

The only important feature of μ is the behavior at the point a in Theorem 16.7.3. Even the hypothesis that the support is compact is superfluous. (Note that the convolution of two distributions with support in a proper convex cone is defined.) In fact, we have

Theorem 16.7.5. *Let* $\mu \in \mathscr{E}'$ *have support in* $\{a\} + \check{\Gamma}^\circ$ *where* $\check{\Gamma}^\circ$ *is a proper convex cone, and assume that* $\mu * E = \delta$ *for some* E *with support in* $\check{\Gamma}^\circ - \{a\}$. *If* $\nu \in \mathscr{D}'$ *has support in* $\{a\} + \check{\Gamma}^\circ$ *and* $a \notin \operatorname{supp}(\nu - \mu)$, *then one can find* F *with support in* $\check{\Gamma}^\circ - \{a\}$ *so that* $\nu * F = \delta$.

Proof. We have

$$\nu * E = \mu * E + (\nu - \mu) * E = \delta - R$$

where $\operatorname{supp} R \subset \check{\Gamma}^\circ$ and $0 \notin \operatorname{supp} R$. It follows that the supports of the iterated convolutions R, $R*R$, $R*R*R, \dots$ are locally finite, so the sum

$$S = \delta + R + R*R + R*R*R + \dots$$

exists and $(\delta - R)*S = \delta$. Hence $\nu * E * S = \delta$ so $F = E*S$ has the required property.

It is easy to show that the conclusion remains valid if we just assume that $\nu - \mu$ is sufficiently smooth near a and not necessarily zero. This provides an abundance of examples of hyperbolic convolution operators which are not differential operators. The case where $\mu = \delta$ and $\mu - \nu$ is a continuous function with support on $\mathbb{R}_+$ is classical, a simple kind of Volterra integral equation. In connection with the mixed problem we have also encountered a less obvious convolution operator in Section 12.9.

Notes

Section 16.1 contains only classical results on subharmonic functions, apart from the asymptotics given in Theorem 16.1.8. This we owe to an oral communication from Arne Beurling who also pointed out that this is a convenient way to Theorem 16.1.9 (due to Titchmarsh [1]). More precise results have been proved by Ahlfors and Heins [1] but we have avoided them by working in the spirit of distribution theory. (Further references to the literature can be found in Hayman and Kennedy [1].) For the class of plurisubharmonic functions studied in Section 16.2 there is a simple asymptotic behavior only at $\complement \mathbb{R}^n$ (see Vauthier [1]) but this is enough for our purposes.

Malgrange [1] initiated a study of convolution equations in $\mathbb{R}^n$. He proved the approximation theorem of Section 16.4 in $\mathbb{R}^n$ and also showed how fundamental solutions can be used to study the inhomogeneous equation. However, it was Ehrenpreis [3] who introduced the notion of invertible distribution $\mu \in \mathscr{E}'$ (and slow decrease of $\hat{\mu}$); he showed that it is equivalent to $\mu * \mathscr{D}'(\mathbb{R}^n) = \mathscr{D}'(\mathbb{R}^n)$. His condition for $\mathscr{D}'_F$ was slightly incorrect and was rectified in Hörmander [14] where the convolution equation was also studied in open subsets of $\mathbb{R}^n$. All this was simplified by the introduction of localizations at infinity in Hörmander [15]. This is reproduced in Section 16.3 with some later improvements. The example given in Theorem 16.3.20 is essentially due to Berenstein and Dostal [1]. A complete proof of the approximation theorem in Section 16.4 was first given in Hörmander

[23]. The proof here is simplified by the use of analytic functionals. Hypoelliptic and hyperbolic convolution equations were first characterized by Ehrenpreis [3, 8]. The hypoelliptic case is illuminated here by the general discussion of singularities in Section 16.3, which also leads to optimal estimates for the singularities.

In the one dimensional case the theory of mean periodic functions of Schwartz [6] is a study of overdetermined systems of homogeneous convolution equations. However, for overdetermined systems with several variables the analogue of Theorem 16.4.1 is not always true: Gurevič [1] has given a system of convolution equations $\mu_j * u = 0, j = 1, \dots, 6$, in $\mathbb{R}^n$, $n > 1$, which has non-trivial solutions but no exponential solution.

As mentioned at the end of Section 16.4 it is usually impossible to find a real analytic solution in $\mathbb{R}^n$ even of a differential equation $P(D)u = f$ where f is real analytic. This was first proved by De Giorgi [2] and Piccinini [1] when $P(D)$ is the heat equation or an elliptic equation in fewer variables. The class of operators $P(D)$ for which this can happen was determined by Hörmander [31].

Appendix A. Some Algebraic Lemmas

The purpose of this appendix is to state with references or proofs the facts concerning polynomials which are needed in the text.

A.1. The Zeros of Analytic Functions

The first lemma is just the implicit function theorem for analytic functions.

Lemma A.1.1. *Let $P(t, z)$ be a polynomial in the $n+1$ variables t and $z = (z_1, \ldots, z_n)$. If $P(t, z) = 0$ but $\partial P(t, z)/\partial t \neq 0$ when $z = z_0$, $t = t_0$, it follows that there is one and only one function $t(z)$ which is analytic in a neighborhood of z_0, equal to t_0 at z_0, and satisfies the equation $P(t(z), z) = 0$.*

Proof. Choose $\varepsilon > 0$ so that $P(t, z_0) \neq 0$ when $0 < |t - t_0| \leqq \varepsilon$. Then $P(t, z) \neq 0$ when $|t - t_0| = \varepsilon$ if $|z - z_0| < \delta$, so

$$t(z) = (2\pi i)^{-1} \int_{|\tau - t_0| = \varepsilon} \tau \partial P(\tau, z)/\partial \tau / P(\tau, z)\, d\tau$$

is the only zero with $|t - t_0| < \varepsilon$ if $|z - z_0| < \delta$. It is obviously an analytic function of z. (See also Section 7.5.)

To examine when the hypotheses of the lemma are fulfilled we introduce the discriminant defined by

(A.1.1) $$\Delta = a_m^{2m-2} \prod_{i<k} (t_i - t_k)^2$$

where $t_1, \ldots, t_m$ are the zeros of

(A.1.2) $$P(t, z) = a_m(z) t^m + a_{m-1}(z) t^{m-1} + \ldots + a_0(z)$$

and we assume $a_m \not\equiv 0$. The discriminant is a polynomial in z. If it is identically 0 then the Euclidean algorithm and Gauss' lemma show that P and $\partial P(t,z)/\partial t$ have a common factor which is a multiple factor of P. (See e.g. van der Waerden [2, § 26, 31].) Hence we have

Lemma A.1.2. *If $P(t,z)$ has no multiple polynomial factors, then the discriminant Δ defined by* (A.1.1) *is a polynomial in z which does not vanish identically. Every point where $a_m \Delta \neq 0$ has a neighborhood V where there are m analytic functions $t_1(z), \dots, t_m(z)$ such that*

$$P(t,z) = a_m(z) \prod_1^m (t - t_j(z)), \qquad z \in V. \tag{A.1.3}$$

We now assume that $n=1$ and study the zeros near 0 when $a_m \Delta = 0$ at 0. Note that there is no zero nearby. Writing $t a_m(z) = w$ we obtain the equation

$$w^m + a_{m-1}(z) w^{m-1} + a_{m-2}(z) a_m(z) w^{m-2} + \dots + a_0(z) a_m(z)^{m-1} = 0 \tag{A.1.4}$$

which is more convenient to study since the leading coefficient is 1. When $z=0$ we have roots $w_1, \dots, w_k$ with multiplicities $m_1, \dots, m_k$ adding up to m. To shorten notation we assume that $w=0$ is a root with multiplicity μ. Then (A.1.4) has μ roots close to 0 for small z, and they are locally analytic functions of z when $z \neq 0$. If we start with one determination of a root $w(z)$ for $0 < z < 2\varepsilon$ we can continue it analytically along the circle $|z| = \varepsilon$ once around the origin. We must return with a root. If it is the original one, then $w(z)$ is bounded and analytic when $0 < |z| < 2\varepsilon$, so the singularity at 0 is removable. Otherwise we continue going around the origin until after $p \leqq \mu$ turns we return to the original value. Then $w(z^p)$ is analytic and bounded when $0 < |z^p| < 2\varepsilon$, hence

$$w(z^p) = \sum_1^\infty c_k z^k.$$

In general we would obtain a constant term also, and returning to the variable t may introduce a finite number of negative powers of z. Thus we have proved:

Lemma A.1.3. *Let $P(t,z)$ be a polynomial in two variables t and z of the form* (A.1.2) *with $m \geqq 1$ and $a_m \not\equiv 0$. Then we have* (A.1.3) *when $0 < |z| < \delta$, where each t_j for some positive integer p is an analytic function of $z^{1/p}$ when $0 < |z^{1/p}| < \delta^{1/p}$, with at most a pole at 0, that is,*

$$t_j(z) = \sum_N^\infty c_k (z^{1/p})^k. \tag{A.1.5}$$

Here N may be a positive or negative integer or 0.

The notation in (A.1.5) is usually simplified to

$$t_j(z)=\sum_N^\infty c_k z^{k/p}$$

but the correct interpretation of this series is given by (A.1.5). The expansion (A.1.5) is called a Puiseux series. Similar expansions involving negative fractional powers of z can of course be given in a neighborhood of infinity. This follows simply by introducing $1/z$ as a new variable instead of z.

An important consequence of the expansion (A.1.5) is that if we choose N so that $c_N \neq 0$, which is possible unless $t_j(z)\equiv 0$, then

$$t_j(z)=c_N z^{N/p}(1+o(1)), \qquad z \to 0, \tag{A.1.6}$$

and similarly when $z\to\infty$. We shall return to this point in Section A.2.

A.2. Asymptotic Properties of Algebraic Functions of Several Variables

Lemma A.1.3 has no analogue for several variables, but it is often possible to obtain a satisfactory substitute by means of the Tarski-Seidenberg theorem. To state this result we need a definition.

Definition A.2.1. *A subset of $\mathbb{R}^n$ is called semi-algebraic if it is a finite union of finite intersections of sets defined by a polynomial equation or inequality.*

Thus one starts with sets of the form

$$\{x;\, P(x)=0\}, \qquad \{x;\, P(x)>0\} \qquad \text{or } \{x;\, P(x)\geqq 0\} \tag{A.2.1}$$

where P is a polynomial, and forms first finite intersections and then finite unions. By definition the union of two semi-algebraic sets is therefore semi-algebraic. The intersection is also semi-algebraic, for if $A=\bigcup A_j$ and $B=\bigcup B_k$, then $A\cap B=\bigcup (A_j\cap B_k)$. Hence the complement of a semi-algebraic set is also semi-algebraic.

Theorem A.2.2 (Tarski-Seidenberg). *If A is a semi-algebraic subset of $\mathbb{R}^{n+m}=\mathbb{R}^n\oplus\mathbb{R}^m$, then the projection A' of A in $\mathbb{R}^m$ is also semi-algebraic.*

In most applications we choose $m=2$ so that A' can be studied by means of the Puiseux series expansions of Section A.1. In the proof of the theorem we may assume that $n=1$ for the general statement follows by iteration of this special case. We denote the variable in $\mathbb{R}$ by x and the variable in $\mathbb{R}^m$ by y. Without restriction we may always assume that A is just an intersection of sets of the form (A.2.1), thus defined by

$$P_1(x,y)>0, \dots, P_r(x,y)\geqq 0, \dots, P_s(x,y)=0$$

say. We want to prove Theorem A.2.2 by induction with respect to the maximum degree of the polynomials occurring here. However, it turns out that this requires that one generalizes the problem by considering all combinations of signs which may occur. We define $\operatorname{sgn} y=0$ if $y=0$ and $\operatorname{sgn} y=\pm 1$ when $y\gtrless 0$.

Let $p_1, \dots, p_s$ be polynomials in one variable x of degree at most m. Let $x_1<x_2<\dots<x_N$ be the points on $\mathbb{R}$ where at least one of them vanishes without vanishing identically. Set $x_0=-\infty$, $x_{N+1}=+\infty$ and $I_k=(x_k, x_{k+1})$, $k=0,\dots,N$. Then $\operatorname{sgn} p_i(x)$ is independent of x when $x\in I_k$ and we denote it by $\operatorname{sgn} p_i(I_k)$. Now we sum up all information about zeros and sign changes in

$$\operatorname{SGN}(p_1,\dots,p_s)=\{\operatorname{sgn} p_i(x_j), \operatorname{sgn} p_i(I_k)\},$$
$$i=1,\dots,s,\ j=1,\dots,N,\ k=0,\dots,N.$$

This is an element in the disjoint union

$$W=\bigcup_{N=0}^{ms}\{-1,0,1\}^{s(2N+1)}.$$

Not all $w\in W$ can occur of course. For example, if $\operatorname{sgn} p_i(x_j)\neq 0$ then

$$\operatorname{sgn} p_i(x_j)=\operatorname{sgn} p_i(I_{j-1})=\operatorname{sgn} p_i(I_j).$$

It is clear that knowing $\operatorname{SGN}(p_1,\dots,p_s)$ we can decide if a system of inequalities $p_1(x)>0,\dots,p_r(x)\geqq 0,\dots,p_s(x)=0$ is satisfied for some x. Indeed, this is equivalent to

$$\operatorname{sgn} p_1(t)=1,\dots,\operatorname{sgn} p_r(t)\neq -1,\dots,\operatorname{sgn} p_s(t)=0$$

for $t=x_j$, some $j=1,\dots,N$, or $t=I_k$, some $k=0,\dots,N$. Theorem A.2.2 is therefore a consequence of

Theorem A.2.2′. *If $P_1(x,y),\dots,P_s(x,y)$ are polynomials of degree at most m with respect to x, then*

$$E=\{y\in\mathbb{R}^n;\ \operatorname{SGN}(P_1(.,y),\dots,P_s(.,y))=w\}$$

is semi-algebraic for every $w\in W$.

The following elementary lemma allows us to prove Theorem A.2.2′ by induction.

Lemma A.2.3. *Let $p_1, \dots, p_s$ be polynomials of degree at most m in one variable. Assume that p_s is exactly of degree $m>0$ and that none is identically 0. If $g_1, \dots, g_s$ are the remainders obtained when p_s is divided by $p_1, \dots, p_{s-1}, p_s'$ then* $\mathrm{SGN}(p_1, \dots, p_{s-1}, p_s', g_1, \dots, g_s) = w$ *determines* $\mathrm{SGN}(p_1, \dots, p_s)$.

Proof. Let $x_1 < \dots < x_N$ be the points on $\mathbb{R}$ where some $p_1, \dots, p_{s-1}, p_s'$, $g_1, \dots, g_s$ is zero without vanishing identically. Knowing w allows us to select the zeros $x_{i_1} < \dots < x_{i_k}$ of $p_1, \dots, p_{s-1}, p_s'$. Since $p_s(x)$ is equal to $g_1(s), \dots,$ or $g_s(x)$ when $p_1(x) = 0, \dots$ or $p_s'(x) = 0$, we can determine $\mathrm{sgn}\, p_s(x_{i_j})$. The sign of the leading coefficient of p_s is $\mathrm{sgn}\, p_s'(I_N)$. In the open intervals bounded by $-\infty, x_{i_1}, \dots, x_{i_k}, +\infty$ the polynomial p_s is monotonic so it has a zero there if and only if p_s has opposite signs at the end points. (The sign at an infinite end point is given by the sign of the highest coefficient.) Thus we can check which intervals contain a zero of p_s. These zeros together with the zeros $x_{i_1}, \dots, x_{i_k}$ which are not just zeros of p_s' give all the relevant zeros of $p_1, \dots, p_s$ and we can read off all the information on signs required to know $\mathrm{SGN}(p_1, \dots, p_s)$. The proof is complete.

Note that we never used the values of $x_1, \dots, x_N$, which we do not know, but only the order relations between them. Replacing $p_1, \dots, p_s$ by $p_1, \dots, p_{s-1}, p_s', g_1, \dots, g_s$ we decrease the number of polynomials of maximal degree m.

Proof of Theorem A.2.2′. By induction we may assume the theorem proved for smaller values of m and also for the same value of m when the number of polynomials of degree m is diminished. For fixed $m_1, \dots, m_s$ the set of all y such that $P_1(x, y), \dots, P_s(x, y)$ are exactly of degree $m_1, \dots, m_s$ with respect to x is obviously semi-algebraic. It suffices to prove that the intersection with E is always semi-algebraic. In doing so we may assume that $m_1, \dots, m_s$ are the orders of $P_1, \dots, P_s$ with respect to x for general y, for higher order terms can otherwise be discarded. Thus

$$P_j(x, y) = q_j(y) x^{m_j} + \text{terms of lower order in } x$$

and we only consider values of y with $q_j(y) \neq 0$ for every j. (Identically vanishing polynomials are omitted after checking that in w their signs are always listed as 0.) Let $m_s = m = \max m_j$, and denote by $g_1, \dots, g_s$ the remainders obtained when P_s is divided by $P_1, \dots, P_{s-1}, P_s'$ and denominators are removed by multiplication with an even power of $q_1, \dots, q_s$. Then

$$\{y \in E;\ q_1(y) \dots q_s(y) \neq 0\}$$

is a finite union of sets of the form

$$\{y;\ q_1(y) \dots q_s(y) \neq 0,\ \mathrm{SGN}(P_1, \dots, P_{s-1}, P_s', g_1, \dots, g_s) = w'\}.$$

This follows from Lemma A.2.3. By the inductive hypothesis this is a semialgebraic set. The proof is complete.

Remark. Theorem A.2.2 can also be stated as follows: If $E \subset \mathbb{R}^{n+m}$ is semi-algebraic, then

$$\{y \in \mathbb{R}^m;\ \exists\, x \in \mathbb{R}^n,\ (x, y) \in E\}$$

is semi-algebraic. Since the complement of a semi-algebraic set is semi-algebraic we can here replace $\exists$ by $\forall$. If $x = (x_1, x_2)$ we may also conclude for example that

$$\{y \in \mathbb{R}^m;\ \forall\, x_1 \in \mathbb{R}^{n_1}\ \exists\, x_2 \in \mathbb{R}^{n_2},\ (x_1, x_2, y) \in E\}$$

is semi-algebraic. We shall make free use of general expressions of this kind below.

Now assume that E is a semi-algebraic set in $\mathbb{R}^{2+n}$ where the coordinates are denoted by (x, y, z); $x, y \in \mathbb{R}$. Define

$$f(x) = \sup\{y;\ \exists z, (x, y, z) \in E\} \tag{A.2.2}$$

with the convention $f(x) = -\infty$ if no such (y, z) exists.

Corollary A.2.4. *If E is a semi-algebraic set in $\mathbb{R}^{2+n}$, then the function f defined by* (A.2.2) *is semi-algebraic, that is, the subgraph*

$$F = \{(x, y);\ y \leqq f(x)\}$$

is semi-algebraic.

Proof. This follows from Theorem A.2.2 since

$$F = \{(x, y), \forall\, \varepsilon > 0\ \exists\, y' > y - \varepsilon\ \exists\, z, (x, y', z) \in E\}.$$

A semi-algebraic function has a very simple structure:

Theorem A.2.5. *If f is a semi-algebraic function on $\mathbb{R}$, then $\mathbb{R}$ can be decomposed into a finite number of intervals (which may be reduced to points) where f is either $+\infty$, $-\infty$ or equal to a continuous algebraic function. If f is finite for large positive x and not identically* 0, *then*

$$f(x) = A x^a (1 + o(1)), \quad x \to +\infty, \tag{A.2.3}$$

where $A \neq 0$ and a is a rational number.

Proof. The projection of the subgraph F of f on $\mathbb{R}$ is semi-algebraic, that is, a finite union of intervals. The complement of the projection is the set where $f=-\infty$, so it is of the same kind. Similarly the complement of the projection of the complement of F on $\mathbb{R}$ is the set where $f=+\infty$. Now consider an open interval I where f is finite. We know that F is defined by say

$$P_1(x, y)\geqq 0, \dots, P_r(x, y)=0. \tag{A.2.4}$$

We can remove from I the finitely many zeros x of the leading coefficient of a $P_j(x, y)$, the discriminant of an irreducible factor or the resultant of two such factors which are not proportional. Let I' be one of the remaining intervals. In I' the equations $P_j(x, y)=0$ with y real define a finite number of functions $y=f_k(x)$, $f_1(x)<f_2(x)<\dots$. The conditions (A.2.4) are either fulfilled in

$$\{(x, y);\ x\in I', f_i(x)<y<f_{i+1}(x)\} \tag{A.2.5}$$

or else they are not valid for any such (x, y), for (A.2.5) defines a connected set where no P_j has a zero. Similarly

$$\{(x, f_j(x));\ x\in I'\} \quad \text{and} \quad \{(x, y);\ x\in I', y<f_1(x)\}$$

is either contained in F or disjoint with F. It follows that $f(x)=f_i(x)$ for a fixed i when $x\in I'$. In particular, this is true in a right semi-infinite interval if $f(x)$ is finite for large positive x, so (A.2.3) follows from (A.1.6).

Corollary A.2.6. *If E is a semi-algebraic set in $\mathbb{R}^{2+n}$ and (A.2.2) is defined and finite for large positive x, then f is identically 0 for large x or else (A.2.3) is valid with $A\neq 0$ and a rational.*

Of course we can replace sup by inf in (A.2.2) and make the same conclusion.

Example A.2.7. If P is a polynomial with $P(\xi)>0$, $\xi\in\mathbb{R}^n$, then

$$P(\xi)>c(1+|\xi|^2)^{-N} \tag{A.2.6}$$

for some $c>0$ and N. In fact,

$$E=\{(x, y, \xi);\ x=1+|\xi|^2, P(\xi)=y\}$$

is semi-algebraic in $\mathbb{R}^{2+n}$, and

$$f(x)=\inf_{1+|\xi|^2=x} P(\xi)=\inf\{y;\ (x, y, \xi)\in E\}.$$

f is a positive continuous function since $P>0$. Hence Corollary A.2.6 gives (A.2.3) when $x\to\infty$, which proves (A.2.6). Note that the example

$P(\xi)=(\xi_1\xi_2-1)^2+\xi_1^2$ shows that (A.2.6) may not be valid for any $N\leqq 0$, for $P(\xi)=\xi_1^2=1/\xi_2^2$ on the hyperbola $\xi_1\xi_2=1$.

It is sometimes important to know not only that the supremum (A.2.2) is semi-algebraic but also that it is attained on a semi-algebraic curve.

Theorem A.2.8. *Let E be a semi-algebraic set in $\mathbb{R}^{2+n}$ such that the range of the projection $E\ni(x,y,z)\to x\in\mathbb{R}$ contains all large positive x. Then one can find Puiseux series $y(x)$, $z(x)$, converging for large positive x, such that $(x,y(x),z(x))\in E$. If $f(x)=\sup\{y;\exists z,(x,y,z)\in E\}$ is finite and the supremum is attained for large positive x, one can take $y(x)=f(x)$.*

Proof. The projection F of E in $\mathbb{R}^2$,

$$F=\{(x,y)\in\mathbb{R}^2;\ (x,y,z)\in E \text{ for some } z\}$$

is semi-algebraic by Theorem A.2.2. The proof of Theorem A.2.5 shows that we can find a finite number of algebraic functions $f_1(x)<f_2(x)<\dots$ defined for $x\geqq x_0$ such that F for $x\geqq x_0$ is the union of a finite number of the curves $\{(x,f_i(x))\}$ and of the strips bounded by them. Hence we can choose an algebraic function $y(x)$ which is either arbitrary or $f_i(x)+cx^{-N}$ for some small c and large N, so that $(x,y(x))\in F$ for $x\geqq x_0$. If the supremum $f(x)$ is finite and attained we can take $y(x)=f(x)$. Now

$$E_1=\{(x,y,z)\in E,\ x\geqq x_0,\ y=y(x)\}$$

is semi-algebraic. The projection on the x axis contains $[x_0,\infty)$. The same argument can therefore be applied to E_1 with z_1 playing the role of y above. In this way we choose $z_1,\dots,z_n$ successively until $z(x)$ is defined so that $(x,y(x),z(x))\in E$. This completes the proof.

Finally we need to know in Section 12.8 that the graph of an analytic continuation of an algebraic function is semi-algebraic:

Theorem A.2.9. *Let $P(t,z)$ be of the form* (A.1.2) *with $\Delta(z)a_m(z)$ not identically* 0. *Let $E\subset\mathbb{C}^{1+n}\times\mathbb{C}^{1+n}$ be the set of all (t_0,z_0,t_1,z_1) such that*

a) $P(t_0,z_0)=P(t_1,z_1)=0$.

b) $\Delta(z)a_m(z)\neq 0$ *if* $z=z(s)=sz_1+(1-s)z_0$, $0\leqq s\leqq 1$.

c) *one can find $t(s)$ analytic when $0\leqq s\leqq 1$ so that $t(0)=t_0$, $t(1)=t_1$ and $P(t(s),z(s))=0$.*

Then E is semi-algebraic.

Proof. The conditions a) and b) have already an algebraic form. (Note that if $w \in \mathbb{C}$ then $w \neq 0$ is equivalent to the real inequality $|w|^2 \neq 0$.) The only problem is to examine c). First we consider the set E_0 defined by means of a), b) and a more restrictive condition than c),

c$_0$) one can find $w \in \mathbb{C}$ such that

i) $\operatorname{Im}(t-T)w \neq 0$ if $p(t, z(s)) = p(T, z(s)) = 0$, $0 \leqq s \leqq 1$ and $t \neq T$.

ii) $\exists t_j^1, \ldots, t_j^m, j=0, 1$, such that $P(t_j^k, z_j) = 0$, $j=0, 1$, $k=1, \ldots, m$, and $\operatorname{Im}(t_j^{k+1} - t_j^k)w > 0$, $j=0, 1$, $k=1, \ldots, m-1$.

iii) $\exists v$, $t_j^v = t_j$, $j=0, 1$.

That c$_0$) implies c) when a) and b) are fulfilled is easy to see. In fact, by b) we have well defined analytic zeros $t^k(s)$ of $P(t, z(s))$ with $t^k(0) = t_0^k$. From i) it follows that $\operatorname{Im}(t^{k+1}(s) - t^k(s))w > 0$ for $0 \leqq s \leqq 1$ since this is true when $s=0$ and no zeros occur. But this determines the order of the zeros so $t^k(1) = t_1^k$ by ii), hence c) is valid by iii). Now E_0 is semi-algebraic by Theorem A.2.2. It is clear that if $(t_0, z_0, t_1, z_1) \in E$ and z_1 is sufficiently close to z_0, then $(t_0, z_0, t_1, z_1) \in E_0$. What remains is to show that we can always choose a bounded number of steps where E and E_0 agree.

Let

$$q(t, s) = a_m(z(s))^N t^{-m} \prod_1^m P(t + t^j(s), z(s))$$

be the polynomial with zeros $t^k(s) - t^j(s)$, $j \neq k$. If N is large enough the coefficients are polynomials in those of P by the main theorem on symmetric functions. ($N = m(m-1)$ is sufficient.) Thus q is a polynomial in s and we have a bound for the degree. Now b) guarantees that we can choose w so that

$$(t_i^j - t_i^k)w \neq \overline{(t_i^{j'} - t_i^{k'})w} \quad \text{if } j \neq k,\ j' \neq k',\ i=0, 1. \tag{A.2.7}$$

In fact, this excludes just a finite number of directions for w. The condition means that $q(t/w, i)$ and $\overline{q(\bar{t}/w, i)}$ have no common zeros, $i = 0, 1$. Thus the resultant R of $q(t/w, s)$ and $\overline{q(\bar{t}/w, s)}$ is not identically 0. It is a polynomial of degree $\leqq k$ in s so it has at most k zeros in $(0, 1)$. Assuming that a) and b) are valid we can therefore find $2k$ points $0 < s_1 < s_2 < \ldots < s_{2k-1} < s_{2k} < 1$ such that $R \neq 0$ outside the intervals (s_{2j-1}, s_{2j}) but these are so small that

$$(t(s_{2j-1}), z(s_{2j-1}), t(s_{2j}), z(s_{2j})) \in E_0$$

anyway. For the other intervals we have condition c$_0$) fulfilled with the chosen value for w. Hence E is the set of all (t_0, z_0, t_1, z_1) satisfying a) and b) such that one can find $s_0, \ldots, s_{2k+1}, T_0, \ldots, T_{2k+1}$ with

$$0=s_0<s_1<\ldots<s_{2k}<s_{2k+1}=1,\ T_0=t_0,\ T_{2k+1}=t_1,$$
$$(T_j, z(s_j), T_{j+1}, z(s_{j+1}))\in E_0,\ j=0,\ldots,2k.$$

By Theorem A.2.2 this proves that E is semi-algebraic.

Notes

The Tarski-Seidenberg theorem was first proved by Tarski [1], but he did not publish it in the mathematical journals. Shortly afterwards Seidenberg [1] gave another proof in a more algebraic spirit. Here we have mainly followed a short manuscript by Cohen [3]. He never published it since it turned out to be close to the original proof by Tarski! In the theory of partial differential equations the Tarski-Seidenberg theorem seems to have been used first in Hörmander [1] in connection with the characterization of hypoelliptic operators. In a completely analogous question concerning hyperbolic operators Gårding [1] had earlier had to use a direct and more complicated algebraic technique. (See the notes for Chapter XII.) A discussion of the Tarski-Seidenberg theorem and its applications can also be found in Gorin [1]. Theorem A.2.9 here is due to Paul Cohen. It was included in Hörmander [21].

Bibliography

Agmon, S.: [1] The coerciveness problem for integro-differential forms. J. Analyse Math. 6, 183–223 (1958).
– [2] Spectral properties of Schrödinger operators. Actes Congr. Int. Math. Nice 2, 679–683 (1970).
– [3] Spectral properties of Schrödinger operators and scattering theory. Ann. Scuola Norm. Sup. Pisa (4) 2, 151–218 (1975).
– [4] Unicité et convexité dans les problèmes différentiels. Sém. Math. Sup. No 13, Les Presses de l'Univ. de Montreal, 1966.
– [5] Lectures on elliptic boundary value problems. van Nostrand Mathematical Studies 2, Princeton, N.J. 1965.
– [6] Problèmes mixtes pour les équations hyperboliques d'ordre supérieur. Coll. Int. CNRS 117, 13–18, Paris 1962.
– [7] Some new results in spectral and scattering theory of differential operators on $\mathbb{R}^n$. Sém. Goulaouic-Schwartz 1978-1979, Exp. II, 1-11.

Agmon, S., A. Douglis and L. Nirenberg: [1] Estimates near the boundary for solutions of elliptic partial differential equations satisfying general boundary conditions. I. Comm. Pure Appl. Math. 12, 623–727 (1959); II. Comm. Pure Appl. Math. 17, 35–92 (1964).

Agmon, S. and L. Hörmander: [1] Asymptotic properties of solutions of differential equations with simple characteristics. J. Analyse Math. 30, 1–38 (1976).

Agranovich, M.S.: [1] Partial differential equations with constant coefficients. Uspehi Mat. Nauk 16:2, 27–94 (1961). (Russian; English translation in Russian Math. Surveys 16:2, 23–90 (1961).)

Ahlfors, L. and M. Heins: [1] Questions of regularity connected with the Phragmén-Lindelöf principle. Ann. of Math. 50, 341–346 (1949).

Airy, G.B.: [1] On the intensity of light in a neighborhood of a caustic. Trans. Cambr. Phil. Soc. 6, 379–402 (1838).

Alinhac, S.: [1] Non-unicité du problème de Cauchy. Ann. of Math. 117, 77–108 (1983).
– [2] Non-unicité pour des opérateurs différentiels à caractéristiques complexes simples. Ann. Sci. École Norm. Sup. 13, 385–393 (1980).
– [3] Uniqueness and non-uniqueness in the Cauchy problem. Contemporary Math. 27, 1–22 (1984).

Alinhac, S. and M.S. Baouendi: [1] Uniqueness for the characteristic Cauchy problem and strong unique continuation for higher order partial differential inequalities. Amer. J. Math. 102, 179–217 (1980).

Alinhac, S. and C. Zuily: [1] Unicité et non-unicité du problème de Cauchy pour des opérateurs hyperboliques à caractéristiques doubles. Comm. Partial Differential Equations 6, 799–828 (1981).

Alsholm, P.K.: [1] Wave operators for long range scattering. Mimeographed report, Danmarks Tekniske Højskole 1975.

Alsholm, P.K. and T. Kato: [1] Scattering with long range potentials. In Partial Diff. Eq., Proc. of Symp. in Pure Math. 23, 393–399. Amer. Math. Soc. Providence, R.I. 1973.

Amrein, W.O., Ph.A. Martin and P. Misra: [1] On the asymptotic condition of scattering theory. Helv. Phys. Acta 43, 313-344 (1970).
Andersson, K.G.: [1] Propagation of analyticity of solutions of partial differential equations with constant coefficients. Ark. Mat. 8, 277-302 (1971).
Andersson, K.G. and R.B. Melrose: [1] The propagation of singularities along glidings rays. Invent. Math. 41, 197-232 (1977).
Arnold, V.I.: [1] On a characteristic class entering into conditions of quantization. Funkcional. Anal. i Priložen. 1, 1-14 (1967) (Russian); also in Functional Anal. Appl. 1, 1-13 (1967).
Aronszajn, N.: [1] Boundary values of functions with a finite Dirichlet integral. Conference on Partial Differential Equations 1954, University of Kansas, 77-94.
– [2] A unique continuation theorem for solutions of elliptic partial differential equations or inequalities of second order. J. Math. Pures Appl. 36, 235-249 (1957).
Aronszajn, N., A. Krzywcki and J. Szarski: [1] A unique continuation theorem for exterior differential forms on Riemannian manifolds. Ark. Mat. 4, 417-453 (1962).
Asgeirsson, L.: [1] Über eine Mittelwerteigenschaft von Lösungen homogener linearer partieller Differentialgleichungen 2. Ordnung mit konstanten Koeffizienten. Math. Ann. 113, 321-346 (1937).
Atiyah, M.F.: [1] Resolution of singularities and division of distributions. Comm. Pure Appl. Math. 23, 145-150 (1970).
Atiyah, M.F. and R. Bott: [1] The index theorem for manifolds with boundary. Proc. Symp. on Differential Analysis, 175-186. Oxford 1964.
– [2] A Lefschetz fixed point formula for elliptic complexes. I. Ann. of Math. 86, 374-407 (1967).
Atiyah, M.F., R. Bott and L. Gårding: [1] Lacunas for hyperbolic differential operators with constant coefficients. I. Acta Math. 124, 109-189 (1970).
– [2] Lacunas for hyperbolic differential operators with constant coefficients. II. Acta Math. 131, 145-206 (1973).
Atiyah, M.F., R. Bott and V.K. Patodi: [1] On the heat equation and the index theorem. Invent. Math. 19, 279-330 (1973).
Atiyah, M.F. and I.M. Singer: [1] The index of elliptic operators on compact manifolds Bull. Amer. Math. Soc. 69, 422-433 (1963).
– [2] The index of elliptic operators. I, III. Ann. of Math. 87, 484-530 and 546-604 (1968).
Atkinson, F.V.: [1] The normal solubility of linear equations in normed spaces. Mat. Sb. 28 (70), 3-14 (1951) (Russian).
Avakumovič, V.G.: [1] Über die Eigenfunktionen auf geschlossenen Riemannschen Mannigfaltigkeiten. Math. Z. 65, 327-344 (1956).
Bang, T.: [1] Om quasi-analytiske funktioner. Thesis, Copenhagen 1946, 101 pp.
Baouendi, M.S. and Ch. Goulaouic: [1] Nonanalytic-hypoellipticity for some degenerate elliptic operators. Bull. Amer. Math. Soc. 78, 483-486 (1972).
Beals, R.: [1] A general calculus of pseudo-differential operators. Duke Math. J. 42, 1-42 (1975).
Beals, R. and C. Fefferman: [1] On local solvability of linear partial differential equations. Ann. of Math. 97, 482-498 (1973).
– [2] Spatially inhomogeneous pseudo-differential operators I. Comm. Pure Appl. Math. 27, 1-24 (1974).
Beckner, W.: [1] Inequalities in Fourier analysis. Ann. of Math. 102, 159-182 (1975).
Berenstein, C.A. and M.A. Dostal: [1] On convolution equations I. In L'anal. harm. dans le domaine complexe. Springer Lecture Notes in Math. 336, 79-94 (1973).
Bernstein, I.N.: [1] Modules over a ring of differential operators. An investigation of the fundamental solutions of equations with constant coefficients. Funkcional. Anal. i Priložen. 5:2, 1-16 (1971) (Russian); also in Functional Anal. Appl. 5, 89-101 (1971).

Bernstein, I.N. and S.I. Gelfand: [1] Meromorphy of the function P^λ. Funkcional. Anal. i Priložen. 3:1, 84–85 (1969) (Russian); also in Functional Anal. Appl. 3, .68–69 (1969).
Bernstein, S.: [1] Sur la nature analytique des solutions des équations aux dérivées partielles du second ordre. Math. Ann. 59, 20–76 (1904).
Beurling, A.: [1] Quasi-analyticity and general distributions. Lectures 4 and 5, Amer. Math. Soc. Summer Inst. Stanford 1961 (Mimeographed).
– [2] Sur les spectres des fonctions. Anal. Harm. Nancy 1947, Coll. Int. XV, 9–29.
– [3] Analytic continuation across a linear boundary. Acta Math. 128, 153–182 (1972).
Björck, G.: [1] Linear partial differential operators and generalized distributions. Ark. Mat. 6, 351–407 (1966).
Björk, J.E.: [1] Rings of differential operators. North-Holland Publ. Co. Math. Library series 21 (1979).
Bochner, S.: [1] Vorlesungen über Fouriersche Integrale. Leipzig 1932.
Boman, J.: [1] On the intersection of classes of infinitely differentiable functions. Ark. Mat. 5, 301–309 (1963).
Bonnesen, T. and W. Fenchel: [1] Theorie der konvexen Körper. Erg. d. Math. u. ihrer Grenzgeb. 3, Springer Verlag 1934.
Bony, J.M.: [1] Une extension du théorème de Holmgren sur l'unicité du problème de Cauchy. C.R. Acad. Sci. Paris 268, 1103–1106 (1969).
– [2] Extensions du théorème de Holmgren. Sém. Goulaouic-Schwartz 1975–1976, Exposé no. XVII.
– [3] Equivalence des diverses notions de spectre singulier analytique. Sém. Goulaouic-Schwartz 1976–1977, Exposé no. III.
Bony, J.M. and P. Schapira: [1] Existence et prolongement des solutions holomorphes des équations aux dérivées partielles. Invent. Math. 17, 95–105 (1972).
Borel, E.: [1] Sur quelques points de la théorie des fonctions. Ann. Sci. École Norm. Sup. 12 (3), 9–55 (1895).
Boutet de Monvel, L.: [1] Comportement d'un opérateur pseudo-différentiel sur une variété à bord. J. Analyse Math. 17, 241–304 (1966).
– [2] Boundary problems for pseudo-differential operators. Acta Math. 126, 11–51 (1971).
– [3] On the index of Toeplitz operators of several complex variables. Invent. Math. 50, 249–272 (1979).
– [4] Hypoelliptic operators with double characteristics and related pseudo-differential operators. Comm. Pure Appl. Math. 27, 585–639 (1974).
Boutet de Monvel, L., A. Grigis and B. Helffer: [1] Parametrixes d'opérateurs pseudo-différentiels à caractéristiques multiples. Astérisque 34–35, 93–121 (1976).
Boutet de Monvel, L. and V. Guillemin: [1] The spectral theory of Toeplitz operators. Ann. of Math. Studies 99 (1981).
Brézis, H.: [1] On a characterization of flow-invariant sets. Comm. Pure Appl. Math. 23, 261–263 (1970).
Brodda, B.: [1] On uniqueness theorems for differential equations with constant coefficients. Math. Scand. 9, 55–68 (1961).
Browder, F.: [1] Estimates and existence theorems for elliptic boundary value problems. Proc. Nat. Acad. Sci. 45, 365–372 (1959).
Buslaev, V.S. and V.B. Matveev: [1] Wave operators for the Schrödinger equation with a slowly decreasing potential. Theor. and Math. Phys. 2, 266–274 (1970). (English translation.)
Calderón, A.P.: [1] Uniqueness in the Cauchy problem for partial differential equations. Amer. J. Math. 80, 16–36 (1958).
– [2] Existence and uniqueness theorems for systems of partial differential equations. Fluid Dynamics and Applied Mathematics (Proc. Symp. Univ. of Maryland 1961), 147–195. New York 1962.

– [3] Boundary value problems for elliptic equations. Outlines of the joint Soviet-American symposium on partial differential equations, 303–304, Novosibirsk 1963.

Calderón, A.P. and R. Vaillancourt: [1] On the boundedness of pseudo-differential operators. J. Math. Soc. Japan 23, 374–378 (1972).

– [2] A class of bounded pseudo-differential operators. Proc. Nat. Acad. Sci. U.S.A. 69, 1185–1187 (1972).

Calderón, A.P. and A. Zygmund: [1] On the existence of certain singular integrals. Acta Math. 88, 85–139 (1952).

Carathéodory, C.: [1] Variationsrechnung und partielle Differentialgleichungen Erster Ordnung. Teubner, Berlin, 1935.

Carleman, T.: [1] Sur un problème d'unicité pour les systèmes d'équations aux dérivées partielles à deux variables indépendentes. Ark. Mat. Astr. Fys. 26B No 17, 1–9 (1939).

– [2] L'intégrale de Fourier et les questions qui s'y rattachent. Publ. Sci. Inst. Mittag-Leffler, Uppsala 1944.

– [3] Propriétés asymptotiques des fonctions fondamentales des membranes vibrantes. C.R. Congr. des Math. Scand. Stockholm 1934, 34–44 (Lund 1935).

Catlin, D.: [1] Necessary conditions for subellipticity and hypoellipticity for the $\bar\partial$ Neumann problem on pseudoconvex domains. In Recent developments in several complex variables. Ann. of Math. Studies 100, 93–100 (1981).

Cauchy, A.: [1] Mémoire sur l'intégration des équations linéaires. C.R. Acad. Sci. Paris 8 (1839). In Œuvres IV, 369–426, Gauthier-Villars, Paris 1884.

Cerezo, A., J. Chazarain and A. Piriou: [1] Introduction aux hyperfonctions. Springer Lecture Notes in Math. 449, 1–53 (1975).

Chaillou, J.: [1] Hyperbolic differential polynomials and their singular perturbations. D. Reidel Publ. Co. Dordrecht, Boston, London 1979.

Charazain, J.: [1] Construction de la paramétrix du problème mixte hyperbolique pour l'équation des ondes. C.R. Acad. Sci. Paris 276, 1213–1215 (1973).

– [2] Formules de Poisson pour les variétés riemanniennes. Invent. Math. 24, 65–82 (1974).

Chazarain, J. and A. Piriou: [1] Introduction à la théorie des équations aux dérivées partielles linéaires. Gauthier-Villars 1981.

Chester, C., B. Friedman and F. Ursell: [1] An extension of the method of steepest descent. Proc. Cambr. Phil. Soc. 53, 599–611 (1957).

Cohen, P.: [1] The non-uniqueness of the Cauchy problem. O.N.R. Techn. Report 93, Stanford 1960.

– [2] A simple proof of the Denjoy-Carleman theorem. Amer. Math. Monthly 75, 26–31 (1968).

– [3] A simple proof of Tarski's theorem on elementary algebra. Mimeographed manuscript, Stanford University 1967, 6 pp.

Colin de Verdière, Y.: [1] Sur le spectra des opérateurs elliptiques à bicharactéristiques toutes périodiques. Comment. Math. Helv. 54, 508–522 (1979).

Cook, J.: [1] Convergence to the Møller wave matrix. J. Mathematical Physics 36, 82–87 (1957).

Cordes, H.O.: [1] Über die eindeutige Bestimmtheit der Lösungen elliptischer Differentialgleichungen durch Anfangsvorgaben. Nachr. Akad. Wiss. Göttingen Math.-Phys. Kl. IIa, No. 11, 239–258 (1956).

Cotlar, M.: [1] A combinatorial inequality and its application to L^2 spaces. Rev. Math. Cuyana 1, 41–55 (1955).

Courant, R. and D. Hilbert: [1] Methoden der Mathematischen Physik II. Berlin 1937.

Courant, R. and P.D. Lax: [1] The propagation of discontinuities in wave motion. Proc. Nat. Acad. Sci. 42, 872–876 (1956).

De Giorgi, E.: [1] Un esempio di non-unicitá della soluzione del problema di Cauchy relativo ad una equazione differenziale lineare a derivate parziali ti tipo parabolico. Rend. Mat. 14, 382–387 (1955).

– [2] Solutions analytiques des équations aux dérivées partielles à coefficients constants. Sém. Goulaouic-Schwartz 1971-1972, Exposé 29.

Deič, V.G., E.L. Korotjaev and D.R. Jafaev: [1] The theory of potential scattering with account taken of spatial anisotropy. Zap. Naučn. Sem. Leningrad Otdel. Mat. Inst. Steklov 73, 35-51 (1977).

Dencker, N.: [1] On the propagation of singularities for pseudo-differential operators of principal type. Ark. Mat. 20, 23-60 (1982).

– [2] The Weyl calculus with locally temperate metrics and weights. Ark. Mat. 24, 59-79 (1986).

Dieudonné, J.: [1] Sur les fonctions continus numériques définies dans un produit de deux espaces compacts. C.R. Acad. Sci. Paris 205, 593-595 (1937).

Dieudonné, J. and L. Schwartz: [1] La dualité dans les espaces $(\mathscr{F})$ et $(\mathscr{L}\mathscr{F})$. Ann. Inst. Fourier (Grenoble) 1, 61-101 (1949).

Dollard, J.D.: [1] Asymptotic convergence and the Coulomb interaction. J. Math. Phys. 5, 729-738 (1964).

– [2] Quantum mechanical scattering theory for short-range and Coulomb interactions. Rocky Mountain J. Math. 1, 5-88 (1971).

Douglis, A. and L. Nirenberg: [1] Interior estimates for elliptic systems of partial differential equations. Comm. Pure Appl. Math. 8, 503-538 (1955).

Duistermaat, J.J.: [1] Oscillatory integrals, Lagrange immersions and unfolding of singularities. Comm. Pure Appl. Math. 27, 207-281 (1974).

Duistermaat, J.J. and V.W. Guillemin: [1] The spectrum of positive elliptic operators and periodic bicharacteristics. Invent. Math. 29, 39-79 (1975).

Duistermaat, J.J. and L. Hörmander: [1] Fourier integral operators II. Acta Math. 128, 183-269 (1972).

Duistermaat, J.J. and J. Sjöstrand: [1] A global construction for pseudo-differential operators with non-involutive characteristics. Invent. Math. 20, 209-225 (1973).

DuPlessis, N.: [1] Some theorems about the Riesz fractional integral. Trans. Amer. Math. Soc. 80, 124-134 (1955).

Egorov, Ju.V.: [1] The canonical transformations of pseudo-differential operators. Uspehi Mat. Nauk 24:5, 235-236 (1969).

– [2] Subelliptic pseudo-differential operators. Dokl. Akad. Nauk SSSR 188, 20-22 (1969); also in Soviet Math. Doklady 10, 1056-1059 (1969).

– [3] Subelliptic operators. Uspehi Mat. Nauk 30:2, 57-114 and 30:3, 57-104 (1975); also in Russian Math. Surveys 30:2, 59-118 and 30:3, 55-105 (1975).

Ehrenpreis, L.: [1] Solutions of some problems of division I. Amer. J. Math. 76, 883-903 (1954).

– [2] Solutions of some problems of division III. Amer. J. Math. 78, 685-715 (1956).

– [3] Solutions of some problems of division IV. Amer. J. Math. 82, 522-588 (1960).

– [4] On the theory of kernels of Schwartz. Proc. Amer. Math. Soc. 7, 713-718 (1956).

– [5] A fundamental principle for systems of linear differential equations with constant coefficients, and some of its applications. Proc. Intern. Symp. on Linear Spaces, Jerusalem 1961, 161-174.

– [6] Fourier analysis in several complex variables. Wiley-Interscience Publ., New York, London, Sydney, Toronto 1970.

– [7] Analytically uniform spaces and some applications. Trans. Amer. Math. Soc. 101, 52-74 (1961).

– [8] Solutions of some problems of division V. Hyperbolic operators. Amer. J. Math. 84, 324-348 (1962).

Enqvist, A.: [1] On fundamental solutions supported by a convex cone. Ark. Mat. 12, 1-40 (1974).

Enss, V.: [1] Asymptotic completeness for quantum-mechanical potential scattering. I. Short range potentials. Comm. Math. Phys. 61, 285-291 (1978).

– [2] Geometric methods in spectral and scattering theory of Schrödinger operators.

In Rigorous Atomic and Molecular Physics, G. Velo and A. Wightman ed., Plenum, New York, 1980–1981 (Proc. Erice School of Mathematical Physics 1980).

Eškin, G.I.: [1] Boundary value problems for elliptic pseudo-differential equations. Moscow 1973; Amer. Math. Soc. Transl. of Math. Monographs 52, Providence, R.I. 1981.

– [2] Parametrix and propagation of singularities for the interior mixed hyperbolic problem. J. Analyse Math. 32, 17–62 (1977).

– [3] General initial-boundary problems for second order hyperbolic equations. In Sing. in Boundary Value Problems. D. Reidel Publ. Co., Dordrecht, Boston, London 1981, 19–54.

– [4] Initial boundary value problem for second order hyperbolic equations with general boundary conditions I. J. Analyse Math. 40, 43–89 (1981).

Fedosov, B.V.: [1] A direct proof of the formula for the index of an elliptic system in Euclidean space. Funkcional. Anal. i Priložen. 4:4, 83–84 (1970) (Russian); also in Functional Anal. Appl. 4, 339–341 (1970).

Fefferman, C.L.: [1] The uncertainty principle. Bull. Amer. Math. Soc. 9, 129–206 (1983).

Fefferman, C. and D.H. Phong: [1] On positivity of pseudo-differential operators. Proc. Nat. Acad. Sci. 75, 4673–4674 (1978).

– [2] The uncertainty principle and sharp Gårding inequalities. Comm. Pure Appl. Math. 34, 285–331 (1981).

Fredholm, I.: [1] Sur l'intégrale fondamentale d'une équation différentielle elliptique à coefficients constants. Rend. Circ. Mat. Palermo 25, 346–351 (1908).

Friedlander, F.G.: [1] The wave front set of the solution of a simple initial-boundary value problem with glancing rays. Math. Proc. Cambridge Philos. Soc. 79, 145–159 (1976).

Friedlander, F.G. and R.B. Melrose: [1] The wave front set of the solution of a simple initial-boundary value problem with glancing rays. II. Math. Proc. Cambridge Philos. Soc. 81, 97–120 (1977).

Friedrichs, K.: [1] On differential operators in Hilbert spaces. Amer. J. Math. 61, 523–544 (1939).

– [2] The identity of weak and strong extensions of differential operators. Trans. Amer. Math. Soc. 55, 132–151 (1944).

– [3] On the differentiability of the solutions of linear elliptic differential equations. Comm. Pure Appl. Math. 6, 299–326 (1953).

– [4] On the perturbation of continuous spectra. Comm. Pure Appl. Math. 1, 361–406 (1948).

Friedrichs, K. and H. Lewy: [1] Über die Eindeutigkeit und das Abhängigkeitsgebiet der Lösungen beim Anfangswertproblem linearer hyperbolischer Differentialgleichungen. Math. Ann. 98, 192–204 (1928).

Fröman, N. and P.O. Fröman: [1] JWKB approximation. Contributions to the theory. North-Holland Publ. Co. Amsterdam 1965.

Fuglede, B.: [1] A priori inequalities connected with systems of partial differential equations. Acta Math. 105, 177–195 (1961).

Gabrielov, A.M.: [1] A certain theorem of Hörmander. Funkcional. Anal. i Priložen. 4:2, 18–22 (1970) (Russian); also in Functional Anal. Appl. 4, 106–109 (1970).

Gårding, L.: [1] Linear hyperbolic partial differential equations with constant coefficients. Acta Math. 85, 1–62 (1951).

– [2] Dirichlet's problem for linear elliptic partial differential equations. Math. Scand. 1, 55–72 (1953).

– [3] Solution directe du problème de Cauchy pour les équations hyperboliques. Coll. Int. CNRS, Nancy 1956, 71–90.

– [4] Transformation de Fourier des distributions homogènes. Bull. Soc. Math. France 89, 381–428 (1961).

– [5] Local hyperbolicity. Israel J. Math. 13, 65–81 (1972).

– [6] Le problème de la dérivée oblique pour l'équation des ondes. C.R. Acad. Sci. Paris 285, 773–775 (1977). Rectification C.R. Acad. Sci. Paris 285, 1199 (1978).

– [7] On the asymptotic distribution of the eigenvalues and eigenfunctions of elliptic differential operators. Math. Scand. 1, 237–255 (1953).

Gårding, L. and J.L. Lions: [1] Functional analysis. Nuovo Cimento N. 1 del Suppl. al Vol. (10)14, 9–66 (1959).

Gårding, L. and B. Malgrange: [1] Opérateurs différentiels partiellement hypoelliptiques et partiellement elliptiques. Math. Scand. 9, 5–21 (1961).

Gask, H.: [1] A proof of Schwartz' kernel theorem. Math. Scand. 8, 327–332 (1960).

Gelfand, I.M. and G.E. Šilov: [1] Fourier transforms of rapidly increasing functions and questions of uniqueness of the solution of Cauchy's problem. Uspehi Mat. Nauk 8:6, 3–54 (1953) (Russian); also in Amer. Math. Soc. Transl. (2) 5, 221–274 (1957).

– [2] Generalized functions. Volume 1: Properties and operations. Volume 2: Spaces of fundamental and generalized functions. Academic Press, New York and London 1964, 1968.

Gevrey, M.: [1] Démonstration du théorème de Picard-Bernstein par la méthode des contours successifs; prolongement analytique. Bull. Sci. Math. 50, 113–128 (1926).

Glaeser, G.: [1] Etude de quelques algèbres Tayloriennes. J. Analyse Math. 6, 1–124 (1958).

Godin, P.: [1] Propagation des singularités pour les opérateurs pseudo-différentiels de type principal à partie principal analytique vérifiant la condition (*P*), en dimension 2. C.R. Acad. Sci. Paris 284, 1137–1138 (1977).

Gorin, E.A.: [1] Asymptotic properties of polynomials and algebraic functions of several variables. Uspehi Mat. Nauk 16:1, 91–118 (1961) (Russian); also in Russian Math. Surveys 16:1, 93–119 (1961).

Grubb, G.: [1] Boundary problems for systems of partial differential operators of mixed order. J. Functional Analysis 26, 131–165 (1977).

– [2] Problèmes aux limites pseudo-différentiels dépendant d'un paramètre. C.R. Acad. Sci. Paris 292, 581–583 (1981).

Grušin, V.V.: [1] The extension of smoothness of solutions of differential equations of principal type. Dokl. Akad. Nauk SSSR 148, 1241–1244 (1963) (Russian); also in Soviet Math. Doklady 4, 248–252 (1963).

– [2] Some classical theorems in spectral theory revisited. Seminar on sing. of sol. of diff. eq., Princeton University Press, Princeton, N.J., 219–259 (1979).

Gudmundsdottir, G.: [1] Global properties of differential operators of constant strength. Ark. Mat. 15, 169–198 (1977).

Guillemin, V.: [1] The Radon transform on Zoll surfaces. Advances in Math. 22, 85–119 (1976).

– [2] Some classical theorems in spectral theory revisited. Seminar on sing. of sol. of diff. eq., Princeton University Press, Princeton, N.J., 219–259 (1979).

– [3] Some spectral results for the Laplace operator with potential on the *n*-sphere. Advances in Math. 27, 273–286 (1978).

Guillemin, V. and D. Schaeffer: [1] Remarks on a paper of D. Ludwig. Bull. Amer. Math. Soc. 79, 382–385 (1973).

Guillemin, V. and S. Sternberg: [1] Geometrical asymptotics. Amer. Math. Soc. Surveys 14, Providence, R.I. 1977.

Gurevič, D.I.: [1] Counterexamples to a problem of L. Schwartz. Funkcional. Anal. i Priložen. 9:2, 29–35 (1975) (Russian); also in Functional Anal. Appl. 9, 116–120 (1975).

Hack, M.N.: [1] On convergence to the Møller wave operators. Nuovo Cimento (10) 13, 231–236 (1959).

Hadamard, J.: [1] Le problème de Cauchy et les équations aux dérivées partielles linéaires hyperboliques. Paris 1932.

Haefliger, A.: [1] Variétés feuilletées. Ann. Scuola Norm. Sup. Pisa 16, 367–397 (1962).

Hanges, N.: [1] Propagation of singularities for a class of operators with double characteristics. Seminar on singularities of sol. of linear partial diff. eq., Princeton University Press, Princeton, N.J. 1979, 113–126.

Hardy, G.H. and J.E. Littlewood: [1] Some properties of fractional integrals. (I) Math. Z. 27, 565–606 (1928); (II) Math. Z. 34, 403–439 (1931–32).

Hausdorff, F.: [1] Eine Ausdehnung des Parsevalschen Satzes über Fourierreihen. Math. Z. 16, 163–169 (1923).

Hayman, W.K. and P.B. Kennedy: [1] Subharmonic functions I. Academic Press, London, New York, San Francisco 1976.

Hedberg, L.I.: [1] On certain convolution inequalities. Proc. Amer. Math. Soc. 36, 505–510 (1972).

Heinz, E.: [1] Über die Eindeutigkeit beim Cauchyschen Anfangswertproblem einer elliptischen Differentialgleichung zweiter Ordnung. Nachr. Akad. Wiss. Göttingen Math.-Phys. Kl. IIa No. 1, 1–12 (1955).

Helffer, B.: [1] Addition de variables et applications à la régularité. Ann. Inst. Fourier (Grenoble) 28:2, 221–231 (1978).

Helffer, B. and J. Nourrigat: [1] Caractérisation des opérateurs hypoelliptiques homogènes invariants à gauche sur un groupe de Lie nilpotent gradué. Comm. Partial Differential Equations 4:8, 899–958 (1979).

Herglotz, G.: [1] Über die Integration linearer partieller Differentialgleichungen mit konstanten Koeffizienten I-III. Berichte Sächs. Akad. d. Wiss. 78, 93–126, 287–318 (1926); 80, 69–114 (1928).

Hersh, R.: [1] Boundary conditions for equations of evolution. Arch. Rational Mech. Anal. 16, 243–264 (1964).

– [2] On surface waves with finite and infinite speed of propagation. Arch. Rational Mech. Anal. 19, 308–316 (1965).

Hirzebruch, F.: [1] Neue topologische Methoden in der algebraischen Geometrie. Springer Verlag, Berlin-Göttingen-Heidelberg 1956.

Hlawka, E.: [1] Über Integrale auf konvexen Körpern. I. Monatsh. Math. 54, 1–36 (1950).

Holmgren, E.: [1] Über Systeme von linearen partiellen Differentialgleichungen. Öfversigt af Kongl. Vetenskaps-Akad. Förh. 58, 91–103 (1901).

– [2] Sur l'extension de la méthode d'intégration de Riemann. Ark. Mat. Astr. Fys. 1, No 22, 317–326 (1904).

Hörmander, L.: [1] On the theory of general partial differential operators. Acta Math. 94, 161–248 (1955).

– [2] Local and global properties of fundamental solutions. Math. Scand. 5, 27–39 (1957).

– [3] On the regularity of the solutions of boundary problems. Acta Math. 99, 225–264 (1958).

– [4] On interior regularity of the solutions of partial differential equations. Comm. Pure Appl. Math. 11, 197–218 (1958).

– [5] On the division of distributions by polynomials. Ark. Mat. 3, 555–568 (1958).

– [6] Differentiability properties of solutions of systems of differential equations. Ark. Mat. 3, 527–535 (1958).

– [7] Definitions of maximal differential operators. Ark. Mat. 3, 501–504 (1958).

– [8] On the uniqueness of the Cauchy problem I, II. Math. Scand. 6, 213–225 (1958); 7, 177–190 (1959).

– [9] Null solutions of partial differential equations. Arch. Rational Mech. Anal. 4, 255–261 (1960).

– [10] Differential operators of principal type. Math. Ann. 140, 124–146 (1960).

– [11] Differential equations without solutions. Math. Ann. 140, 169–173 (1960).

– [12] Hypoelliptic differential operators. Ann. Inst. Fourier (Grenoble) 11, 477–492 (1961).

– [13] Estimates for translation invariant operators in L^p spaces. Acta Math. 104, 93–140 (1960).
– [14] On the range of convolution operators. Ann. of Math. 76, 148–170 (1962).
– [15] Supports and singular supports of convolutions. Acta Math. 110, 279–302 (1963).
– [16] Pseudo-differential operators. Comm. Pure Appl. Math. 18, 501–517 (1965).
– [17] Pseudo-differential operators and non-elliptic boundary problems. Ann. of Math. 83, 129–209 (1966).
– [18] Pseudo-differential operators and hypoelliptic equations. Amer. Math. Soc. Symp. on Singular Integrals, 138–183 (1966).
– [19] An introduction to complex analysis in several variables. D. van Nostrand Publ. Co., Princeton, N.J. 1966.
– [20] Hypoelliptic second order differential equations. Acta Math. 119, 147–171 (1967).
– [21] On the characteristic Cauchy problem. Ann. of Math. 88, 341–370 (1968).
– [22] The spectral function of an elliptic operator. Acta Math. 121, 193–218 (1968).
– [23] Convolution equations in convex domains. Invent. Math. 4, 306–317 (1968).
– [24] On the singularities of solutions of partial differential equations. Comm. Pure Appl. Math. 23, 329–358 (1970).
– [25] Linear differential operators. Actes Congr. Int. Math. Nice 1970, 1, 121–133.
– [26a] The calculus of Fourier integral operators. Prospects in math. Ann. of Math. Studies 70, 33–57 (1971).
– [26] Fourier integral operators I. Acta Math. 127, 79–183 (1971).
– [27] Uniqueness theorems and wave front sets for solutions of linear differential equations with analytic coefficients. Comm. Pure Appl. Math. 24, 671–704 (1971).
– [28] A remark on Holmgren's uniqueness theorem. J. Diff. Geom. 6, 129–134 (1971).
– [29] On the existence and the regularity of solutions of linear pseudo-differential equations. Ens. Math. 17, 99–163 (1971).
– [30] On the singularities of solutions of partial differential equations with constant coefficients. Israel J. Math. 13, 82–105 (1972).
– [31] On the existence of real analytic solutions of partial differential equations with constant coefficients. Invent. Math. 21, 151–182 (1973).
– [32] Lower bounds at infinity for solutions of differential equations with constant coefficients. Israel J. Math. 16, 103–116 (1973).
– [33] Non-uniqueness for the Cauchy problem. Springer Lecture Notes in Math. 459, 36–72 (1975).
– [34] The existence of wave operators in scattering theory. Math. Z. 146, 69–91 (1976).
– [35] A class of hypoelliptic pseudo-differential operators with double characteristics. Math. Ann. 217, 165–188 (1975).
– [36] The Cauchy problem for differential equations with double characteristics. J. Analyse Math. 32, 118–196 (1977).
– [37] Propagation of singularities and semiglobal existence theorems for (pseudo-)differential operators of principal type. Ann. of Math. 108, 569–609 (1978).
– [38] Subelliptic operators. Seminar on sing. of sol. of diff. eq. Princeton University Press, Princeton, N.J., 127–208 (1979).
– [39] The Weyl calculus of pseudo-differential operators. Comm. Pure Appl. Math. 32, 359–443 (1979).
– [40] Pseudo-differential operators of principal type. Nato Adv. Study Inst. on Sing. in Bound. Value Problems. Reidel Publ. Co., Dordrecht, 69–96 (1981).
– [41] Uniqueness theorems for second order elliptic differential equations. Comm. Partial Differential Equations 8, 21–64 (1983).
– [42] On the index of pseudo-differential operators. In Elliptische Differentialgleichungen Band II, Akademie-Verlag, Berlin 1971, 127–146.
– [43] L^2 estimates for Fourier integral operators with complex phase. Ark. Mat. 21, 297–313 (1983).
– [44] On the subelliptic test estimates. Comm. Pure Appl. Math. 33, 339–363 (1980).

Hurwitz, A.: [1] Über die Nullstellen der Bessel'schen Funktion. Math. Ann. 33, 246–266 (1889).

Iagolnitzer, D.: [1] Microlocal essential support of a distribution and decomposition theorems - an introduction. In Hyperfunctions and theoretical physics. Springer Lecture Notes in Math. 449, 121-132 (1975).

Ikebe, T.: [1] Eigenfunction expansions associated with the Schrödinger operator and their applications to scattering theory. Arch. Rational Mech. Anal. 5, 1-34 (1960).

Ikebe, T. and Y. Saito: [1] Limiting absorption method and absolute continuity for the Schrödinger operator. J. Math. Kyoto Univ. 12, 513-542 (1972).

Ivrii, V.Ja: [1] Sufficient conditions for regular and completely regular hyperbolicity. Trudy Moskov. Mat. Obšč. 33, 3-65 (1975) (Russian); also in Trans. Moscow Math. Soc. 33, 1-65 (1978).

– [2] Wave fronts for solutions of boundary value problems for a class of symmetric hyperbolic systems. Sibirsk. Mat. Ž. 21:4, 62-71 (1980) (Russian); also in Sibirian Math. J. 21, 527-534 (1980).

– [3] On the second term in the spectral asymptotics for the Laplace-Beltrami operator on a manifold with boundary. Funkcional. Anal. i Priložen. 14:2, 25-34 (1980) (Russian); also in Functional Anal. Appl. 14, 98-106 (1980).

Ivrii, V.Ja and V.M. Petkov: [1] Necessary conditions for the correctness of the Cauchy problem for non-strictly hyperbolic equations. Uspehi Mat. Nauk 29:5, 3-70 (1974) (Russian); also in Russian Math. Surveys 29:5, 1-70 (1974).

Iwasaki, N.: [1] The Cauchy problems for effectively hyperbolic equations (general case). J. Math. Kyoto Univ. 25, 727–743 (1985).

Jauch, J.M. and I.I. Zinnes: [1] The asymptotic condition for simple scattering systems. Nuovo Cimento (10) 11, 553-567 (1959).

Jerison D. and C.E. Kenig: [1] Unique continuation and absence of positive eigenvalues for Schrödinger operators. Ann. of Math. 121, 463–488 (1985).

John, F.: [1] On linear differential equations with analytic coefficients. Unique continuation of data. Comm. Pure Appl. Math. 2, 209-253 (1949).

– [2] Plane waves and spherical means applied to partial differential equations. New York 1955.

– [3] Non-admissible data for differential equations with constant coefficients. Comm. Pure Appl. Math. 10, 391-398 (1957).

– [4] Continuous dependence on data for solutions of partial differential equations with a prescribed bound. Comm. Pure Appl. Math. 13, 551-585 (1960).

– [5] Linear partial differential equations with analytic coefficients. Proc. Nat. Acad. Sci. 29, 98-104 (1943).

Jörgens, K. and J. Weidmann: [1] Zur Existenz der Wellenoperatoren. Math. Z. 131, 141–151 (1973).

Kashiwara, M.: [1] Introduction to the theory of hyperfunctions. In Sem. on microlocal analysis, Princeton Univ. Press, Princeton, N.J., 1979, 3-38.

Kashiwara, M. and T. Kawai: [1] Microhyperbolic pseudo-differential operators. I. J. Math. Soc. Japan 27, 359–404 (1975).

Kato, T.: [1] Growth properties of solutions of the reduced wave equation with a variable coefficient. Comm. Pure Appl. Math. 12, 403–425 (1959).

Keller, J.B.: [1] Corrected Bohr-Sommerfeld quantum conditions for nonseparable systems. Ann. Physics 4, 180–188 (1958).

Kitada, H.: [1] Scattering theory for Schrödinger operators with long-range potentials. I: Abstract theory. J. Math. Soc. Japan 29, 665-691 (1977). II: Spectral and scattering theory. J. Math. Soc. Japan 30, 603-632 (1978).

Knapp, A.W. and E.M. Stein: [1] Singular integrals and the principal series. Proc. Nat. Acad. Sci. U.S.A. 63, 281-284 (1969).

Kohn, J.J.: [1] Harmonic integrals on strongly pseudo-convex manifolds I, II. Ann. of Math. 78, 112-148 (1963); 79, 450–472 (1964).

– [2] Pseudo-differential operators and non-elliptic problems. In Pseudo-differential operators, CIME conference, Stresa 1968, 157–165. Edizione Cremonese, Roma 1969.

Kohn, J.J. and L. Nirenberg: [1] On the algebra of pseudo-differential operators. Comm. Pure Appl. Math. 18, 269–305 (1965).

– [2] Non-coercive boundary value problems. Comm. Pure Appl. Math. 18, 443–492 (1965).

Kolmogorov, A.N.: [1] Zufällige Bewegungen. Ann. of Math. 35, 116–117 (1934).

Komatsu, H.: [1] A local version of Bochner's tube theorem. J. Fac. Sci. Tokyo Sect. I-A Math. 19, 201–214 (1972).

– [2] Boundary values for solutions of elliptic equations. Proc. Int. Conf. Funct. Anal. Rel. Topics, 107–121. University of Tokyo Press, Tokyo 1970.

Kreiss, H.O.: [1] Initial boundary value problems for hyperbolic systems. Comm. Pure Appl. Math. 23, 277–298 (1970).

Krzyżański, M. and J. Schauder: [1] Quasilineare Differentialgleichungen zweiter Ordnung vom hyperbolischen Typus. Gemischte Randwertaufgaben. Studia Math. 6, 162–189 (1936).

Kumano-go, H.: [1] Factorizations and fundamental solutions for differential operators of elliptic-hyperbolic type. Proc. Japan Acad. 52, 480–483 (1976).

Kuroda, S.T.: [1] On the existence and the unitary property of the scattering operator. Nuovo Cimento (10) 12, 431–454 (1959).

Lascar, B. and R. Lascar: [1] Propagation des singularités pour des équations hyperboliques à caractéristiques de multiplicité au plus double et singularités Masloviennes II. J. Analyse Math. 41, 1–38 (1982).

Lax, A.: [1] On Cauchy's problem for partial differential equations with multiple characteristics. Comm. Pure Appl. Math. 9, 135–169 (1956).

Lax, P.D.: [2] On Cauchy's problem for hyperbolic equations and the differentiability of solutions of elliptic equations. Comm. Pure Appl. Math. 8, 615–633 (1955).

– [3] Asymptotic solutions of oscillatory initial value problems. Duke Math. J. 24, 627–646 (1957).

Lax, P.D. and L. Nirenberg: [1] On stability for difference schemes: a sharp form of Gårding's inequality. Comm. Pure Appl. Math. 19, 473–492 (1966).

Lebeau, G.: [1] Fonctions harmoniques et spectre singulier. Ann. Sci. École Norm. Sup. (4) 13, 269–291 (1980).

Lelong, P.: [1] Plurisubharmonic functions and positive differential forms. Gordon and Breach, New York, London, Paris 1969.

– [2] Propriétés métriques des variétés définies par une équation. Ann. Sci. École Norm. Sup. 67, 22–40 (1950).

Leray, J.: [1] Hyperbolic differential equations. The Institute for Advanced Study, Princeton, N.J. 1953.

– [2] Uniformisation de la solution du problème linéaire analytique de Cauchy près de la variété qui porte les données de Cauchy. Bull. Soc. Math. France 85, 389–429 (1957).

Lerner, N.: [1] Unicité de Cauchy pour des opérateurs différentiels faiblement principalement normaux. J. Math. Pures Appl. 64, 1–11 (1985).

Lerner, N. and L. Robbiano: [1] Unicité de Cauchy pour des opérateurs de type principal. J. Analyse Math. 44, 32–66 (1984/85).

Levi, E.E.: [1] Caratterische multiple e problema di Cauchy. Ann. Mat. Pura Appl. (3) 16, 161–201 (1909).

Levinson, N.: [1] Transformation of an analytic function of several variables to a canonical form. Duke Math. J. 28, 345–353 (1961).

Levitan, B.M.: [1] On the asymptotic behavior of the spectral function of a self-adjoint differential equation of the second order. Izv. Akad. Nauk SSSR Ser. Mat. 16, 325–352 (1952).

– [2] On the asymptotic behavior of the spectral function and on expansion in eigenfunctions of a self-adjoint differential equation of second order II. Izv. Akad. Nauk SSSR Ser. Mat. 19, 33–58 (1955).

Lewy, H.: [1] An example of a smooth linear partial differential equation without solution. Ann. of Math. 66, 155–158 (1957).
– [2] Extension of Huyghen's principle to the ultrahyperbolic equation. Ann. Mat. Pura Appl. (4) 39, 63–64 (1955).
Lions, J.L.: [1] Supports dans la transformation de Laplace. J. Analyse Math. 2, 369–380 (1952–53).
Lions, J.L. and E. Magenes: [1] Problèmes aux limites non homogènes et applications I–III. Dunod, Paris, 1968–1970.
Łojasiewicz, S.: [1] Sur le problème de division. Studia Math. 18, 87–136 (1959).
Lopatinski, Ya.B.: [1] On a method of reducing boundary problems for a system of differential equations of elliptic type to regular integral equations. Ukrain. Mat. Ž. 5, 123–151 (1953). Amer. Math. Soc. Transl. (2) 89, 149–183 (1970).
Ludwig, D.: [1] Exact and asymptotic solutions of the Cauchy problem. Comm. Pure Appl. Math. 13, 473–508 (1960).
– [2] Uniform asymptotic expansions at a caustic. Comm. Pure Appl. Math. 19, 215–250 (1966).
Luke, G.: [1] Pseudodifferential operators on Hilbert bundles. J. Differential Equations 12, 566–589 (1972).
Malgrange, B.: [1] Existence et approximation des solutions des équations aux dérivées partielles et des équations de convolution. Ann. Inst. Fourier (Grenoble) 6, 271–355 (1955–56).
– [2] Sur une classe d'opérateurs différentiels hypoelliptiques. Bull. Soc. Math. France 85, 283–306 (1957).
– [3] Sur la propagation de la régularité des solutions des équations à coefficients constants. Bull. Math. Soc. Sci. Math. Phys. R.P. Roumanie 3 (53), 433–440 (1959).
– [4] Sur les ouverts convexes par rapport à un opérateur différentiel. C. R. Acad. Sci. Paris 254, 614–615 (1962).
– [5] Sur les systèmes différentiels à coefficients constants. Coll. CNRS 113–122, Paris 1963.
– [6] Ideals of differentiable functions. Tata Institute, Bombay, and Oxford University Press 1966.
Mandelbrojt, S.: [1] Analytic functions and classes of infinitely differentiable functions. Rice Inst. Pamphlet 29, 1–142 (1942).
– [2] Séries adhérentes, régularisations des suites, applications. Coll. Borel, Gauthier-Villars, Paris 1952.
Martineau, A.: [1] Les hyperfonctions de M. Sato. Sém. Bourbaki 1960–1961, Exposé No 214.
– [2] Le "edge of the wedge theorem" en théorie des hyperfonctions de Sato. Proc. Int. Conf. Funct. Anal. Rel. Topics, 95–106. University of Tokyo Press, Tokyo 1970.
Maslov, V.P.: [1] Theory of perturbations and asymptotic methods. Moskov. Gos. Univ., Moscow 1965 (Russian).
Mather, J.: [1] Stability of C^∞ mappings: I. The division theorem. Ann. of Math. 87, 89–104 (1968).
Melin, A.: [1] Lower bounds for pseudo-differential operators. Ark. Mat. 9, 117–140 (1971).
– [2] Parametrix constructions for right invariant differential operators on nilpotent groups. Ann. Global Analysis and Geometry 1, 79–130 (1983).
Melin, A. and J. Sjöstrand: [1] Fourier integral operators with complex-valued phase functions. Springer Lecture Notes in Math. 459, 120–223 (1974).
– [2] Fourier integral operators with complex phase functions and parametrix for an interior boundary value problem. Comm. Partial Differential Equations 1:4, 313–400 (1976).
Melrose, R.B.: [1] Transformation of boundary problems. Acta Math. 147, 149–236 (1981).

– [2] Equivalence of glancing hypersurfaces. Invent. Math. 37, 165–191 (1976).
– [3] Microlocal parametrices for diffractive boundary value problems. Duke Math. J. 42, 605–635 (1975).
– [4] Local Fourier-Airy integral operators. Duke Math. J. 42, 583–604 (1975).
– [5] Airy operators. Comm. Partial Differential Equations 3:1, 1–76 (1978).
– [6] The Cauchy problem for effectively hyperbolic operators. Hokkaido Math. J. To appear.
– [7] The trace of the wave group. Contemporary Math. 27, 127–167 (1984).
Melrose, R.B. and J. Sjöstrand: [1] Singularities of boundary value problems I, II. Comm. Pure Appl. Math. 31, 593–617 (1978); 35, 129–168 (1982).
Mihlin, S.G.: [1] On the multipliers of Fourier integrals. Dokl. Akad. Nauk SSSR 109, 701–703 (1956) (Russian).
Mikusiński, J.: [1] Une simple démonstration du théorème de Titchmarsh sur la convolution. Bull. Acad. Pol. Sci. 7, 715–717 (1959).
– [2] The Bochner integral. Birkhäuser Verlag, Basel and Stuttgart 1978.
Minakshisundaram, S. and Å. Pleijel: [1] Some properties of the eigenfunctions of the Laplace operator on Riemannian manifolds. Canad. J. Math. 1, 242–256 (1949).
Mizohata, S.: [1] Unicité du prolongement des solutions des équations elliptiques du quatrième ordre. Proc. Jap. Acad. 34, 687–692 (1958).
– [2] Systèmes hyperboliques. J. Math. Soc. Japan 11, 205–233 (1959).
– [3] Note sur le traitement par les opérateurs d'intégrale singulière du problème de Cauchy. J. Math. Soc. Japan 11, 234–240 (1959).
– [4] Solutions nulles et solutions non analytiques. J. Math. Kyoto Univ. 1, 271–302 (1962).
– [5] Some remarks on the Cauchy problem. J. Math. Kyoto Univ. 1, 109–127 (1961).
Møller, C.: [1] General properties of the characteristic matrix in the theory of elementary particles. I. Kongl. Dansk. Vidensk. Selsk. Mat.-Fys. Medd. 23, 2–48 (1945).
Morrey, C.B.: [1] The analytic embedding of abstract real-analytic manifolds. Ann. of Math. 68, 159–201 (1958).
Morrey, C.B. and L. Nirenberg: [1] On the analyticity of the solutions of linear elliptic systems of partial differential equations. Comm. Pure Appl. Math. 10, 271–290 (1957).
Moyer, R.D.: [1] Local solvability in two dimensions: Necessary conditions for the principle-type case. Mimeographed manuscript, University of Kansas 1978.
Müller, C.: [1] On the behaviour of the solutions of the differential equation $\Delta U = F(x, U)$ in the neighborhood of a point. Comm. Pure Appl. Math. 7, 505–515 (1954).
Münster, M.: [1] On A. Lax's condition of hyperbolicity. Rocky Mountain J. Math. 8, 443–446 (1978).
– [2] On hyperbolic polynomials with constant coefficients. Rocky Mountain J. Math. 8, 653–673 (1978).
von Neumann, J. and E. Wigner: [1] Über merkwürdige diskrete Eigenwerte. Phys. Z. 30, 465–467 (1929).
Nirenberg, L.: [1] Remarks on strongly elliptic partial differential equations. Comm. Pure Appl. Math. 8, 648–675 (1955).
– [2] Uniqueness in Cauchy problems for differential equations with constant leading coefficients. Comm. Pure Appl. Math. 10, 89–105 (1957).
– [3] A proof of the Malgrange preparation theorem. Liverpool singularities I. Springer Lecture Notes in Math. 192, 97–105 (1971).
– [4] On elliptic partial differential equations. Ann. Scuola Norm. Sup. Pisa (3) 13, 115–162 (1959).
– [5] Lectures on linear partial differential equations. Amer. Math. Soc. Regional Conf. in Math., 17, 1–58 (1972).

Nirenberg, L. and F. Treves: [1] Solvability of a first order linear partial differential equation. Comm. Pure Appl. Math. 16, 331-351 (1963).

– [2] On local solvability of linear partial differential equations. I. Necessary conditions. II. Sufficient conditions. Correction. Comm. Pure Appl. Math. 23, 1-38 and 459-509 (1970); 24, 279-288 (1971).

Nishitani, T.: [1] Local energy integrals for effectively hyperbolic operators I, II. J. Math. Kyoto Univ. 24, 623-658 and 659-666 (1984).

Noether, F.: [1] Über eine Klasse singulärer Integralgleichungen. Math. Ann. 82, 42-63 (1921).

Olejnik, O.A.: [1] On the Cauchy problem for weakly hyperbolic equations. Comm. Pure Appl. Math. 23, 569-586 (1970).

Olejnik, O.A. and E.V. Radkevič: [1] Second order equations with non-negative characteristic form. In Matem. Anal. 1969, ed. R.V. Gamkrelidze, Moscow 1971 (Russian). English translation Plenum Press, New York-London 1973.

Oshima, T.: [1] On analytic equivalence of glancing hypersurfaces. Sci. Papers College Gen. Ed. Univ. Tokyo 28, 51-57 (1978).

Palamodov, V.P.: [1] Linear differential operators with constant coefficients. Moscow 1967 (Russian). English transl. Grundl. d. Math. Wiss. 168, Springer Verlag, New York, Heidelberg, Berlin 1970.

Paley, R.E.A.C. and N. Wiener: [1] Fourier transforms in the complex domain. Amer. Math. Soc. Coll. Publ. XIX, New York 1934.

Pederson, R.: [1] On the unique continuation theorem for certain second and fourth order elliptic equations. Comm. Pure Appl. Math. 11, 67-80 (1958).

– [2] Uniqueness in the Cauchy problem for elliptic equations with double characteristics. Ark. Mat. 6, 535-549 (1966).

Peetre, J.: [1] Théorèmes de régularité pour quelques classes d'opérateurs différentiels. Thesis, Lund 1959.

– [2] Rectification à l'article "Une caractérisation abstraite des opérateurs différentiels". Math. Scand. 8, 116-120 (1960).

– [3] Another approach to elliptic boundary problems. Comm. Pure Appl. Math. 14, 711-731 (1961).

– [4] New thoughts on Besov spaces. Duke Univ. Math. Series I. Durham, N.C. 1976.

Persson, J.: [1] The wave operator and P-convexity. Boll. Un. Mat. Ital. (5) 18-B, 591-604 (1981).

Petrowsky, I.G.: [1] Über das Cauchysche Problem für Systeme von partiellen Differentialgleichungen. Mat. Sb. 2 (44), 815-870 (1937).

– [2] Über das Cauchysche Problem für ein System linearer partieller Differentialgleichungen im Gebiete der nichtanalytischen Funktionen. Bull. Univ. Moscow Sér. Int. 1, No. 7, 1-74 (1938).

– [3] Sur l'analyticité des solutions des systèmes d'équations différentielles. Mat. Sb. 5 (47), 3-70 (1939).

– [4] On the diffusion of waves and the lacunas for hyperbolic equations. Mat. Sb. 17 (59), 289-370 (1945).

– [5] Some remarks on my papers on the problem of Cauchy. Mat. Sb. 39 (81), 267-272 (1956). (Russian.)

Pham The Lai: [1] Meilleures estimations asymptotiques des restes de la fonction spectrale et des valeurs propres relatifs au laplacien. Math. Scand. 48, 5-31 (1981).

Piccinini, L.C.: [1] Non surjectivity of the Cauchy-Riemann operator on the space of the analytic functions on $\mathbb{R}^n$. Generalization to the parabolic operators. Bull. Un. Mat. Ital. (4) 7, 12-28 (1973).

Pliš, A.: [1] A smooth linear elliptic differential equation without any solution in a sphere. Comm. Pure Appl. Math. 14, 599-617 (1961).

– [2] The problem of uniqueness for the solution of a system of partial differential equations. Bull. Acad. Pol. Sci. 2, 55-57 (1954).

– [3] On non-uniqueness in Cauchy problem for an elliptic second order differential equation. Bull. Acad. Pol. Sci. 11, 95–100 (1963).

Poincaré, H.: [1] Sur les propriétés du potentiel et les fonctions abeliennes. Acta Math. 22, 89–178 (1899).

Povzner, A.Ya.: [1] On the expansion of arbitrary functions in characteristic functions of the operator $-\Delta u+cu$. Mat. Sb. 32 (74), 109–156 (1953).

Radkevič, E.: [1] A priori estimates and hypoelliptic operators with multiple characteristics. Dokl. Akad. Nauk SSSR 187, 274–277 (1969) (Russian); also in Soviet Math. Doklady 10, 849–853 (1969).

Ralston, J.: [1] Solutions of the wave equation with localized energy. Comm. Pure Appl. Math. 22, 807–823 (1969).

– [2] Gaussian beams and the propagation of singularities. MAA Studies in Mathematics 23, 206–248 (1983).

Reed, M. and B. Simon: [1] Methods of modern mathematical physics. III. Scattering theory. Academic Press 1979.

Rempel, S. and B.-W. Schulze: [1] Index theory of elliptic boundary problems. Akademie-Verlag, Berlin (1982).

de Rham, G.: [1] Variétés différentiables. Hermann, Paris, 1955.

Riesz, F.: [1] Sur l'existence de la dérivée des fonctions d'une variable réelle et des fonctions d'intervalle. Verh. Int. Math. Kongr. Zürich 1932, I, 258–269.

Riesz, M.: [1] L'intégrale de Riemann-Liouville et le problème de Cauchy. Acta Math. 81, 1–223 (1949).

– [2] Sur les maxima des formes bilinéaires et sur les fonctionnelles linéaires. Acta Math. 49, 465–497 (1926).

– [3] Sur les fonctions conjuguées. Math. Z. 27, 218–244 (1928).

– [4] Problems related to characteristic surfaces. Proc. Conf. Diff. Eq. Univ. Maryland 1955, 57–71.

Rothschild, L.P.: [1] A criterion for hypoellipticity of operators constructed from vector fields. Comm. Partial Differential Equations 4:6, 645–699 (1979).

Saito, Y.: [1] On the asymptotic behavior of the solutions of the Schrödinger equation $(-\Delta+Q(y)-k^2)V=F$. Osaka J. Math. 14, 11–35 (1977).

– [2] Eigenfunction expansions for the Schrödinger operators with long-range potentials $Q(y)=O(|y|^{-\varepsilon})$, $\varepsilon>0$. Osaka J. Math. 14, 37–53 (1977).

Sakamoto, R.: [1] E-well posedness for hyperbolic mixed problems with constant coefficients. J. Math. Kyoto Univ. 14, 93–118 (1974).

– [2] Mixed problems for hyperbolic equations I. J. Math. Kyoto Univ. 10, 375–401 (1970), II. J. Math. Kyoto Univ. 10, 403–417 (1970).

Sato, M.: [1] Theory of hyperfunctions I. J. Fac. Sci. Univ. Tokyo I, 8, 139–193 (1959).

– [2] Theory of hyperfunctions II. J. Fac. Sci. Univ. Tokyo I, 8, 387–437 (1960).

– [3] Hyperfunctions and partial differential equations. Proc. Int. Conf. on Funct. Anal. and Rel. Topics, 91–94, Tokyo University Press, Tokyo 1969.

– [4] Regularity of hyperfunction solutions of partial differential equations. Actes Congr. Int. Math. Nice 1970, 2, 785–794.

Sato, M., T. Kawai and M. Kashiwara: [1] Hyperfunctions and pseudodifferential equations. Springer Lecture Notes in Math. 287, 265–529 (1973).

Schaefer, H.H.: [1] Topological vector spaces. Springer Verlag, New York, Heidelberg, Berlin 1970.

Schapira, P.: [1] Hyperfonctions et problèmes aux limites elliptiques. Bull. Soc. Math. France 99, 113–141 (1971).

– [2] Propagation at the boundary of analytic singularities. Nato Adv. Study Inst. on Sing. in Bound. Value Problems. Reidel Publ. Co., Dordrecht, 185–212 (1981).

– [3] Propagation at the boundary and reflection of analytic singularities of solutions of linear partial differential equations. Publ. RIMS, Kyoto Univ., 12 Suppl., 441–453 (1977).

Schechter, M.: [1] Various types of boundary conditions for elliptic equations. Comm. Pure Appl. Math. 13, 407–425 (1960).
– [2] A generalization of the problem of transmission. Ann. Scuola Norm. Sup. Pisa 14, 207–236 (1960).
Schwartz, L.: [1] Théorie des distributions I, II. Hermann, Paris, 1950–51.
– [2] Théorie des noyaux. Proc. Int. Congr. Math. Cambridge 1950, I, 220–230.
– [3] Sur l'impossibilité de la multiplication des distributions. C. R. Acad. Sci. Paris 239, 847–848 (1954).
– [4] Théorie des distributions à valeurs vectorielles I. Ann. Inst. Fourier (Grenoble) 7, 1–141 (1957).
– [5] Transformation de Laplace des distributions. Comm. Sém. Math. Univ. Lund, Tome suppl. dédié à Marcel Riesz, 196–206 (1952).
– [6] Théorie générale des fonctions moyenne-périodiques. Ann. of Math. 48, 857–929 (1947).
Seeley, R.T.: [1] Singular integrals and boundary problems. Amer. J. Math. 88, 781–809 (1966).
– [2] Extensions of C^∞ functions defined in a half space. Proc. Amer. Math. Soc. 15, 625–626 (1964).
– [3] A sharp asymptotic remainder estimate for the eigenvalues of the Laplacian in a domain of $\mathbb{R}^3$. Advances in Math. 29, 244–269 (1978).
– [4] An estimate near the boundary for the spectral function of the Laplace operator. Amer. J. Math. 102, 869–902 (1980).
– [5] Elliptic singular integral equations. Amer Math. Soc. Symp. on Singular Integrals, 308–315 (1966).
Seidenberg, A.: [1] A new decision method for elementary algebra. Ann. of Math. 60, 365–374 (1954).
Shibata, Y.: [1] E-well posedness of mixed initial-boundary value problems with constant coefficients in a quarter space. J. Analyse Math. 37, 32–45 (1980).
Siegel, C.L.: [1] Zu den Beweisen des Vorbereitungssatzes von Weierstrass. In Abhandl. aus Zahlenth. u. Anal., 299–306. Plenum Press, New York 1968.
Sjöstrand, J.: [1] Singularités analytiques microlocales. Prépublications Université de Paris-Sud 82-03.
– [2] Analytic singularities of solutions of boundary value problems. Nato Adv. Study Inst. on Sing. in Bound. Value Prob., Reidel Publ. Co., Dordrecht, 235–269 (1981).
– [3] Parametrices for pseudodifferential operators with multiple characteristics. Ark. Mat. 12, 85–130 (1974).
– [4] Propagation of analytic singularities for second order Dirichlet problems I, II, III. Comm. Partial Differential Equations 5:1, 41–94 (1980), 5:2, 187–207 (1980), and 6:5, 499–567 (1981).
– [5] Operators of principal type with interior boundary conditions. Acta Math. 130, 1–51 (1973).
Sobolev, S.L.: [1] Méthode nouvelle à résoudre le problème de Cauchy pour les équations linéaires hyperboliques normales. Mat. Sb. 1 (43), 39–72 (1936).
– [2] Sur un théorème d'analyse fonctionnelle. Mat. Sb. 4 (46), 471–497 (1938). (Russian; French summary.) Amer. Math. Soc. Transl. (2) 34, 39–68 (1963).
Sommerfeld, A.: [1] Optics. Lectures on theoretical physics IV. Academic Press, New York, 1969.
Stein, E.M.: [1] Singular integrals and differentiability properties of functions. Princeton Univ. Press 1970.
Sternberg, S.: [1] Lectures on differential geometry. Prentice-Hall Inc. Englewood Cliffs, N.J., 1964.
Stokes, G.B.: [1] On the numerical calculation of a class of definite integrals and infinite series. Trans. Cambridge Philos. Soc. 9, 166–187 (1850).

Svensson, L.: [1] Necessary and sufficient conditions for the hyperbolicity of polynomials with hyperbolic principal part. Ark. Mat. 8, 145-162 (1968).

Sweeney, W.J.: [1] The *D*-Neumann problem. Acta Math. 120, 223-277 (1968).

Szegö, G.: [1] Beiträge zur Theorie der Toeplitzschen Formen. Math. Z. 6, 167-202 (1920).

Täcklind, S.: [1] Sur les classes quasianalytiques des solutions des équations aux dérivées partielles du type parabolique. Nova Acta Soc. Sci. Upsaliensis (4) 10, 1-57 (1936).

Tarski, A.: [1] A decision method for elementary algebra and geometry. Manuscript, Berkeley, 63 pp. (1951).

Taylor, M.: [1] Gelfand theory of pseudodifferential operators and hypoelliptic operators. Trans. Amer. Math. Soc. 153, 495-510 (1971).

– [2] Grazing rays and reflection of singularities of solutions to wave equations. Comm. Pure Appl. Math. 29, 1-38 (1976).

– [3] Diffraction effects in the scattering of waves. In Sing. in Bound. Value Problems, 271-316. Reidel Publ. Co., Dordrecht 1981.

– [4] Pseudodifferential operators. Princeton Univ. Press, Princeton, N.J., 1981.

Thorin, O.: [1] An extension of a convexity theorem due to M. Riesz. Kungl Fys. Sällsk. Lund. Förh. 8, No 14 (1939).

Titchmarsh, E.C.: [1] The zeros of certain integral functions. Proc. London Math. Soc. 25, 283-302 (1926).

Treves, F.: [1] Solution élémentaire d'équations aux dérivées partielles dépendant d'un paramètre. C. R. Acad. Sci. Paris 242, 1250-1252 (1956).

– [2] Thèse d'Hörmander II. Sém. Bourbaki 135, 2e éd. (Mai 1956).

– [3] Relations de domination entre opérateurs différentiels. Acta Math. 101, 1-139 (1959).

– [4] Opérateurs différentiels hypoelliptiques. Ann. Inst. Fourier (Grenoble) 9, 1-73 (1959).

– [5] Local solvability in L^2 of first order linear PDEs. Amer. J. Math. 92, 369-380 (1970).

– [6] Fundamental solutions of linear partial differential equations with constant coefficients depending on parameters. Amer. J. Math. 84, 561-577 (1962).

– [7] Un théorème sur les équations aux dérivées partielles à coefficients constants dépendant de paramètres. Bull. Soc. Math. France 90, 473-486 (1962).

– [8] A new method of proof of the subelliptic estimates. Comm. Pure Appl. Math. 24, 71-115 (1971).

– [9] Introduction to pseudodifferential and Fourier integral operators. Volume 1: Pseudodifferential operators. Volume 2: Fourier integral operators. Plenum Press, New York and London 1980.

Vauthier, J.: [1] Comportement asymptotique des fonctions entières de type exponentiel dans $\mathbb{C}^n$ et bornées dans le domaine réel. J. Functional Analysis 12, 290-306 (1973).

Vekua, I.N.: [1] Systeme von Differentialgleichungen erster Ordnung vom elliptischen Typus und Randwertaufgaben. Berlin 1956.

Veselič, K. and J. Weidmann: [1] Existenz der Wellenoperatoren für eine allgemeine Klasse von Operatoren. Math. Z. 134, 255-274 (1973).

– [2] Asymptotic estimates of wave functions and the existence of wave operators. J. Functional Analysis. 17, 61-77 (1974).

Višik, M.I.: [1] On general boundary problems for elliptic differential equations. Trudy Moskov. Mat. Obšč. 1, 187-246 (1952) (Russian). Also in Amer. Math. Soc. Transl. (2) 24, 107-172 (1963).

Višik, M.I. and G.I. Eškin: [1] Convolution equations in a bounded region. Uspehi Mat. Nauk 20:3 (123), 89-152 (1965). (Russian.) Also in Russian Math. Surveys 20:3, 86-151 (1965).

– [2] Convolution equations in a bounded region in spaces with weighted norms. Mat. Sb. 69 (111), 65–110 (1966) (Russian). Also in Amer. Math. Soc. Transl. (2) 67, 33–82 (1968).
– [3] Elliptic convolution equations in a bounded region and their applications. Uspehi Mat. Nauk 22:1 (133), 15–76 (1967). (Russian.) Also in Russian Math. Surveys 22:1, 13–75 (1967).
– [4] Convolution equations of variable order. Trudy Moskov. Mat. Obšč. 16, 25-50 (1967). (Russian.) Also in Trans. Moskov. Mat. Soc. 16, 27–52 (1967).
– [5] Normally solvable problems for elliptic systems of convolution equations. Mat. Sb. 74 (116), 326-356 (1967). (Russian.) Also in Math. USSR-Sb. 3, 303-332 (1967).
van der Waerden, B.L.: [1] Einführung in die algebraische Geometrie. Berlin 1939.
– [2] Algebra I-II. 4. Aufl. Springer Verlag, Berlin-Göttingen-Heidelberg 1959.
Wang Rou-hwai and Tsui Chih-yung: [1] Generalized Leray formula on positive complex Lagrange-Grassmann manifolds. Res. Report, Inst. of Math., Jilin Univ. 8209, 1982.
Warner, F.W.: [1] Foundations of differentiable manifolds and Lie Groups. Scott, Foresman and Co., Glenview, Ill., London, 1971.
Weinstein, A.: [1] The order and symbol of a distribution. Trans. Amer. Math. Soc. 241, 1-54 (1978).
– [2] Asymptotics of eigenvalue clusters for the Laplacian plus a potential. Duke Math. J. 44, 883–892 (1977).
– [3] On Maslov's quantization condition. In Fourier integral operators and partial differential equations. Springer Lecture Notes in Math. 459, 341-372 (1974).
Weyl, H.: [1] The method of orthogonal projection in potential theory. Duke Math. J. 7, 411–444 (1940).
– [2] Die Idee der Riemannschen Fläche. 3. Aufl., Teubner, Stuttgart, 1955.
– [3] Über gewöhnliche Differentialgleichungen mit Singularitäten und die zugehörigen Entwicklungen willkürlicher Funktionen. Math. Ann. 68, 220–269 (1910).
– [4] Das asymptotische Verteilungsgesetz der Eigenwerte linearer partieller Differentialgleichungen (mit einer Anwendung auf die Theorie der Hohlraumstrahlung). Math. Ann. 71, 441–479 (1912).
Whitney, H.: [1] Analytic extensions of differentiable functions defined in closed sets. Trans. Amer. Math. Soc. 36, 63–89 (1934).
Widom, H.: [1] Eigenvalue distribution in certain homogeneous spaces. J. Functional Analysis 32, 139–147 (1979).
Yamamoto, K.: [1] On the reduction of certain pseudo-differential operators with non-involution characteristics. J. Differential Equations 26, 435–442 (1977).
Zeilon, N.: [1] Das Fundamentalintegral der allgemeinen partiellen linearen Differentialgleichung mit konstanten Koeffizienten. Ark. Mat. Astr. Fys. 6, No 38, 1-32 (1911).
Zerner, M.: [1] Solutions de l'équation des ondes présentant des singularités sur une droite. C. R. Acad. Sci. Paris 250, 2980–2982 (1960).
– [2] Solutions singulières d'équations aux dérivées partielles. Bull. Soc. Math. France 91, 203-226 (1963).
– [3] Domaine d'holomorphie des fonctions vérifiant une équation aux dérivées partielles. C. R. Acad. Sci. Paris 272, 1646-1648 (1971).
Zuily, C.: [1] Uniqueness and non-uniqueness in the Cauchy problem. Progress in Math. 33, Birkhäuser, Boston, Basel, Stuttgart 1983.
Zygmund, A.: [1] On a theorem of Marcinkiewicz concerning interpolation of operators. J. Math. Pures Appl. 35, 223–248 (1956).

Index

(See also the index in Volume I.)

Index of Notation

(See also the index of notation in Volume I.)

Spaces of Functions and Distributions

Special Symbols

Printing: Krips bv, Meppel
Binding: Litges & Dopf, Heppenheim